Lecture Notes in Computer Science 1290

Edited by G. Goos, J. Hartmanis and J. van Leeuwen

Springer

Berlin
Heidelberg
New York
Barcelona
Budapest
Hong Kong
London
Milan
Paris
Santa Clara
Singapore
Tokyo

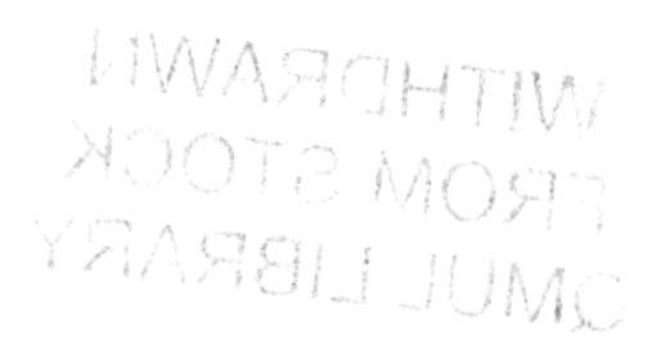

Eugenio Moggi Giuseppe Rosolini (Eds.)

Category Theory and Computer Science

7th International Conference, CTCS '97
Santa Margherita Ligure
Italy, September 4-6, 1997
Proceedings

Volume Editors

Eugenio Moggi
Università di Genova
Dipartimento di Informatica e Scienze dell'Informazione
Via Dodecaneso 35, I-16146 Genova, Italy
E-mail: moggi@disi.unige.it

Giuseppe Rosolini
Università di Genova
Dipartimento di Matematica
Via Dodecaneso 35, I-16146 Genova, Italy
E-mail: rosolini@disi.unige.it

Cataloging-in-Publication data applied for

Die Deutsche Bibliothek - CIP-Einheitsaufnahme

Category theory and computer science : 7th international conference ; proceedings / CTCS '97, S. Margherita Ligure, Italy, September 4 - 6, 1997. Eugenio Moggi ; Giuseppe Rosolini (ed.). - Berlin ; Heidelberg ; New York ; Barcelona ; Budapest ; Hong Kong ; London ; Milan ; Paris ; Santa Clara ; Singapore ; Tokyo : Springer, 1997
(Lecture notes in computer science ; Vol. 1290)
ISBN 3-540-63455-X

CR Subject Classification (1991): F.3, F.4.1, D.2.1, D.3.1, D.3.3

ISSN 0302-9743
ISBN 3-540-63455-X Springer-Verlag Berlin Heidelberg New York

Printed in Germany

Typesetting: Camera-ready by author
SPIN 10546414 06/3142 – 5 4 3 2 1 0 Printed on acid-free paper

Preface

Key features of category theory are its ability to draw precise and often unexpected bridges among different areas of computer science and mathematics, and to provide a few generic principles for organizing mathematical theories.

The papers of this volume were presented at the seventh meeting of the Biennial Conference on Category Theory and Computer Science held in S. Margherita Ligure, 4-6 September 1997. They address issues in: reasoning principles for types, rewriting, semantics, and structuring of logical systems. The conference programme had four invited speakers: John Baez, Richard Bird, Barry Jay, and Gordon Plotkin. Some of them have kindly contributed to this volume with introductory and overview papers related to their talks. Proceedings of previous meetings are also published in this series (Volumes 240, 283, 389, 530, and 953).

The organizers would like to thank the Italian CNR (Gruppo Nazionale Informatica Matematica, Comitato Nazionale Scienza e Tecnologia dell'Informazione) and the University of Genova (Ufficio Pubblicazioni e Relazioni Esterne) for their financial support.

Genova, June 1997

Eugenio Moggi
Giuseppe Rosolini

Programme Committee

S. Abramsky, Edinburgh (UK)
P.-L. Curien, LIENS (France)
P. Dybjer, Chalmers (Sweden)
P. Johnstone, Cambridge (UK)
G. Longo, LIENS (France)
G. Mints, Stanford (USA)
J. Mitchell, Stanford (USA)
E. Moggi, Genova (Italy)
A. Pitts, Cambridge (UK)
A. Poigne, GMD (Germany)
G. Rosolini, Genova (Italy)
D. Rydeheard, Manchester (UK)
F-J. de Vries, ETL (Japan)

Additional Referees

Gianna Bellè
Gianluca Cattani
Maura Cerioli
Gang Chen
Thierry Coquand
Pierre Damphousse
Marcelo Fiore
Giovanna Guerrini
Claudio Hermida
Barnaby Hilken
Patrik Jansson
Barry Jay
Ole Hoegh Jensen
Achim Jung
Stefan Kahrs
Richard Kennaway
Yoshiki Kinoshita
Kazem Lellahi
Paul-Andre Mellies
Catuscia Palamidessi
Gianna Reggio
Peter Sewell
Valentin Shehtman
Harold Simmons
Alex Simpson

Table of Contents

An Introduction to n-Categories

John C. Baez

Department of Mathematics, University of California,
Riverside, California 92521 USA

Abstract. An n-category is some sort of algebraic structure consisting of objects, morphisms between objects, 2-morphisms between morphisms, and so on up to n-morphisms, together with various ways of composing them. We survey various concepts of n-category, with an emphasis on 'weak' n-categories, in which all rules governing the composition of j-morphisms hold only up to equivalence. (An n-morphism is an equivalence if it is invertible, while a j-morphism for $j < n$ is an equivalence if it is invertible up to a $(j+1)$-morphism that is an equivalence.) We discuss applications of weak n-categories to various subjects including homotopy theory and topological quantum field theory, and review the definition of weak n-categories recently proposed by Dolan and the author.

1 Introduction

Very roughly, an n-category is an algebraic structure consisting of a collection of 'objects', a collection of 'morphisms' between objects, a collection of '2-morphisms' between morphisms, and so on up to n, with various reasonable ways of composing these j-morphisms. A 0-category is just a set, while a 1-category is just a category. Recently n-categories for arbitrarily large n have begun to play an increasingly important role in many subjects. The reason is that they let us *avoid mistaking isomorphism for equality.*

In a mere set, elements are either the same or different; there is no more to be said. In a category, objects can be different but still 'the same in a way'. In other words, they can be unequal but still isomorphic. Even better, we can explicitly keep track of the way they are the same: the isomorphism itself. This more nuanced treatment of 'sameness' is crucial to much of mathematics, physics, and computer science. For example, it underlies the modern concept of symmetry: since an object can be 'the same as itself in different ways', it has a symmetry group, its group of automorphisms. Unfortunately, in a category this careful distinction between equality and isomorphism breaks down when we study the morphisms. Morphisms in a category are either the same or different; there is no concept of isomorphic morphisms. In a 2-category this is remedied by introducing 2-morphisms between morphisms. Unfortunately, in a 2-category we cannot speak of isomorphic 2-morphisms. To remedy this we need the notion of 3-category, and so on.

The plan of this paper is as follows. We do not begin by defining n-categories. Many definitions have been proposed. So far, all of them are a bit complicated.

Ultimately a number of them should turn out to be equivalent, but this has not been shown yet. In fact, the correct sense of 'equivalence' here is a rather subtle issue, intimately linked with n-category theory itself. Thus before mastering the details of any particular definition, it is important to have a sense of the issues involved. Section 2 starts with a rough sketch of various approaches to defining n-categories. Section 3 describes how n-categories are becoming important in a variety of fields, and how they should allow us to formalize some previously rather mysterious analogies between different subjects. Section 4 sketches a particular definition of 'weak n-categories' due to Dolan and the author [5]. In the Conclusions we discuss the sense in which various proposed definitions of weak n-category should be equivalent.

2 Various Concepts of n-Category

To start thinking about n-categories it is helpful to use pictures. We visualize the objects as 0-dimensional, i.e., points. We visualize the morphisms as 1-dimensional, i.e., intervals, or more precisely, arrows going from one point to another. In this picture, composition of morphisms corresponds to gluing together an arrow $f: x \to y$ and an arrow $g: y \to z$ to obtain an arrow $fg: x \to z$:

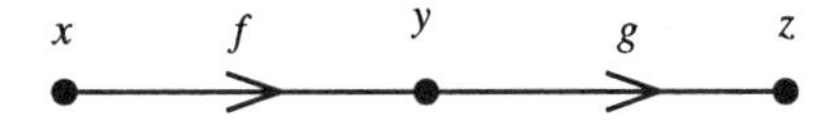

Note that while the notation fg for the composite of $f: x \to y$ and $g: y \to z$ is somewhat nonstandard, it fits the picture better than the usual notation.

Continuing on in this spirit, we visualize the 2-morphisms as 2-dimensional, and compose 2-morphisms in a way that corresponds to gluing together 2-dimensional shapes. Of course, we should choose some particular shapes for our 2-morphisms. For example, we could use a 'bigon':

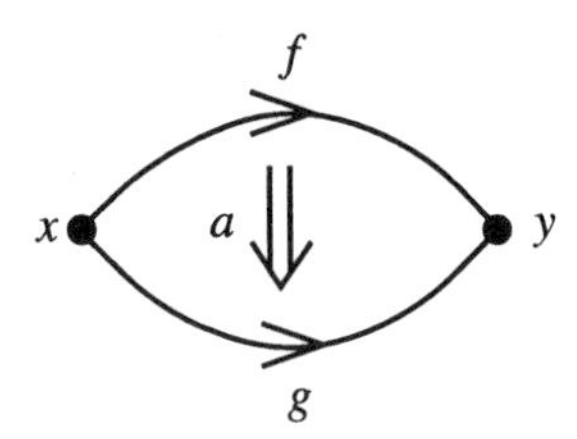

as the shape of a 2-morphism $a: f \Rightarrow g$ between morphisms $f, g: x \to y$ with the same source and target. This is the sort of 2-morphism used in the standard definitions of 'strict 2-categories' [21] — usually just called 2-categories — and the somewhat more general 'bicategories' [11]. There are two geometrically natural ways to compose 2-morphisms shaped like bigons. First, given 2-morphisms $a: f \Rightarrow g$ and $b: g \Rightarrow h$ as below, we can 'vertically' compose them to obtain a 2-morphism $a \cdot b: f \Rightarrow g$:

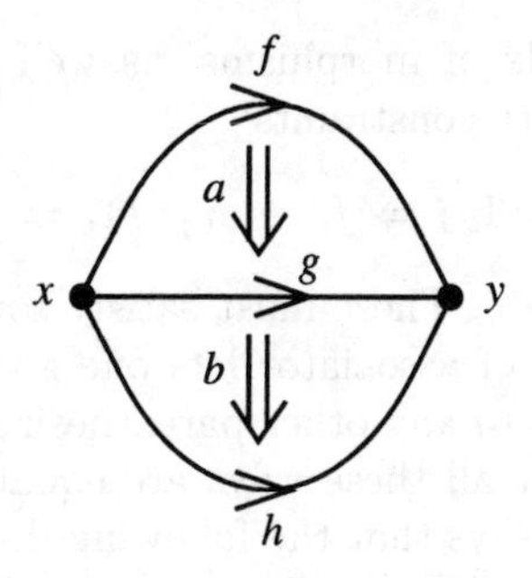

Second, given 2-morphisms $a: f \Rightarrow g$ and $b: h \Rightarrow i$ as below, we can 'horizontally' compose them to obtain a 2-morphism $ab: fh \Rightarrow gi$:

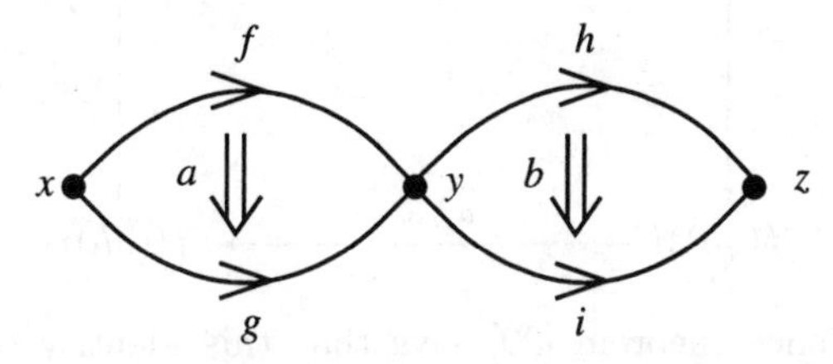

The definition of a strict 2-category is easy to state. The objects and morphisms must satisfy the usual rules holding in a category, while horizontal and vertical composition satisfy some additional axioms: vertical and horizontal composition are associative, for each morphism f there is a 2-morphism $1_f: f \Rightarrow f$ that is an identity for vertical composition, and for each identity morphism 1_x the 2-morphism 1_{1_x} is also an identity for horizontal composition. Finally, we require the following 'interchange law' relating vertical and horizontal composition:

$$(a \cdot b)(c \cdot d) = (ac) \cdot (bd)$$

whenever either side is well-defined. This makes the following composite 2-morphism unambiguous:

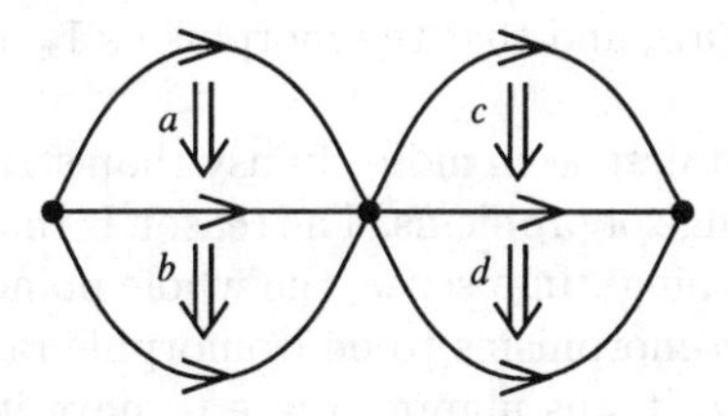

We can think of it either as the result of first doing two vertical composites and then one horizontal composite, or as the result of first doing two horizontal composites and then one vertical composite.

The definition of a bicategory is similar, but instead of requiring that the associativity and unit laws for morphisms hold 'on the nose' as equations, one requires merely that they hold up to isomorphism. Thus one has invertible 'associator' 2-morphisms

$$a_{f,g,h}: (fg)h \Rightarrow f(gh)$$

for every composable triple of morphisms, as well as invertible 2-morphisms called 'left and right identity constraints'

$$l_f : 1_x f \Rightarrow f, \qquad r_f : f 1_y \Rightarrow f$$

for every morphism $f: x \to y$. These must satisfy some equations of their own. For example, repeated use of associator lets one go from any parenthesization of a product of morphisms to any other parenthesization, but one can do so in many ways. To ensure that all these ways are equal, one imposes the Stasheff 'pentagon identity', which says that the following diagram commutes:

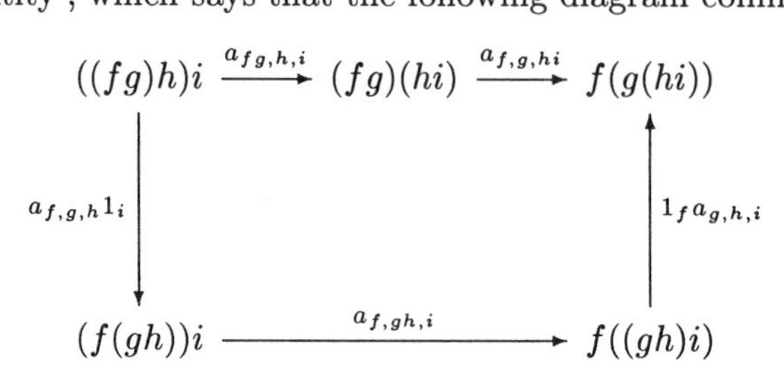

Mac Lane's coherence theorem [29] says that this identity suffices. Similarly, given morphisms $f: x \to y$ and $g: y \to z$, one requires that the following triangle commute:

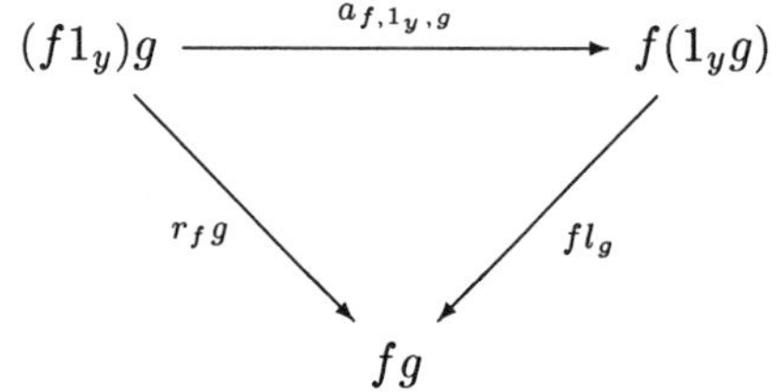

One also requires that the associators and unit constraints are natural with respect to their arguments. Also, as with strict 2-categories, one requires that vertical composition be associative, that vertical and horizontal composition satisfy the interchange law, and that the morphisms 1_x are identities for vertical composition.

While bicategories at first seem more clumsy than strict 2-categories, they are applicable to a wider range of problems. The reason is that frequently 'everything is only true up to something'. In a sense, the whole point of introducing $(n+1)$-morphisms is to allow n-morphisms to be isomorphic rather than merely equal. From this point of view, it was inappropriate to have imposed equational laws between 1-morphisms in the definition of a strict 2-category, and the definition of bicategory corrects this problem. This is known as 'weakening'.

To see some bicategories that are not strict 2-categories, consider bicategories with one object. Given a bicategory C with one object x, we can form a category $\tilde{C}$ whose objects are the morphisms of C and whose morphisms are the 2-morphisms of C. This is a special sort of category: we can 'multiply' the objects of $\tilde{C}$, since they are really just morphisms in C from x to itself. We call this sort of category — one that is really just a bicategory with one object —

a 'weak monoidal category'. We can do the same thing starting with a strict 2-category and get a 'strict monoidal category'.

The category Set becomes a weak monoidal category if we multiply sets using the Cartesian product. However, it is not a strict monoidal category! The reason is that the Cartesian product is not strictly associative:

$$(X \times Y) \times Z \neq X \times (Y \times Z).$$

To see this, one needs to pry into the set-theoretic definition of ordered pairs. The usual von Neumann definition is $(x, y) = \{\{x\}, \{x, y\}\}$, and using this, we clearly do not have strict associativity for the Cartesian product. Instead, we have associativity *up to a specified isomorphism*, the associator:

$$a_{X,Y,Z}: (X \times Y) \times Z \to X \times (Y \times Z)$$

which satisfies the pentagon identity. This is a typical example of how the bicategories found 'in nature' tend not to be strict 2-categories.

So far we have considered only bigons as possible shapes for 2-morphisms, but there are many other choices. For example, we might wish to use triangles going from a pair of morphisms $f: x \to y$, $g: y \to z$ to a morphism $h: x \to z$:

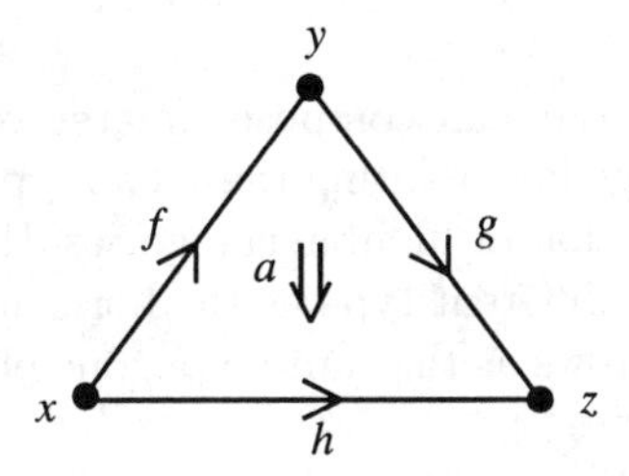

There is no good way to glue together triangles of this type to form other triangles of this type, but if we also allow the 'reverse' sort of triangle going from a single morphism to a pair:

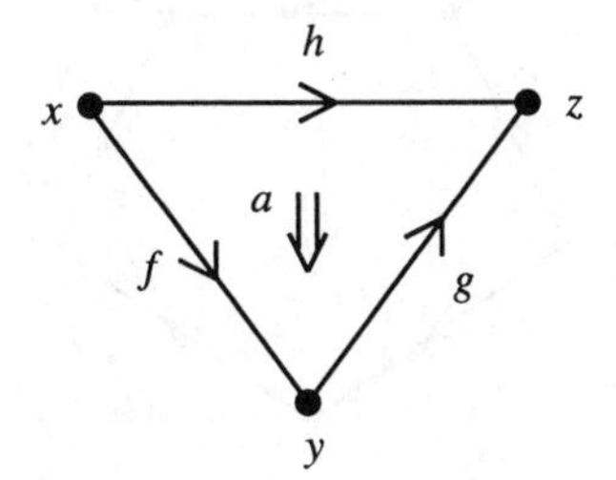

then there are various ways to glue together 3 triangles to form a larger one. For example:

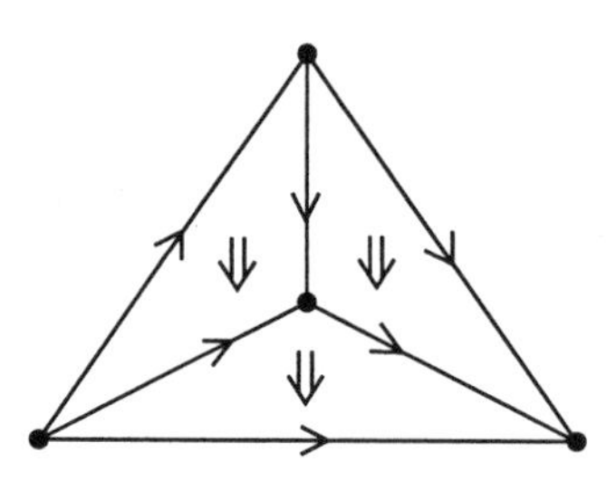

One can do quite a bit of topology in a purely combinatorial way using triangles and their higher-dimensional analogues, called simplices [31]. In these applications one often assumes that all j-morphisms are invertible, at least in some weakened sense. In the 2-dimensional case, this motivates the idea of 'reversing' a triangle.

Another approach would be to use squares going from a pair of morphisms $f: x \to y$, $g: y \to z$ to a pair $h: x \to w$, $i: w \to z$:

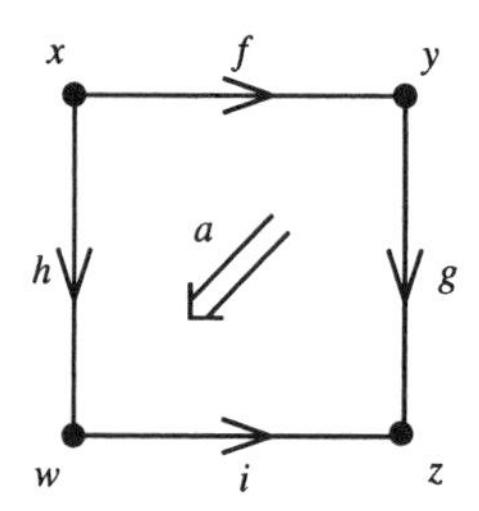

Much as with bigons, one can compose squares vertically and horizontally and require an 'interchange law' relating these two types of composition. This is the idea behind the definition of 'double categories' [16, 21], where the vertical arrows are treated as of a different type as the horizontal ones. If one treats the vertical and horizontal arrows as the same type, one obtains a theory equivalent to that of strict 2-categories.

Alternatively, one might argue that the business of picking a particular shape of 2-morphism as 'basic' is somewhat artificial. One might instead allow all possible polygons as shapes for 2-morphisms. The idea would be to use polygons whose boundary is divided into two parts having the arrows consistently oriented:

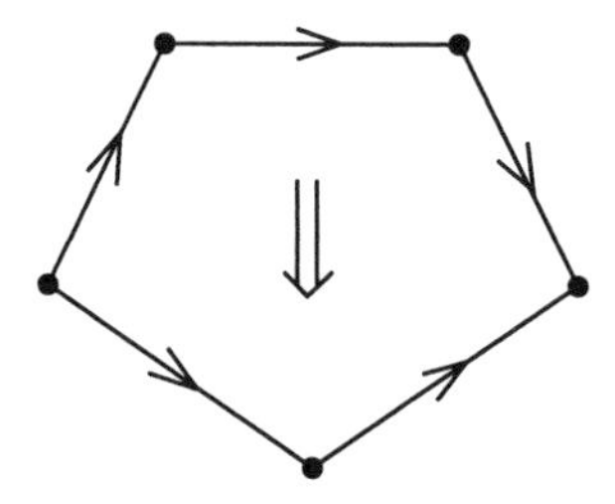

There are many ways to compose such polygons. However, while this approach might seem more general, one can actually define and work with these more general polygons within the theory of strict 2-categories [25, 34, 35].

Yet another approach would be to use only polygons having many 'infaces' but only one 'outface', like this:

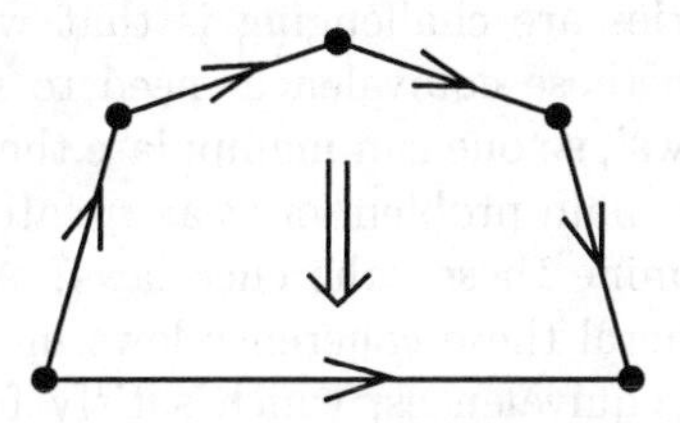

As we shall see, this has certain advantages of its own.

What about n-categories for higher n? In general, j-morphisms can be visualized as j-dimensional solids, and part of their boundary represents the source while the rest represents the target. However, as n increases one can imagine more and more definitions of 'n-category', because there are more and more choices for the shapes of the j-morphisms. The higher-dimensional analogues of bigons are called 'globes'. Globes are the basic shape in the traditional approach to 'strict n-categories', often simply called 'n-categories' [17, 15]. There has also been a lot of work on globular weak n-categories, such as Gordon, Power and Street's 'tricategories' [23], Trimble's 'tetracategories' [43], and Batanin's 'weak ω-categories' [8], which can have j-morphisms of arbitrarily high dimension. The higher-dimensional analogues of triangles, namely simplices, are used in the 'Kan complexes' favored by topologists [31], as well as in Street's 'simplicial weak ω-categories' [39] and Lawrence's 'n-algebras' [28]. The higher-dimensional analogues of squares, namely cubes, are used in Ehresmann's 'n-tuple categories' [16], as well as the work of Brown and his collaborators [13]. Finally, Dolan and the author [5] have given a definition of 'weak n-categories' based on some new shapes called 'opetopes'. We describe these in Section 4.

In addition to the issue of shapes for j-morphisms, there is the issue of the laws that composition operations should satisfy. Most importantly, there is the distinction between 'strict' and 'weak' approaches. In the 'strict' approach, composition of j-morphisms satisfies equational laws for all j. The philosophy behind the 'weak' approach is that equations should hold only at the top level, between n-morphisms. Laws concerning j-morphisms for $j < n$ should always be expressed as $(j+1)$-morphisms, or more precisely, 'equivalences'. Roughly, the idea here is that an equivalence between $(n-1)$-morphisms is an invertible n-morphism, while an equivalence between j-morphisms for lesser j is recursively defined as a $(j+1)$-morphism that is invertible *up to equivalence.*

Strict n-categories are fairly well-understood [15], but the interesting and challenging sort of n-categories are the weak ones. Weak n-categories are interesting because these are the ones that tend to arise naturally in applications. The reason for this is simple yet profound. Equations of the form $x = x$ are completely useless. All interesting equations are of the form $x = y$. Equations of this form can always be viewed as asserting the existence of a reversible sort of computation transforming x to y. In n-categorical terms, they assert the existence of an equivalence $f\colon x \to y$. To face up to this fact, it is helpful to systematically avoid equational laws and work explicitly with equivalences, instead. This leads naturally to working with weak n-categories, and eventually weak ω-categories.

The reason n-categories are challenging is that when equational laws are replaced by equivalences, these equivalences need to satisfy new laws of their own, called 'coherence laws', so one can manipulate them with some of the same facility as equations. The main problem of weak n-category theory is: how does one systematically determine these coherence laws? A systematic approach is necessary, because in general these coherence laws must themselves be treated not as equations but as equivalences, which satisfy further coherence laws of their own, and so on! This quickly becomes very bewildering if one proceeds on an ad hoc basis.

For example, suppose one tries to write down definitions of 'globular weak n-categories', that is, weak n-categories in the approach where the j-morphisms are shaped like globes. These are usually called categories, bicategories, tricategories, tetracategories, and so on. The definition of a category is quite concise; the most complicated axiom is the associative law $(fg)h = f(gh)$. As we have seen, in the definition of a bicategory this law is replaced by a 2-morphism, the associator, which in turn satisfies the pentagon identity. In the definition of a tricategory, the pentagon identity is replaced by a 3-isomorphism satisfying a coherence law which is best depicted using a 3-dimensional commutative diagram in the shape of the 3-dimensional 'associahedron'. In the definition of a tetracategory, this becomes a 3-morphism which satisfies a coherence law given by the 4-dimensional associahedron. In fact, the associahedra of all dimensions were worked out by Stasheff [38] in 1963 using homotopy theory. However, there are other sequences of coherence laws to worry about, spawned by the equational laws of the form $1f = f = f1$, and also the interchange laws governing the various higher-dimensional analogues of 'vertical' and 'horizontal' composition.

At this point the reader can be forgiven for wondering if the rewards of setting up a theory of weak n-categories really justify the labor involved. Before proceeding, let us describe some of the things n-categories should be good for.

3 Applications of n-Categories

One expects n-categories to show up in any situation where there are things, processes taking one thing to another, 'meta-processes' taking one process to another, 'meta-meta-processes', and so on. Clearly computer science is deeply concerned with such situations. Unfortunately, the author is not competent to discuss applications to this subject! Some other places where applications are evident include:

1. n-category theory
2. homotopy theory
3. topological quantum field theory

The first application is circular, but not viciously so. The point is that the study of n-categories leads to applications of $(n+1)$-category theory. The other two applications may sound abstruse and specialized, but there is a good reason for discussing them here. Pure n-category theory treats the most general iterated notion of process. Homotopy theory limits its attention to processes that are 'invertible', at least up to equivalence. Topological quantum field theory focuses attention on processes which have 'adjoints' or 'duals'. While generally not invertible even up to equivalence, such processes are reversible in a broader sense (the classic examples from category theory being adjoint functors). In what follows we briefly summarize all three applications in turn.

3.1 n-Category Theory

While self-referential, this application is perhaps the most fundamental. A 0-category is just a *set.* When one studies sets one is naturally led to consider the set of all sets. However, this turns out to be a bad thing to do, not merely because of Russell's paradox (which is easily sidestepped), but because one is interested not just in sets but also in the functions between them. What is interesting is thus the *category* of all sets, Set.

This category is in some sense the primordial category. Indeed, the Yoneda embedding theorem shows how every category can be thought of as a category of 'sets with extra structure' — at least modulo some technical issues of 'size'. However, when we study categories of sets with extra structure, it turns out to be worthwhile to develop category theory as a subject in its own right. In addition to categories and functors, natural transformations play a crucial role here. Thus one is led to study the *2-category* of all categories, Cat. This 2-category has categories as objects, functors between categories as morphisms, and natural transformations between functors as 2-morphisms.

The ladder of n-categories continues upwards in this way. For each n there is an $(n + 1)$-category of all n-categories, nCat. To really understand n-categories we need to understand this $(n + 1)$-category. Eventually this requires an understanding of $(n+1)$-categories in general, which then leads us to define $(n+1)$Cat.

There are some curious subtleties worth noting here, though. The 2-category Cat happens to be a strict 2-category. We could think of it as a bicategory if we wanted, but weakening happens not to be needed here, since functors compose associatively 'on the nose', not just up to a natural transformation. Using the fact that Cat is the primordial 2-category, one can show that every bicategory is equivalent to a strict 2-category in a certain precise sense. Technically speaking, one proves this using the Yoneda embedding for bicategories [23].

The fact that every weak 2-category can be 'strictified' seems to have held back work on weak n-categories: it raised the hope that every weak n-category might be equivalent to a strict one. It turned out, however, that the strict and weak approaches diverge as we continue to ascend the ladder of n-categories. On the one hand, we can always construct a strict $(n + 1)$-category of strict n-categories. On the other hand, we can construct a weak $(n + 1)$-category of weak n-categories. The latter is *not* equivalent to a strict $(n + 1)$-category for $n \geq 2$.

Consider the case $n = 2$. On the one hand, we can form a strict 3-category 2Cat whose objects are strict 2-categories. We can visualize a strict 2-category as a bunch of points, arrows and bigons. For simplicity, let us consider a very small 2-category C with just one interesting 2-morphism:

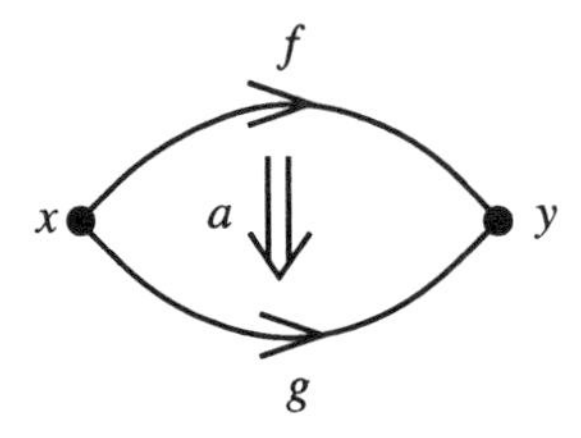

(We have not drawn the identity morphisms and 2-morphisms.) The morphisms in 2Cat are called '2-functors'. A 2-functor $F: C \to D$ sends objects to objects, morphisms to morphisms, and 2-morphisms to 2-morphisms, strictly preserving all structure. We can visualize F as creating a picture of the 2-category C in the 2-category D:

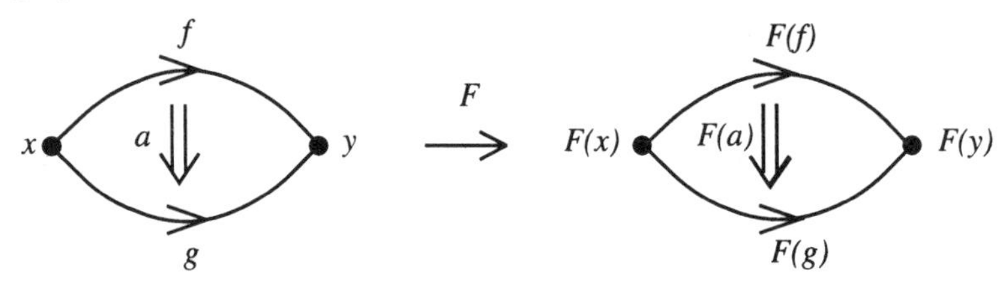

The 2-morphisms in 2Cat are called 'natural transformations'. A natural transformation $A: F \Rightarrow G$ between 2-functors $F, G: C \to D$ sends each object in C to a morphism in D and each morphism in C to a 2-morphism in D, and satisfies some conditions similar to those in the definition of a natural transformation between functors. We can visualize A as a prism going from one picture of C in D to another, built using commutative squares:

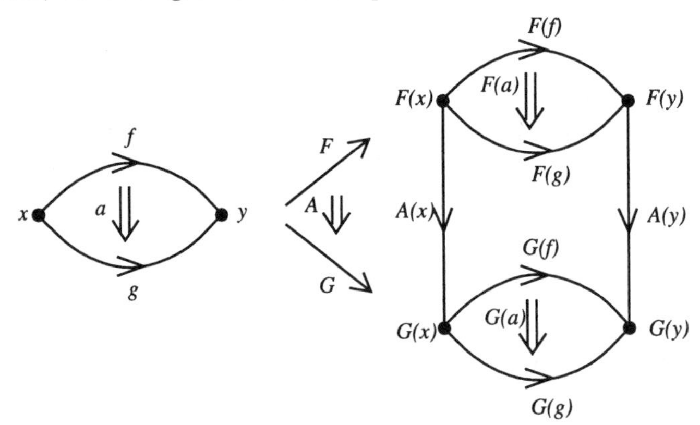

Finally, the 3-morphisms are called 'modifications'. A modification M from a natural transformation $A: F \Rightarrow G$ to a natural transformation $B: F \Rightarrow G$ sends each object $x \in C$ to a 2-morphism $M(x): A(x) \Rightarrow B(x)$ in D, in a manner satisfying some naturality conditions. We can visualize a modification M as follows:

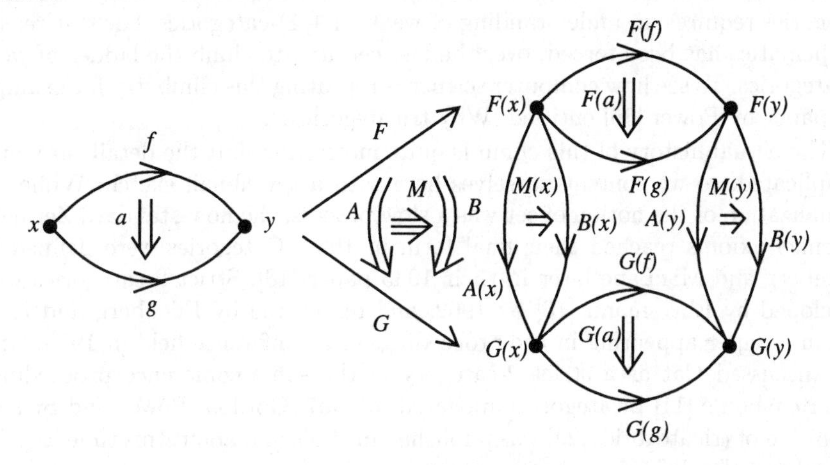

Note how the n-dimensionality of an n-category leads naturally to the $(n+1)$-dimensionality of nCat.

If instead we adopt the weak approach, we can form a tricategory Bicat whose objects are bicategories [23]. The morphisms in Bicat are called 'pseudofunctors'. A pseudofunctor $F: C \to D$ need not strictly preserve all the structure. For example, given morphisms $f: x \to y$ and $g: y \to z$ in C, we do not require that $F(fg) = F(f)F(g)$. Instead, we require only that the two sides are isomorphic by a specified 2-morphism, which in turn must satisfy some coherence laws. The 2-morphisms in Bicat are called 'pseudonatural transformations'. In these, the squares that had to commute in the definition of a natural transformation need only commute up to a 2-isomorphism satisfying certain coherence laws. The 3-morphisms in Bicat are called 'modifications'. Here there is no room for weakening, since a modification sends each object in C to a 2-morphism in D, and the only sort of laws that 2-morphisms can satisfy in a bicategory are equational laws.

One can show that Bicat is not equivalent to a strict 3-category, so we really need the more general notion of tricategory. Or do we? One might argue that we needed tricategories only because we made the mistake of not strictifying our bicategories. After all, every bicategory is equivalent to a strict 2-category. Perhaps if we replaced every bicategory with an equivalent strict 2-category, we could work in the strict 3-category 2Cat and never need to think about Bicat.

Alas, while superficially plausible, this line of argument is naive. We have said that every bicategory C is equivalent to a strict 2-category C'. But what does 'equivalent' mean here, precisely? It means 'equivalent, as an object of Bicat'. In other words, for every bicategory C there is a strict 2-category C' and a pseudofunctor $F: C \to C'$ that is invertible up to a pseudonatural transformation that is invertible up to a modification that is invertible! In practice, therefore, the business of strictifying bicategories requires a solid understanding of Bicat as the full-fledged tricategory it is.

There is much more to say about this subject, but the basic point is that, like it or not, sets are weak 0-categories, and a deep understanding of weak n-categories requires an understanding of weak $(n+1)$-categories. For this reason mathematics has been forced, over the last century, to climb the ladder of weak n-categories. To see how computer science is repeating this climb, try for example the paper by Power [36] entitled 'Why tricategories?'.

The actual history of this climb is quite interesting, but the details are quite complicated, so we content ourselves here with a thumbnail sketch. While the formalization of the notion of set was a slow process, the now-standard Zermelo-Fraenkel axioms reached their final form in 1922. Categories were defined by Eilenberg and MacLane later in their 1945 paper [18]. Strict 2-categories were developed by Ehresmann [16] by 1962, and reinvented by Eilenberg and Kelly [17] in a paper appearing in the proceedings of a conference held in 1965. Gray [20] discussed Cat as a strict 2-category in the same conference proceedings, and Bénabou's [11] bicategories appeared in 1967. Gordon, Power and Street's definition of tricategories [23] was published in 1995, and about this time Trimble formulated the definition of tetracategories [43].

Subsequent work has concentrated on radically accelerating this process by defining weak n-categories for all n simultaneously. Actually, Street [39] proposed a simplicial definition of weak n-categories for all n in 1987, but this appears not to have been seriously studied, perhaps in part because it came too early! Starting in 1995, Dolan and the author gave a definition of weak n-categories using 'opetopic sets' [4, 5], Tamsamani gave a definition using 'multisimplicial sets' [41, 42], and Batanin gave definition of globular weak ω-categories [8, 9, 40]. Dolan and the author have constructed the weak $(n+1)$-category of their n-categories, and Simpson [37] has constructed the weak $(n+1)$-category of Tamsamani's n-categories. Now the focus is turning towards working with these different definitions and seeing whether they are equivalent. We return to this last issue in the Conclusions.

3.2 Homotopy Theory

A less inbred application of n-category theory is to the branch of algebraic topology known as homotopy theory. In fact, many of our basic insights into n-categories come from this subject. The reason is not far to seek. Topology concerns the category Top whose objects are topological spaces and whose morphisms are continuous maps. Unfortunately, there is no useful classification of topological spaces up to isomorphism — an isomorphism in Top being called a 'homeomorphism'. When topologists realized this, they retreated to the goal of classifying spaces up to various coarser equivalence relations. Homotopy theory is all about properties that are preserved by continuous deformations. More precisely, given spaces $X, Y \in$ Top and maps $F, G\colon X \to Y$, one defines a 'homotopy' from F to G to be a map $H\colon [0,1] \times x \to y$ with

$$H(0,\cdot) = F, \qquad H(1,\cdot) = G.$$

Homotopy theory studies properties of maps that are preserved by homotopies. Thus two spaces X and Y are 'the same' for the purposes of homotopy theory, or more precisely 'homotopy equivalent', if there are maps $F: X \to Y$, $G: Y \to X$ which are inverses *up to homotopy.*

In fact, what we have done here is made Top into a 2-category whose objects are spaces, whose morphisms are maps between spaces, and whose 2-morphisms are homotopies between maps. This allows us to replace the categorical concept of isomorphism between spaces by the more flexible 2-categorical concept of equivalence. However, work on homotopy theory soon led to the study of 'higher homotopies'. Since a homotopy is itself a map, the concept of a homotopy between homotopies makes perfect sense, and we may iterate this indefinitely. This amounts to treating Top as an n-category for arbitrarily large n, or for that matter, as an ω-category.

It is worthwhile pondering how the seemingly innocuous category Top became an ω-category. The key trick was to use the unit interval $[0, 1]$ to define higher-level morphisms. The reason this trick works is that the unit interval resembles an *arrow* going from 0 to 1. One could say that the abstract arrow we use in category theory is a kind of metaphor for the unit interval — or conversely, that the unit interval we use in topology is a kind of metaphor for the process of going from 'here' to 'there'. However, unlike the most general sort of abstract arrow, the unit interval has a special feature: we can go from 1 to 0 as easily as we can go from 0 to 1.

Taking advantage of this insight, Grothendieck [24] proposed thinking of homotopy theory as a branch of n-category theory, as follows. We should be able to associate to any space X a weak ω-category $\Pi(X)$ whose objects are points $x \in X$, whose morphisms are paths (maps $F: [0, 1] \to X$) going from one point to another, whose 2-morphisms are certain paths of paths, and so on. Due to the special feature of the unit interval, every j-morphism in this ω-category should be an equivalence. We call this special sort of ω-category an 'ω-groupoid', since a category with all morphisms invertible is called a groupoid.

Grothendieck also argued that conversely, we should be able to obtain a topological space $N(G)$ from any weak ω-groupoid G, essentially by taking seriously the picture we can draw with points for objects of G, intervals for morphisms of G, and so on. By this means we should be able to obtain weak ω-functors $\Pi: \mathrm{Top} \to \omega\mathrm{Gpd}$ and $N: \omega\mathrm{Gpd} \to \mathrm{Top}$. Using these, we should be able to show that that the weak ω-categories Top and ωGpd are equivalent, as objects of ωCat. In short, homotopy theory is another word for the study of ω-groupoids!

There many ways to try to realize this program, a number of which have already obtained results. It is well-known that all of homotopy theory can be done purely combinatorially using 'Kan complexes' [31], which may be regarded as simplicial weak ω-categories. Brown, Higgins, Loday, and collaborators have developed a variety of approaches using cubes [13]. Kapranov and Voevodsky [26] have shown that homotopy theory is in principle equivalent to the study of their '∞-groupoids'. Tamsamani has also shown that his approach to weak n-categories reduces to homotopy theory in the n-groupoid case [42].

Many homotopy theorists might doubt the importance of seeing homotopy theory as a branch of n-category theory. In a sense, they already implicitly know many of the lessons n-category theory has to offer: the idea of replacing equations by equivalences, the importance of 'homotopies between homotopies', and the crucial importance of coherence laws. Eventually n-category should be able to help homotopy theory in its treatment of morphisms that are not equivalences. In the short term, however, the question is not what n-categories can do for homotopy theory, but what homotopy theory can do for n-categories.

In fact, many ideas in n-category theory have already had their origin in homotopy theory. A good example is Stasheff's work on the associahedron [38]. Recall that a 'monoid' is a set equipped with an associative product and multiplicative unit, while a 'topological monoid' is a monoid equipped with a topology for which the product is continuous. Stasheff wanted to uncover the homotopy-invariant structure contained in a topological monoid. Suppose X is a topological monoid and Y is a space equipped with a homotopy equivalence to X. What sort of structure does Y inherit from X? Clearly we can use the homotopy equivalence to transport the product and unit from X to Y, obtaining a product and unit on Y satisfying the laws of a monoid *up to homotopy*. For example, the two maps

$$F, G\colon Y \times Y \times Y \to Y$$

given by

$$F(y_1, y_2, y_3) = (y_1 y_2) y_3$$
$$G(y_1, y_2, y_3) = y_1 (y_2 y_3)$$

need not be equal, but there is a homotopy between them, the 'associator'. Stasheff showed that this associator satisfies the pentagon identity up to homotopy, and that this homotopy satisfies a coherence law of its own, again up to homotopy, and so on ad infinitum. By working out these coherence laws in detail, he discovered the associahedron. Later the associahedron turned out to be relevant to weak ω-categories in general. Part of the reason is that we can think of the space Y above as a special sort of ω-category. A monoid can be thought of as a category with one object, by viewing the monoid elements as morphisms from this object to itself. Similarly, we can view Y as a weak ω-category with one object, points of Y as morphisms from this object to itself, paths between these as 2-morphisms, and so on.

3.3 Topological Quantum Field Theory

In physics, interest in n-categories was sparked by developments in relating topology and quantum field theory [22]. One can roughly date the beginning of this story to 1985, when Jones came across a wholly unexpected invariant of knots while studying some operator algebras invented by von Neumann in his work on the mathematical foundations of quantum theory. Soon this 'Jones polynomial' was generalized to a family of knot invariants. It was then realized that these generalizations could be systematically derived from algebraic structures known as 'quantum groups', first invented by Drinfel'd and collaborators in their work on exactly soluble 2-dimensional field theories [14, 30]. This approach involved 2-dimensional pictures of knots. The story became even more exciting when Witten came up with a manifestly 3-dimensional approach to the new knot invariants, deriving them from a quantum field theory in 3-dimensional spacetime now known as Chern-Simons-Witten theory. This approach also gave invariants of 3-dimensional manifolds.

These developments exposed a deep but mysterious unity in what at first might seem like disparate branches of algebra, topology, and quantum physics. Interestingly, it appears that the roots of this unity lie in certain aspects of n-category theory. In fact, this is the main reason for the author's interest in n-categories: it seems that a good theory of weak n-categories is needed as a framework for the mathematics that will be able to reduce the currently rather elaborate subject of 'topological quantum field theory' to its simple essence. Having explained this at length elsewhere [3], we limit our remarks here to a few key points.

Quantum physics relies crucially on the theory of Hilbert spaces. For simplicity, we limit our attention here to the finite-dimensional case, defining a 'Hilbert space' to be a finite-dimensional complex vector space H equipped with an 'inner product'

$$\langle \cdot, \cdot \rangle : H \times H \to \mathbb{C}$$

which is linear in the second argument, conjugate-linear in the first, and satisfies $\langle \psi, \phi \rangle = \overline{\langle \phi, \psi \rangle}$ for all $\psi, \phi \in H$ and $\langle \psi, \psi \rangle > 0$ for all nonzero $\psi \in H$. The inner product allows us to define the norm of a vector $\psi \in H$ by

$$\|\psi\| = \langle \psi, \psi \rangle^{1/2},$$

but its main role in physics is to compute amplitudes. States of a quantum system are described by vectors with norm 1. If one places a quantum system in the state ψ, and then does an experiment to see if it is in some state ϕ, the probability that the answer is 'yes' equals

$$|\langle \phi, \psi \rangle|^2.$$

This is automatically a real number between 0 and 1. However, when one delves deeper into the theory, it appears that even more fundamental than the probability is the 'amplitude'

$$\langle \phi, \psi \rangle,$$

which is of course a complex number.

The role of the inner product in quantum physics has always been a source of puzzles to those with an interest in the philosophical foundations of the subject. Complex amplitudes lack the intuitive immediacy of probabilities. From the category-theoretic point of view, part of the problem is to understand the category of Hilbert spaces. The objects are Hilbert spaces, but what are the morphisms? Typically morphisms are required to preserve all the structure in sight. This suggests taking the morphisms to be linear operators preserving the inner product. However, other linear operators are also important. Particularly in topological quantum field theory, there are good reasons to take *all* linear operators as morphisms. However, if we define a category Hilb this way, then Hilb is equivalent to the category Vect of complex vector spaces. This then raises the question: how does Hilb really differ from Vect, if as categories they are equivalent?

Luckily, quantum theory suggests an answer to this question. Given any linear operator $F: H \to H'$ between Hilbert spaces, we may define the 'adjoint' $F^*: H' \to H$ to be the unique linear operator with

$$\langle F\phi, \psi \rangle = \langle \phi, F^* \psi \rangle$$

for all $\phi \in H$, $\psi \in H'$. This sort of adjoint is basic to quantum theory. From the category-theoretic point of view, the role of the adjoint is to make Hilb into a '$*$-category': a category C equipped with a contravariant functor $*: C \to C$ fixing objects and satisfying $*^2 = 1_C$. While Hilb and Vect are equivalent as categories, only Hilb is a $*$-category.

This is particularly important in topological quantum field theory. The mysterious relationships between topology, algebra and physics exploited by this subject amount in large part to the existence of interesting functors from various topologically defined categories to the category Hilb. These topologically defined categories are always $*$-categories, and the really interesting functors from them to Hilb are always '$*$-functors', functors preserving the $*$-structure. Physically, the $*$ operation corresponds to *reversing the direction of time.* For example, there is a $*$-category whose objects are collections of points and whose morphisms are 'tangles':

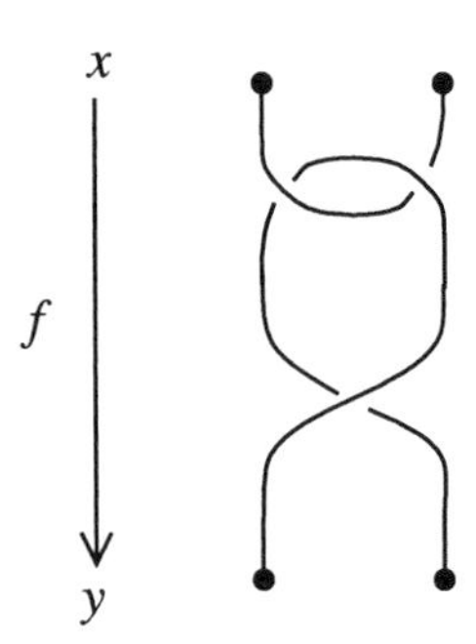

We can think of this morphism $f: x \to y$ as representing the trajectories of a collection of particles and antiparticles, where particles and antiparticles can be created or annihilated in pairs. Reversing the direction of time, we obtain the 'dual' morphism $f^*: y \to x$:

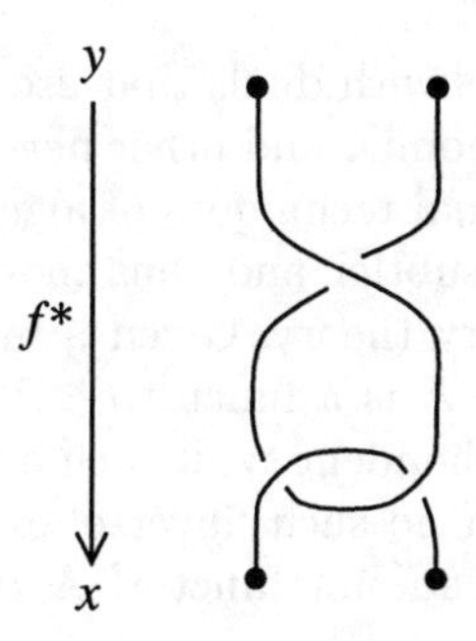

This morphism is not the inverse of f, since the composite ff^* is a nontrivial tangle:

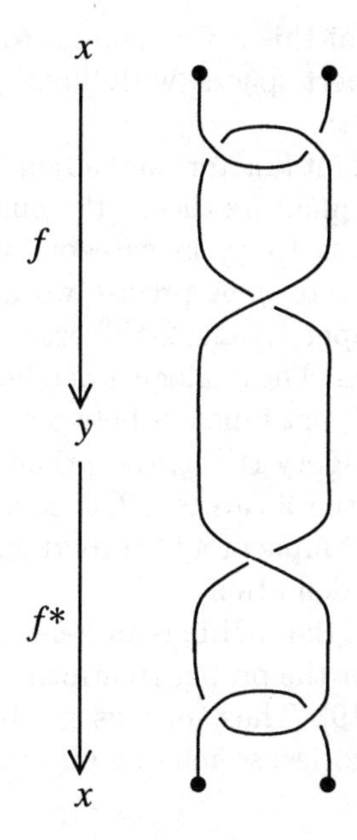

Any groupoid becomes a $*$-category if we set $f^* = f^{-1}$ for every morphism f, but the most interesting $*$-categories in topological quantum field theory are not groupoids.

The above example involves 1-dimensional curves in 3-dimensional spacetime. More generally, topological quantum field theory studies n-dimensional manifolds embedded in $(n + k)$-dimensional spacetime, which in the $k \to \infty$ limit appear as 'abstract' n-dimensional manifolds. It appears that these are best described using certain 'n-categories with duals', meaning n-categories in which every j-morphism f has a dual f^*. Unfortunately, so far the details have only been worked out in certain low-dimensional cases [2, 6]. The main problem is that the notion of 'n-category with duals' is only beginning to be understood.

One class of n-categories with duals should be the n-groupoids; this would explain many relationships between topological quantum field theory and homotopy theory [33]. However, the novel aspects of topological quantum field theory should arise from n-categories with duals that are not n-groupoids. Indeed, this explains why the Jones polynomial and other new knot invariants were not discovered earlier using traditional techniques of algebraic topology.

The idea that duals are subtler and thus more interesting than inverses is already familiar from category theory. Given a functor $F: C \to D$, the correct sort of weakened 'inverse' to F is a functor $G: D \to C$ such that FG and GF are naturally isomorphic to the identity; if such a functor G exists then F is an equivalence. However, even if no such 'inverse' exists, the functor F may have a kind of 'dual', namely an adjoint functor! A right adjoint $F^*: C \to D$, for example, would satisfy:

$$\hom(Fx, y) \cong \hom(x, F^*y)$$

for all $x \in C$, $y \in D$. Note that this is very similar to the definition of the adjoint of a linear map between Hilbert spaces, with 'hom' playing the role of the inner product.

The analogy between adjoint functors and adjoint linear operators relies upon a deeper analogy: just as in quantum theory the inner product $\langle \phi, \psi \rangle$ represents the *amplitude* to pass from ϕ to ψ, in category theory $\hom(x, y)$ represents the *set of ways* to go from x to y. A precise working out of this analogy can be found in the author's paper [2] on '2-Hilbert spaces'. These are to Hilbert spaces as categories are to sets. The analogues of adjoint linear operators between Hilbert spaces are certain adjoint functors between 2-Hilbert spaces. Just as the primordial example of a category is Set, the primordial example of a 2-Hilbert space is Hilb. Also, just as the 2-category Cat is a 3-category, it appears that the 2-category 2Hilb is an example of a '3-Hilbert space' — a concept which has not yet been given a proper definition.

More generally, it appears that nHilb is an n-category with duals, and that 'n-Hilbert spaces' are needed for the proper treatment of n-dimensional topological quantum field theories [3, 19]. Thus, just as mathematics has been forced to ascend the ladder of n-categories, so may be physics!

4 A Definition of Weak n-Category

As discussed in Section 2, any definition of n-categories involves a choice of the basic shapes of j-morphisms and a choice of allowed ways to glue them together. Any definition of weak n-categories also requires a careful treatment of coherence laws. In what follows we present a definition of weak n-categories in which all these issues are handled in a tightly linked way. In this definition, the basic shapes of j-morphisms are the j-dimensional 'opetopes'. The allowed ways of gluing together the j-dimensional opetopes correspond precisely to the $(j+1)$-dimensional opetopes. Moreover, the coherence laws satisfied by composition correspond to still higher-dimensional opetopes!

Before going into the details, let us give a rough sketch of this works. First consider some low-dimensional opetopes. The only 0-dimensional opetope is the point:

There is no way to glue together 0-dimensional opetopes. The only 1-dimensional opetope is the interval, or more precisely the arrow:

The allowed ways of gluing together 1-dimensional opetopes are given by the 2-dimensional opetopes. The first few 2-dimensional opetopes are as follows:

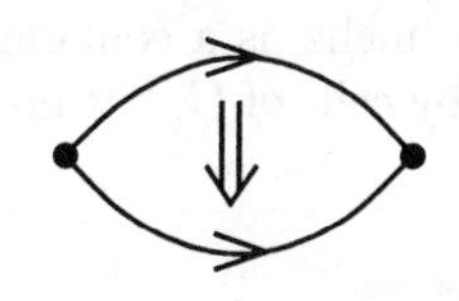
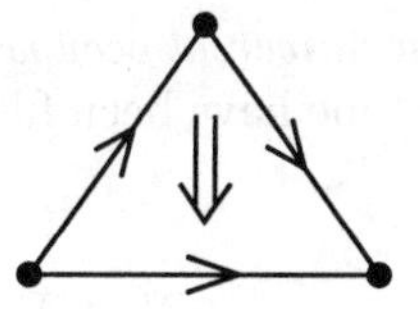
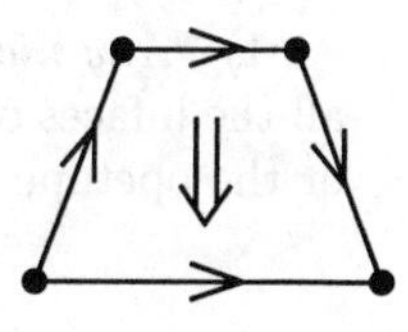

For any $k \geq 0$, there is a 2-dimensional opetope with k 'infaces' and one 'outface'. (We are glossing over some subleties here; for reasons noted later, there are really $k!$ such opetopes.)

The allowed ways of gluing together 2-dimensional opetopes are given by the 3-dimensional opetopes. There are many of these; a simple example is as follows:

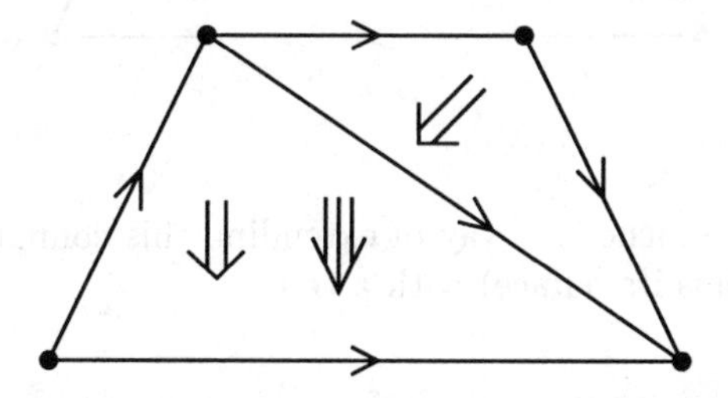

This may be a bit hard to visualize, but it depicts a 3-dimensional shape whose front consists of two 3-sided 'infaces', and whose back consists of a single 4-sided 'outface'. We have drawn double arrows on the infaces but not on the outface. Note that while this shape is topologically a ball, it cannot be realized as a polyhedron with planar faces. This is typical of opetopes.

In general, a $(n+1)$-dimensional opetope has any number of infaces and exactly one outface: the infaces are n-dimensional opetopes glued together in a tree-like pattern, while the outface is a single n-dimensional opetope. For example, the 3-dimensional opetope above corresponds to the following tree:

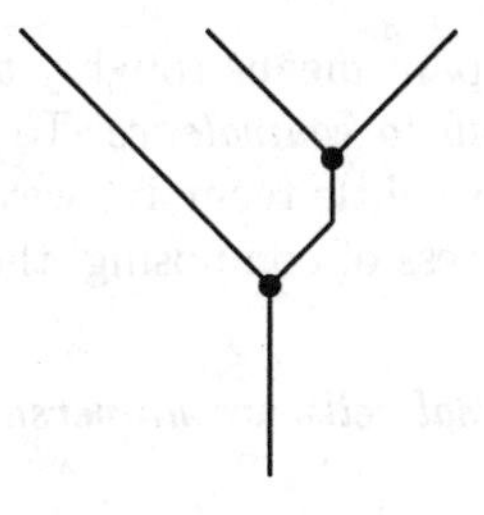

The two triangular infaces of the opetope correspond to the two nodes in this tree. This is a rather special tree; in general, we allow nonplanar trees with any number of nodes and any number of edges coming into each node.

In our approach, a weak n-category is a special sort of 'opetopic set'. Basically, an opetopic set is a set of 'cells' shaped like opetopes, such that any face of any cell is again a cell. In a weak n-category, the j-dimensional cells play the role of j-morphisms. An opetopic set C is an n-category if it satisfies the following two properties:

1) *"Any niche has a universal occupant."* A 'niche' is a configuration where all the infaces of an opetope have been filled in by cells of C, but not the outface or the opetope itself:

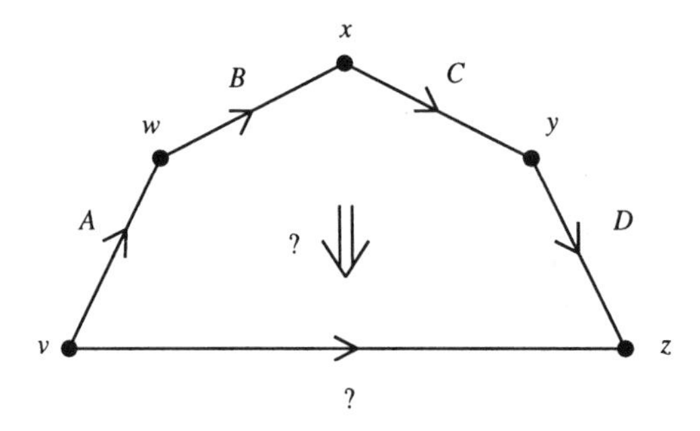

An 'occupant' of the niche is a way of extending this configuration by filling in the opetope (and thus its ouface) with a cell:

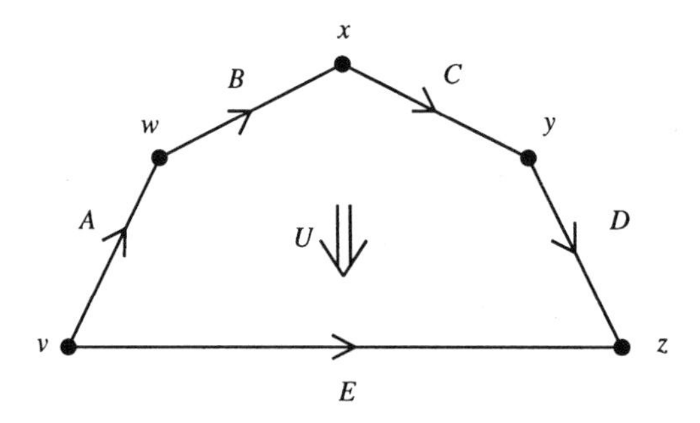

The 'universality' of an occupant means roughly that every other occupant factors through the given one *up to equivalence.* To make this precise we need to define universality in a rather subtle recursive way. We may think of a universal occupant of a niche as 'a process of composing' the infaces, and its outface as 'a composite' of the infaces.

2) *"Composites of universal cells are universal."* Suppose that U, V, and W below are universal cells:

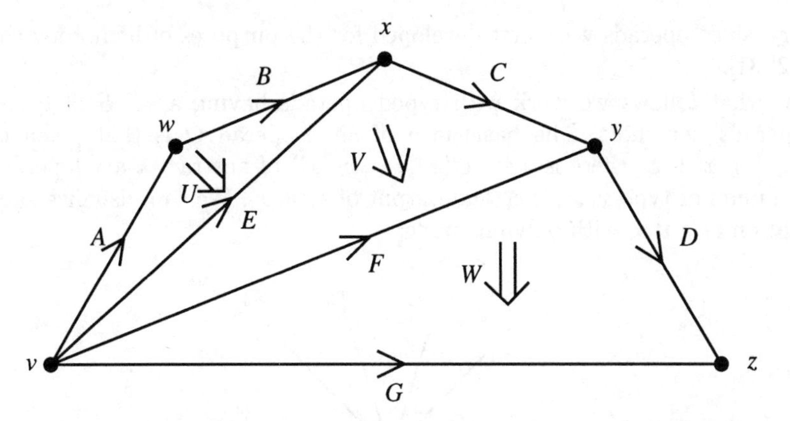

Then we can compose them, and we are guaranteed that their composite is again universal, and thus that the outface G is a composite of the cells A, B, C, D. Note that a process of composing U, V, W is described by a universal occupant of a niche of one higher dimension.

Note that in this approach to weak n-categories, composition of cells is not an operation in the traditional sense: the composite is defined by a universal property, and is thus unique only *up to equivalence.* Only at the top level, for the n-cells in an n-category, is the composite truly unique. This may seem odd at first glance, but in fact it closely reflects actual mathematical practice. For example, it is unnatural to think of the Cartesian product as an operation on Set. We can do it, but there are as many ways to do so as there are ways to define ordered pairs in set theory; there is certainly nothing sacred about the von Neumann definition. If we arbitrarily choose a way, we can think of Set as a weak monoidal category, i.e., a bicategory with one object. However, we can avoid this arbitrariness if we define the Cartesian product of sets by a universal property, using the category-theoretic concept of 'product'. Then the product of sets is only defined up to a natural isomorphism, and Set becomes our sort of weak 2-category with one object. In this approach, all the necessary coherence laws *follow automatically* from the universal property defining the product.

In the following sections we first review the theory of operads, and then explain how this theory can be used to define the opetopes. After a brief discussion of some notions related to opetopic sets, we give the definition of 'universal occupant' of a niche, and then the definition of weak n-category. At various points we skim over technical details; these can all be found in our paper [5].

4.1 Operads

To describe the opetopes we need to specify exactly which tree-like patterns we can use to glue together the opetopes of a given dimension; these are the opetopes of the next higher dimension. For this we use the theory of 'operads'. An operad is a gadget consisting of abstract k-ary 'operations' and various ways of composing them, and the n-dimensional opetopes will be the operations of a certain operad. This is another example of how n-category theory is indebted to homotopy theory, since operads were first developed for the purposes of homotopy theory [1, 12, 31].

In what follows we work with typed operads having a set S of types, or 'S-operads' for short. The basic idea of an S-operad O is that given types $x_1, \ldots, x_k, x' \in S$, there is a set $O(x_1, \ldots, x_k; x')$ of abstract k-ary 'operations' with inputs of type $x_1, \ldots, x_k$ and output of type x'. We can visualize such an operation as a tree with only one node:

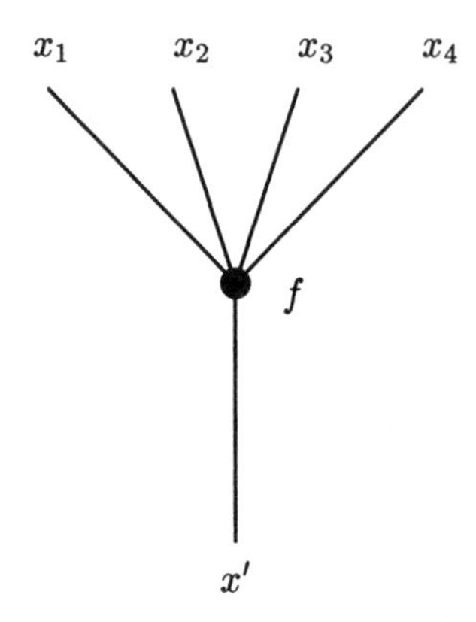

In an operad, we can get new operations from old ones by composing them, which we can visualize in terms of trees as follows:

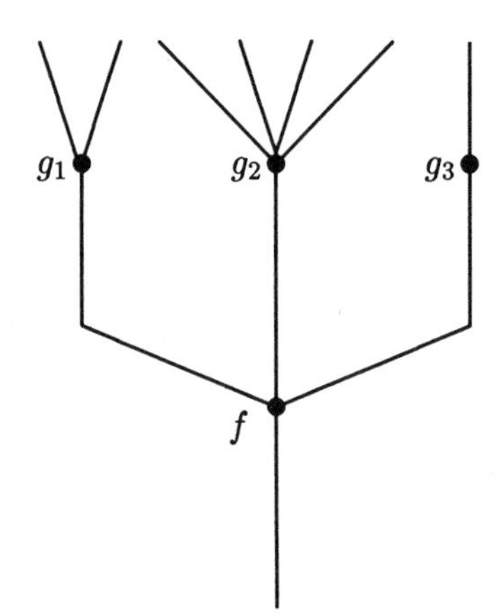

We can also obtain new operations from old by permuting arguments, and there is a unary 'identity' operation of each type. Finally, we demand a few plausible axioms: the identity operations act as identities for composition, permuting arguments is compatible with composition, and composition is 'associative', making composites of the following sort well-defined:

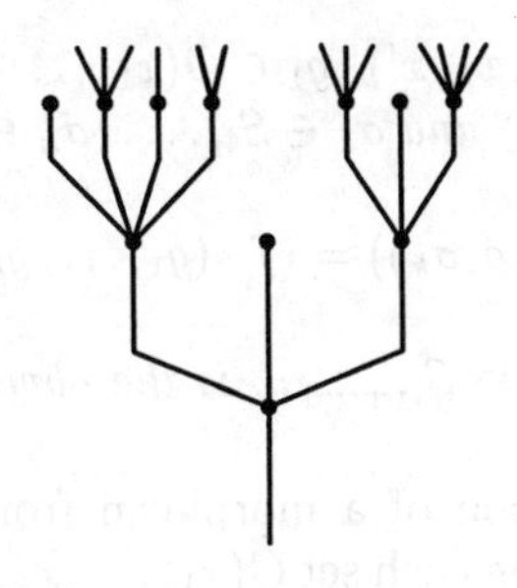

Formally, we have:

Definition 1. *For any set S, an 'S-operad' O consists of*

1. *for any $x_1, \ldots, x_k, x' \in S$, a set $O(x_1, \ldots, x_k; x')$*
2. *for any $f \in O(x_1, \ldots, x_k; x')$ and any $g_1 \in O(x_{11}, \ldots, x_{1i_1}; x_1), \ldots,$ $g_k \in O(x_{k1}, \ldots, x_{ii_k}; x_k)$, an element*

$$f \cdot (g_1, \ldots, g_k) \in O(x_{11}, \ldots, x_{1i_1}, \ldots\ldots, x_{k1}, \ldots, x_{ii_k}; x')$$

3. *for any $x \in S$, an element $1_x \in O(x; x)$*
4. *for any permutation $\sigma \in S_k$, a map*

$$\sigma: O(x_1, \ldots, x_k; x') \to O(x_{\sigma(1)}, \ldots, x_{\sigma(k)}; x')$$
$$f \mapsto f\sigma$$

such that:

(a) *whenever both sides make sense,*

$$f \cdot (g_1 \cdot (h_{11}, \ldots, h_{1i_1}), \ldots, g_k \cdot (h_{k1}, \ldots, h_{ki_k})) =$$

$$(f \cdot (g_1, \ldots g_k)) \cdot (h_{11}, \ldots, h_{1i_1}, \ldots\ldots, h_{k1}, \ldots, h_{ki_k})$$

(b) *for any $f \in O(x_1, \ldots, x_k; x')$,*

$$f = 1_{x'} \cdot f = f \cdot (1_{x_1}, \ldots, 1_{x_k})$$

(c) *for any $f \in O(x_1, \ldots, x_k; x')$ and $\sigma, \sigma' \in S_k$,*

$$f(\sigma\sigma') = (f\sigma)\sigma'$$

(d) *for any $f \in O(x_1, \ldots, x_k; x')$, $\sigma \in S_k$, and $g_1 \in O(x_{11}, \ldots, x_{1i_1}; x_1), \ldots,$ $g_k \in O(x_{k1}, \ldots, x_{ki_k}; x_k)$,*

$$(f\sigma) \cdot (g_{\sigma(1)}, \ldots, g_{\sigma(k)}) = (f \cdot (g_1, \ldots, g_k))\, \rho(\sigma),$$

where $\rho: S_k \to S_{i_1 + \cdots + i_k}$ is the obvious homomorphism.

(e) *for any* $f \in O(x_1, \ldots, x_k; x')$, $g_1 \in O(x_{11}, \ldots, x_{1i_1}; x_1), \ldots,$ $g_k \in O(x_{k1}, \ldots, x_{ki_k}; x_k)$, *and* $\sigma_1 \in S_{i_1}, \ldots, \sigma_k \in S_{i_k}$,

$$(f \cdot (g_1\sigma_1, \ldots, g_k\sigma_k)) = (f \cdot (g_1, \ldots, g_k))\, \rho'(\sigma_1, \ldots, \sigma_k),$$

where $\rho' \colon S_{i_1} \times \cdots \times S_{i_k} \to S_{i_1+\cdots+i_k}$ *is the obvious homomorphism.*

There is an obvious notion of a morphism from an S-operad O to an S-operad O': a function mapping each set $O(x_1, \ldots, x_k, x')$ to the corresponding set $O'(x_1, \ldots, x_k, x')$, preserving composition, identities, and the symmetric group actions. An important example is an 'algebra' of an operad, in which its abstract operations are represented as actual functions:

Definition 2. *For any S-operad O, an 'O-algebra' A consists of:*

1. *for any $x \in S$, a set $A(x)$.*
2. *for any $f \in O(x_1, \ldots, x_k; x')$, a function*

$$\alpha(f) \colon A(x_1) \times \cdots \times A(x_k) \to A(x')$$

such that:

(a) *whenever both sides make sense,*

$$\alpha(f \cdot (g_1, \ldots, g_k)) = \alpha(f)(\alpha(g_1) \times \cdots \times \alpha(g_k))$$

(b) *for any $x \in C$, $\alpha(1_x)$ acts as the identity on $A(x)$*

(c) *for any $f \in O(x_1, \ldots, x_k, x')$ and $\sigma \in S_k$,*

$$\alpha(f\sigma) = \alpha(f)\sigma,$$

where $\sigma \in S_k$ acts on the function $\alpha(f)$ on the right by permuting its arguments.

We can think of an operad as a simple sort of theory, and its algebras as models of this theory. Thus we can study operads either 'syntactically' or 'semantically'. To describe an operad syntactically, we list:

1. the set S of *types*,
2. the sets $O(x_1, \ldots, x_k; x')$ of *operations*,
3. the set of all *reduction laws* saying that some composite of operations (possibly with arguments permuted) equals some other operation.

This is like a presentation in terms of generators and relations, with the reduction laws playing the role of relations. On the other hand, to describe an operad semantically, we describe its algebras.

4.2 Opetopes

The following fact is the key to defining the opetopes. Let O be an S-operad, and let $\mathrm{elt}(O)$ be the set of all operations of O.

Theorem 3. *There is an* $\mathrm{elt}(O)$*-operad* O^+ *whose algebras are* S*-operads over* O*, i.e.,* S*-operads equipped with a homomorphism to* O*.*

We call O^+ the 'slice operad' of O. One can describe O^+ syntactically as follows:

1. The types of O^+ are the operations of O.
2. The operations of O^+ are the reduction laws of O.
3. The reduction laws of O^+ are the ways of combining reduction laws of O to give other reduction laws.

The 'level-shifting' going on here as we pass from O to O^+ captures the process by which equational laws are promoted to equivalences and these equivalences satisfy new coherence laws of their own. In this context, the new laws are just *the ways of combining the old laws.*

Note that we can iterate the slice operad construction. Let O^{n+} denote the result of applying the slice operad construction n times to the operad O if $n \geq 1$, or just O itself if $n = 0$.

Definition 4. *An* n*-dimensional 'O-opetope' is a type of* O^{n+}*, or equivalently, if* $n \geq 1$*, an operation of* $O^{(n-1)+}$*.*

In particular, we define an n-dimensional 'opetope' to be an n-dimensional O-opetope when O is the simplest operad of all:

Definition 5. *The 'initial untyped operad'* I *is the* S*-operad with:*

1. *only one type:* $S = \{x\}$
2. *only one operation, the identity operation* $1 \in O(x; x)$
3. *all possible reduction laws*

Semantically, I is the operad whose algebras are just sets.

The opetopes emerge from I as follows. The 0-dimensional opetopes are the types of I, but there is only one type, so there is only one 0-dimensional opetope:

x

The 1-dimensional opetopes are the types of I^+, or in other words, the operations of I. There is only one operation of I, the identity operation, so there is only one 1-dimensional opetope:

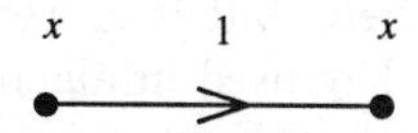

The 2-dimensional opetopes are the types of I^{++}, or in other words, the operations of I^+, which are the reduction laws of I. These reduction laws all state that the identity operation composed with itself k times equals itself. This leads to 2-dimensional opetopes with k infaces and one outface, as follows:

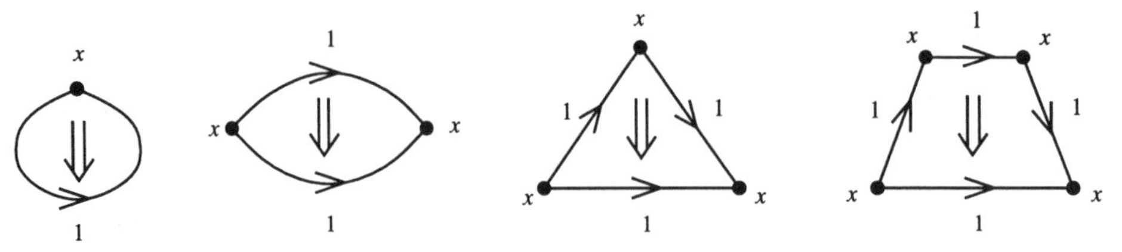

Actually there are $k!$ different 2-dimensional opetopes with k infaces, since the permutation group S_k acts freely on the set of k-ary operations of I^+. We could keep track of these by labelling the infaces with some permutation of k distinct symbols.

The 3-dimensional opetopes are the types of I^{+++}, or in other words, the operations of I^{++}, which are the reduction laws of I^+. These state that some composite of 2-dimensional opetopes equals some other 2-dimensional opetope. This leads to 3-dimensional opetopes like the following:

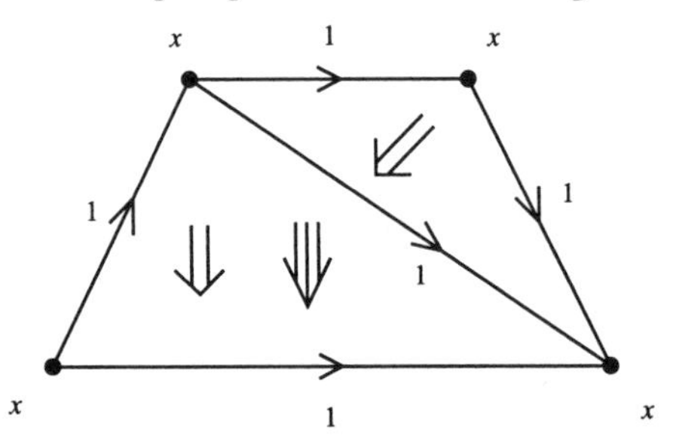

In general, the $(n+1)$-dimensional opetopes describe all possible ways of composing n-dimensional opetopes. Since the $(n+1)$-dimensional opetopes are the operations of the operad I^{n+}, all allowed ways of composing them can be described by trees. One can use this to describe the opetopes of all dimensions using 'metatree notation'. In this notation, an n-dimensional opetope is represented as a list of n labelled trees. This notation nicely handles the nuances of how the permutation group S_k acts on the set of opetopes with k infaces.

4.3 Opetopic sets

A weak n-category will be an 'opetopic set' with certain properties. An opetopic set consists of collections of 'cells' of different shapes, one collection for each opetope. The face of any cell is again a cell, and one can keep track of this using 'face maps' going between the collections of cells. These maps also satisfy certain relations. We omit the details here, but it is worth noting that all this can be handled nicely using a trick widely used in algebraic topology [31]. The idea is that there is a category Op whose objects are opetopes. The morphisms in this category describe how one opetope is included as a specified face of another. An opetopic set may then be defined as a contravariant functor $S\colon \mathrm{Op} \to \mathrm{Set}$. Such a functor assigns to each opetope t a set $S(t)$ of cells of that shape, or 't-cells'. Moreover, if $f: s \to t$ is a morphism in Op describing how s is a particular face of t, the face map $S(f): S(t) \to S(s)$ describes how each t-cell of S has a given s-cell as this particular face.

For the definition of a weak n-category we need some terminology concerning opetopic sets. If $j \geq 1$, we may schematically represent a j-dimensional cell x in an opetopic set as follows:

$$(a_1, \ldots, a_k) \xrightarrow{\quad x \quad} a'$$

Here $a_1, \ldots, a_k$ are the infaces of x and a' is the outface of x; all these are cells of one lower dimension. A configuration just like this, satisfying all the incidence relations satisfied by the boundary of a cell, but with x itself missing:

$$(a_1, \ldots, a_k) \xrightarrow{\quad ? \quad} a'$$

is called a 'frame'. A 'niche' is like a frame with the outface missing:

$$(a_1, \ldots, a_k) \xrightarrow{\quad ? \quad} ?$$

Similarly, a 'punctured niche' is like a frame with the outface and one inface missing:

$$(a_1, \ldots, a_{i-1}, ?, a_{i+1}, \ldots, a_k) \xrightarrow{\quad ? \quad} ?$$

If one of these configurations (frame, niche, or punctured niche) can be extended to an actual cell, the cell is called an 'occupant' of the configuration. Occupants of the same frame are called 'frame-competitors', while occupants of the same niche are called 'niche-competitors'.

4.4 Universality

The only thing we need now to define the notion of weak n-category is the concept of a 'universal' occupant of a niche. This is also the subtlest aspect of the whole theory. We explain it briefly here, but it seems that the only way to really understand it is to carefully work through examples.

Before confronting the precise definition of universality, it is important to note that the main role of universality is to define the notion of 'composite':

Definition 6. *Given a universal occupant u of a j-dimensional niche:*

$$(a_1, \ldots, a_k) \xrightarrow{\quad u \quad} b$$

we call b a 'composite' of $(a_1, \ldots, a_k)$.

It is also important to keep in mind the role played by cells of different dimensions. In our framework an n-category usually has cells of arbitrarily high dimension. For $j \leq n$ the j-dimensional cells play the role of j-morphisms, while for $j > n$ they play the role of 'equations', 'equations between equations', and so on. The definition of universality depends on n in a way that has the following effects. For $j \leq n$ there may be many universal occupants of a given j-dimensional niche, which is why we speak of 'a' composite rather than 'the' composite. There is at most one occupant of any given $(n+1)$-dimensional niche, which is automatically universal. Thus composites of n-cells are unique, and we may think of the universal occupant of an $(n+1)$-dimensional niche as an equation saying that the composite of the infaces equals the outface. For $j > n+1$ there is exactly one occupant of each j-dimensional frame, indicating that the composite of the equations corresponding to the infaces equals the equation corresponding to the outface.

The basic idea of universality is that a j-dimensional niche-occupant is universal if all of its niche-competitors factor through it uniquely, *up to equivalence.* For $j \geq n+1$ this simply amounts to saying that each niche has a unique occupant, while for $j = n$ it means that each niche has an occupant through which all of its niche-competitors factor uniquely. In general, we require that composition with a universal niche-occupant set up a 'balanced punctured niche' of one higher dimension. Heuristically, one should think of a balanced punctured niche as defining an equivalence between occupants of its outface and occupants of its missing outface.

Definition 7. *A j-dimensional niche-occupant:*

$$(c_1, \ldots, c_k) \xrightarrow{\quad u \quad} d$$

is said to be 'universal' if and only if $j > n$ and u is the only occupant of its niche, or $j \leq n$ and for any frame-competitor d' of d, the $(j+1)$-dimensional punctured niche

$$\begin{array}{c} ((c_1, \ldots, c_k) \xrightarrow{u} d,\ d \xrightarrow{?} d') \\ \Big\downarrow {\scriptstyle ?} \\ (c_1, \ldots, c_k) \xrightarrow{?} d' \end{array}$$

and its mirror-image version

$$\begin{array}{c}(d \xrightarrow{?} d',\ (c_1, \ldots, c_k) \xrightarrow{u} d) \\ \Big\downarrow ? \\ (c_1, \ldots, c_k) \xrightarrow{?} d'\end{array}$$

are balanced.

Finally we must define the concept of 'balanced punctured niche'. The reader may note that the first numbered condition in the following definition generalizes the concept of an 'essentially surjective' functor, while the second generalizes the concept of a 'fully faithful' functor.

Definition 8. *An m-dimensional punctured niche:*

$$(a_1, \ldots, a_{i-1}, ?, a_{i+1}, \ldots, a_k) \xrightarrow{?} ?$$

is said to be 'balanced' if and only if $m > n+1$ or:

1. *any extension*

$$(a_1, \ldots, a_{i-1}, ?, a_{i+1}, \ldots, a_k) \xrightarrow{?} b$$

   extends further to:

$$(a_1, \ldots, a_{i-1}, a_i, a_{i+1}, \ldots, a_k) \xrightarrow{u} b$$

   with u universal in its niche, and

2. *for any occupant*

$$(a_1, \ldots, a_{i-1}, a_i, a_{i+1}, \ldots, a_k) \xrightarrow{u} b$$

   universal in its niche, and frame-competitor a_i' of a_i, the $(m+1)$-dimensional punctured niche

$$\begin{array}{c}(a_i' \xrightarrow{?} a_i,\ (a_1, \ldots, a_{i-1}, a_i, a_{i+1}, \ldots, a_k) \xrightarrow{u} b) \\ \Big\downarrow ? \\ (a_1, \ldots, a_{i-1}, a_i', a_{i+1}, \ldots, a_k) \xrightarrow{?} b\end{array}$$

and its mirror-image version

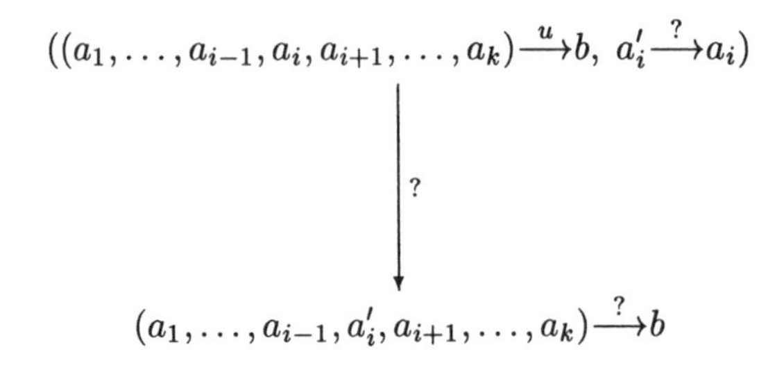

are balanced.

Note that while the definitions of 'balanced' and 'universal' call upon each other recursively, there is no bad circularity. Using these definitions, it is easy to define a weak n-category. While the definition below does not explicitly depend on n, it depends on n through the definition of 'universal' niche-occupant.

Definition 9. *A 'weak n-category' is an opetopic set such that 1) every niche has a universal occupant, and 2) composites of universal cells are universal.*

5 Conclusions

The above definition of weak n-category is really a beginning, rather than an end. We turn the reader to the papers by Dolan and the author [5] and the forthcoming work of Hermida, Makkai and Power for more. The weak $(n+1)$-category of weak n-categories is beginning to be understood; the generalizations of n-categories where we replace I by an arbitrary operad have also turned out to be very interesting. However, before the really interesting applications of n-category theory can be worked out, there is still much basic work to be done.

In particular, it is important to compare various different definitions of weak n-category, so that the subject does not fragment. As one might expect, the question of when two definitions of weak n-category are 'equivalent' is rather subtle. This question seems to have first been seriously pondered by Grothendieck [24], who proposed the following solution. Suppose that for all n we have two different definitions of weak n-category, say 'n-category$_1$' and 'n-category$_2$'. Then we should try to construct the $(n+1)$-category$_1$ of all n-categories$_1$ and the $(n+1)$-category$_1$ of all n-categories$_2$ and see if these are equivalent as objects of the $(n+2)$-category$_1$ of all $(n+1)$-categories$_1$. If so, we may say the two definitions are equivalent as seen from the viewpoint of the first definition.

There are some touchy points here worth mentioning. First, there is considerable freedom of choice involved in constructing the two $(n+1)$-categories$_1$ in question; one should do it in a 'reasonable' way, but this is not necessarily easy. Secondly, there is no guarantee that we might not get a different answer for the question if we reversed the roles of the two definitions. Nonetheless, it should be interesting to compare different definitions of weak n-category in this way.

A second solution is suggested by homotopy theory, which again comes to the rescue. Many different approaches to homotopy theory are in use, and though superficially very different, there is a well-understood sense in which they are fundamentally the same. Different approaches use objects from different categories to represent topological spaces, or more precisely, the homotopy-invariant information in topological spaces, called their 'homotopy types'. These categories are not equivalent, but each one is equipped with a class of morphisms playing the role of homotopy equivalences. Given a category C equipped with a specified class of morphisms called 'equivalences', under mild assumptions one can adjoin inverses for these morphisms, and obtain a category called the 'homotopy category' of C. Two categories with specified equivalences may be considered the same for the purposes of homotopy theory if their homotopy categories are equivalent in the usual sense of category theory. Homotopy theorists have proved that all the popular approaches to homotopy theory are the same in this sense [10].

The same strategy should be useful in n-category theory. Any definition of weak n-category should come along with a definition of an 'n-functor' for which there is a category with weak n-categories as objects and n-functors as morphisms, and there should be a specified class of n-functors called 'equivalences'. This allows the construction a homotopy category of n-categories. Then, for two definitions of weak n-category to be considered equivalent, we require that their homotopy categories be equivalent.

Dolan and the author have constructed the homotopy category of their n-categories, and Simpson [37] has constructed the homotopy category of Tamsamani's n-categories. Now we need machinery to check whether these homotopy categories, and those corresponding to other definitions, are equivalent. Once these preliminary chores are completed, there should be many exciting things we can do with n-categories.

References

1. J. F. Adams, *Infinite Loop Spaces*, Princeton U. Press, Princeton, 1978.
2. J. Baez, Higher-dimensional algebra II: 2-Hilbert spaces, to appear in *Adv. Math.*, preprint available online as q-alg/9609018 and at http://math.ucr.edu/home/baez/
3. J. Baez and J. Dolan, Higher-dimensional algebra and topological quantum field theory, *Jour. Math. Phys.* **36** (1995), 6073-6105.
4. J. Baez and J. Dolan, letter to R. Street, Nov. 30, 1995, corrected version as of Dec. 3, 1995 available at http://math.ucr.edu/home/baez/
5. J. Baez and J. Dolan, Higher-dimensional algebra III: n-Categories and the algebra of opetopes, to appear in *Adv. Math.*, preprint available online as q-alg/9702014 and at http://math.ucr.edu/home/baez/
6. J. Baez and L. Langford, 2-Tangles, preprint available online as q-alg/9703033 and at http://math.ucr.edu/home/baez/
7. J. Baez and M. Neuchl, Higher-dimensional algebra I: Braided monoidal 2-categories, *Adv. Math.* **121** (1996), 196-244.

8. M. Batanin, On the definition of weak ω-category, Macquarie Mathematics Report No. 96/207.
9. M. Batanin, Monoidal globular categories as a natural environment for the theory of weak n-categories, available at http://www-math.mpce.mq.edu.au/~mbatanin/papers.html
10. H.-J. Baues, Homotopy types, in *Handbook of Algebraic Topology*, ed. I. M. James, Elsevier, New York, 1995
11. J. Bénabou, Introduction to bicategories, Springer Lecture Notes in Mathematics **47**, New York, 1967, pp. 1-77.
12. J. M. Boardman and R. M. Vogt, *Homotopy invariant structures on topological spaces*, Lecture Notes in Mathematics 347, Springer, Berlin, 1973.
13. R. Brown, From groups to groupoids: a brief survey, *Bull. London Math. Soc.* **19** (1987), 113-134.
14. V. Chari and A. Pressley, *A Guide to Quantum Groups*, Cambridge U. Press, 1994.
15. S. E. Crans, On combinatorial models for higher dimensional homotopies, Ph.D. thesis, University of Utrecht, Utrecht, 1991.
16. C. Ehresmann, *Catégories et Structures*, Dunod, Paris, 1965.
17. S. Eilenberg and G. M. Kelly, Closed categories, in *Proceedings of the Conference on Categorical Algebra*, eds. S. Eilenberg *et al*, Springer, New York, 1966.
18. S. Eilenberg and S. Mac Lane, General theory of natural equivalences, *Trans. Amer. Math. Soc.* **58** (1945), 231-294.
19. D. Freed, Higher algebraic structures and quantization, *Commun. Math. Phys.* **159** (1994), 343-398.
20. J. Gray, *Formal Category Theory: Adjointness for 2-Categories*, Springer Lecture Notes in Mathematics **391**, Berlin, 1974.
21. G. Kelly and R. Street, Review of the elements of 2-categories, Springer Lecture Notes in Mathematics **420**, Berlin, 1974, pp. 75-103.
22. T. Kohno, ed., *New Developments in the Theory of Knots*, World Scientific, Singapore, 1990.
23. R. Gordon, A. J. Power, and R. Street, Coherence for tricategories, *Memoirs Amer. Math. Soc.* **117** (1995) Number 558.
24. A. Grothendieck, Pursuing stacks, unpublished manuscript, 1983, distributed from UCNW, Bangor, United Kingdom.
25. M. Johnson, The combinatorics of n-categorical pasting, *Jour. Pure Appl. Alg.* **62** (1989), 211-225.
26. M. Kapranov and V. Voevodsky, ∞-groupoids and homotopy types, *Cah. Top. Geom. Diff.* **32** (1991), 29-46.
27. M. Kapranov and V. Voevodsky, 2-Categories and Zamolodchikov tetrahedra equations, in *Proc. Symp. Pure Math.* **56** Part 2 (1994), AMS, Providence, pp. 177-260.
28. R. Lawrence, Triangulation, categories and extended field theories, in *Quantum Topology*, eds. R. Baadhio and L. Kauffman, World Scientific, Singapore, 1993, pp. 191-208.
29. S. Mac Lane, Natural associativity and commutativity, *Rice U. Studies* **49** (1963), 28-46.
30. S. Majid, *Foundations of Quantum Group Theory*, Cambridge U. Press, Cambridge, 1995.
31. J. P. May, *Simplicial Objects in Algebraic Topology*, Van Nostrand, Princeton, 1968.
32. J. P. May, *The Geometry of Iterated Loop Spaces*, Lecture Notes in Mathematics 271, Springer, Berlin, 1972.

33. T. Porter, TQFTs from homotopy n-types, to appear in *J. London Math. Soc.*, currently available at http://www.bangor.ac.uk/~mas013/preprint.html
34. A. J. Power, A 2-categorical pasting theorem, *Jour. Alg.* **129** (1990), 439-445.
35. A. J. Power, An n-categorical pasting theorem, Springer Lecture Notes in Mathematics **1488**, New York,
36. A. J. Power, Why tricategories?, *Info. Comp.* **120** (1995), 251-262.
37. C. Simpson, A closed model structure for n-categories, internal Hom, n-stacks and generalized Seifert-Van Kampen, preprint available as alg-geom/9704006.
38. J. D. Stasheff, Homotopy associativity of H-spaces I & II, *Trans. Amer. Math. Soc.* **108** (1963), 275-292, 293-312.
39. R. Street, The algebra of oriented simplexes, *Jour. Pure Appl. Alg.* **49** (1987), 283-335.
40. R. Street, The role of Michael Batanin's monoidal globular categories, available at www-math.mpce.mq.edu.au/~coact/street_nw97.ps
41. Z. Tamsamani, Sur des notions de ∞-categorie et ∞-groupoide non-strictes via des ensembles multi-simpliciaux, preprint available as alg-geom/9512006.
42. Z. Tamsamani, Equivalence de la théorie homotopique des n-groupoides et celle des espaces topologiques n-tronqués, preprint available as alg-geom/9607010.
43. T. Trimble, The definition of tetracategory (handwritten diagrams, August 1995).

Allegories as a Basis for Algorithmics

Richard S. Bird

Programming Research Group, Oxford University
Wolfson Building, Parks Road, Oxford, OX1 3QD, UK

1 Introduction

Imagine a calculus designed both for the calculation of individual algorithms and the study of algorithmic strategies in general; what should it be like? It seems curious that while the mathematics for the analysis of algorithms is well-developed, including laws for the manipulation of binomial identities, recurrence relations, generating functions, and so on (see e.g. [9, 10]), the development of a similar kind of mathematics for the construction of algorithms is still regarded as being unimportant by many algorithm designers. Instead, algorithmic principles – such as divide and conquer, branch and bound, and so on – continue to be explained by example, and individual algorithms are justified with a degree of formality far below that contained in, say, the simplification of a binomial formula.

A calculus for algorithmics is not quite the same thing as a methodology of programming, since the objectives are different. What is sought is not an industrial-strength theory for the production of reliable software, but simply a theory capable of expressing and reasoning about algorithms at a high level of abstraction. There is, or should be, an emphasis on calculation, i.e. equational or inequational reasoning with formal expressions. Equational logic is simple, and the ability to prove something by calculation is an important indicator of the maturity of a mathematical theory. One may shy away from giving the details of a tedious calculation, or relegate it to a machine, but the existence of a routine calculation is a testament to the success of a theory, not its failure. Nevertheless, notational issues are important; a calculation should be capable of being conducted in both a succinct and transparent notation, designed for human rather than machine consumption.

Our aim in this talk is to argue that one suitable calculus for algorithmics is provided by the theory of allegories. Allegories, invented by Freyd and Ščedrov[8], describe a categorical calculus of relations. A calculus of relations is important in the calculation of algorithms because one needs a degree of freedom in specification and proof that a calculus of functions alone cannot provide. A categorical calculus is appropriate for at least three related reasons. Firstly, category theory provides a unifying framework for algorithmic ideas; secondly, notational clutter is reduced by emphasising point-free compositional reasoning with arrows; thirdly, category theory provides a simple device, the functor, for describing datatypes concisely. As a result, algorithmic strategies and solutions can be formulated without reference to particular datatypes. Specific algorithms for specific problems are obtained by instantiating functor parameters in a suitable manner. This style of program derivation has been called *generic* or *polytypic* program derivation (e.g. see [3, 12, 15]). There is also some indication (e.g. see [7]) that the category itself can usefully be regarded as a parameter of the theory; instantiating the category in different ways can then lead to different computational paradigms, e.g. functional, declarative, state-based, or reactive algorithms.

Our aim in what follows is to give an overview of the use of allegories in the study of algorithmics. The notions of category, functor, natural transformation, and so on, are taken to be understood. A fuller account, from first principles, appears in [2]. This text is, in part, a summary of the work by a number of researchers (including Backhouse, Cockett, de Moor, Fokkinga, Hutton, Jay, Jeuring, Malcolm, Meertens, Meijer, Möller, and others) pursuing similar objectives. In particular, it was Backhouse [1] and de Moor [6] who first proposed using a relational theory of datatypes for program construction, and de Moor who first suggested using an allegorical basis.

2 An example

Since our primary interest is in practical problem solving rather than in foundations *per se*, we will set the scene with a specific computational problem. The problem will be used to motivate ideas and to introduce the style of expression we have in mind.

The *rally driver's* problem is described as follows: imagine a rally taking place in some desert. Along the route are placed n petrol dumps, and at dump $j\,(1 \leq j \leq n)$ is a quantity p_j of petrol. The distance from dump j to dump $j+1$ (or to the finish if $j = n$) is a known quantity d_j, where both p_j and d_j are measured in terms of the same unit. Suppose that a driver is at the start of the rally with p_0 petrol in the tank, and the distance to the first dump is d_0 units. Assuming for simplicity that the car's tank can contain an arbitrarily large amount of petrol, and that the driver can certainly reach the finish by stopping at every dump to pick up more petrol, identify a minimum number of stops necessary to complete the rally. For example, take the following instance in which $n = 4$:

```
          4          8          3          7          2
START ------> 1 -------> 2 -------> 3 ------> 4 -------> FINISH
  10          5         10          9         8
```

In an optimal solution the driver has to make two stops, either at dumps 1 and 2, or at dumps 1 and 3. Dump 1 has to be included, otherwise the driver will never be able to reach point 2 of the route. The general constraint, of course, is that the driver should never run out of petrol between dumps. The rally driver's problem has been around for about ten years (in fact, it was originally called the Mark Thatcher problem after the son of a British prime minister who did indeed get lost in a desert), but it was only recently that it was realised (by Sharon Curtis, see [5]) that there is a simple greedy solution.

To specify the problem formally, we need to choose a suitable representation for feasible solutions. One possibility is to use a subsequence of 1 to n. Such a subsequence, $i_1, i_2, \ldots, i_k$ say, is *safe* if, setting $i_0 = 0$ and $i_{k+1} = n+1$, we have

$$p_{i_0} + p_{i_1} + \cdots + p_{i_j} \geq d_0 + d_1 + \cdots + d_{i_{j+1}-1}$$

for each $j \; (0 \leq j \leq k)$. In particular, it is assumed that the sequence $1, 2, \ldots, n$, which corresponds to stopping at every dump, is safe.

With this definition, the problem can be specified in the following way:

$$\begin{aligned} M &= \mathit{min}\; R \cdot \Lambda(\mathit{safe} \cdot \mathit{subseq}) \\ R &= \mathit{length}^{\circ} \cdot \mathit{leq} \cdot \mathit{length}. \end{aligned}$$

These equations define two relations, M and R, in terms of other relations and functions. Both M and R have type

$$\mathit{List\ Int} \leftarrow \mathit{List\ Int}.$$

We write $R : A \leftarrow B$ in preference to $R : B \rightarrow A$ because the typing rule for composition takes the smoother form: if $R : A \leftarrow B$ and $S : B \leftarrow C$, then $R \cdot S : A \leftarrow C$. We think of relations and functions as taking arguments on the right and delivering results on the left; for example, $\mathit{safe} \cdot \mathit{subseq}$ applied to an argument x delivers a safe subsequence of x. Thus, $x \; \mathit{subseq} \; y$ holds if x is a subsequence of y. The relation safe is a subrelation of the identity relation (and is called a *coreflexive*); it holds for a list x if x is a safe sequence. The operator Λ (called *power transpose*) converts a relation into the corresponding set-valued function, and $\mathit{min}\; R : A \leftarrow P(A)$ is the relation defined by $x \, (\mathit{min}\; R) \, xs$ if x is a minimum element of the set xs under the relation R. The particular choice of R described above defines $x \, R \, y$ if the length of x is less than or equal to (leq) the length of y. The converse of a relation S is denoted by S°. In words, M holds between x and $[1, 2, \ldots, n]$ if x is a shortest safe subsequence of $[1, 2, \ldots, n]$.

To appreciate why relations are used in the specification, consider an alternative version that uses only functions:

$$m = \mathit{minlength} \cdot \mathit{filter\ safe} \cdot \mathit{subseqs}.$$

In this version, *subseqs* returns a set of all subsequences of the input, and *filter safe* retains only the safe subsequences. So far, the new specification is equivalent to the earlier one. The difference arises with *minlength*, which is a function rather than a relation. Consequently, m is a function and the task of program derivation is simply to find a more efficient way of computing m, given the assumption that the input sequence is itself safe. Use of *minlength* rather than a relation $\mathit{min}\; R$ constrains both the specification and any subsequent development. In general, it is simply not possible to get to all the possible programs for computing an optimal solution if one adopts a functional approach: one needs the degree of freedom provided by relations in order not to restrict access to possible solutions.

Under a relational approach, the programmer's task is to obtain a suitable *refinement* of M. In the present case, this means finding some function f satisfying $f \cdot safe \subseteq M \cdot safe$. The restriction to safe inputs is necessary for the problem to have a solution. Of course, the programmer has also to ensure that values of f can be computed in an acceptably efficient manner. Reasoning with $\subseteq$ (i.e. inequational reasoning) rather than with $(=)$ enables the programmer to exploit the extra degree of freedom provided by relations.

The example hopefully demonstrates why a calculus of relations is appropriate as a basis for deriving purely functional algorithms, but does not yet show why such a calculus should be categorical. The major reason concerns datatypes and their descriptions as initial algebras of functors. Consider the definition of the relation *subseq*. We can imagine the datatype of lists being introduced by the following declaration:

$$\textbf{data}\ List\ A = nil \mid cons\,(A \times List\ A).$$

In functional programming there is a function *fold* associated with *List A* that captures definition by structural recursion of functions that take lists as arguments:

$$\begin{aligned} fold\,(c,f)\ nil &= c \\ fold\,(c,f)\,(cons\,(a,x)) &= f\,(a, fold\,(c,f)\ x). \end{aligned}$$

Using an extension of *fold* to relations, we can now define *subseq*:

$$subseq = fold\,(nil, cons \cup snd).$$

The projection function *snd* is defined by $snd\,(a,x) = x$. The definition of *subseq* constructs an arbitrary subsequence by considering each element of the input (from right to left) and either accepting it (via *cons*), or rejecting it (via *snd*). Of course, this is not the only way to define *subseq*, and an alternative will be considered later on.

It is clear now that one can view the rally driver's problem as an instance of a more general problem of the form

$$\begin{aligned} M &: B \leftarrow List\ A \\ M &= min\ R \cdot \Lambda(ok \cdot fold\ S). \end{aligned}$$

But one can go further and remove the dependence on lists and the particular function *fold* described above. As we will show below, a function *fold* can be defined for any datatype that can be described as an initial algebra, so the expression above can be thought of as having an additional parameter, the datatype over which the *fold* is taken. As a result, the specification of M becomes truly generic and one can regard the task of refining M as providing an algorithmic scheme for a whole class of algorithms. Put another way, algorithmic strategies are embodied in general methods for tackling such problems. In a similar way, one can study problems, such as

$$\begin{aligned} M &= min\ R \cdot \Lambda(fold\ S \cdot (fold\ T)^{\circ}) \\ M &= min\ R \cdot \Lambda(safe \cdot S^{*}), \end{aligned}$$

that capture other general forms of specification.

3 Datatypes as initial algebras

The standard presentation (e.g. [11, 14]) of parameterised datatypes as initial algebras uses functors of type $\mathbf{C} \leftarrow \mathbf{C} \times \mathbf{C}$ over a suitable base category $\mathbf{C}$. For example, the datatype declaration

$$\mathbf{data}\ \mathit{List}\ A = \mathit{nil} \mid \mathit{cons}\,(A \times \mathit{List}\ A)$$

is associated with the functor F (called the *base* functor of the datatype) defined by

$$F(A, B) = 1 + A \times B$$
$$F(f, g) = id_1 + f \times g.$$

In part, the type declaration asserts the existence of an isomorphism

$$\mathit{List}\ A \cong F(A, \mathit{List}\ A)$$

for each A, and names $[\mathit{nil}, \mathit{cons}]_A : \mathit{List}\ A \leftarrow F(A, \mathit{List}\ A)$ as the witness of the isomorphism. (The case construction uses square brackets, so that if $f : A \leftarrow B$ and $g : A \leftarrow C$, then $[f, g] : A \leftarrow B + C$ applies f to left components and g to right components.)

The type declaration also asserts that the arrow $[\mathit{nil}, \mathit{cons}]_A$ is an initial object in a category whose objects are (F, A)-algebras and whose arrows are (F, A)-homomorphisms. An (F, A)-algebra is an arrow of $\mathbf{C}$ of type $B \leftarrow F(A, B)$, and a (F, A)-homomorphism $h : f \leftarrow g$, where $f : B \leftarrow F(A, B)$ and $g : C \leftarrow F(A, C)$, is an arrow $h : B \leftarrow C$ of $\mathbf{C}$ satisfying

$$h \cdot g = f \cdot F(id_A, h).$$

The initiality of $[\mathit{nil}, \mathit{cons}]_A$ means that for every (F, A)-algebra f there is a unique (F, A)-homomorphism $h : f \leftarrow [\mathit{nil}, \mathit{cons}]_A$. This homomorphism is denoted by $\mathit{fold}\, f$ (which, strictly speaking, should be decorated by both F and A).

Generalising F, and replacing $\mathit{List}\ A$ by $\mathit{Type}\ A$ and $[\mathit{nil}, \mathit{cons}]$ by

$$\alpha_A : \mathit{Type}\ A \leftarrow F(A, \mathit{Type}\ A),$$

we obtain that the arrow $\mathit{fold}\, f$ is characterised by the universal property

$$h = \mathit{fold}\, f \equiv h \cdot \alpha_A = f \cdot F(id_A, h).$$

The type constructor *Type* can be made into a functor by defining

$$\mathit{Type}\, f = \mathit{fold}\, (\alpha \cdot F(f, id)).$$

For example, $\mathit{List}\, f$ is the standard function $\mathit{map}\, f$ of functional programming. Functors introduced in this way will be called *type* functors. Appealing to the universal property of *fold*, we obtain for $f : A \leftarrow B$ that

$$\mathit{Type}\, f \cdot \alpha_B = \alpha_A \cdot F(f, \mathit{Type}\, f),$$

so $\alpha : \mathit{Type} \leftarrow G$, where $G(A, B) = F(A, \mathit{Type}\, A)$, is a natural transformation.

The standard presentation is sufficient for most of the datatype declarations commonly found in functional programming; in particular, it suffices for the so-called *regular* datatypes. A regular datatype is one whose base functor F can be built from constant functors, the identity and projection functors, coproduct $(+)$ and product $(\times)$, and type functors. However, the standard presentation cannot deal with type declarations such as the following:

$$\mathbf{data}\ \mathit{Nest}\, A = \mathit{box}\, A \mid \mathit{wrap}\, (\mathit{Nest}\, (\mathit{List}\, A))$$

For example, *Nest Int* contains the following elements:

$$\begin{array}{l} \mathit{box}\, 3 \\ \mathit{wrap}\, (\mathit{box}\, [1, 2, 3]) \\ \mathit{wrap}\, (\mathit{wrap}\, (\mathit{box}\, [[1], [2, 3]])) \end{array}$$

Such datatype declarations are legal in many functional languages, though the Hindley-Milner type system currently imposed on these languages prevents one from implementing many useful functions over such datatypes. On the other hand, it is becoming apparent that non-regular datatypes are useful in a number of applications (e.g. see [4, 17]). In particular, [4] makes use of the non-regular datatype

$$\mathbf{data}\ \mathit{Bush}\, A = \mathit{null} \mid \mathit{fork}\, (A \times \mathit{Bush}\, (\mathit{Bush}\, A) \times \mathit{Bush}\, (\mathit{Bush}\, A))$$

to represent a generalisation of the trie data structure.

It is not yet clear what is the right way to set up a functorial semantics, and appropriate language implementations, of non-regular datatypes that includes regular datatypes as a special case. One possible approach is to use second-order functors with type

$$\mathit{Nat}(\mathbf{C}) \leftarrow \mathit{Nat}(\mathbf{C}) \times \mathit{Nat}(\mathbf{C}).$$

The category $\mathit{Nat}(\mathbf{C})$ has functors of type $\mathbf{C} \leftarrow \mathbf{C}$ as objects, and natural transformations of $\mathbf{C}$ as arrows. For example, the second-order functor Δ associated with *Nest*, is defined by

$$\begin{aligned} \Delta(F, G)(A) &= F(A) + G(\mathit{List}\, A) \\ \Delta(F, G)(f) &= F(f) + G(\mathit{List}\, f) \\ \Delta(\eta, \psi)_A &= \eta_A + \psi_{\mathit{List}\, A}. \end{aligned}$$

Under this scheme, the *fold* function takes natural transformations to natural transformations. The standard presentation of regular datatypes, and the standard uses of *fold* on such types, can be recovered using the fact that every arrow of $\mathbf{C} \leftarrow \mathbf{C}$ extends to a natural transformation between constant functors.

4 Allegories

We will give the barest outline of allegories; fuller details can be found in [8] or [2]. An allegory is a category endowed with additional structure: one can compare two arrows having the same source and target under a partial order $\subseteq$, take the intersection of two such arrows with $\cap$, and take an arbitrary arrow to its converse with a unary operator $(\,)^{\circ}$. The properties of these operations are what one would expect, except possibly for an additional law, called the *modular* law, that states

$$R \cdot S \cap T \subseteq R \cdot (S \cap R^{\circ} \cdot T),$$

for all arrows R, S and T of the same type. By convention, composition binds tighter than any other operator.

An allegory **A** has three subcategories of special interest, the categories formed by taking just: (1) the *simple* arrows, also called partial functions; (2) the *entire* arrows, also called total relations; (3) the arrows that are both simple and entire, that is, functions. The subcategory of simple and entire arrows of an allegory **A** is denoted by $Fun(\mathbf{A})$. For simplicity, the arrows of an allegory will be called relations, and the arrows of $Fun(\mathbf{A})$ will be called functions.

Allegories are very general, and extra conditions have to be imposed to get closer to set-theoretic relations. In particular, an allegory is *tabular* if for every relation $R : A \leftarrow B$ there exist a pair of functions $f : A \leftarrow C$ and $g : B \leftarrow C$ such that $R = f \cdot g^{\circ}$ and $f^{\circ} \cdot f \cap g^{\circ} \cdot g = id$. The allegory **Rel** of sets and relations is tabular. The existence of tabulations makes it possible to mimic pointwise proofs in an allegorical setting.

An allegory is *unitary* if it possesses a *unit*, namely, an object U such that id_U is the largest arrow of type $U \leftarrow U$ and such that for all A there is an entire arrow of type $U \leftarrow A$. In a tabular allegory, a unit is a terminal object in the subcategory of functions.

An allegory is a *power* allegory if it contains a power object $P(A)$ for every object A, with an associated membership relation $\in : A \leftarrow P(A)$, and if for every $R : A \leftarrow B$ there is a function $\Lambda R : P(A) \leftarrow B$ such that

$$f = \Lambda R \equiv \in \cdot f = R.$$

It is immediate from this universal property that Λ sets up an isomorphism between relations and set-valued functions.

An allegory is *locally complete* if one can take the union of any set of relations with the same source and target. Given a locally-complete allegory, we can define the operation of (right) division on two arrows R and S with the same target by

$$R/S = \bigcup\{X \mid X \cdot S \subseteq R\}.$$

Hence we have the universal property

$$X \subseteq R/S \equiv X \cdot S \subseteq R$$

for all X. The division operator gives us a way of expressing properties involving universal quantification. In particular, the relation *min* R is defined by

$$\begin{aligned} min\ R &: A \leftarrow P(A) \\ min\ R &= \in \cap (R/\ni), \end{aligned}$$

where $\ni$ is an abbreviation for $\in^\circ$. This definition says that $x\,(min\ R)\,xs$ if x is a member of xs (i.e. $x \in xs$), and if xRy for all y in xs (i.e. $x\,(R/\ni)\,xs$).

The additional operations and features provided by tabular, unitary, locally complete, power allegories are necessary in an adequate calculus for the specification and derivation of programs. On occasions, though surprisingly rarely, one also needs the assumption that the allegory is *Boolean*, so that the complementation operator is available for service. We will suppose in what follows that $\mathbf{A}$ is an allegory with the additional features described above.

The problem, however, is that useful categorical definitions of datatypes are not possible in an allegorical setting. Since each allegory is identical to its opposite, dual categorical constructs coincide. In particular, products coincide with coproducts, which is not what one wants in a sensible theory of datatypes.

One solution is to define all relevant datatype constructions in $\mathbf{C} = Fun(\mathbf{A})$ and then extend them in some canonical way to $\mathbf{A}$. The power functor P and all regular functors of $\mathbf{C}$ (though not all functors) can be extended in a unique way to *monotonic* functors of $\mathbf{A}$. A bifunctor F is monotonic if

$$R_1 \subseteq S_1 \ \wedge\ R_2 \subseteq S_2 \Rightarrow F(R_1, R_2) \subseteq F(S_1, S_2).$$

In particular, if $Fun(\mathbf{A})$ has products, then $\mathbf{A}$ contains a monotonic extension, and this extension – called *relational* product and also denoted by $\times$ – is adequate for programming purposes. Similarly, a terminal object of $Fun(\mathbf{A})$ translates to a unit of $\mathbf{A}$.

If F is a monotonic bifunctor that has an initial algebra α in the subcategory $Fun(\mathbf{A})$ of a power allegory $\mathbf{A}$, then α is initial in the whole allegory $\mathbf{A}$. In fact,

$$fold\ R = \in \cdot fold\,(\Lambda(R \cdot F(id, \in))).$$

As we have seen in the case of *subseq*, the ability to define relational folds gives a very useful increase in descriptive power.

5 Algorithmics

To give a flavour of the kinds of calculation that can be performed in the calculus outlined above, consider an optimisation problem of the form

$$M = min\ R \cdot \Lambda(safe \cdot S^*),$$

where R is some given preorder, *safe* is coreflexive, and $S : Type\ A \leftarrow Type\ A$ for some datatype *Type*. The relation S^*, the reflexive transitive closure of S, can be defined by a universal property, or as the least solution (under $\subseteq$) for

X of the equation $X = id \cup X \cdot S$, or in terms of relational folds. Consider the datatype

$$\textbf{data}\ Iterate\ A = stop\ A \mid again\ (Iterate\ A)$$

The function $fold\,[id, id]$ takes an element $again^n\ (stop\ x)$ of $iterate\ A$ and returns x. The relation $fold\,[id, S]$ applied to the same value returns a y such that $y\ S^n\ x$. Hence

$$S^* = fold\,[id, S] \cdot (fold\,[id, id])^{\circ}.$$

Expressions of the form $fold\ S \cdot (fold\ T)^{\circ}$ are called *hylomorphisms* ([16]) and are important in a systematic account of algorithmics because they capture the notion of a divide and conquer scheme. In fact, $X = fold\ S \cdot (fold\ T)^{\circ}$ if and only if X is the least solution of the recursion equation

$$X = S \cdot F(id, X) \cdot T^{\circ},$$

where F is the base functor associated with the $fold$. Practically every relation of interest in programming can be defined as a relational hylomorphism.

Consider now the relation $lim\ S$ defined by

$$lim\ S = notdom\ S \cdot S^*,$$

where the coreflexive $notdom\ S$ holds for values not in the domain of S. In words, $lim\ S$ iterates S for as long as possible. If S is a well-founded relation, then $lim\ S$ is entire. A full account of the properties of $lim\ S$ is given in [5]. In a sense, the relation lim generalises $fold$ in the same way that **while** statements generalise **for** statements in imperative programming. The analogy is not quite accurate, since $lim\ S$ can be defined as a hylomorphism; the real generalisation is the ability to consider the converse of $fold$ operations in a relational setting.

If S is well-founded, $S \subseteq R$, and $safe \cdot S \subseteq S \cdot safe$, then

$$min\ R \cdot \Lambda lim\ (safe \cdot S) \cdot safe \subseteq M \cdot safe.$$

In words, an optimal solution can be obtained by iterating $safe \cdot S$ for as long as possible and in all possible ways, and then taking a minimum value under R in the set of possible results. The formal proof of this result is omitted. One can construct examples to show that each of the three conditions is necessary.

Now set $T = safe \cdot S$ and consider the relation $G = min\ Q \cdot \Lambda T$, where Q is another given preorder. The relation G represents a greedy step in the calculation of $lim\ T$, in which a minimum under Q is chosen from among the possible results returned by $safe \cdot S$ applied to a given input. Under what conditions do we have

$$lim\ G \subseteq min\ R \cdot \Lambda lim\ T \quad ? \tag{1}$$

This inclusion captures the idea of refining M by a greedy algorithm.

The following calculation gives the highlights of one possible answer. First, the definition of *min R* gives rise to the following universal property:

$$X \subseteq min\ R \cdot \Lambda T \equiv X \subseteq T \ \wedge \ X \subseteq R/T^{\circ}$$

Hence (1) holds if and only if

$$lim\ G \subseteq lim\ T \tag{2}$$

$$lim\ G \subseteq R/(lim\ T)^{\circ}. \tag{3}$$

Using the fact that *lim G* is the least solution for X of the inequation

$$notdom\ G \cup X \cdot G \subseteq X,$$

(the proof is an appeal to the Knaster-Tarski theorem on least fixed points), inclusions (2) and (3) follow from:

$$notdom\ G \ \cup\ lim\ T \cdot G \subseteq lim\ T \tag{4}$$

$$notdom\ G \ \cup\ R/(lim\ T)^{\circ} \cdot G \subseteq R/(lim\ T)^{\circ}. \tag{5}$$

Using the universal properties of division and union, we obtain that (4) and (5 are equivalent to:

$$notdom\ G \subseteq lim\ T \tag{6}$$

$$notdom\ G \cdot (lim\ T)^{\circ} \subseteq R \tag{7}$$

$$lim\ T \cdot G \subseteq lim\ T \tag{8}$$

$$R/(lim\ T)^{\circ} \cdot G \cdot (lim\ T)^{\circ} \subseteq R. \tag{9}$$

Inclusions (6) and (7) hold if $dom\ G = dom\ S$, for then we get

$$notdom\ G \ = \ notdom\ S \ \subseteq \ notdom\ T$$

since $T \subseteq S$. Hence

$$notdom\ G \ \subseteq \ notdom\ T \cdot T^{*} \ = \ lim\ T$$

and

$$\begin{aligned} notdom\ G \cdot (lim\ T)^{\circ} &\subseteq notdom\ T \cdot (lim\ T)^{\circ} \\ &= (lim\ T \cdot notdom\ T)^{\circ} \\ &= notdom\ T \\ &\subseteq R. \end{aligned}$$

The last step holds because R is assumed to be reflexive, i.e. $id \subseteq R$. Inclusion (8) holds because $G \subseteq T$ and so

$$lim\ T \cdot G \ \subseteq \ lim\ T \cdot T \ \subseteq \ lim\ T.$$

Finally, (9) holds under the assumption that

$$G \cdot (lim\ T)^{\circ} \subseteq (lim\ T)^{\circ} \cdot R$$

because then we obtain

$$\begin{aligned} R/(lim\ T)^{\circ} \cdot G \cdot (lim\ T)^{\circ} &\subseteq R/(lim\ T)^{\circ} \cdot (lim\ T)^{\circ} \cdot R \\ &\subseteq R \cdot R \subseteq R. \end{aligned}$$

The last step holds because R is assumed to be transitive, i.e. $R \cdot R \subseteq R$.

One could go on in a similar vein, constructing yet more stronger conditions that imply (1). The two we have at the moment are that $dom\ G = dom\ S$ (in words, a greedy step is possible when any step is), and that

$$G \cdot (lim\ T)^{\circ} \subseteq (lim\ T)^{\circ} \cdot R.$$

In words, a greedy first step will never lead to a worse solution than some other step.

The calculation described above is typical of manipulations in a calculus of relations. It consists of equational and inequational reasoning with relations, using universal properties of operators, explicit definitions of subsidiary relations, properties of folds and converses of folds, and so on.

6 An instantiation

Consider the rally driver's problem again. Previously, we defined *subseq* to be the relation $fold\ [nil, cons \cup snd]$, but an alternative is to define $subseq = delete^*$, where $x\ delete\ y$ if x is the sequence y with one element removed. Formally, we can define

$$delete = cat \cdot (id \times tail) \cdot cat^{\circ},$$

where $cat : List \leftarrow List \times List$ concatenates two lists, and $tail : List \hookleftarrow List$ removes the first element of a list. The definition is: $tail = snd \cdot cons^{\circ}$. The relation *tail* is simple but not entire. Installing the alternative definition of *subseq*, the rally driver's problem is to refine

$$min\ R \cdot \Lambda(safe \cdot delete^*),$$

where $R = length^{\circ} \cdot leq \cdot length$, to a function that can be implemented with acceptable efficiency.

We can apply the greedy theory of the previous section to the problem. Firstly, it is clear that *delete* is well founded, and that

$$\begin{aligned} safe \cdot delete &\subseteq delete \cdot safe \\ delete &\subseteq R. \end{aligned}$$

The second inclusion is intuitively obvious, and the first inclusion hold because adding a stop to a safe subsequence cannot possibly result in an unsafe sequence. It follows that

$$min\ R \cdot \Lambda(lim\ (safe \cdot delete)) \cdot safe \subseteq min\ R \cdot \Lambda(safe \cdot delete^*) \cdot safe.$$

Next, choose Q to be the relation

$$Q = reserve^\circ \cdot geq \cdot reserve,$$

where $geq = leq^\circ$ and *reserve* computes the total amount of petrol left in the tank at the end of the rally. With this definition of Q, the relation

$$G = min\ Q \cdot \Lambda(safe \cdot delete)$$

removes a stop such that the resulting subsequence remains safe and the reserve at the end of the rally is maximised. Equivalently, G misses out a dump with a smallest amount of petrol among all possible misses that give a safe result. It is clear that a greedy step is possible if any step is, so we have only the second condition to check. This condition says that if G identifies i to be a best dump to miss out, and $[j_1, j_2, \ldots, j_k]$ is a safe solution, then there is another safe solution incorporating i that is of no greater length. If $i = j_p$ for some p, then there is nothing to prove. Otherwise, define j_p to be j_k if $j_k < i$, and the smallest integer in $[j_1, \ldots, j_k]$ greater than i, otherwise. The claim is that

$$[j_1, \ldots, j_{p-1}, i, j_{p+1}, \ldots, j_k]$$

is also a safe subsequence (which may be improved further by missing out additional stops). Briefly, replacing any stop to the left of i by i also gives a safe subsequence, as does replacing the first stop to the right of i by i. Formal verification of this claim is omitted. By sorting the dumps in ascending order of the amount of petrol they provide, the greedy algorithm can be implemented to run in $O(n \log n)$ steps. This solution is a dual version of one first discovered by Sharon Curtis in her D.Phil thesis; Curtis's algorithm proceeds by adding stops to an initially empty sequence, while the above algorithm removes stops from an initially full sequence.

The rally driver's problem is interesting for a number of reasons. First, it is an example of a greedy solution that does not fall into the class of matroid or greedoid structures (see [13]); in this respect it is more akin to Huffman coding. Secondly, starting with the definition of *subseq* in terms of *fold* does not lead (at least, not easily) to the solution above. One cannot build an optimal solution incrementally from right to left, or from left to right. On the other hand, the rally driver's problem as expressed in the form

$$min\ R \cdot \Lambda(safe \cdot fold\ S),$$

was one of a set of programming problems that helped Oege de Moor and myself to develop a useful theory of *thinning* algorithms. Indeed, it appears as an exercise (Exercise 8.30) in our book. Thinning algorithms are useful for other optimisation problems, such as the knapsack problem, that cannot be solved by a greedy algorithm.

Acknowledgements I would like to thank Roland Backhouse for pointing out a number of errors in an earlier draft, and Oege de Moor for many technical insights.

References

1. R.C. Backhouse and P.F. Hoogendijk. Elements of a relational theory of datatypes. In B. Möller, H. Partsch, and S. Schuman, editors, *Formal Program Development*, volume LNCS 755, pages 7–42. Springer-Verlag, 1993.
2. R. Bird and O. de Moor. *Algebra of Programming*. International Series in Computing Science. Prentice Hall, 1996.
3. R. S. Bird, P. F. Hoogendijk, and O. De Moor. Generic programming with relations and functors. *Journal of Functional Programming*, 6(1):1–28, 1996.
4. R.H. Connelly and F. Lockwood Morris. A generalisation of the trie data structure. *Mathematical Structures in Computer Science*, 5(3):381–418, 1995.
5. S. Curtis. *A relational approach to optimization problems*. D.Phil. thesis, Computing Laboratory, Oxford, UK, 1996.
6. O. de Moor. *Categories, Relations and Dynamic Programming*. D.Phil. thesis, Computing Laboratory, Oxford, UK, 1992. published in *Mathematical Structures in Computer Science*, volume 4, pp 33–70 (1994).
7. O. de Moor. An exercise in polytypic programming: Repmin. Programming Research Group, Oxford University, 1996.
8. P. J. Freyd and A. Ščedrov. *Categories, Allegories*, volume 39 of *Mathematical Library*. North-Holland, 1990.
9. R.L. Graham, D.E. Knuth, and O. Patashnik. *Concrete Mathematics*. Addison-Wesley, 1989.
10. D.H. Greene and D.E. Knuth. *Mathematics for the Analysis of Algorithms*. Progress in Computer Science. Birkhäuser, 1982.
11. T. Hagino. *Category theoretic approach to data types*. PhD thesis, Laboratory for Foundations of Computer Science, University of Edinburgh, UK, 1987. Technical Report ECS-LFCS-87-38.
12. J. Jeuring. Polytypic pattern matching. In S. Peyton Jones, editor, *Functional Programming and Computer Architecture*, pages 238–248. Association for Computing Machniery, 1995.
13. B. Korte, L. Lovasz, and R. Schrader. *Greedoids*, volume 4 of *Algorithms and Combinatorics*. Springer-Verlag, 1991.
14. E.G. Manes and M.A. Arbib. *Algebraic Approaches to Program Semantics*. Texts and Monographs in Computing Science. Springer-Verlag, 1986.
15. L. Meertens. Calculate polytypically! In *PLILP Conference*, Lecture Notes in Computer Science. Springer-Verlag, 1996.
16. E. Meijer. *Calculating Compilers*. Ph.D thesis, University of Nijmegen, The Netherlands, 1992.
17. C. Okasaki. *Purely Functional Data Structures*. Ph.D thesis, School of Computer Science, Carnegie Mellon University, 1996.

Separating Shape from Data

C.B. Jay

School of Computing Sciences, University of Technology, Sydney,
P.O. Box 123, Broadway NSW 2007, Australia.
email: cbj@socs.uts.edu.au
www: http://linus.socs.uts.edu.au/~cbj

Shape theory [Jay95] gives a precise categorical account of how data is stored within data structures, or shapes. This separation of shape (of type S) and data (given by a list of A) can be represented by a pullback

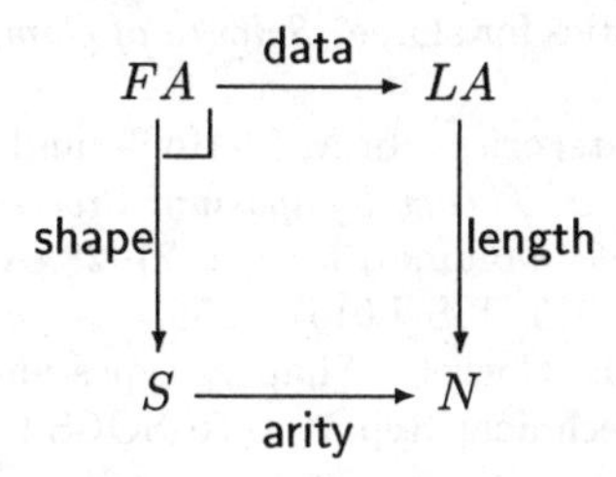

that corresponds to a cartesian natural transformation data : $F \Rightarrow L$. The pullback expresses the property that the only link between the shape and the data is that there be exactly the right amount of data to fill the shape. This definition includes the usual inductive types, such as lists and trees, and also types of (regular and irregular) arrays, but excludes sets, co-inductive types and general function types. Sets are excluded because their shape (cardinality) depends on the equality of elements, while co-inductive types and functions represent data processes, rather than data storage.

The simplicity of the description above suggests that a rather weak categorical setting, let us say a *data category*, may suffice for expressing the key properties of data types. Though some preliminary results are presented in [Jay96], many open problems remain.

The shape-data separation supports novel approaches to computation. One application is *shape polymorphism*, first proposed in [JC94], in which programs can be applied to arguments of different (types and) shapes, e.g. to a list, tree or matrix. The latest work in this line is Functorial ML [BJM96], which extends core ML with a class of functors for representing shapes. Shape polymorphism is similar in outlook to *polytypic* programming [JJ96]. The key difference is that shape polymorphism is *parametric*, i.e. the execution of shape polymorphic operations does not rely on type information, whereas polytypism uses type information to specialise its algorithms.

Another application is to *shape analysis*. Whereas shape polymorphism tries to delay commitment to a choice of shape, shape analysis tries to infer shapes as soon as possible, preferably during compilation. Its most obvious application is to the detection of array bound errors; VEC[JS96] is an experimental functional language in which all array bound errors are detected statically. Deeper applications are concerned with memory management, especially as it applies to parallel programming. It can be used with a *cost monad* to estimate parallel execution times [JCSS97]. Future work aims to use this cost model to automate parallelisation of programs. Other possible applications include an account of *adaptive polymorphism* of object-oriented programs [Lie96, JN96].

Shape theory provides a distinctive approach to data types that opens up both theoretical challenges and new computational possibilities.

References

[BJM96] G. Bellé, C. B. Jay, and E. Moggi. Functorial ML. In *PLILP '96*, volume 1140 of *LNCS*, pages 32–46. Springer Verlag, 1996. TR SOCS-96.08.

[Jay95] C.B. Jay. A semantics for shape. *Science of Computer Programming*, 25:251–283, 1995.

[Jay96] C.B. Jay. Data categories. In M.E. Houle and P. Eades, editors, *Computing: The Australasian Theory Symposium Proceedings, Melbourne, Australia, 29–30 January, 1996*, volume 18, pages 21–28. Australian Computer Science Communications, 1996. ISSN 0157–3055.

[JC94] C.B. Jay and J.R.B. Cockett. Shapely types and shape polymorphism: Extended version. Technical Report UTS-SOCS-94-??, University of Technology, Sydney, 1994.

[JCSS97] C.B. Jay, M.I. Cole, M. Sekanina, and P. Steckler. Costing distribution of a parallel functional language, 1997. accepted as distinguished paper for *Europar'97*.

[JN96] C.B. Jay and J. Noble. Shaping object-oriented programs. Technical report, UTS, 1996.

[JS96] C.B. Jay and M. Sekanina. Shape checking of array programs. Technical Report 96.09, University of Technology, Sydney, 1996.

[JJ96] J. Jeuring and P. Jansson. Polytypic programming. In J. Launchbury, E. Meijer, and T. Sheard, editors, *Advanced Functional Programming, Second International School*, pages 68–114. Springer-Verlag, 1996. LNCS 1129.

[Lie96] Karl Lieberherr. *Adaptive Object-Oriented Software: The Demeter Method with Propagation Patterns.* PWS Publishing Company, 1996.

Axiomatic Rewriting Theory III

A Factorisation Theorem in Rewriting Theory

Paul-André Melliès
LFCS, University of Edinburgh
<paulm@dcs.ed.ac.uk>

Abstract. Some computations on a symbolic term M are more judicious than others, for instance the ***leftmost outermost*** derivations in the λ-calculus. In order to characterise generically that kind of judicious computations, [M] introduces the notion of ***external*** derivations in its axiomatic description of Rewriting Systems: a derivation $e : M \longrightarrow P$ is said to be external when the derivation $e; f : M \longrightarrow Q$ is standard whenever the derivation $f : P \longrightarrow Q$ is standard.
In this article, we show that in every Axiomatic Rewriting System [M,1] every derivation $d : M \longrightarrow Q$ can be factorised as an external derivation $e : M \longrightarrow P$ followed by an ***internal*** derivation $m : P \longrightarrow Q$. Moreover, this epi-mono factorisation is functorial (i.e there is a nice diagram) in the sense of Freyd and Kelly [FK].
Conceptually, the factorisation property means that the efficient part of a computation can always be separated from its junk. Technically, the property is the key step towards our illuminating interpretation of Berry's stability (semantics) as a syntactic phenomenon (rewriting). In fact, contrary to the usual Lévy derivation spaces, the external derivation spaces enjoy meets.

1 Motivations on two syntactic λ-calculi

There are algebraic reasons behind the confluence of the λ-calculus: [Lé] shows that the category $[\mathcal{C}_\lambda]$ of ***derivations*** up to ***redex-permutation*** enjoys pushouts. Our first move is to recall the construction of $[\mathcal{C}_\lambda]$.

THE CONSTRUCTION OF $[\mathcal{C}_\lambda]$. The λ-calculus generates a transition graph $\mathcal{G}_\lambda$ whose vertices are the λ-terms up to α-conversion, and whose arrows $M \longrightarrow P$ are the β-redexes from M to P. We recall that a *β-redex* $M \longrightarrow P$ is a couple (M, o, P) where o is the occurrence in M where the β-reduction occurs. The category $\underline{\mathcal{C}}_\lambda$ is the ***free*** category on this graph $\mathcal{G}_\lambda$:

1. its objects are the vertices of $\mathcal{G}_\lambda$,
2. its morphisms from M to P are the sequences $(M_1, r_1, M_2, \cdots, M_n, r_n, M_{n+1})$ where M_j's are vertices and r_j's β-redexes such that $M_1 = M$ and $M_{n+1} = P$ and $r_j : M_j \longrightarrow M_{j+1}$ for every j. A morphism in $\underline{\mathcal{C}}_\lambda$ is called a ***derivation***,
3. composition of derivation is just concatenation. If $d : M \longrightarrow P$ and $e : P \longrightarrow Q$, we write $d; e$ their composite from M to Q. In particular, the notation $r_1; \cdots; r_n$ denotes the derivation $(M_1, r_1, M_2, \cdots, M_n, r_n, M_{n+1})$.

The crux in the construction of $[\mathcal{C}_\lambda]$ is to identify the derivations from M to P with the same computational content but different reduction orders. To do this, Lévy introduces a ***permutation*** equivalence $\equiv$ which identifies the different developments of a set $\{u, v\}$ of coinitial β-redexes. We recall that a ***development*** of a set of coinitial β-redexes is a derivation which sequentialises their simultaneous reduction. In [Lé], the sequentialisation is formalised with a notion of ***residual*** which permits to trace β-redexes in the course of computation.

A ***redex permutation*** in the λ-calculus can be of two species whose paradigms are permutations (1) and (2):

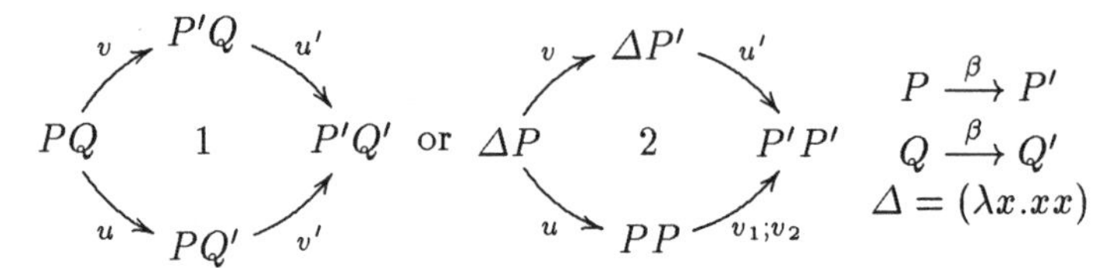

In the first permutation, the permuted β-redexes u and v are ***disjoint***. In the second one, the β-redex u ***contains*** the other β-redex v. Anticipating Section 2, the permutation (1) from $v; u'$ to $u; v'$ is declared ***disjoint*** because it permutes disjoint redexes, and the permutation (2) from $v; u'$ to $u; v_1; v_2$ ***standardising*** because it permutes the outer redex u before the inner redex v.

Lévy defines the ***permutation equivalence relation*** $\equiv$ as the least equivalence relation

1. which identifies the different developments of $\{u, v\}$ for every couple (u, v) of coinitial β-redexes,
2. which is closed under composition: if $f \equiv g$ for $f, g : P \longrightarrow Q$ then $d_1; f; d_2 \equiv d_1; g; d_2$ for every $d_1 : M \longrightarrow P$ and $d_2 : Q \longrightarrow N$.

In other words, two derivations f and g are identified by $\equiv$ when a sequence of permutations operates on f (to and fro) and transforms it into g.
The category $[\mathcal{C}_\lambda]$ is defined as the quotient of $\underline{\mathcal{C}}_\lambda$ by the equivalence relation $\equiv$:

1. its objects are the λ-terms up to α-conversion,
2. its morphisms $M \longrightarrow P$ are the derivations up to the permutation equivalence $\equiv$.

The category $[\mathcal{C}_\lambda]$ is not just a preorder category. The λ-term $I(Ia)$ for $I = \lambda x.x$ can be rewritten in two different ways to Ia, and these two computations are mirrored in the category $[\mathcal{C}_\lambda]$.

PUSHOUTS. [Lé] shows that $[\mathcal{C}_\lambda]$ enjoys the ***pushout*** property. This result can be read as follows: if $M \xrightarrow{f} P$ and $M \xrightarrow{g} Q$ in $[\mathcal{C}_\lambda]$, then there exists two morphisms f' and g' such that the following diagram commutes in $[\mathcal{C}_\lambda]$:

$$\begin{array}{ccc} M & \xrightarrow{g} & Q \\ {\scriptstyle f}\downarrow & & \vdots\,{\scriptstyle f'} \\ P & \underset{g'}{\cdots\!\!\rightarrow} & N \end{array}$$

and furthermore, if two morphisms $f; h_1$ and $g; h_2$ from M to O are equal in $[\mathcal{C}_\lambda]$ (equivalence by permutation in $\underline{\mathcal{C}}_\lambda$), there exists a unique morphism $h : N \longrightarrow O$ (uniqueness up to $\equiv$ in $\underline{\mathcal{C}}_\lambda$) such that the following diagram commutes in $[\mathcal{C}_\lambda]$:

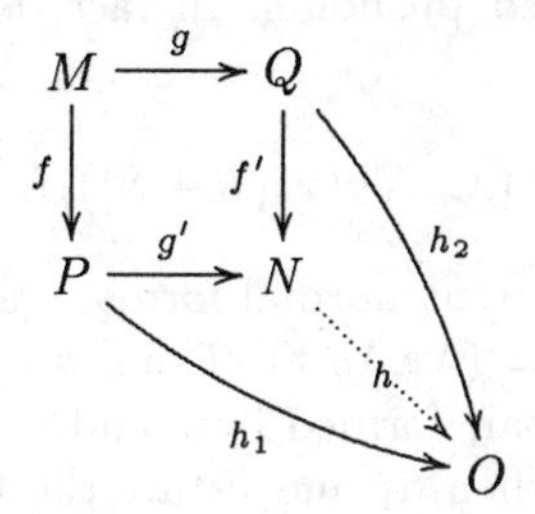

Clearly, the confluence property of the λ-calculus (the so-called Church-Rosser property) is a direct consequence of the existence of pushouts in $[\mathcal{C}_\lambda]$.

PLACARD. We claim that this pushout property is one of the most important results obtained in the field of rewriting theory. From our point of view, it justifies all further attempts to understand rewriting from a structural or algebraic point of view.

Robustness is naturally reinforced by genericity: the same pushout construction is possible on any ***orthogonal*** rewriting system, with an analogous notion of permutation equivalence, see [HL]. For more information on the later developments of this result (standardisation, normalisation, optimality), have a look at [Lé,HL,K,Bo,Ba,GLM,CK,M].

MOTIVATIONS. This paper is motivated by a negative observation: there seems to be no simple adaptation of the pushout property to ***non orthogonal*** (in particular non confluent) rewriting systems. Consider for instance the λ_+-calculus, a λ-calculus enriched with the operator $+$ and the two rewrite rules:

$$P + Q \longrightarrow P \qquad P + Q \longrightarrow Q$$

The λ_+-calculus is not confluent. Nevertheless, a notion of permutation equivalence $\equiv$ can be defined in the way of [Bo,CK,M] with a permutation on *non-conflicting* redexes. Hence, a category $[\mathcal{C}_+]$ of λ_+-derivations up to $\equiv$ (or Lévy permutation classes) can be constructed. Clearly, by non confluence, this category does not enjoy pushouts. Consequently, it is natural to ask whether $[\mathcal{C}_+]$ enjoys a weaker notion of pushouts: ***bounded pushouts***.

BOUNDED PUSHOUTS. A ***span*** in a category $\mathcal{C}$ is a couple (f, g) of coinitial morphisms. A span $P \xleftarrow{f} M \xrightarrow{g} Q$ is ***bounded*** in $\mathcal{C}$ when there exists two morphisms $h_1 : P \longrightarrow O$ and $h_2 : Q \longrightarrow O$ such that $f; h_1 = g; h_2$:

$$\begin{array}{ccc} M & \xrightarrow{g} & Q \\ {\scriptstyle f}\downarrow & & \vdots {\scriptstyle h_2} \\ P & \xrightarrow[h_1]{\cdots} & O \end{array}$$

A category enjoys ***bounded pushouts*** when every bounded span has a pushout.

A COUNTER-EXAMPLE. We adapt an example of [Lé] and show that the category $[\mathcal{C}_+]$ does not enjoy bounded pushouts. In fact, let $P \overset{r}{\longleftarrow} M \overset{s}{\longrightarrow} Q$ be the following span in $[\mathcal{C}_+]$:

$$(\lambda x.Ka(xbR))K \overset{r}{\longleftarrow} (\lambda x.Ka(xb(R+S)))K \overset{s}{\longrightarrow} (\lambda x.Ka(xbS))K$$

where R and S are two different normal forms, r and s are the two conflicting $+$-redexes from M, and $K = (\lambda x.\lambda y.x)$. The span $P \overset{r}{\longleftarrow} M \overset{s}{\longrightarrow} Q$ is bounded in $[\mathcal{C}_+]$ despite the critical pair formed by r and s. In fact, it is bounded twice. On one hand $r; h_1 \equiv s; h_2$ when h_1 and h_2 are the two following derivations:

$$h_1 : P \longrightarrow (\lambda x.a)K \qquad h_2 : Q \longrightarrow (\lambda x.a)K$$

On the other hand $r; i_1 \equiv s; i_2$ when i_1 and i_2 are the two following derivations:

$$i_1 : P \longrightarrow Ka(KbR) \longrightarrow Kab \qquad i_2 : P \longrightarrow Ka(KbS) \longrightarrow Kab$$

So, we obtain two different ***commutative*** diagrams in $[\mathcal{C}_+]$:

$$\begin{array}{ccc} M & \xrightarrow{g} & Q \\ {\scriptstyle f}\downarrow & & \downarrow{\scriptstyle h_2} \\ P & \xrightarrow[h_1]{} & (\lambda x.a)K \end{array} \qquad\qquad \begin{array}{ccc} M & \xrightarrow{g} & Q \\ {\scriptstyle f}\downarrow & & \downarrow{\scriptstyle i_2} \\ P & \xrightarrow[i_1]{} & Kab \end{array}$$

Observe that each derivation h_1, h_2, i_1 and i_2 represents a ***minimal*** procedure to erase R or S. What minimal means here is that any (hypothetical) pushout diagram:

$$\begin{array}{ccc} M & \xrightarrow{g} & Q \\ {\scriptstyle f}\downarrow & PO & \downarrow{\scriptstyle f'} \\ P & \xrightarrow[g']{} & N \end{array}$$

would induce commutative diagrams:

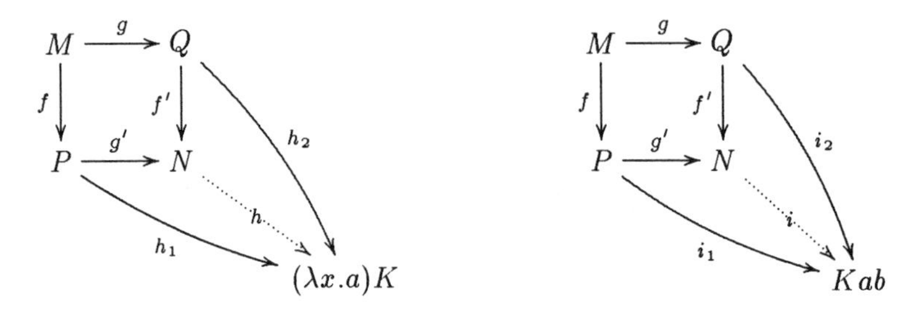

where h and i would be isos, with the consequence that $(\lambda x.a)K$ and Kab are isomorphic in $[\mathcal{C}_+]$, which is impossible because the identity morphisms are the only isos in $[\mathcal{C}_+]$. We conclude that $P \stackrel{r}{\longleftarrow} M \stackrel{s}{\longrightarrow} Q$ is a bounded span without a pushout [1] and that the category $[\mathcal{C}_+]$ does not enjoy bounded pushouts.

An aside: The category specialist unhappy with the notion of bounded pushouts should verify that the slice category $([\mathcal{C}_+]{\downarrow}a)$ does not enjoy pushouts either. This will convince him that the malicious phenomenon above cannot be overcome with a categorically more satisfactory apparatus than bounded pushouts [2].

DIRECTIONS. Well, observing that the category $[\mathcal{C}_+]$ does not enjoy bounded pushouts is a bit depressing, but there is still some hope that an interesting ***subcategory*** of $[\mathcal{C}_+]$ enjoys them. In particular, our counter-example arises from the possibility of erasing two ***unnecessary*** and ***conflicting*** derivations r and s. What would happen if we restrict our investigation to a class of ***necessary*** derivations in $[\mathcal{C}_+]$?

At this point, we decide to extend our scope (forget the λ_+-calculus!) and develop our analysis in the framework of ***Axiomatic Rewriting Systems*** (ARS). Introduced in [1], ARSs are transition systems (or abstract rewriting systems) equipped with a notion of concurrence and duplication in the spirit of Concurrent Transition Systems [S]. A wide range of orthogonal and non orthogonal Rewriting Systems can be modelled as ARSs, in particular first order term rewriting systems, the λ-calculus, the λ_+-calculus, the call-by-value λ-calculus, the $\lambda\sigma$-calculus, combinatory reduction systems, interaction nets or the π-calculus. Because we work in ARSs from now, all theorems in the sequel apply generically to any of these calculi. In particular, we define in every ARS a subcategory $[\mathcal{E}]$ of external derivations and prove that it enjoys the bounded pushout property — thus solving in every ARS the problem we motivated in the λ_+-calculus.

STRUCTURE OF THE PAPER. Section 2 introduces Axiomatic Rewriting Systems and explains the derived notions of ***standard*** and ***external*** derivation. The categories $[\mathcal{C}]$, $[\mathcal{E}]$ and $[\mathcal{M}]$ are defined. A summary of the paper's results is provided at the end of the section. Section 3 introduces a 2-categorical notion of ***oriented 2-pushouts***, a precious tool to prove subsequently [section 3.2] that $[\mathcal{E}]$ enjoys bounded pushouts in every Lévy permutation category $[\mathcal{C}]$ constructed from an ARS $(\mathcal{G}, \rhd)$, [section 3.3] that $[\mathcal{E}]$ and $[\mathcal{M}]$ are orthogonal subcategories of $[\mathcal{C}]$. An ingenious characterisation of the ***factorisation systems*** defined in [FK] is given in Section 4 and subsequently applied to prove that $([\mathcal{E}], [\mathcal{M}])$ is a factorisation system in every Lévy permutation category $[\mathcal{C}]$ constructed from an ARS.

---

[1] Another simpler example suggested by Vincent van Oostrom (private communication) is the bounded span $F(A,B) \longleftarrow F(A,A) \longrightarrow F(A,C)$ in the first order rewriting system whose rules are $A \longrightarrow B$, $A \longrightarrow C$, $F(B,x) \longrightarrow B$ and $F(C,x) \longrightarrow C$.

[2] In posets, bounded pushouts correspond to bounded binary joins $x \vee y$ and pushouts in $([\mathcal{C}_+]{\downarrow}a)$ to binary joins $x \vee_a y$ in the principal ideal $\{x, x \leq a\}$. The first notion is stronger than the second one since -1- the existence of $x \vee y$ implies the existence of $x \vee_a y$ whenever $x \leq a$ and $y \leq a$, with in that case $x \vee_a y = x \vee y$, and -2- when $x \vee y$ does not exist the existence and value of $x \vee_a y$ may depend on a.

2 Axiomatic Rewriting Theory

In Section 1, we shew that the Lévy pushout construction is difficult to extend from orthogonal to non orthogonal rewriting systems. The example of the λ_+-calculus bears evidence that only a subcategory of well-behaved computations can enjoy bounded pushouts.

Here, we generalise our prospect and introduce Axiomatic Rewriting Systems (ARS) to describe there -1- the generic construction of a category $[\mathcal{C}]$ of derivations up to Lévy permutation equivalence, generalising in this the constructions above of $[\mathcal{C}_\lambda]$ and $[\mathcal{C}_+]$, -2- the generic standardisation theorem obtained in [GLM,M,1], -3- the generic construction of a subcategory $[\mathcal{E}]$ composed of the ***external*** morphisms in $[\mathcal{C}]$.

ARS. An Axiomatic Rewriting System is defined in [1] as a couple $(\mathcal{G}, \triangleright)$ composed of:

1. a graph $\mathcal{G} = (T, R, \partial_0, \partial_1)$ where T is a set of ***terms***, R is a set of ***redexes***, $\partial_0 : R \longrightarrow T$ and $\partial_1 : R \longrightarrow T$ are respectively the source and target functions. We write $M \xrightarrow{u} N$ when $\partial_0 u = M$ and $\partial_1 u = N$,
2. a binary relation $\triangleright$ between coinitial and cofinal paths of $\mathcal{G}$.

We recall that a ***path*** in a graph $\mathcal{G}$ is a sequence

$$(M_1, u_1, M_2, ..., M_m, u_m, M_{m+1})$$

where $M_i \xrightarrow{u_i} M_{i+1}$ for every $i \in [1...m]$. When $m = 0$, the path (M_1) is said to be ***empty***. Two paths $(M_1, u_1, ..., u_m, M_m)$ and $(N_1, v_1, ..., v_n, N_n)$ are coinitial (resp. cofinal) when $M_1 = N_1$ (resp. $M_m = N_n$). In the sequel, paths are also called ***derivations*** to follow the Rewriting terminology.

Many concrete Rewriting Systems can be modelled as an ARS $(\mathcal{G}, \triangleright)$. For instance, the λ-calculus:

- the graph $\mathcal{G}$ is the transition graph of the calculus, in that case $\mathcal{G}_\lambda$,
- the relation $\triangleright$ mirrors the ***oriented*** redex permutations of the calculus. By oriented permutations we mean either (1) disjoint permutations from $u; v'$ to $v; u'$, (2) standardising permutations from $v; u'$ to $u; f$ (where the derivation f reduces the copies of v through u). In this case, (1) and (2) are mirrored by relations $v; u' \triangleright u; v'$ and $v; u' \triangleright u; f$ in the ARS $(\mathcal{G}, \triangleright)$.

Many important concepts in Rewriting Theory are expressed in $(\mathcal{G}, \triangleright)$ without referring to the concrete underlying calculus. For instance, a ***disjoint*** (resp. ***standardising***) permutation is formally defined in $(\mathcal{G}, \triangleright)$ as a couple (f, g) of (coinitial & cofinal) derivations such that $f \triangleright g$ and $g \triangleright f$ (resp. $f \triangleright g$ but not $g \triangleright f$). A ***one-step*** oriented permutation from d to e is defined in $(\mathcal{G}, \triangleright)$ as a quadruple (d_1, f, g, d_2) such that $f \triangleright g$, $d = d_1; f; d_2$ and $e = d_1; g; d_2$.

Thence, every ARS $(\mathcal{G}, \triangleright)$ gives rise to a 2-category $\mathcal{C}$ whose carrier is $\underline{\mathcal{C}}$, the free category on $\mathcal{G}$:

1. $\mathcal{C}$'s objects are the vertices of the graph $\mathcal{G}$,
2. its morphisms $f : P \longrightarrow Q$ are the derivations from P to Q,
3. its 2-cells $\alpha : f \Rightarrow g$ are the sequences of one-step oriented permutations, up to disjoint permutations [3].

Lévy permutation equivalence $\equiv$ can also be expressed in $(\mathcal{G}, \rhd)$ (or in the 2-category $\mathcal{C}$) as the least equivalence relation containing $\Rightarrow$. In fact, the equivalence classes of $\equiv$ correspond exactly to the connected components of the hom-categories $\mathcal{C}(P, Q)$. The category $[\mathcal{C}]$ is thence defined as the quotient of $\mathcal{C}$ by $\equiv$. The canonical functor $[\cdot]$ from the free category $\underline{\mathcal{C}}$ to the category $[\mathcal{C}]$ transports a derivation d to its Lévy class $[d]$:

$$[\cdot] : \underline{\mathcal{C}} \longrightarrow [\mathcal{C}]$$

STANDARD DERIVATIONS. A derivation d in $(\mathcal{G}, \rhd)$ is called ***standard*** when no standardising permutation appears from d after any sequence of disjoint permutations. The derivation d is thus a $\Rightarrow$-normal form up to disjoint permutations: if $\alpha : d \Rightarrow e$ then α is a sequence of disjoint permutations.

Ten elementary axioms are introduced in [1] to establish the ***standardisation*** theorem which states that there exists in every Lévy class $\equiv$ a unique standard derivation, up to disjoint permutations. Every standard derivation is therefore a canonical representative of its Lévy permutation class.

Assumption: The ten axioms are so important that we integrate them in the definition of ARSs and consider from now that every ARS $(\mathcal{G}, \rhd)$ verifies them.

The standardisation theorem can also be expressed as a property of the 2-category $\mathcal{C}$.

Theorem 1 ([M,1]). *Every 2-category $\mathcal{C}$ constructed from an ARS $(\mathcal{G}, \rhd)$ can be enriched to a **standardisation** 2-category $(\mathcal{C}, \Downarrow)$ — see definition 2.*

Definition 2. A ***standardisation*** 2-category $\mathcal{C}$ is a 2-category equipped with an unary operator $\Downarrow$ on morphisms such that:

1. *$\forall e : P \longrightarrow Q$, if $d \equiv e$ then there exists a unique 2-cell $\alpha : e \Rightarrow \Downarrow_d$,*
2. *if there is a 2-cell $\alpha : \Downarrow_d \Rightarrow e$ then α is a 2-iso: $\Downarrow_d \simeq e$.*

*To express this in the categorical idiom, $\Downarrow_d$ is a **strong terminal object** in the 2-connected component of $\mathcal{C}(P, Q)$ which contains d.*

We recall from [1] that two derivations $f, g : P \longrightarrow Q$ are 2-isomorphic in $\mathcal{C}$ (i.e there exist 2-isos $\alpha : f \Rightarrow g$ and $\alpha^{-1} : g \Rightarrow f$) if and only if there is a sequence of ***disjoint*** permutations from P to Q. We write $f \simeq g$ in that case. Observe that assertion 1. in Definition 2 implies that $\Downarrow_d$ is unique in its connected component, up to disjoint permutations: if $d \equiv e$ then $\Downarrow_d \simeq \Downarrow_e$.

[3] To be honest, up to disjoint permutations and a bit more, see [1] for details.

We usually call $\underline{\mathcal{C}}$ the derivation category, $\mathcal{C}$ the 2-category and $[\mathcal{C}]$ the Lévy permutation category of the ARS $(\mathcal{G}, \triangleright)$ they mirror.

External derivations. That a standard derivation stands among the very best computations in its Lévy permutation class does not mean that it is judicious at all. In the λ-calculus, the derivation $Ka(Ix) \longrightarrow Kax$ is standard but cannot be judicious because its Lévy class itself is not judicious. We have to find a stronger criterion than standardness to characterise the good computations of a calculus. [M,2] propounds the following criterion. A derivation $e : M \longrightarrow P$ is called ***external*** when the derivation $e; f : M \longrightarrow Q$ is ***standard*** whenever $f : P \longrightarrow Q$ is ***standard***.

In the λ-calculus for instance, the head-redex $(\lambda x.M)N \longrightarrow M[x := N]$ is external but not the redex $u : I((\lambda x.M)N) \longrightarrow I(M[x := N])$ because the derivation $u; f$ where $f : I(M[x := N]) \longrightarrow M[x := N]$ is standard:

$$u; f : \; I((\lambda x.M)N) \longrightarrow I(M[x := N]) \xrightarrow{f} M[x := N]$$

can be standardised to:

$$\Downarrow_{u;f} : \; I((\lambda x.M)N) \longrightarrow (\lambda x.M)N \longrightarrow M[x := N]$$

We observe in [M,2] that the composite $d; e : M \longrightarrow Q$ of two external derivations $d : M \longrightarrow P$ and $e : P \longrightarrow Q$ is also external (immediate from the definition). Consequently, the external derivations form a subcategory $\mathcal{E}$ of the derivation category $\underline{\mathcal{C}}$. Another point: In every Axiomatic Rewriting System, external derivations are standard, therefore all standard normalising derivations are in $\mathcal{E}$ (very easy). For these two reasons, we claim in [M,2] that $\mathcal{E}$ is ***the*** category of well-behaved computations (but are you convinced?)

External vs internal. However, the category $\mathcal{E}$ is a subcategory of $\underline{\mathcal{C}}$ and some translation is required to transport it to a subcategory of $[\mathcal{C}]$. The category $[\mathcal{E}]$ image of $\mathcal{E}$ under $[\cdot]$ is called the subcategory of *external* morphisms in $[\mathcal{C}]$.

The notion of external derivation has a dual. A derivation $m : M \longrightarrow Q$ is ***internal*** when the derivation $[e] : M \longrightarrow P$ is iso whenever $e \in \mathcal{E}$ and $m \equiv e; f$. To express this another way, a derivation is internal when it contains no external derivation up to $\equiv$. $\mathcal{M}$ denotes the ***class*** of internal derivations (unfortunately we do not know yet that $\mathcal{M}$ is a category, this will be proved in the sequel). Of course, the image $[\mathcal{M}]$ of $\mathcal{M}$ under $[\cdot]$ is a class of morphisms in $[\mathcal{C}]$.

Summary of the results. In this paper, we show that $[\mathcal{E}]$ enjoys bounded pushouts in every ARS and consequently solve the problem opened at the end of Section 1. To speak the truth, we prove something more fundamental perhaps. We show that $([\mathcal{E}], [\mathcal{M}])$ forms a ***factorisation system*** in the sense of Freyd and Kelly [FK]. This robust property means:

1. that $[\mathcal{E}]$ and $[\mathcal{M}]$ are categories,
2. that every morphism f can be factored as $f = e; m$ with $e \in [\mathcal{E}]$ and $m \in [\mathcal{M}]$,

3. that this factorisation is functorial: if $(e;m);v = u;(f;n)$ where $e, f \in [\mathcal{E}]$ and $m, n \in [\mathcal{M}]$, there is a unique w rendering commutative the diagram:

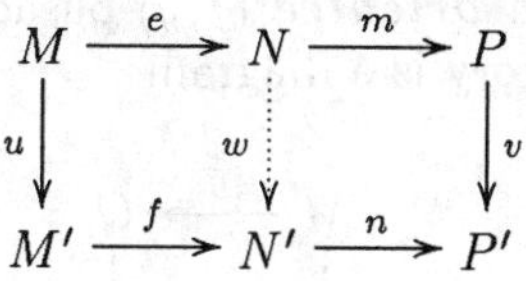

$$\begin{array}{ccccc} M & \xrightarrow{e} & N & \xrightarrow{m} & P \\ {\scriptstyle u}\downarrow & & \vdots {\scriptstyle w} & & \downarrow{\scriptstyle v} \\ M' & \xrightarrow{f} & N' & \xrightarrow{n} & P' \end{array}$$

We mention some important consequences of the factorisation theorem:

- if $e_1; e_2 \in [\mathcal{E}]$ then $e_2 \in [\mathcal{E}]$,
- if $m_1; m_2 \in [\mathcal{M}]$ then $m_1 \in [\mathcal{M}]$,
- $[\mathcal{E}]$ is closed under pushouts and $[\mathcal{M}]$ is closed under pullbacks,
- the fibered coproduct of $e_\alpha : A \longrightarrow B_\alpha$ is in $[\mathcal{E}]$ if each e_α is in $[\mathcal{E}]$.

ORTHOGONALITY. One guiding idea in [FK] is that two morphisms can be ***orthogonal*** in a category $\mathcal{C}$. A morphism e is orthogonal to a morphism m when for every commutative diagram:

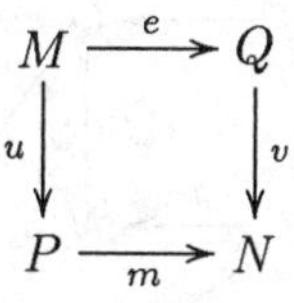

$$\begin{array}{ccc} M & \xrightarrow{e} & Q \\ {\scriptstyle u}\downarrow & & \downarrow{\scriptstyle v} \\ P & \xrightarrow[m]{} & N \end{array}$$

there is a ***unique*** morphism w rendering the diagram:

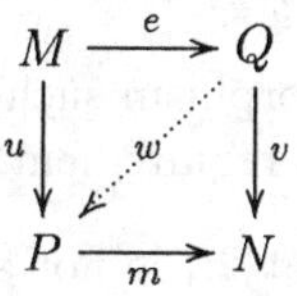

$$\begin{array}{ccc} M & \xrightarrow{e} & Q \\ {\scriptstyle u}\downarrow & \swarrow{\scriptstyle w} & \downarrow{\scriptstyle v} \\ P & \xrightarrow[m]{} & N \end{array}$$

commutative: $e; w = u$ and $w; f = v$. In that case, we write $e \perp m$. Beware: the relation $\perp$ is not symmetric.

If $\mathcal{X}$ is a class of morphim in a category $\mathcal{C}$, we write $\mathcal{X}^{\downarrow}$ the set of morphisms m such that $x \perp m$ for every $x \in \mathcal{X}$, and $\mathcal{X}^{\uparrow}$ the set of morphisms e such that $e \perp x$ for every $x \in \mathcal{X}$. We say that two classes $\mathcal{X}$ and $\mathcal{Y}$ of morphisms of $\mathcal{C}$ are orthogonal when $\mathcal{X}^{\downarrow} \supset \mathcal{Y}$ or (equivalently) when $\mathcal{X} \subset \mathcal{Y}^{\uparrow}$.

3 A 2-categorical proof that $[\mathcal{E}]$ enjoys bounded pushouts

This section is concerned with a proof that $[\mathcal{E}]$ enjoys bounded pushouts. The result is obtained from 2-categorical considerations on ***oriented*** pushout diagrams in the 2-category $\mathcal{C}$ constructed from $(\mathcal{G}, \triangleright)$. This detour through 2-categorical techniques should not be a surprise. It testifies that the standardisation structures behind the construction of the Lévy category $[\mathcal{C}]$ cannot be neglected during the analysis of non orthogonal ARSs.

3.1 Oriented 2-pushouts

ORIENTED PUSHOUTS. An ***oriented*** (f,g)-pushout diagram (written OPO in the diagrams) in a 2-category is a diagram

$$\begin{array}{ccc} M & \xrightarrow{g} & Q \\ {\scriptstyle f}\downarrow & OPO & \downarrow{\scriptstyle f'} \\ P & \xrightarrow[g']{} & N \end{array}$$

such that:

1. there is a cell $\alpha : g; f' \Rightarrow f; g'$,
2. for every two morphisms $h_1 : P \longrightarrow O$ and $h_2 : Q \longrightarrow O$, if there is a cell $\beta : g; h_2 \Rightarrow f; h_1$, then there exists a morphism $h : N \longrightarrow O$ and a cell $\gamma : h_2 \Rightarrow f'; h$ such that $h_1 \equiv g'; h$.

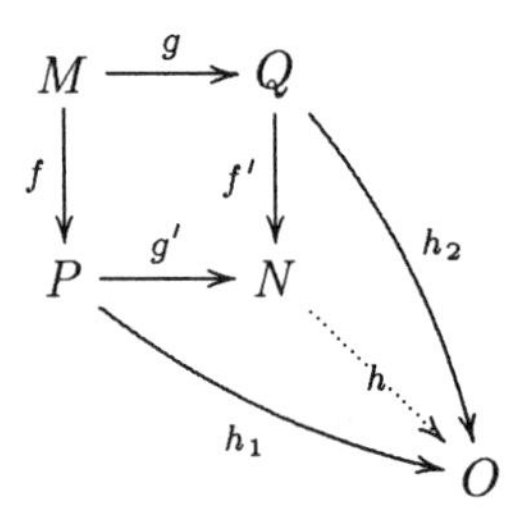

3. Moreover, if h' is another morphism such that $h_1 \equiv g'; h'$ and $h_2 \equiv f'; h'$, then $h \equiv h'$. We call this last requirement the ***universality*** condition.

Observe that the definition (point 2.) is not symmetric. This justifies our taxonomy of *oriented* pushouts: in general, a (f,g)-pushout is not a (g,f)-pushout.

Lemma 3 (Horizontal Pasting). .

$$\text{If } \begin{array}{ccc} M_1 & \xrightarrow{g_1} & M_2 \\ {\scriptstyle f}\downarrow & OPO & \downarrow{\scriptstyle f'} \\ P_1 & \xrightarrow[g_1']{} & P_2 \end{array} \text{ and } \begin{array}{ccc} M_2 & \xrightarrow{g_2} & M_3 \\ {\scriptstyle f'}\downarrow & OPO & \downarrow{\scriptstyle f''} \\ P_2 & \xrightarrow[g_2']{} & P_3 \end{array} \text{ then } \begin{array}{ccc} M_1 & \xrightarrow{g_1;g_2} & M_3 \\ {\scriptstyle f}\downarrow & OPO & \downarrow{\scriptstyle f''} \\ P_1 & \xrightarrow[g_1';g_2']{} & P_3 \end{array}$$

Proof. We use the traditional technique of diagram chasing. Let us call (A) and (B) the two first O-pushout diagrams.

First of all, observe (by cell composition) that $g_1; g_2; f'' \Rightarrow f; g_1; g_2$.

Then, let there be two morphisms $h_1 : P_1 \longrightarrow O$ and $h_2 : M_3 \longrightarrow O$ and a cell $\beta : g_1; g_2; h_2 \Rightarrow f; h_1$. We will show the existence of a morphism $h : P_3 \longrightarrow O$ and a cell $\gamma : h_2 \Rightarrow f''; h$ such that $h_1 \equiv g_1'; g_2'; h$. By diagram chasing on (A), there is a morphism $i : P_2 \longrightarrow O$ and a cell $\gamma_1 : g_2; h_2 \Rightarrow f'; i$ such that $h_1 \equiv g_1'; i$. The existence of γ_1 permits to chase on (B) and deduce that there is a morphism $h : P_3 \longrightarrow O$ and a cell $\gamma_2 : h_2 \Rightarrow f''; h$ such that $i \equiv g_2'; h$. The morphism h then verifies the two expected conditions: $h_1 \equiv g_1'; g_2'; h$ and $\gamma = \gamma_2 : h_2 \Rightarrow f''; h$.

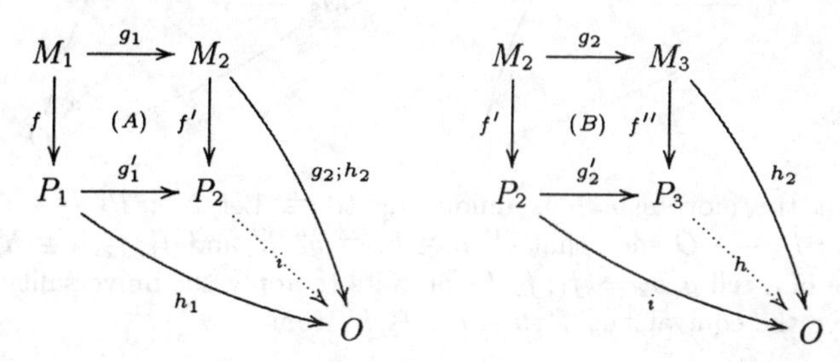

We show the universality condition on h. Let h' be another morphism such that $h_1 \equiv g_1'; g_2'; h'$ and $h_2 \equiv f''; h'$. The cell $\beta : g_1; g_2; h_2 \Rightarrow f; h_1$ permits to apply the universality condition on (A) and deduce $g_2'; h \equiv i \equiv g_2'; h'$ from the relations:

$$g_1'; (g_2'; h) \equiv h_1 \equiv g_1'; (g_2'; h')$$
$$f'; (g_2'; h) \equiv g_2; (f''; h) \equiv g_2; h_2 \equiv g_2; (f''; h') \equiv f'; (g_2'; h')$$

Because $g_2; h_2 \Rightarrow f'; i$, we can apply the universality condition on (B) and deduce from

$$\begin{array}{ll} g_2'; h \equiv g_2'; h' & \text{proved above} \\ f''; h \equiv h_2 \equiv f''; h' & \text{hypothesis} \end{array}$$

that $h \equiv h'$. We conclude. ∎

Lemma 4 (Vertical Pasting). *Suppose that $(\mathcal{C}, \Downarrow)$ is a standardisation 2-category whose morphism f_2 is external.*

$$\text{If } \begin{array}{ccc} M_1 & \xrightarrow{g} & P_1 \\ {\scriptstyle f_1}\downarrow & OPO & \downarrow{\scriptstyle f_1'} \\ M_2 & \xrightarrow[g']{} & P_2 \end{array} \text{ and } \begin{array}{ccc} M_2 & \xrightarrow{g'} & P_2 \\ {\scriptstyle f_2}\downarrow & OPO & \downarrow{\scriptstyle f_2'} \\ M_3 & \xrightarrow[g'']{} & P_3 \end{array} \text{ then } \begin{array}{ccc} M_1 & \xrightarrow{g} & P_1 \\ {\scriptstyle f_1;f_2}\downarrow & OPO & \downarrow{\scriptstyle f_1';f_2'} \\ M_3 & \xrightarrow[g'']{} & P_3 \end{array}$$

Proof. Less traditional than lemma 3 because we use the fact that f_2 is external. Let us call (A) and (B) the two first O-pushout diagrams.

First of all, observe that $g; f_1'; f_2' \Rightarrow f_1; f_2; g''$ by vertical composition of the cells underlying (A) and (B).

Suppose the existence of two morphisms $h_1 : M_3 \longrightarrow O$ and $h_2 : P_1 \longrightarrow O$ and of a cell $\beta : g; h_2 \Rightarrow f_1; f_2; h_1$. We construct a morphism $h : P_3 \longrightarrow O$ and a cell $\gamma : h_2 \Rightarrow f_1'; f_2'; h$ such that $h_1 \equiv g''; h$. By chasing on (A), there is a morphism $i : P_2 \longrightarrow O$ and a cell $\gamma_1 : h_2 \Rightarrow f_1'; i$ such that $f_2; h_1 \equiv g'; i$. Because f_2 is external, the morphism $f_2; \Downarrow_{h_1}$ is standard, hence $g'; i \Rightarrow f_2; \Downarrow_{h_1}$. This allows to chase on (B) and deduce that there is a morphism $h : P_3 \longrightarrow O$ and a cell $\gamma_2 : i \Rightarrow f_2'; h$ such that $\Downarrow_{h_1} \equiv g''; h$. We deduce from the equivalence $h_1 \equiv \Downarrow_{h_1}$ that h verifies the existential conditions of oriented 2-pushouts: $h_1 \equiv g''; h$ and $\gamma : h_2 \Rightarrow f_1'; f_2'; h$, with γ the vertical composite of the cells γ_1 and γ_2.

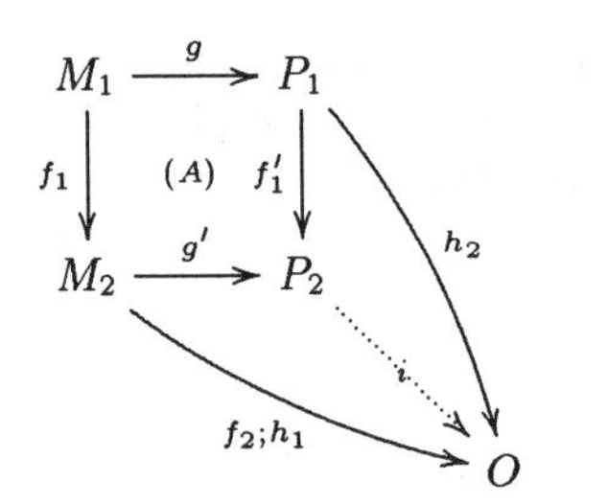

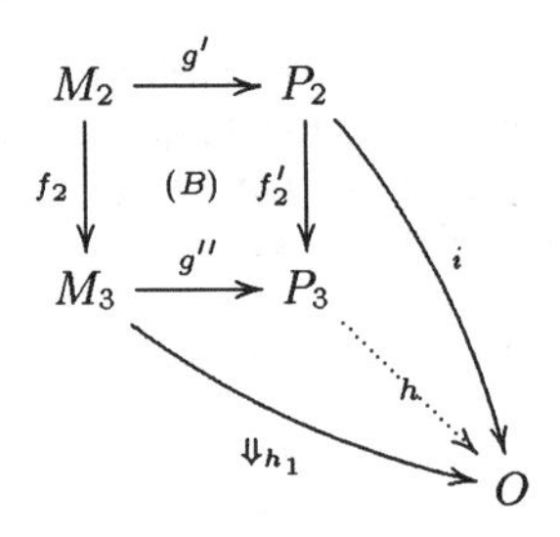

We show that the morphism h is unique up to $\equiv$. Let $h' : P_3 \longrightarrow O$ be another morphism $h' : P_3 \longrightarrow O$ such that $g'';h \equiv h_1 \equiv g'';h'$ and $f_1';f_2';h \equiv h_2 \equiv f_1';f_2';h'$. The existence of a cell $g;h_2 \Rightarrow f_1;f_2;h_1$ permits to apply the universality condition on (A) and derive the equivalence $f_2';h \equiv i \equiv f_2';h'$ from

$$g';(f_2';h) \equiv f_2;(g'';h) \equiv f_2;(g'';h') \equiv g';(f_2';h') \quad \text{and} \quad f_1';(f_2';h) \equiv f_1';(f_2';h')$$

Because $g';i \Rightarrow f_2;\Downarrow_{h_1}$, we can apply the universality condition on (B) to establish $h \equiv h'$ from

$$g'';h \equiv h_1 \equiv g'';h' \quad \text{and} \quad f_2';h \equiv f_2';h'$$

We conclude. ■

Lemma 5. *Let $(\mathcal{C}, \Downarrow)$ be a standardisation 2-category. If $e : M \longrightarrow P$ is external in an oriented (e, f)-pushout diagram of the form*

$$\begin{array}{ccc} M & \xrightarrow{f} & Q \\ {\scriptstyle e}\downarrow & OPO & \downarrow{\scriptstyle e'} \\ P & \xrightarrow[f']{} & N \end{array}$$

then $e' : Q \longrightarrow N$ is external too.

Proof. Easy. Let $h : N \longrightarrow O$ be any morphism. We will show that $e';\Downarrow_h \simeq \Downarrow_{e';h}$ and conclude. Consider the following diagram:

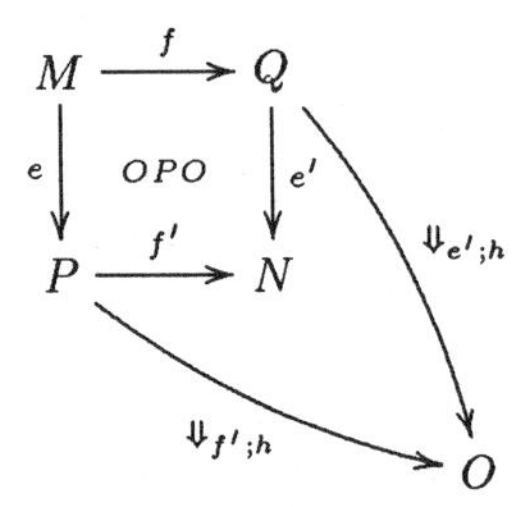

The series of equivalence

$$f;\Downarrow_{e';h} \equiv f;e';h \equiv e;f';h$$

implies the existence of a cell $f;\Downarrow_{e';h} \Rightarrow \Downarrow_{e;f';h}$. Because the derivation e is external, $\Downarrow_{e;f';h} \simeq e;\Downarrow_{e';h}$ and we deduce the existence of a cell

$$\gamma : f;\Downarrow_{e';h} \Rightarrow e;\Downarrow_{f';h}$$

The existence of γ allows to chase on the O-pushout diagram and deduce the existence of a morphism i such that

$$\Downarrow_{e';h} \Rightarrow e';i \qquad \Downarrow_{f';h} \equiv f';i \tag{1}$$

The equivalence $e';h \equiv e';i$ and $f';h \equiv f';i$ follow (1). By universality, the equivalence $h \equiv i$ follows. The definition of a standardisation 2-category tells that $\Downarrow_h$ is ***terminal*** in its connected component in $\mathcal{C}(N, O)$, in particular that there exists a cell $i \Rightarrow \Downarrow_h$. We deduce from this and (1) that

$$\Downarrow_{e';h} \Rightarrow e';i \Rightarrow e';\Downarrow_h$$

The definition of a standardisation 2-category tells also that $\Downarrow_{e';h}$ is ***strong*** in its connected component in $\mathcal{C}(Q, O)$, thus that $\Downarrow_{e';h} \simeq e';\Downarrow_h$. We conclude. ■

3.2 Consequence 1 on ARSs: The subcategory [$\mathcal{E}$] enjoys bounded pushouts

We apply the 2-categorical results of section 3.1 to ARSs. First of all, we import a lemma from [M]:

Lemma 6. *Let $\mathcal{C}$ be the 2-category of an ARS $(\mathcal{G}, \rhd)$. Suppose that $P \xleftarrow{r} M \xrightarrow{f} Q$ is a span in $\mathcal{C}$ and that r is a redex. If there exists two derivations h_1, h_2 and a cell $\gamma : f; h_2 \Rightarrow r; h_1$, then there exists an oriented 2-pushout diagram:*

$$\begin{array}{ccc} M & \xrightarrow{f} & Q \\ {\scriptstyle r}\downarrow & OPO & \downarrow{\scriptstyle r'} \\ P & \xrightarrow[f']{} & N \end{array}$$

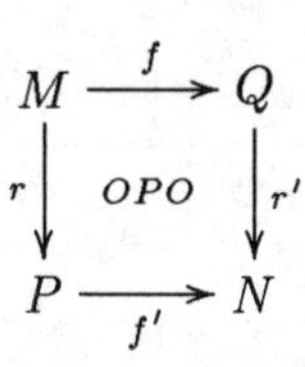

where r' is either a redex or an empty derivation.

Proof. The property is a consequence of two results of [M]: the ***transitivity*** lemma (see lemma 4.21) and the ***left-simplification*** or ***epi*** theorem (see theorem 4.58). ■

We use the Vertical Pasting lemma 4 and lemma 6 to prove the following theorem, a key step towards theorem 8.

Theorem 7. *Let $\mathcal{C}$ be the 2-category of an ARS $(\mathcal{G}, \rhd)$, and let $P \xleftarrow{e} M \xrightarrow{f} Q$ be a span in $\mathcal{C}$. Suppose that e is external: $e \in \mathcal{E}$, and that there exists two derivations h_1 and h_2 such that $e; h_1 \equiv f; h_2$. Then, there is an oriented 2-pushout diagram in $\mathcal{C}$:*

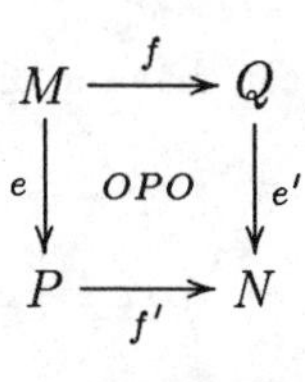

$$\begin{array}{ccc} M & \xrightarrow{f} & Q \\ {\scriptstyle e}\downarrow & OPO & \downarrow{\scriptstyle e'} \\ P & \xrightarrow[f']{} & N \end{array}$$

Moreover, the morphism e' is external: $e' \in \mathcal{E}$.

Proof. By induction on the number of rewrite steps in the external derivation e. First of all, the equivalence $e; h_1 \equiv f; h_2$ and $e \in \mathcal{E}$ implies that $f; h_2 \Rightarrow e; \Downarrow_{h_1}$. Suppose that $e = r; E$ for a redex r. The existence of a cell $f; h_2 \Rightarrow r; (E; \Downarrow_{h_1})$ permits to apply lemma 6 and deduce that there is an oriented (r, f)-pushout. In particular, there exists a morphism $h : N' \longrightarrow O$ such that $h_2 \Rightarrow r'; h$ and $h \equiv E; \Downarrow_{h_1}$:

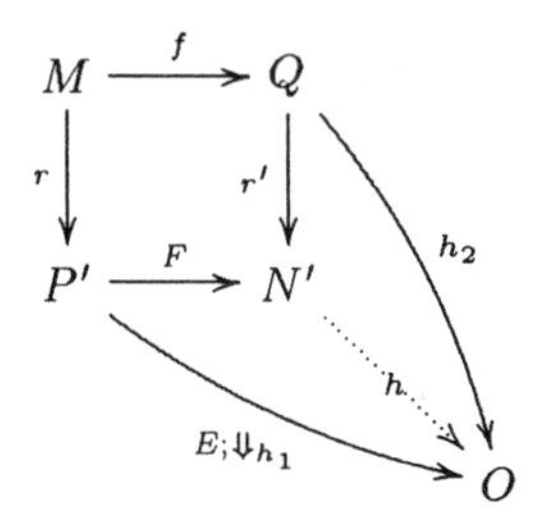

Because $e \in \mathcal{E}$, the ***right decomposition*** lemma, see [M,2], establishes that $E \in \mathcal{E}$. By externality of E, $F; h \Rightarrow E; \Downarrow_{h_1}$. By induction hypothesis, there is an oriented (E, F)-pushout diagram, which by lemma 4 can be pasted to the OPO-diagram above. We conclude.

Externality of e' is a corollary of lemma 5. ∎

Theorem 8 (bounded pushouts). *Let $[\mathcal{C}]$ be the Lévy permutation category of an ARS $(\mathcal{G}, \rhd)$ and let $P \xleftarrow{e} M \xrightarrow{f} Q$ be a bounded span in $[\mathcal{C}]$. If e is external: $e \in [\mathcal{E}]$, then there exists a pushout diagram of the form:*

$$\begin{array}{ccc} M & \xrightarrow{f} & Q \\ {\scriptstyle e}\downarrow & PO & \downarrow{\scriptstyle e'} \\ P & \xrightarrow[f']{} & N \end{array}$$

More over, the morphism $e' : Q \longrightarrow N$ is external: $e' \in [\mathcal{E}]$.

Proof. Fairly simple with theorem 7. The equality $e; h_1 = f; h_2$ in $[\mathcal{C}]$ implies that $E; H_1 \equiv F; H_2$ for derivations E, F, H_1 and H_2 in the permutation classes e, f, h_1 and h_2. Moreover, since $e \in [\mathcal{E}]$, the derivation E can be chosen external: $E \in \mathcal{E}$. This establishes the existence of a cell $F; H_2 \Rightarrow E; \Downarrow_{H_1}$. Henceforth, there is an oriented (E, F)-pushout diagram in $\mathcal{C}$:

$$\begin{array}{ccc} M & \xrightarrow{F} & Q \\ {\scriptstyle E}\downarrow & OPO & \downarrow{\scriptstyle E'} \\ P & \xrightarrow[F']{} & N \end{array}$$

By externality of E, it should be clear that the diagram:

$$\begin{array}{ccc} M & \xrightarrow{f} & Q \\ {\scriptstyle e}\downarrow & OPO & \downarrow{\scriptstyle e'} \\ P & \xrightarrow[f']{} & N \end{array}$$

is a pushout in $[\mathcal{C}]$ for e' and f' the permutation classes of E' and F'. Observe that e' is external: $e' \in [\mathcal{E}]$, because, by theorem 7, $E' \in \mathcal{E}$. We conclude. ■

3.3 Consequence 2 on ARSs: the classes $[\mathcal{E}]$ and $[\mathcal{M}]$ are orthogonal in $[\mathcal{C}]$

We use theorem 8 to prove that external morphisms and internal morphisms are orthogonal in $[\mathcal{C}]$.

Corollary 9 (orthogonality). *Let $[\mathcal{C}]$ be the Lévy permutation category of an ARS $(\mathcal{G}, \rhd)$. If $e \in [\mathcal{E}]$ and $m \in [\mathcal{M}]$ then $e \perp m$.*

Proof. Let the following diagram be commutative in $[\mathcal{C}]$:

$$\begin{array}{ccc} M & \xrightarrow{e} & Q \\ {\scriptstyle u}\downarrow & & \downarrow{\scriptstyle v} \\ P & \xrightarrow[m]{} & O \end{array}$$

By theorem 8 there is a pushout diagram $e; u' = u; e'$, and therefore a commutative diagram:

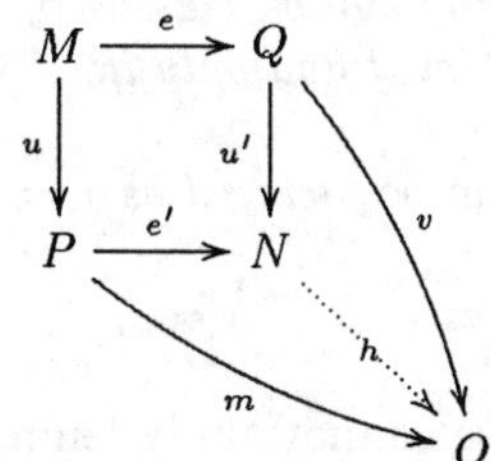

with $e' \in [\mathcal{E}]$ and $m = e'; h$. Because m is internal and e' is external by theorem 8, the morphism e' is an iso. This establishes that there is a pushout diagram of the form (we write $w = u'; (e')^{-1}$).

$$\begin{array}{ccc} M & \xrightarrow{e} & Q \\ {\scriptstyle u}\downarrow & PO & \downarrow{\scriptstyle w} \\ P & \underset{\mathrm{id}_P}{=\!=} & P \end{array}$$

Observe that $w; m = v$ and that $w = w'$ whenever $e; w' = u$. The property $e \perp m$ follows. ■

4 The factorisation theorem

4.1 Three equivalent definitions

This section is concerned with various (three) equivalent definitions of a factorisation system. The two first definitions appear in a seminal paper by Freyd and

Kelly, see [FK]. The third definition was specially designed [4] to apply on ARSs and solve our specific rewriting problem: to prove that $([\mathcal{E}], [\mathcal{M}])$ is a factorisation system of the Lévy permutation category $[\mathcal{C}]$.

REMARK. Sections 4.1 and 4.2 use $\mathcal{C}$ to denote a category, $\mathcal{E}$ and $\mathcal{M}$ to denote two classes of morphisms in $\mathcal{C}$. In fact, we reuse the notations of [FK] and forget for some time (except when explicitly mentionned) our rewriting theoretic meanings of $\mathcal{E}$ and $\mathcal{M}$.

FACTORISATION SYSTEM. A factorisation system on a category $\mathcal{C}$ is a pair $(\mathcal{E}, \mathcal{M})$ such that:

1. *every morphism f in $\mathcal{C}$ can be factored as $f = e; m$ with $e \in \mathcal{E}$ and $m \in \mathcal{M}$,*
2. *$\mathcal{E} \subset \mathcal{M}^{\uparrow}$ and $\mathcal{M} \subset \mathcal{E}^{\downarrow}$,*
3. *both $\mathcal{E}$ and $\mathcal{M}$ contain the isos and are closed under composition.*

SECOND DEFINITION. Unfortunately, we cannot use directly the definition above: we do not know yet that our class $[\mathcal{M}]$ of internal morphisms in $[\mathcal{C}]$ is a category. There is another possibility: to use the following lemma, also appearing in [FK].

Lemma 10 ([FK]). *Let $\mathcal{E}$ and $\mathcal{M}$ be classes of morphisms in a category $\mathcal{C}$. $(\mathcal{E}, \mathcal{M})$ is a factorisation system if and only if:*

1. *every morphism f in $\mathcal{C}$ can be factored as $f = e; m$ with $e \in \mathcal{E}$ and $m \in \mathcal{M}$,*
2. *$\mathcal{E} = \mathcal{M}^{\uparrow}$ and $\mathcal{M} = \mathcal{E}^{\downarrow}$.*

THIRD DEFINITION. Again, we cannot apply lemma 10 on ARSs, this time because we do not know if the equalities $[\mathcal{E}] = [\mathcal{M}]^{\uparrow}$ and $[\mathcal{M}] = [\mathcal{E}]^{\downarrow}$ hold in our rewriting categories. So, we take the opportunity to establish another characterisation of factorisation systems, see section 4.2 for a proof.

Lemma 11 (characterisation). *Let $\mathcal{E}$ and $\mathcal{M}$ be classes of morphisms in a category $\mathcal{C}$. $(\mathcal{E}, \mathcal{M})$ is a factorisation system if and only if:*

1. *every morphism f in $\mathcal{C}$ can be factored as $f = e; m$ with $e \in \mathcal{E}$ and $m \in \mathcal{M}$,*
2. *$\mathcal{E} \subset \mathcal{M}^{\uparrow}$ and $\mathcal{M} \subset \mathcal{E}^{\downarrow}$,*
3. *$e \in \mathcal{E}$ and j iso imply that $e; j \in \mathcal{E}$,*
4. *$m \in \mathcal{M}$ and j iso imply that $j; m \in \mathcal{M}$.*

To our knowledge, lemma 11 is the weakest characterisation of factorisation systems in the literature. Section 4.3 applies the characterisation to establish that $([\mathcal{E}], [\mathcal{M}])$ is a factorisation system of the Lévy permutation category $[\mathcal{C}]$.

[4] I believed for some time that the third characterisation was mine. In fact, the result was already known to some people as I could check later on the category mailing list, look at [AHS,DT,T] for more information.

4.2 A useful characterisation of factorisation systems

This section is devoted to the proof of lemma 11. First of all, we prove that simple and nice result:

Lemma 12. *Let e be epi or m mono or $e \perp m$. Then*

$$e; f \perp m \Rightarrow f \perp m$$

Proof. suppose that $e; f \perp m$. Let the diagram:

$$\begin{array}{ccc} X & \xrightarrow{f} & Y_1 \\ {\scriptstyle u}\downarrow & & \downarrow{\scriptstyle v} \\ Y_2 & \xrightarrow[m]{} & Z \end{array}$$

be commutative. We may left compose it with e to obtain a diagram:

$$\begin{array}{ccccc} X_0 & \xrightarrow{e} & X & \xrightarrow{f} & Y_1 \\ & & {\scriptstyle u}\downarrow & & \downarrow{\scriptstyle v} \\ & & Y_2 & \xrightarrow[m]{} & Z \end{array}$$

By $e; f \perp m$, there exists a morphism $w : Y_1 \longrightarrow Y_2$ such that $e; f; w = e; u$ and $w; m = v$. We initiate a case study to show that $f; w = u$:

1. if the morphism e is epi: the two arrows $f; w$ and u are equal,
2. if the morphism m is mono: $f; (w; m) = f; v = u; m$ implies that $f; w = u$,
3. if the morphism e is orthogonal to m, then consider the commutative diagram

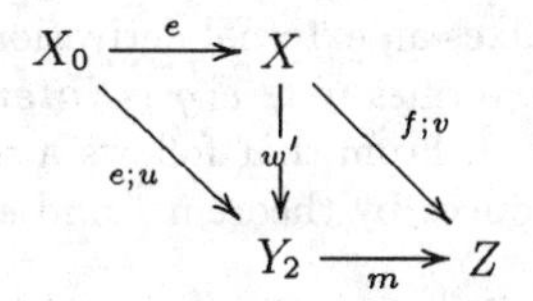

   and conclude that $u = w' = f; w$ from the unicity of the arrow w'.

We show the unicity of w: suppose that $f; x = u$ and $x; m = v$ for a morphism $x : X \longrightarrow Y_2$; then $e; f; w' = e; u$ and therefore $x = w$ by definition of $e; f \perp m$. Consequently the unicity of w is trivial.

We conclude that $e \perp m$. ■

Lemma 12 may be dualised:

Lemma 13. *Let m be mono or e be epi or $e \perp m$. Then*

$$e \perp f; m \Rightarrow e \perp f$$

Lemma 14 (Characterisation lemma 11). *Let $\mathcal{E}$ and $\mathcal{M}$ be two classes of morphisms in a category $\mathcal{C}$. The couple $(\mathcal{E}, \mathcal{M})$ is a factorisation system if and only if:*

1. *every morphism f in $\mathcal{C}$ can be factored as $f = e; m$ with $e \in \mathcal{E}$ and $m \in \mathcal{M}$,*
2. *$\mathcal{E} \subset \mathcal{M}^{\uparrow}$ and $\mathcal{M} \subset \mathcal{E}^{\downarrow}$,*
3. *$e \in \mathcal{E}$ and j iso imply that $e; j \in \mathcal{E}$,*
4. *$m \in \mathcal{M}$ and j iso imply that $j; m \in \mathcal{M}$.*

Proof. If $(\mathcal{E}, \mathcal{M})$ is a factorisation system then the four assertions are true. We will prove that $\mathcal{M}^{\uparrow} \subset \mathcal{E}$ and $\mathcal{E}^{\downarrow} \subset \mathcal{M}$ to apply lemma 10 and show the converse. Let $t \in \mathcal{M}^{\uparrow}$. This morphism may be factored in $t = e; m$ with $e \in \mathcal{E}$ and $m \in \mathcal{M}$. Since $e \in \mathcal{E}$ and $\mathcal{E} \subset \mathcal{M}^{\uparrow}$ we obtain that $e \in \mathcal{M}^{\uparrow}$ and hence e is orthogonal to any morphism in $\mathcal{M}$. That $m \in \mathcal{M}^{\uparrow}$ follows lemma 12.

In particular, the morphism m is orthogonal to itself and therefore m is iso. By hypothesis 4., the morphism $t = e; m$ is in $\mathcal{E}$. We conclude that $\mathcal{E} = \mathcal{M}^{\uparrow}$. The other equation $\mathcal{M} = \mathcal{E}^{\downarrow}$ is proved dually. Lemma 10 shows that $(\mathcal{E}, \mathcal{M})$ is a factorisation system. ■

4.3 Consequence: The external-internal factorisation theorem in every ARS

It is now easy to apply lemma 14 and prove the factorisation theorem 15 on every Lévy permutation category $[\mathcal{C}]$ given by an ARS $(\mathcal{G}, \triangleright)$. Point 2. of lemma 14 is corollary 9. Point 3. and 4. are immediate consequences of the fact that the only isos in $[\mathcal{C}]$ are the identities.

Thus, we only need to prove the first point of the lemma, which means here that every morphism F in $[\mathcal{C}]$ can be factored as $E; M$ with $E \in [\mathcal{E}]$ and $M \in [\mathcal{M}]$. Because every morphism F in $[\mathcal{C}]$ is of the form $[f]$ for f in $\underline{\mathcal{C}}$, we only need to prove that every derivation f in $\underline{\mathcal{C}}$ is Lévy equivalent to some $e; m$ for $e \in \mathcal{E}$ and $m \in \mathcal{M}$.

The ***length*** of a derivation d is its number of rewrite steps. Let n be the length of $\Downarrow_f$. We say that a derivation $e : M \longrightarrow Q$ is a ***prefix*** of $d : M \longrightarrow Q$ (up to $\simeq$) when there exists $h : P \longrightarrow Q$ such that $d \simeq e; h$. The length of a prefix of $\Downarrow_f$ being less than n, there exists among its prefixes an external derivation e of ***maximal length***. We show that the derivation g which verifies $f \simeq e; g$ is ***internal***. Suppose that $[g] = [e']; [h]$ for some external derivation e'. From that follows a series of equivalence $g \equiv e'; h$ and $f \equiv e; g \equiv (e; e'; h)$ which induces by theorem 1 and $e, e' \in \mathcal{E}$ that

$$\Downarrow_f \simeq \Downarrow_{e;e';h} \simeq e; \Downarrow_{e';h} \simeq (e; e'); \Downarrow_h$$

But $\mathcal{E}$ is a subcategory of $\underline{\mathcal{C}}$ and therefore $e; e'$ is an external prefix of $\Downarrow_f$. By definition of e, the derivations e and $e; e'$ have the same length, hence e' is iso (=empty). Thus, $g \in \mathcal{M}$. Point 1. follows the $[\mathcal{E}][\mathcal{M}]$-factorisation $F = [e]; [g]$ of $F = [f]$.

Theorem 15 (factorisation). *The couple $([\mathcal{E}], [\mathcal{M}])$ is a factorisation system in every Lévy permutation category $[\mathcal{C}]$ constructed from an Axiomatic Rewriting System $(\mathcal{G}, \triangleright)$.*

In particular, every Axiomatic Rewriting System $(\mathcal{G}, \triangleright)$ (and related categories $\underline{\mathcal{C}}$, $[\mathcal{C}]$, $\mathcal{E}$, $[\mathcal{E}]$, $\mathcal{M}$, $[\mathcal{M}]$) verifies the three following properties:

1. the class $\mathcal{M}$ of internal derivations is closed under composition,
2. every morphism f in $[\mathcal{C}]$ (= permutation class) can be factored as $f = e; m$ with $e \in [\mathcal{E}]$ and $m \in [\mathcal{M}]$,

3. this factorisation is functorial: if $(e;m);v = u;(f;n)$ where $e, f \in [\mathcal{E}]$ and $m, n \in [\mathcal{M}]$, there is a unique w in $[\mathcal{C}]$ rendering commutative the diagram:

$$\begin{array}{ccccc} M & \xrightarrow{e} & N & \xrightarrow{m} & P \\ {\scriptstyle u}\downarrow & & \vdots{\scriptstyle w} & & \downarrow{\scriptstyle v} \\ M' & \xrightarrow{f} & N' & \xrightarrow{n} & P' \end{array}$$

5 Conclusion

Duplication is the source of many conceptual and technical difficulties in rewriting theory. To use an ***external-internal*** factorisation and establish a non-duplicating realm of ***external*** computations is a new[5] and promising idea. With standardisation, it is one of the very few generic techniques available on non confluent rewriting systems.

The factorisation theorem is the first explicit borrowing from category theory in the author's axiomatic exploration of rewriting systems. The robustness of the property should convince other authors that categorical concepts can be fruitfully imported to rewriting theory. It also justifies the abstract approach to rewriting developed in [M], in particular the definition of external derivations.

Let us conclude the paper with a short survey of the interesting properties of the category $[\mathcal{E}]$ of external derivations, interwoven with future directions.

- two external derivations e_1 and e_2 equivalent by permutation: $e_1 \equiv e_2$ are also equivalent by ***disjoint*** permutation: $e_1 \simeq e_2$,
- in the category $[\mathcal{E}]$, every morphism e is an epi: if $e; e_1 \simeq e; e_2$, then $e_1 \simeq e_2$; and a mono: if $e_1; e \simeq e_2; e$, then $e_1 \simeq e_2$,
- every slice category $(M{\downarrow}[\mathcal{E}])$ is a partial order with bounded joins ***and*** bounded meets. Reckon that the slice categories $(M{\downarrow}[\mathcal{C}_\lambda])$ called ***derivation spaces*** in [Lé,HL] are only $\sqcup$-semi-lattices because they do not have meets. The existence of bounded meets in $(M{\downarrow}[\mathcal{E}])$ shall be used in [4] to establish a generic ***syntactic*** stability theorem.
- in important calculi like orthogonal first-order rewriting systems [HL], the λ-calculus [Lé] and the $\lambda\sigma$-calculus [M,2], every external strategy normalises. In particular, the ***external*** derivations from a ***normalising*** term M form a ***complete*** rewriting system (i.e. confluent and strongly normalising). This observation should help to apply homological techniques, see [La,LP], to these frameworks.

Acknowledgements

I would like to thank Marcelo Fiore for pointing [FK] at an early stage of this work. Many thanks also to Walter Tholen, Hans Porst and Paul Taylor for helping my reference search on the category mailing list.

[5] But ideas of that nature already appear in a paper written by Boudol in the mid'80s, see [Bo], where every computation is projected to its ***needed*** component.

References

[AHS] J. Adamek, H. Herrlich, G. Strecker, "*Abstract and Concrete Categories*", Wiley Interscience in Pure and Applied Mathematics, 1990.

[Ba] H. Barendregt, "*The Lambda Calculus: Its Syntax and Semantics*". North Holland, 1985.

[Bo] G. Boudol, "Computational semantics of term rewriting systems". Algebraic methods in Semantics, Maurice Nivat and John C. Reynolds (eds), Cambridge University Press, 1985.

[CK] D. Clark and R. Kennaway, "Event structures and non-orthogonal term graph rewriting". To appear in MSCS, 1996.

[DT] D. Dikranjan and W. Tholen, "*Categorical Structure of Closure Operators. With Applications to Topology, Algebra and Discrete Mathematics*", Kluwer Academic Publishers (Dordrecht, Boston, London), 1995.

[FK] P.J. Freyd, G.M. Kelly, "Categories of continuous functors, I". *Journal of Pure and Applied Algebra* 2, pp 169—191, 1972.

[GLM] G. Gonthier, J-J. Lévy, P-A. Melliès, "An abstract standardisation theorem". Seventh Annual IEEE Symposium on Logic In Computer Science, August 1992.

[HL] G. Huet, J-J. Lévy, "Call by Need Computations in Non-Ambiguous Linear Term Rewriting Systems". Rapport de recherche INRIA 359, 1979. Republished as "Computations in orthogonal rewriting systems, I and II", in Jean-Louis Lassez and Gordon Plotkin, editors, *Computational logic, essays in honor of Alan Robinson*, pages 395—443. MIT Press, Cambridge, Massachussets, 1991.

[K] J. W. Klop, "*Combinatory Reduction Systems*". PhD thesis, Rijksuniversiteit Utrecht, *Mathematics Centre Tract*, volume 127, June 1980.

[La] Y. Lafont, "A new finiteness condition for monoids presented by complete rewriting systems (after Craig C. Squier)". Journal of Pure and Applied Algebra 98, 1995.

[LP] Y. Lafont, Alain Prouté, "Church-Rosser property and homology of monoids". Mathematical Structures in Computer Science 1, 297 - 326, 1991.

[Lé] J-J. Lévy, "Réductions correctes et optimales dans le λ-calcul". Thèse de Doctorat d'Etat, Université Paris VII, 1978.

[Mc] S. Mac Lane, "*Categories for the working mathematician*", Springer Verlag, 1971.

[M] P-A. Melliès, "Description abstraite des systèmes de réécriture", Thèse de l'Université Paris VII, December 1996.

[1] P-A. Melliès, "Axiomatic Rewriting Theory I: An axiomatic standardisation theorem", in preparation.

[2] P-A. Melliès, "Axiomatic Rewriting Theory II: The lambda-sigma-calculus has the finite cone property", presented at the School on Rewriting and Type Theory, Glasgow, September 1996. Submitted to publication. Available by `ftp` at `http://www.dcs.ed.ac.uk/home/paulm/`.

[4] P-A. Melliès, "Axiomatic Rewriting Theory IV: The fundamental theorem of rewriting theory", in preparation.

[S] E. Stark, "Concurrent Transition Systems", Journal of Theoretical Computer Science, vol. 64, May 1989.

[T] P. Taylor, "*Practical Foundations*". Cambridge Studies in Advanced Mathematics, 1997 (in preparation).

Monads and Modular Term Rewriting

Christoph Lüth[1] and Neil Ghani[2]

[1] Bremen Institute for Safe Systems, FB 3, Universität Bremen
Postfach 330440, 28334 Bremen
cxl@informatik.uni-bremen.de

[2] The School Of Computer Science
University of Birmingham, Birmingham, England
nxg@cs.bham.ac.uk

Abstract. Monads can be used to model term rewriting systems by generalising the well-known equivalence between universal algebra and monads on the category **Set**. In [Lü96], this semantics was used to give a purely categorical proof of the modularity of confluence for the disjoint union of term rewriting systems. This paper provides further support for monadic semantics of rewriting by giving a categorical proof of the most general theorem concerning the modularity of strong normalisation. In the process, we improve upon the technical aspects of earlier work.

1 Introduction

Term rewriting systems (TRSs) are widely used throughout computer science as they provide an abstract model of computation while retaining a relatively simple syntax and semantics. Reasoning about large term rewriting systems can be very difficult and an alternative is to define structuring operations which build large term rewriting systems from smaller ones. Of particular interest is whether key properties are *modular*, that is when does a structured term rewriting system inherit properties from its components?

Although most properties are not in general modular, there are a number of results in the literature providing sufficient conditions for the modularity of key properties. Research originally focussed on the *disjoint union*, for which confluence is modular (Toyama's Theorem), whereas strong normalisation is not. However, strong normalisation is modular under a variety conditions, such as both systems are not collapsing (i.e. contain no *collapsing rules*) [Rus87], both systems are not *duplicating* [Rus87], one system is neither duplicating nor collapsing [Mid89], or both systems are *simplifying* [KO92]. Modularity results for conditional term rewriting systems and unions which permit limited sharing of term constructors are rather unsatisfactory and tend to require rather strong syntactic conditions. Overall, although many specific modularity results are known, what is lacking is a coherent framework which explains the underlying principles behind these results.

We believe that part of the problem is the overly concrete, syntactic nature of term rewriting and that a more abstract semantics is needed. *Abstract*

Reduction Systems provide a semantics for term rewriting systems using relations, but relations do not posses enough structure to adequately model key concepts such as *substitution, context, layer structure* etc. Thus the relational model is used mainly as an organisational tool with the difficult results proved directly in the syntax. Category theory has been used to provide a semantics for term rewriting systems at an intermediate level of abstraction between the actual syntax and the relational model, using structures such as *2-categories* [RS87,See87], *Sesqui-categories* [Ste94] or *confluent categories* [Jay90]. However, despite some one-off results [Gha95,RS87], these approaches have failed to make a lasting impact on term rewriting.

An alternative approach starts from the observation that the categorical treatment of universal algebra is based on the idea of a monad on the category **Set**. Since term rewriting systems can be regarded as a generalisation of universal algebra it is natural to model a term rewriting system by a monad over a more structured base category. The basic theory of monads over categories with more structure than **Set** has been developed by Kelly and Power [KP93] and forms the theoretical basis of this research.

Monads offer a general methodology for the study of modularity in term rewriting. Firstly, one proves that the semantics is *compositional* wrt. the structuring operation in question. For the disjoint union of term rewriting systems, this means proving that if Θ is a term rewriting system and T_Θ is its semantics, then $\mathsf{T}_{\Theta_1+\Theta_2} \cong \mathsf{T}_{\Theta_1} + \mathsf{T}_{\Theta_2}$. Next we express the action of the monad representing the combined term rewriting system as a pointwise colimit over the base category $\mathsf{T}_{\Theta_1+\Theta_2}(X) = \mathit{colim}\mathcal{D}_X$. Finally we prove that if the objects of $\mathcal{D}_X$ satisfy the desired property, then so does $\mathit{colim}\,\mathcal{D}_X$.

This methodology is particularly pleasing as the diagram $\mathcal{D}_X$ abstractly represents the fundamental concept in modular term rewriting of the *layer structure* on terms. In addition, the conditions on the use of variables which occur in the literature arise naturally as conditions on the units of the component monads. In [Lü96], Lüth used this approach to give an entirely categorical proof of Toyama's theorem. This paper proves the most general result for the modularity of strong normalisation for disjoint unions and also improves some technical aspects of earlier work.

The paper is divided as follows. Section 2 motivates the use of monads as models of term rewriting systems by recalling the equivalence between universal algebra and finitary monads on **Set**. Section 3 formally introduces term rewriting systems, and Section 4 defines the monadic semantics for term rewriting systems. Section 5 shows how disjoint unions of term rewriting systems are treated semantically while sections 6 and 7 contain the actual modularity results. Section 8 finishes with directions for further research. We would like to thank Don Sannella, Stefan Kahrs and John Power for many stimulating discussions. Glory, glory to the Hibees!

2 Universal Algebra and Monads

Definition 1 (Monad). A *monad* $\mathsf{T} = \langle T, \eta, \mu \rangle$ on a category $\mathcal{C}$ is given by an endofunctor $T : \mathcal{C} \to \mathcal{C}$, called the *action*, and two natural transformations, $\eta : \mathbf{1}_\mathcal{C} \Rightarrow T$, called the *unit*, and $\mu : TT \Rightarrow T$, called the *multiplication* of the monad, satisfying the *monad laws*: $\mu{\cdot}T\eta = \mathbf{1}_\mathcal{C} = \mu{\cdot}\eta_T$, and $\mu{\cdot}T\mu = \mu{\cdot}\mu_T$.

The monadic approach to term rewriting generalises the well known equivalence between (finitary) monads on the category **Set** and universal algebra. Thus, in order to motivate our constructions, we begin with a brief account of this equivalence. However, since this material is standard category theory, we omit most proofs and instead refer the reader to the standard references ([Man76], [Rob94] and [Mac71, Section VI]).

Every algebraic theory defines a monad on **Set** whose action maps a set to the free algebra over this set. The unit maps a variable to the associated term, while the multiplication describes the process of substitution. The monad laws ensure that substitution behaves correctly, i.e. substitution is associative and the variables are left and right units. Thus monads form an abstract calculus for equational reasoning where *variables, substitution* and *term algebra* (represented by the unit, multiplication and action of the monad) are the primitive concepts. We now make these ideas precise.

Definition 2 (Signature). A (single-sorted) *signature* consists of a function $\Sigma : \mathbb{N} \to \mathbf{Set}$. The set of *n-ary operators* of Σ is defined $\Sigma_n \stackrel{def}{=} \Sigma(n)$

Definition 3 (Term Algebra). Given a signature Σ and a set of variables X, the *term algebra* $T_\Sigma(X)$ is defined inductively:

$$\frac{x \in X}{{'x} \in T_\Sigma(X)} \qquad \frac{f \in \Sigma_n \quad t_1, \ldots t_n \in T_\Sigma(X)}{f(t_1, \ldots, t_n) \in T_\Sigma(X)}$$

Quotes are used to distinguish a variable $x \in X$ from the term ${'x} \in T_\Sigma(X)$. This will be important when analysing the layer structure on terms formed from the disjoint union of two signatures. A term of the form ${'x}$ will be called a *term variable* while all other terms are called *compound terms*. An element of $T_\Sigma(X)$ will be called a term built over X.

Lemma 4. *The map $X \mapsto T_\Sigma(X)$ defines a monad T_Σ on* **Set**.

Lemma 4 generalises to many-sorted signatures— if S is a set of sorts, then an S-sorted signature defines a monad on $\mathbf{Set}^S$. Monads arising via the term algebra construction satisfy an important continuity condition, namely they are *finitary*. To understand this condition, observe that the term algebra $T_\Sigma(X)$ built over on an infinite set X of variables can be given as

$$T_\Sigma(X) = \bigcup_{X_0 \subset X \text{ is finite}} T_\Sigma(X_0)$$

This equation holds because all the operators in Σ have a finite arity and thus a term built over X can only contain a finite number of variables — such terms are therefore built over a finite subset of X. Categorically this is expressed by saying the functor T_Σ is *finitary*:

Definition 5 (Finitary Monads). A functor is *finitary* iff it preserves filtered colimits [Mac71]. A monad is *finitary* iff its action is finitary.

Lemma 6. *If Σ is a signature, then T_Σ is finitary [Rob94, Lemma 1.7].*

One can consider signatures with operations of infinite arities in which case the associated monad satisfies a suitably generalised definition of finitariness. All monads we shall consider are finitary — an example of a monad which isn't finitary is the powerset monad on **Set** which forms powersets of arbitrary large sets and hence has "operations" of arbitrary large arity. Monads also model *algebraic theories*:

Definition 7 (Equations and Algebraic Theories). Given a signature Σ,a *Σ-equation* is of the form $X \vdash t = s$ where X is a set and $t, s \in T_\Sigma(X)$. An *algebraic theory* $\langle \Sigma, E \rangle$ consists of a signature Σ and a set E of Σ-equations.

The term algebra construction generalises from signatures to algebraic theories by mapping a set X to the term algebra quotiented by the equivalence relation generated from the equations and hence we again obtain a finitary monad over **Set**. The category of algebras of this monad is equivalent to the category of models of $\mathcal{A}$, justifying the correctness of the monadic semantics: "universal algebra is the study of finitary monads over **Set**" [Man76].

One key property of this monadic semantics for algebraic theories is that it is *compositional.* For the disjoint union of algebraic theories, this means $\mathsf{T}_{\mathcal{A}_1+\mathcal{A}_2} \cong \mathsf{T}_{\mathcal{A}_1} + \mathsf{T}_{\mathcal{A}_2}$. This compositionality property is established by showing that every finitary monad arises from an algebraic theory called the *internal language* of the monad.

Definition 8 (Internal Signature). The *internal signature* of a finitary monad $\mathsf{S} = \langle S, \eta, \mu \rangle$ on **Set** is given by

$$\Sigma_S(n) \stackrel{def}{=} \bigcup_{card(X)=n} S(X)$$

We can define a map $\varepsilon_{S,X} : T_{\Sigma_S}(X) \to SX$ interpreting terms from $T_{\Sigma_S}(X)$ in SX. We say that a monad S *admits* an equation (X, l, r) where $l, r \in T_{\Sigma_S}(X)$, written $S \models_X l = r$, if $\varepsilon_{S,X}(l) = \varepsilon_{S,X}(r)$. The set of equations admitted by S, written $\mathcal{E}_S$, is defined as $\mathcal{E}_S \stackrel{def}{=} \{(X, l, r) \mid S \models_X l = r\}$.

Definition 9 (Internal Language). The internal language of a finitary monad S on **Set** is given by $\mathcal{L}_S \stackrel{def}{=} \langle \Sigma_S, \mathcal{E}_S \rangle$.

Crucially, these constructions are adjoint: that is, there is an adjunction $\mathsf{T} \dashv \mathcal{L} : \mathbf{AlgTh} \to \mathbf{Mon}_{Fin}(\mathbf{Set})$ where the categories **AlgTh** of algebraic theories and $\mathbf{Mon}_{Fin}(\mathbf{Set})$ of finitary monads on **Set** are appropriately defined [BW85]. The evaluation ε_S is the counit of this adjunction. Since T is left adjoint, it preserves colimits and hence the semantics is compositional.

In summary, monads provide a semantics for algebraic theories with the concepts of term-algebra, variable and substitution taken as primitive. This semantics is compositional, allowing us to reason about the disjoint union of algebraic theories in terms of the component theories.

3 Term Rewriting Systems

We now briefly review the theory of term rewriting systems — further details may be found in [Klo92]. First, fix a countably infinite set V of variables.

Definition 10 (Term Rewriting Systems). A *term rewriting system* $\Theta = \langle \Sigma, R \rangle$ consists of a signature Σ and a set R of Σ-rewrite rules of the form $r : t \to s$ where $t, s \in T_\Sigma(V)$.

A rewrite rule $r : t \to s$ gives rise to the *one-step reduction* relation $C[\sigma(t)] \to_r C[\sigma(s)]$, where $C[\,]$ is a context and σ is a substitution. The one-step reduction relation $\to_R$ of a term rewriting system $\Theta = \langle \Sigma, R \rangle$ is defined as the union of $\{\to_r\}_{r \in R}$, while the *many-step reduction relation*, denoted $\twoheadrightarrow_R$, is the transitive-reflexive closure of the one-step reduction relation.

A rewrite rule $r : t \to s$ is called *expanding* if t is a variable, and *collapsing* if s is a variable. It is said to *introduce variables* if there is a variable occuring in s which does not occur in t, and be *duplicating* if a variable occurs more often in s than in t. Traditionally, rewrite rules are not to allowed to be expanding or variable-introducing, but semantically these restrictions are unnatural and hence omitted. The two key properties of term rewriting systems are *confluence* and *strong normalisation*.

Definition 11 (Confluence and SN). A term rewriting system is *confluent* iff $\forall x, y_1, y_2 . \, x \twoheadrightarrow_R y_1 \wedge x \twoheadrightarrow_R y_2 \; \exists z . \, y_1 \twoheadrightarrow_R z \wedge y_2 \twoheadrightarrow_R z$. It is *strongly normalising* (SN, terminating, Noetherian) iff there is no infinite sequence $x_1 \to_R x_2 \to_R x_3 \to_R \ldots$.

A term rewriting system which is both confluent and SN is called *complete*. Modular term rewriting studies how properties of large term rewriting systems are inherited from their component systems. The key definitions are

Definition 12 (Disjointness and Modularity). Given two term rewriting systems $\Theta_1 = \langle \Sigma_1, R_1 \rangle$ and $\Theta_2 = \langle \Sigma_2, R_2 \rangle$ their *disjoint union* is defined as $\langle \Sigma_1 + \Sigma_2, R_1 + R_2 \rangle$. A property P is *modular* if the disjoint union of Θ_1 and Θ_2 satisfies P iff Θ_1 satisfies P and Θ_2 satisfies P.

4 Monads as Models of Term Rewriting Systems

Our semantics for term rewriting systems generalises the treatment of algebraic theories as finitary monads over the category **Set**. We regard term rewriting systems as a generalised signature and hence its semantics is naturally given by a monad over a more structured base category. The choice of the base category depends on the specific aspects of rewriting one is interested in. We start by using the category **Pre** of preorders as a base category because definition 14 is notationally easier[1], although later we switch to **Cat**.

Kelly and Power [KP93] have shown how algebraic theories can be generalised to categories other than **Set** in such a way that the theory of section 2 can be developed at this more abstract level. This general theory requires the arity of operations, variables and term algebra to have the same structure (sets in the case of algebraic theories, preorders here) so as to allow a uniform treatment of term formation by the multiplication of the monad. Thus, each rewrite rule must be given an arity which is a *preorder* and the term algebra construction must map a *preorder* of variables to a *preorder* of rewrites. This leads to a more general form of rewrite rules:

Definition 13 (Generalised Rewrite Rules). A *generalised rewrite rule* in a signature Σ is a triple (X, l, r), written as $(X \vdash l \rightarrow r)$, where $X = (X_0, \rightarrow_X)$ is a finite preorder and $l, r \in T_\Sigma(X_0)$ are terms.

Thus, in order to instantiate a generalised rewrite rule (X, l, r), one must not only supply terms for the free variables of the rule, but these terms must have rewrites between them which conform to the order structure of X — see rule [INST] of Definition 14. The traditional rewrite rules of definition 10 are of course generalised rewrite rules whose the arities are discrete.

In universal algebra, each signature Σ defines a functor T_Σ whose action is to map a set X to the term algebra $T_\Sigma(X)$ built using the operators of Σ as term constructors and the elements of X as variables. The equivalent construction for term rewriting systems is called a *term reduction algebra*:

Definition 14 (Term Reduction Algebra). Given a term rewriting system $\Theta = \langle \Sigma, R \rangle$ and a preorder $X = (X_0, \rightarrow_X)$, the *term reduction algebra* $T_\Theta(X)$ is the smallest preorder $\rightarrow_{T_\Theta(X)}$ on the terms $T_\Sigma(X_0)$ satisfying the following inference rules (where $t[t_1, \ldots, t_n]$ is the substitution of the n vari-

[1] The term algebra construction for categories is technically more complicated as extra equations are required to ensure the term algebra is a category.

ables in $t \in T_\Omega(Y)$ with terms $t_1, \ldots, t_n$):

$$[\textsc{Var}] \quad \frac{x \rightarrow_X y}{{}'x \rightarrow_{T_\Theta(X)} {}'y}$$

$$[\textsc{Pre}] \quad \frac{t_1 \rightarrow_{T_\Theta(X)} s_1, \ldots, t_n \rightarrow_{T_\Theta(X)} s_n}{f(t_1, \ldots, t_n) \rightarrow_{T_\Theta(X)} f(s_1, \ldots, s_n)} \; f \in \Sigma_n$$

$$[\textsc{Inst}] \quad \frac{\begin{array}{l}(Y \vdash l \rightarrow r) \in R,\; Y = (\{y_1, \ldots, y_n\}, \rightarrow_Y) \\ \forall i, j = 1, \ldots, n.\; y_i \rightarrow_Y y_j \;\Rightarrow\; t_i \rightarrow_{T_\Theta(X)} t_j\end{array}}{l[t_1, \ldots, t_n] \rightarrow_{T_\Theta(X)} r[t_1, \ldots, t_n]} \; t_1, \ldots, t_n \in T_\Sigma(X)$$

So the term reduction algebra $T_\Theta(X)$ has as objects the terms which can be built over X and has as rewrites the transitive-reflexive closure of the union of the rewrites of Θ and the rewrites of X closed under the term constructors. This construction defines a monad and this semantics is compositional.

Lemma 15. *The map $X \mapsto T_\Theta(X)$ defines a finitary,* **Pre***-enriched monad* T_Θ*. Furthermore, this semantics is compositional.*

Proof. The proofs follow those for algebraic theories. See [Lü97]. □

Our construction of the coproduct of two monads in Section 5 requires the following technical properties.

Definition 16 (Regular Monads). A monad T is *regular* if

1) the action T preserves *weakly filtered colimits* (colimits of weakly filtered diagrams) where a diagram D is weakly filtered if for all $i, j \in D$, there is a $k \in D$ and morphisms $m : i \rightarrow k$, $n : j \rightarrow k$;
2) the unit is a mono (i.e. every component of the unit is a mono).

Lemma 17. *For a term rewriting system Θ, the monad T_Θ is regular.*

Proof. That the unit is a monomorphism is easy to see. To show that T_Θ preserves weakly filtered diagrams, it is sufficient to show that T_Θ preserves both filtered colimits (because the underlying monad on **Set** is finitary, see lemma 6), and coequalizers (because it does not identify any terms). □

Enriched Monads

The crucial insight behind the constructions of the previous section is the proper *enrichment* [Kel82]. In particular, the base category $\mathcal{A}$ has to be enriched over a closed monoidal category $\mathcal{V}$. Further, $\mathcal{A}$ and $\mathcal{V}$ have to be locally finitely presentable, i.e. have a small set $\mathcal{N}$ of objects representing isomorphism classes of *finitely presentable objects* [KP93]; for **Set**, $\mathcal{N}$ is the natural numbers and the finitely presentable objects are the finite sets.

In the enriched setting, a signature over $\mathcal{A}$ is a map $\Sigma : \mathcal{N} \to \mathcal{A}$, giving for every $c \in \mathcal{N}$ the operations of arity c. The term algebra is then given by a functor $T_\Sigma : \mathcal{A} \to \mathcal{A}$ which maps an $\mathcal{A}$-object of variables X to the $\mathcal{A}$-object of Σ-terms constructed over X. Formally, $T_\Sigma(X)$ is defined as the colimit in $\mathcal{A}$ of the chain $T_0(X) \hookrightarrow T_1(X) \hookrightarrow \ldots$ where $T_0(X) = X$ and

$$T_{n+1}(X) = X + \sum_{c \in \mathcal{N}} [c, T_n(X)] \otimes \Sigma(c) \tag{1}$$

Note how the closed structure over which $\mathcal{A}$ enriches occurs in equation 1. We think of $T_{n+1}(X)$ as the terms of depth $n+1$, constructed as operations of arity c applied to c-objects of terms of depth n. Our models of term rewriting systems arise when we take $\mathcal{A} = \mathcal{V} = \mathbf{Pre}$ with the usual cartesian closed structure providing the enrichment.

Each of the rules of Definition 14 arises as a special case of equation 1. For instance the rule [VAR] stems from the inclusion of X in $T_1(X)$, while specialising equation 1 to the declaration of rewrite rules gives the following equivalent formulation of [INST]

$$\frac{\theta \in \mathbf{Pre}(Y, T_\Theta(X)) \quad (Y \vdash l \to r) \in R}{\theta(l) \to_{T_\Theta(X)} \theta(r)} \ [\text{INST'}]$$

We can also specialise equation 1 to the declaration of term constructors and hence obtain an equivalent formulation of rule [PRE].

Of course, one can vary not only the base category, but also the choice of the monoidal structure. There are in fact two monoidal closed structures over **Cat** — the usual cartesian structure and a monoidal structure which has as objects the same objects as the cartesian product but whose morphisms are alternating sequences of morphisms from each category. Categories enriched over this alternative monoidal structure are called *Sesqui-categories* and have been used as alternative models for term rewriting [Ste94] since they have a categorical notion of "length". It is our intention to use this observation to compare the Sesqui-category approach within our monadic framework.

Monadic Versions of Rewriting Concepts

In the remainder of the paper, we give semantic proofs of modularity results for confluence and strong normalisation. The first step is to define confluence and strong normalisation for arbitrary monads, and show that these definitions coincide with the traditional definitions of section 3.

Definition 18 (Confluence for Monads). A category $\mathcal{C}$ is *confluent* if for any two morphisms $\alpha : x \to x_1, \beta : x \to x_2$ there are morphisms $\gamma : x_1 \to z, \delta : x_2 \to z$ such that $\gamma \cdot \alpha = \delta \cdot \beta$. A monad $\mathsf{T} = \langle T, \eta, \mu \rangle$ on **Cat** is *confluent* if $T\mathcal{X}$ is confluent whenever $\mathcal{X}$ is.

Definition 19 (SN for Monads). A category $\mathcal{C}$ is *SN*, written $\mathcal{C} \models \text{SN}$, if its *underlying order* $R^{-}(\mathcal{C})$ is SN, where $R^{-}(\mathcal{C})$ is defined as follows:

$$R^{-}(\mathcal{C}) \stackrel{def}{=} (|\mathcal{C}|, \{x > y \mid \exists \alpha : x \to y \wedge \alpha \neq \mathbf{1}_x\}) \tag{2}$$

A monad T on **Cat** is strongly normalising if $T\mathcal{X} \models \text{SN}$ whenever $\mathcal{X} \models \text{SN}$.

The definition of a confluent category is different from Stell's [Ste94] which does not require the completions to form a commuting diagram; it is used by Jay [Jay90] but his confluent functors have a different intention and hence only require the identity and composition to be preserved up to having a common reduct.

Lemma 20. *A TRS Θ is confluent iff T_Θ is a confluent monad. Similarly, Θ is SN iff T_Θ is SN.*

Proof. If X is a preorder and Θ is a term rewriting system, then $\mathsf{T}_\Theta(X)$ is the transitive-reflexive closure of the union of the one-step reduction relation $\to_R$ and the closure of the variable rewrites in X under application of operations. Thus lemma 20 amounts to proving that the addition of variable rewrites to a term rewriting system does not change its properties. See [Lü97]. □

Collapsing and expanding rewrites also have categorical formulations.

Definition 21 (Non-Expanding/Non-Collapsing Monads). A functor $F : \mathcal{X} \to \mathcal{Y}$ is *non-expanding*, if for all objects $x \in \mathcal{X}$ and all morphisms $\alpha : Fx \to y'$ in $\mathcal{Y}$ there is a morphism $\beta : x \to y$ in $\mathcal{X}$ such that $F\beta = \alpha$. A monad $\mathsf{T} = \langle T, \eta, \mu \rangle$ on **Cat** is non-expanding if all components of the unit η are non-expanding, and the action preserves non-expanding functors, i.e. if $F : \mathcal{X} \to \mathcal{Y}$ is non-expanding, then so is TF.

A functor $F : \mathcal{X} \to \mathcal{Y}$ is *non-collapsing*, if F^{op} is non-expanding. A monad $\mathsf{T} = \langle T, \eta, \mu \rangle$ on **Cat** is non-collapsing if all components of the unit are non-collapsing, and the action preserves non-collapsing functors

One may easily verify that a term rewriting system Θ is non-expanding (non-collapsing) iff T_Θ is non-expanding(non-collapsing).

5 A Monadic Approach to Modularity

We have given a semantics to term rewriting systems in terms of monads on **Cat**. By lemma 20 we can reason about the disjoint union of Θ_1 and Θ_2 by reasoning about its semantics $\mathsf{T}_{\Theta_1+\Theta_2}$, which by lemma 15 is isomorphic to the coproduct of $\mathsf{T}_{\Theta_1} + \mathsf{T}_{\Theta_2}$. This section gives a pointwise construction of the coproduct of two regular monads as the colimit of a diagram. Since this diagram is built solely from the component monads, we can reason about the coproduct monad in terms of the component monads.

Consider terms built in the disjoint union of two signatures Ω, Σ. Such terms have an inherent notion of *layer*, that is one can decompose a term constructed from symbols in the union of two disjoint signatures into a term constructed from symbols in only one signature and strictly smaller subterms whose head symbol is from the other signature. Thus terms built from operations of $\Omega + \Sigma$ are contained in

$$\begin{aligned} T_{\Omega+\Sigma}(X) = & X + T_\Omega(X) + T_\Sigma(X) + T_\Omega T_\Sigma(X) + T_\Sigma T_\Omega(X) + \\ & T_\Omega T_\Sigma T_\Omega(X) + T_\Sigma T_\Omega T_\Sigma(X) + \ldots \end{aligned} \tag{3}$$

However this disjoint union is too large as each component of the sum in equation 3 contains a separate copy of the variables X. Therefore this sum is quotiented by taking the colimit of a diagram including all arrows formed using the units and multiplications of the monads. Formally, let $\mathsf{T}_1, \mathsf{T}_2$ be two regular monads on **Cat**, let $\mathcal{L} \stackrel{def}{=} \{1,2\}$, and define $W \stackrel{def}{=} \mathcal{L}^*$ to be the words over $\mathcal{L}$, and for $w \in W$, $T^w : \mathbf{Cat} \to \mathbf{Cat}$ by $T^\varepsilon \stackrel{def}{=} 1_{\mathbf{Cat}}$ and $T^{jv} \stackrel{def}{=} T_j T^v$ where $j \in \mathcal{L}, v \in W$. As notational shortcuts, we also define the natural transformations $\eta^u_{i,v} \stackrel{def}{=} T^u(\eta_{i,T^v})$ and $\mu^u_{j,v} \stackrel{def}{=} T^u(\mu_{j,T^v})$ for $u, v \in W, j \in \mathcal{L}$.

Definition 22 (The Colimit Diagram $\mathcal{D}_\mathcal{X}$). For every category $\mathcal{X}$, the diagram $\mathcal{D}_\mathcal{X}$ has as objects the categories $T^w(\mathcal{X})$ for $w \in W$, and as edges:

$$(\eta^u_{i,v})_\mathcal{X} : T^{uv}(\mathcal{X}) \to T^{uiv}(\mathcal{X}) \qquad (\mu^u_{j,v})_\mathcal{X} : T^{ujjv}(\mathcal{X}) \to T^{ujv}(\mathcal{X})$$

Lemma 23. *The map on categories $\mathcal{X} \mapsto$ colim $D_\mathcal{X}$ extends to a monad which is the coproduct of the monads T_1 and T_2.*

Proof. Functoriality follows from the universal property of the colimit, and by the fact that all arrows in the diagram are natural transformations. The unit is simply the inclusion of $\mathcal{X}$ into the colimiting object. The multiplication uses the fact that the diagram $D_\mathcal{X}$ is weakly filtered, and hence preserved by the two functors T_1, T_2. The monad laws and universal property follow from various diagram chases (see [Lü97] for details). □

Note that $D_\mathcal{X}$ is not filtered, since there is e.g. no arrow in the diagram which makes $\eta_1^{12} \cdot \eta_2^1$ and $\eta_{2,1} \cdot \eta_{1,21}$ equal.

Analysing the Coproduct Monad

By the dual of Theorem 2 in [Mac71, pg. 109], every colimit can be expressed via coproducts and coequalizers. In particular, the colimit of $D_\mathcal{X}$ is given by the coequalizer of Diagram 4, where on the left side, for every morphism

$$\coprod_{d: T^u\mathcal{X} \to T^v\mathcal{X} \in D_\mathcal{X}} T^u\mathcal{X} \underset{G}{\overset{F}{\rightrightarrows}} \coprod_{w \in W} T^w\mathcal{X} \tag{4}$$

$d : T^u\mathcal{X} \to T^v\mathcal{X}$ in $D_\mathcal{X}$ (with $u, v \in W$) there is a component $T^u\mathcal{X}$ in the coproduct, and F and G are defined as $F(T^u\mathcal{X}) \stackrel{def}{=} \iota_u(T^u\mathcal{X})$, $G(T^u\mathcal{X}) \stackrel{def}{=} \iota_v(d(T^u\mathcal{X}))$ where ι_u and ι_v are injections into the coproduct on the right.

Lemma 24. *Given two functors $F, G : \mathcal{X} \to \mathcal{Y}$, their coequalizer is a functor $Q : \mathcal{Y} \to \mathcal{Z}$, where $\mathcal{Z}$ is defined as follows:*

1) *The objects are the objects of $\mathcal{Y}$, quotiented by the equivalence closure $\equiv$ of the relation $\sim$ defined as $x \sim y \Leftrightarrow \exists z \in \mathcal{X} \,.\, Fz = x, Gz = y$.*
2) *Morphisms are sequences $<f_1, \ldots, f_n>$ of morphisms $f_i \in \mathcal{Y}(x_i, y_i)$ such that $y_i \equiv x_{i+1}$, quotiented by the smallest equivalence relation $\equiv$ compatible with composition in $\mathcal{Y}$ s.t. $<f, g> \equiv <g \cdot f>$ if f, g are composable in $\mathcal{Y}$, and $<Fh> \equiv <Gh>$ for all morphisms h in $\mathcal{X}$.*

Deciding the Equivalence: Normal Forms

The terms of the disjoint union of two monads are equivalence classes of objects from $\coprod_{w \in W} T^w \mathcal{X}$. In this section, we improve upon the presentation of [Lü96] by introducing a pair of reduction systems which reduce the objects and morphisms of $\coprod_{w \in W} T^w \mathcal{X}$ to a unique normal form, deciding this equivalence. We stress that these constructions occur at the level of regular monads and nowhere do we use the fact that these monads arise from term rewriting systems.

Definition 25 (The Reduction System $\to_{Ob}$). Define the following reduction systems on the objects of $\coprod_{w \in W} T^w \mathcal{X}$:

$$\begin{aligned}
\to_\mu &\stackrel{def}{=} \{x \to_\mu \mu^u_{j,v}(x) \mid u, v \in W, j \in \mathcal{L}\} \\
\to_\eta &\stackrel{def}{=} \{\eta^u_{j,v}(x) \to_\eta x \mid u, v \in W, j \in \mathcal{L}\} \\
\to_{Ob} &\stackrel{def}{=} \to_\eta \cup \to_\mu
\end{aligned}$$

We show that $\to_{Ob}$, is complete and hence obtain a decision procedure for the associated equality. First, define the *rank* of a term $t \in T^w \mathcal{X}$ as $\text{rank}(t) \stackrel{def}{=} |w|$ (where $|w|$ is the length of the word w).

Lemma 26. *$\twoheadrightarrow_{Ob}$ is complete.*

Proof. For each reduction step $t \to_{Ob} u$, the rank of t is strictly greater than that of u and hence $\to_{Ob}$ is SN. For confluence, we refer to lemma 13 of [Lü96]. Clause (i) of that lemma implies confluence of $\to_\mu$, and clause (ii) implies confluence of $\to_\eta$. Clause (iii) implies that $\to_\eta$ and $\to_\mu$ commute and hence $\twoheadrightarrow_{Ob}$ is confluent. The cited proof also elucidates the necessity for the units η_1, η_2 to be monomorphisms. □

Since $\twoheadrightarrow_{Ob}$ is complete, every object in $t \in \coprod_{w \in W} T^w \mathcal{X}$ reduces to a unique normal form which we denote NF(t). This forms a decision procedure for the equivalence of the objects:

Lemma 27. *Given $t, t' \in \coprod_{w \in W} T^w \mathcal{X}$, $Qt = Qt'$ iff $NF(t) = NF(t')$.*

Proof. $\mathrm{NF}(t) = \mathrm{NF}(t')$ iff t and t' are related is the equational theory on $\coprod_{w \in W} T^w\mathcal{X}$ generated by $\twoheadrightarrow_{Ob}$. This theory is clearly the same as that induced by the coequalizer of diagram 4. □

We now consider morphisms in the coequalizer of diagram 4. Since such morphisms are sequences of morphisms in $\coprod_{w \in W} T^w\mathcal{X}$, we start by considering the normal forms of morphisms in $\coprod_{w \in W} T^w\mathcal{X}$.

Definition 28 (The Reduction System $\rightarrow_{Mor}$). Define the reduction systems on the morphisms of $\coprod_{w \in W} T^w\mathcal{X}$:

$$\begin{aligned}
\rightarrow_\mu &\stackrel{def}{=} \{\alpha \rightarrow_\mu \mu^u_{j,v}(\alpha) \mid u, v \in W, j \in \mathcal{L}\} \\
\rightarrow_\eta &\stackrel{def}{=} \{\eta^u_{j,v}(\alpha) \rightarrow_\eta \alpha \mid u, v \in W, j \in \mathcal{L}\} \\
\rightarrow_{Mor} &\stackrel{def}{=} \rightarrow_\eta \cup \rightarrow_\mu
\end{aligned}$$

Lemma 29. *$\rightarrow_{Mor}$ is complete, and every morphism α in $T^w\mathcal{X}$ reduces to a unique normal form NF(α) s.t. for all β, $Q\alpha = Q\beta$ iff NF(α) = NF(β).*

Proof. Analogously to lemma 26 and 27. □

The mapping of terms and morphisms to their normal form can not be extended to a functor, since the presence of non-expanding and non-collapsing rewrites means that the normal form need not preserve the source and target of a morphism. For example given a rewrite α : $\texttt{'}x \rightarrow \texttt{G('}x\texttt{)}$ in $T_1(X)$, then $\mathrm{NF}(\alpha) = \alpha$ which is in $T_1(X)$ while $\mathrm{NF}(\texttt{'}x) = x$.

Definition 30 (Layer-Expanding and Layer-Collapsing). Let $\alpha : s \rightarrow t$ be in $T^w\mathcal{X}$, and $\mathrm{NF}(\alpha) : s' \rightarrow t'$ its normal form. Then α is called *layer-collapsing (layer-expanding) in* T_j if there are $u, v \in W, j \in \mathcal{L}$ and $y \in T^{uv}\mathcal{X}$ s.t. $t' = \eta^u_{j,v}(y)$ $(s' = \eta^u_{j,v}(y))$.

Lemma 31. *$\alpha : s \rightarrow t$ in $T^w\mathcal{X}$ is layer-expanding (layer-collapsing) iff for NF(α) : $s' \rightarrow t'$, $s' \neq$ NF(s) ($t' \neq$ NF(t)).*

Proof. We can apply $\rightarrow_\mu$ to a morphism $\alpha : x \rightarrow y$ iff we can apply it to its source x and target y. It is feasible that we can apply $\rightarrow_\eta$ to x (or y) but not to α; namely, if $x = \eta^u_{j,v}(x')$ (or $y = \eta^u_{j,v}(y')$) and for all $\beta : x' \rightarrow y'$, $\eta^u_{j,v}(\beta) \neq \alpha$ (and so $\alpha \not\rightarrow_\eta \beta$), but then α is layer-expanding (or layer-collapsing). □

Note that a rewrite can be expanding or collapsing in both systems at the same time. Further note that there can only be layer-expanding (collapsing) rewrites in T_j if T_j is expanding (collapsing).

For sequences $<\alpha_1, \ldots, \alpha_n>$ in the colimit, we do not really need to decide the equality on them, but merely want to reason about their length (in the light of lemma 39 below). Hence we introduce the notion of minimal length:

Definition 32 (Minimal Length). A sequence $A = <\alpha_1, \ldots, \alpha_n>$ is of *minimal length* iff all elements are normal forms: $\forall i = 1, \ldots, n.\, \alpha_i = \mathrm{NF}(\alpha_i)$, and no equivalent sequence is shorter: $B \equiv A \Rightarrow |A| \leq |B|$.

For an example, consider the two monads given by the following two term rewriting systems $R_1 = \{\texttt{F(F('}x\texttt{))} \to \texttt{H('}x\texttt{)}\}, R_2 = \{\texttt{G('}y\texttt{)} \to \texttt{'}y\}$ Then there are reductions $\alpha_1 : \texttt{F('G('F('}x\texttt{)))} \to \texttt{F(''F('}x\texttt{))}$ in $T_1T_2T_1(X)$ and $\alpha_2 : \texttt{F(F('}x\texttt{))} \to \texttt{H('}x\texttt{)}$ in $T_1(X)$. Since $\mathrm{NF}(\texttt{F(''F('}x\texttt{)))} = \texttt{F(F('}x\texttt{))}$, one can form a sequence $<\alpha_1, \alpha_2>$ although α_1 and α_2 are not composable as they inhabit different components of $\coprod_{w \in W} T^w \mathcal{X}$. This situation (and its dual, where α_2 would be layer-expanding) is prototypical, as the following lemma shows:

Lemma 33. *A sequence $A = <\alpha_1, \ldots, \alpha_n>$ is of minimal length iff for all $l = 1, \ldots, n$, $\alpha_l : x_l \to y_l$ is in normal form, and for all $k = 1, \ldots, n-1$:*

- *α_k is layer-collapsing, with $y_k = \eta^w_{j,v}(z)$ $(w, v \in W, j \in \mathcal{L})$, and there are $r, s \in W, i \in \mathcal{L}$ s.t. $w = ri$, $v = is$, $i \neq j$, $x_{k+1} = \mu^r_{i,s}(z)$ and for all $\beta : z \to z'$, $\mu^r_{i,s}(\beta) \neq \alpha_{k+1}$;*
- *or α_{k+1} is layer-expanding, with $x_{k+1} = \eta^w_{j,v}(z)$ $(w, v \in W, j \in \mathcal{L})$, and there are $r, s \in W, i \in \mathcal{L}$ s.t. $w = ri$, $v = is$, $i \neq j$, $y_k = \mu^r_{i,s}(z)$ and for all $\beta : z' \to z$, $\mu^r_{i,s}(\beta) \neq \alpha_k$.*

Then (α_k, α_{k+1}) are called an incomposable pair.

Proof. A is of minimal length iff we cannot compose α_k and α_{k+1}, and there are no β_k, β_{k+1} which are composable and equivalent to α_k, α_{k+1}. In particular, $y_k \neq x_{k+1}$ but $\mathrm{NF}(y_k) = \mathrm{NF}(x_{k+1})$, hence $\mathrm{NF}(y_k) \neq y_k$ or $x_{k+1} \neq \mathrm{NF}(x_{k+1})$. By lemma 31,$\alpha_k$ is layer-collapsing iff $\mathrm{NF}(y_k) \neq y_k$. Then the second part of the first clause ensures that there is no β' which is composable with α_k s.t. $Q(\beta) = Q(\alpha_{k+1})$. The second clause is the dual of the first (with $x_{k+1} \neq \mathrm{NF}(x_{k+1})$). □

We close this section by showing that given two monad morphisms $\kappa : \mathsf{T}_1 \to \mathsf{S}_1$, $\lambda : \mathsf{T}_2 \to \mathsf{S}_2$, their coproduct $\kappa + \lambda : \mathsf{T}_1 + \mathsf{T}_2 \to \mathsf{S}_1 + \mathsf{S}_2$ preserves the normal form with respect to the two reduction systems $\twoheadrightarrow_{Ob}, \twoheadrightarrow_{Mor}$ above. Intuitively $\kappa + \lambda$ replaces every T_1 layer with its image in S_1 under κ and similarly for T_2 layers. Formally the components of $\kappa + \lambda$ are constructed by defining the obvious cone over the diagram whose colimit defines $(\mathsf{T}_1 + \mathsf{T}_2)(X)$. It is however not the case that $\kappa + \lambda$ preserves sequences of minimal length, since in general the conditions of lemma 33 are not preserved.

Lemma 34. *Given $\kappa : \mathsf{T}_1 \to \mathsf{S}_1$, $\lambda : \mathsf{T}_2 \to \mathsf{S}_2$ which are epi, let $M \stackrel{def}{=} \kappa + \lambda$. Then $M(NF(t)) = NF(M(t))$ for $t \in T^w \mathcal{X}$, and $M(NF(\alpha)) = NF(M(\alpha))$ for $\alpha : s \to t$ in $T^w \mathcal{X}$.*

Proof. By induction on the derivations $t \twoheadrightarrow_{Ob} \mathrm{NF}(t)$, $\alpha \twoheadrightarrow_{Mor} \mathrm{NF}(\alpha)$. Essentially, whenever we can reduce $t \rightarrow_{Ob} t'$, then we can reduce $M(t) \rightarrow_{Ob} M(t')$ (by naturality of the unit and multiplication of T_1 and T_2, and κ and λ being monad morphisms.) □

6 Modularity of Confluence

We now prove the modularity of confluence. The first step towards proving confluence is to find conditions under which a functor preserves confluence.

Definition 35 (One-Step Completion). Given a functor $Q : \mathcal{Y} \rightarrow \mathcal{Z}$, the category $\mathcal{Y}$ has the *one-step completion property with respect to* Q, written $\mathcal{Y} \models_Q \diamond$, if for all morphisms $\alpha : x \rightarrow x', \beta : y \rightarrow y'$ in $\mathcal{Y}$ such that $Qx = Qy$ there are morphisms $\gamma : v \rightarrow v', \delta : w \rightarrow w'$ in $\mathcal{Y}$ such that $Q\gamma \cdot Q\alpha = Q\delta \cdot Q\beta$.

Lemma 36. *Let $Q : \mathcal{Y} \rightarrow \mathcal{Z}$ be the coequalizer of two functors $F, G : \mathcal{X} \rightarrow \mathcal{Y}$ in* **Cat**. *If $\mathcal{Y}$ is confluent and $\mathcal{Y} \models_Q \diamond$, then $\mathcal{Z}$ is confluent.*

Proof. Given two morphisms $\alpha = [<\alpha_1, \dots, \alpha_n>]$ and $\beta = [<\beta_1, \dots, \beta_m>]$ in $\mathcal{Z}$ with the same source. Then (since $\mathcal{Y} \models_Q \diamond$) there are β_1', α_1' such that $Q(\beta_1') \cdot Q(\alpha_1) = Q(\alpha_1') \cdot Q(\beta_1)$. By induction on the length n and m of α and β, respectively, we obtain completions $\alpha' \stackrel{def}{=} [<\alpha_1^{(m)}, \dots, \alpha_n^{(m)}>]$, $\beta' \stackrel{def}{=} [<\beta_1^{(n)}, \dots, \beta_m^{(n)}>]$ such that $\beta' \cdot \alpha = \alpha' \cdot \beta$. □

To prove that the coequalizer of diagram 4 is confluent we show that $\coprod_{w \in W} T^w X \models_Q \diamond$ where Q is the coequalising functor. In [Lü96], this was done using a *witness relation.* Here, the witnesses are replaced by the conceptually simpler normal forms.

Lemma 37. *The coproduct of two confluent, non-expanding, regular monads is confluent.*

Proof. We first show that if $\mathcal{X}$ is confluent then $\coprod_{w \in W} T^w \mathcal{X} \models_Q \diamond$. Since $\mathcal{X}$, T_1 and T_2 are confluent and coproducts in **Cat** preserve confluence, $\coprod_{w \in W} T^w \mathcal{X}$ is confluent. Given $\alpha : x \rightarrow x'$ and $\beta : y \rightarrow y'$ in $\coprod_{w \in W} T^w \mathcal{X}$ such that $Qx = Qy$, by lemma 31 ($\mathsf{T}_1, \mathsf{T}_2$ are non-expanding) $\mathrm{NF}(\alpha) : \mathrm{NF}(x) \rightarrow x_0$ and $\mathrm{NF}(\beta) : \mathrm{NF}(y) \rightarrow y_0$. Further, since $Qx = Qy$ by lemma 27 $\mathrm{NF}(x) = \mathrm{NF}(y)$. Hence there are completions $\gamma : x_0 \rightarrow z_0$ and $\delta : y_0 \rightarrow z_0$ s.t. $\gamma \cdot \mathrm{NF}(\alpha) = \delta \cdot \mathrm{NF}(\beta)$, and hence $Q\gamma \cdot Q\alpha = Q\delta \cdot Q\beta$.

Then by lemma 36 the colimit of diagram 4 is confluent if $\mathcal{X}$ is confluent, and so the coproduct monad is confluent. □

The modularity of confluence for TRSs follows easily:

Theorem 38 (Toyama). *Confluence is modular for non-expanding TRSs.*

Proof. Let Θ_1 and Θ_2 be confluent TRSs. By lemma 17 the monads T_{Θ_1} and T_{Θ_2} are regular, non-expanding and by lemma 20 confluent, and so is their coproduct (lemma 37). By lemma 15 this coproduct models the disjoint union of Θ_1 and Θ_2 and hence by lemma 20 this TRS is confluent. □

7 Modularity of Strong Normalisation

As mentioned in the introduction, strong normalisation (SN) is *not* a modular property for the disjoint union of term rewriting systems. We will below find conditions under which the disjoint union of two SN monads cannot be strongly normalising, from which several conditions under which the union is SN will be derived. This is an adaption of the minimal counterexamples technique in [Gra92]. We first need a criterion to determine when the disjoint union of two monads is not SN.

Lemma 39. *Given monads $\mathsf{T}_1, \mathsf{T}_2$ on* **Cat** *s.t. $\mathsf{T}_1 \models SN, \mathsf{T}_2 \models SN$, then $\mathsf{T}_1 + \mathsf{T}_2 \not\models SN$ iff for all $n \in \mathbb{N}$ there is a sequence A of minimal length s.t. $|A| > n$ (called an* infinite sequence*).*

Proof. The lemma follows from the observation that every sequence $A = \langle\alpha_1, \ldots, \alpha_n\rangle$ (with $\alpha_i : x_i \to y_i$) of minimal length gives rise to a sequence $[x_1] < [x_2] < \ldots [y_n]$ in the underlying order $R^-(T\mathcal{X})$ of the coproduct monad at $\mathcal{X}$. See [Lü97] for the details. □

A term rewriting system Θ is called *strongly normalising under deterministic collapses* (SNDC or $\mathcal{C}_\mathcal{E}$-terminating) [Ohl94,Gra92] if it is SN and the disjoint union $\Theta + \mathcal{C}_\mathcal{E}$ is SN, where $\mathcal{C}_\mathcal{E}$ is the term rewriting system $\mathcal{C}_\mathcal{E} \stackrel{def}{=} \{\mathtt{G}('x, 'y) \to 'x, \mathtt{G}('x, 'y) \to 'y\}$. A recent term rewriting result is that the disjoint union is not SN if one system is SNDC and the other collapsing. The term rewriting proof is a rather intricate encoding construction. In this setting, the proof is far simpler: we find a monad $\mathsf{T}_\perp$ representing $\mathcal{C}_\mathcal{E}$ and then analyse its combination with T_1. This combination will be obtained by a universal property of $\mathsf{T}_\perp$. We first define SNDC for monads, and show this definition is equivalent to the one used in term rewriting:

Definition 40 (SNDC). A monad T on **Cat** is called *strongly normalising under deterministic collapses*, $T \models \text{SNDC}$ if $T \models \text{SN} \wedge T + T_\perp \models \text{SN}$

Lemma 41. *The term rewriting system Θ is strongly normalising under deterministic collapses iff. the monad T_Θ is.*

Proof. Using lemma 39, we must show there is an infinite reduction in $\Theta + \mathcal{C}_\mathcal{E}$ iff there is an infinite sequence of minimal length in $T_1 + T_\perp$. One direction is easy, since every non-identity rewrite $\alpha : s \to t$ in $T_1 + T_\perp$ gives rise to at least one rewrite step in $\Theta + \mathcal{C}_\mathcal{E}$. For the other direction, we draw upon [Gra92, Lemma 2], which shows that an infinite derivation in $\Theta + \mathcal{C}_\mathcal{E}$ contains infinitely many rewrites which satisfy the criteria of lemma 33[2]. □

[2] Namely, they are destructive at level 2.

Definition 42 (A Monad called $\mathsf{T}_\perp$). The monad $\mathsf{T}_\perp = \langle T_\perp, \eta_\perp, \mu_\perp \rangle$ on **Cat** is defined as follows: it maps a category $\mathcal{X}$ to the category $T_\perp(\mathcal{X})$, which has as objects $|T_\perp \mathcal{X}| \stackrel{def}{=} \{\perp\} + |\mathcal{X}|$ and as morphisms

$$T_\perp(\mathcal{X})(x, y) \stackrel{def}{=} \begin{cases} \{!_y\} & \text{if } x = \perp \\ \emptyset & \text{if } x \neq \perp, y = \perp \\ \mathcal{X}(x, y) & \text{otherwise} \end{cases}$$

with the evident composition (for $f : x \to y$, $f \cdot !_x = !_y$ etc.). For a functor $F : \mathcal{X} \to \mathcal{Y}$, $T_\perp(F)$ maps $\perp$ to $\perp$, and x (with $x \in \mathcal{X}$) to Fx in $T_\perp(\mathcal{Y})$, and similarly on the morphisms. The unit $\eta_{\perp,\mathcal{X}} : \mathcal{X} \to T_\perp \mathcal{X}$ is the injection of the category $\mathcal{X}$ into $T_\perp \mathcal{X}$, and the multiplication $\mu_{\perp,\mathcal{X}} : T_\perp T_\perp \mathcal{X} \to T_\perp \mathcal{X}$ identifies the two adjoined objects.

From the term rewriting point, this monad can be seen as representing the system $\mathcal{C}_\mathcal{E}$.[3] From the categorical point of view, this monad freely adjoins an initial object $\perp$ to a category. This monad is terminal amongst non-expanding monads on **Cat**:

Lemma 43. *For any non-expanding monad* T *on* **Cat**, *there is a unique monad morphism* $!_T : \mathsf{T} \to \mathsf{T}_\perp$.

Proof. For a category $\mathcal{X}$, $!_{T,\mathcal{X}} : T\mathcal{X} \to T_\perp \mathcal{X}$ is defined on objects as

$$!_{T,\mathcal{X}}(x) \stackrel{def}{=} \begin{cases} x_0 & \text{if } x = \eta(x_0) \\ \perp & \text{otherwise} \end{cases}$$

and similarly on the morphisms. This is the only definition which makes $!_T$ a monad morphism. Note $!_{T,\mathcal{X}}$ is only a functor if T is non-expanding. □

By lemma 43, for two monads T_1 and T_2, there is a monad morphism $1{+}! : \mathsf{T}_1 + \mathsf{T}_2 \to \mathsf{T}_1 + \mathsf{T}_\perp$, substituting all compound terms from T_2 in $\mathsf{T}_1 + \mathsf{T}_2$ with the object $\perp$ from $T_\perp$. The proof of our main result proceeds as follows: since $\mathsf{T}_1 + \mathsf{T}_2$ is not SN, there is an infinite sequence of minimal length, A. We consider the image of A under the monad morphism $M \stackrel{def}{=} 1{+}! : \mathsf{T}_1 + \mathsf{T}_2 \to \mathsf{T}_1 + T_\perp$. From lemma 33, we know when a sequence has minimal length. We will show that the monad morphism preserves these properties, so there will be an infinite sequence of minimal length in $\mathsf{T}_1 + T_\perp$ as well, showing that T_1 is not SNDC. Recall from lemma 33 the notion of incomposable pairs. Assuming both monads to be non-expanding, only the second case of lemma 33 applies:

Lemma 44. *Given an incomposable pair* (α, β) *where* α *is layer-collapsing in* T_2, *then* $(M\alpha, M\beta)$ *is an incomposable pair as well.*

[3] Although of course $\mathsf{T}_\perp$ is not the monad $\mathsf{T}_{\mathcal{C}_\mathcal{E}}$ given by that system because of its multiplication.

Proof. By lemma 34, $M(\mathrm{NF}(\alpha)) = \mathrm{NF}(M\alpha)$ (as both 1 and ! are epi), so $M(\alpha) : x_1 \to y_1, M(\beta) : x_2 \to y_2$ is in normal form. The other conditions follow since M is given by a cone morphism ν between the two cones over the diagrams defining the colimits: since $\eta^v_{2,w}$ and $\mu^r_{1,s}$ are morphisms of the diagram, ν preserves them. □

Theorem 45. *Given two regular, non-expanding, SN monads* T_1 *and* T_2 *on* **Cat**, *if* $\mathsf{T}_1 + \mathsf{T}_2 \not\models$ *SN, then either* $\mathsf{T}_1 \not\models$ *SNDC and* T_2 *is collapsing or vice versa.*

Proof. By lemma 39, there is an infinite sequence $A = <\alpha_1, \ldots, \alpha_n \ldots>$ of minimal length in $\mathsf{T}_1 + \mathsf{T}_2$. By lemma 33, all α_i in A have to be collapsing, so at least one of T_1 or T_2 is collapsing. Further, there are infinitely many rewrites collapsing in T_1, or infinitely many rewrites collapsing in T_2. Wolg. assume the latter, and consider the sequence MA in $\mathsf{T}_1 + \mathsf{T}_\perp$. If we compose all $M\alpha_i$ and $M\alpha_{i+1}$ which can be composed, we obtain a sequence A' of minimal length which is equivalent to MA, but by lemma 44, if α_i is collapsing in T_2, $(M\alpha_i, M\alpha_{i+1})$ will remain an incomposable pair, so A' will be infinite as well. Hence, by lemma 39, $\mathsf{T}_1 + \mathsf{T}_\perp$ is not strongly normalising, so $\mathsf{T}_1 \not\models$ SNDC. □

Theorem 45 has a host of interesting corollaries such as:

Corollary 46. *The following modularity results follow from Theorem 45:*

1) SN is modular for non-collapsing systems.
2) SN is modular for non-duplicating systems.
3) SN is modular if one system is non-collapsing and non-duplicating.
4) SN is modular for simplifying systems.

Proof. The first is obvious. For the rest, non-duplicating and simplifying systems are strongly normalising under deterministic collapses [Gra92]. □

Hence, all of the conditions listed in the introduction follow as corollaries from Theorem 45. [Ohl94] contains further derived criteria.

8 Conclusions and Further Work

We have shown how monads can be used to give a semantics to term rewriting systems by generalising the well-known equivalence between universal algebra and finitary monads on **Set**. Monads are well suited to the study of modular term rewriting as the key concepts have concise monadic formulations. We believe this paper provides ample justification for these claims, and further for the more general claim that category theory provides a useful level of abstraction for the study of rewriting.

We propose to extend this work and tackle open problems in modular term rewriting. Firstly, monads can be used to model more general notions of term rewriting for which current modularity results are less than satisfactory. In particular research on modularity for conditional term rewriting is at an advanced stage. Another area where significant problems remain is that of modularity for unions which permit limited forms of sharing. Categorically, these unions are modelled by push-outs which again have a compositional semantics. This observation allows us to apply the methodology outlined in the introduction to study modularity for these more general structuring operations.

References

[BW85] M. Barr and C. Wells. *Toposes, Triples and Theories.* Springer 1985.

[Gha95] N. Ghani. *Adjoint Rewriting.* PhD thesis, University of Edinburgh, 1995.

[Gra92] B. Gramlich. Generalized sufficient conditions for modular termination of rewriting. In *Proc. 3rd ICALP*, LNCS 632, pages 53–68. Springer, 1992.

[Jay90] C. B. Jay. Modelling reductions in confluent categories. In *Proc. Durham Symposium on Applications of Categories in Computer Science*, 1990.

[Kel82] G. M. Kelly. *Basic Concepts of Enriched Category Theory*, LMS Lecture Notes 64. Cambridge University Press, 1982.

[Klo92] J. W. Klop. Term rewriting systems. In S. Abramsky et.al., eds., *Handbook of Logic in Computer Science* Vol. 2, pages 1–116. OUP, 1992.

[KO92] M. Kurihara and A. Ohuchi. Modularity of simple termination of term rewriting systems with shared constructors. *TCS* 103:273– 282, 1992.

[KP93] G. M. Kelly and A. J. Power. Adjunctions whose counits are coequalizers, and presentations of finitary monads. *JPAA* 89:163– 179, 1993.

[Lü96] C. Lüth. Compositional term rewriting: An algebraic proof of Toyama's theorem. In *RTA '96*, LNCS 1103, pages 261– 275, Springer Verlag, 1996.

[Lü97] C. Lüth. *Categorical Term Rewriting: Monads and Modularity.* PhD thesis, University of Edinburgh, 1997. Forthcoming.

[Mac71] S. Mac Lane. *Categories for the Working Mathematician.* Springer 1971.

[Man76] E. G. Manes. *Algebraic Theories*, Springer Verlag, 1976.

[Mid89] A. Middeldorp. A sufficient condition for the termination of the direct sum of term rewriting systems. In *Proc. 4th LICS*, p. 396–401. June 1989.

[Ohl94] E. Ohlebusch. On the modularity of termination of term rewriting systems. *TCS* 136:333– 360, 1994.

[Rob94] E. Robinson. Variations on algebra: monadicity and generalisations of equational theories. Tech. Rep. 6/94, Sussex Univ. Comp. Sci., 1994.

[RS87] D. E. Rydeheard and J. G. Stell. Foundations of equational deduction: A categorical treatment of equational proofs and unification algorithms. In *CTCS '87*, LNCS 283, pages 114– 139. Springer Verlag, 1987.

[Rus87] M. Rusinowitch. On the termination of the direct sum of term-rewriting systems. *Information Processing Letters*, 26(2):65–70, 1987.

[See87] R. A. G. Seely. Modelling computations: A 2-categorical framework. In *Proc. 2nd LICS*, pages 65–71, 1987.

[Ste94] J. G. Stell. Modelling term rewriting systems by Sesqui-categories. Technical Report TR94-02, Keele Unversity, January 1994.

A 2-Categorical Presentation of Term Graph Rewriting*

A. Corradini[1], F. Gadducci[2,3]

[1] CWI, Kruislaan 413, 1098 SJ Amsterdam, The Netherlands, (andrea@cwi.nl).
[2] TUB, Fachbereich 13 Informatik, Franklinstraße 28/29, 10587 Berlin, Germany, (gfabio@cs.tu-berlin.de).
[3] Corresponding author: Phone: +49-30-31473557, Fax: +49-30-31423516.

Abstract. It is well-known that a term rewriting system can be faithfully described by a cartesian 2-category, where horizontal arrows represent terms, and cells represent rewriting sequences. In this paper we propose a similar, original 2-categorical presentation for *term graph rewriting*. Building on a result presented in [8], which shows that term graphs over a given signature are in one-to-one correspondence with arrows of a gs-monoidal category freely generated from the signature, we associate with a term graph rewriting system a gs-monoidal 2-category, and show that cells faithfully represent its rewriting sequences. We exploit the categorical framework to relate term graph rewriting and term rewriting, since gs-monoidal (2-)categories can be regarded as "weak" cartesian (2-)categories, where certain (2-)naturality axioms have been dropped.

Keywords: term graph rewriting, directed acyclic graphs, categorical models, 2-categories.

1 Introduction

The classical theory of *term graph rewriting* studies the issue of representing finite terms as directed, acyclic graphs, and of modeling term rewriting via graph rewriting (among the many contributions to this theory, we mention [28, 2, 13, 10, 1]; see also the book [27] and the references therein). The main advantage of using graphs is that the sharing of common subterms can be represented explicitly, avoiding to copy a subterm when applying a rewrite rule with more than one occurrence of a variable in its right-hand side. Furthermore the rewriting process is speeded up, because the rewriting steps do not have to be repeated for each copy of an identical subterm. For these reasons term graph rewriting is often used, for example, in the implementation of functional languages [24].

* Research partly supported by the EC TMR Network GETGRATS (General Theory of Graph Transformation Systems) through the Technical University of Berlin. A. Corradini is on leave from Dipartimento di Informatica, Pisa, Italy, and he is also partly supported by the EC Fixed Contribution Contract n. EBRFMBICT960840.

The relationship between classical term rewriting and term graph rewriting has been studied in many papers in the literature, including those mentioned above. All the authors agree that term graph rewriting is a sort of term rewriting with explicit sharing of subterms. However, in our opinion, there is an unsatisfactory gap between the achievements of the theory of term rewriting and of term graph rewriting, respectively. On the one hand, indeed, the terms over a signature Σ have an elegant characterization, using Universal Algebra techniques, as elements of the free algebra generated by a signature, or also, using more categorical tools, as the arrows of a suitable cartesian category; and term rewriting sequences over a given system can be defined not only in the usual "operational" way, using the notions of redex and substitution, but also more abstractly as the cells of a suitable cartesian 2-category. Such a categorical account has the advantage of being independent of representation details, and of stressing the intrinsic algebraic structure of terms and their rewriting; in fact, on the basis of the mentioned results one can safely claim that the essential structure of the collection of terms over a signature is "cartesianity".

On the other hand, only more concrete and operational presentations of term graphs and their rewriting are available in literature. For example, term graphs have been defined, among other ways, as directed (acyclic) graphs satisfying a number of constraints [2], as suitable labeled hypergraphs called *jungles* [13], or as sets of recursive equations [1]. None of these (essentially equivalent) representations is able to highlight the algebraic structure of term graphs, and a consequence of this is that the simple correspondence they have with terms is often made more obscure by the need of encoding and decoding functions. In fact, for each of the mentioned representations of term graphs suitable functions have been defined that, for example, extract from a term graph the term it represents (via some "unraveling"), or, viceversa, that from a given term return a term graph representing it (and exhibiting some minimal or maximal sharing).

The goal of this paper (together with its predecessor [8]) is to fill the gap described above, by proposing an original, clean categorical description of term graph rewriting. More precisely, while [8] has shown that term graphs over a signature correspond one-to-one to arrows of a *gs-monoidal* category, building on that result we show here that term graph rewriting sequences (satisfying a mild restriction) are faithfully represented by cells of a gs-monoidal 2-category. Summing up, one could say that "gs-monoidality characterizes term graph rewriting", in the same sense as "cartesianity characterizes term rewriting".

Before explaining concisely the main contribution of the paper, it is worth recalling that the terms over a given signature Σ can be regarded as the arrows of a cartesian category (called the *algebraic theory* of Σ) freely generated (in a suitable way) by Σ (see e.g. [21, 18]). Such a category has (underlined) natural numbers as objects, and its generators are arrows like $g : \underline{n} \to \underline{1}$, where g is an operator of arity n in Σ; in this category arrows from $\underline{n}$ to $\underline{m}$ are in one-to-one correspondence with m-tuples of terms over n variables. Furthermore, a term rewrite rule $R = l \to r$ (where l and r are two terms over Σ) can be represented as a *cell*, i.e., a vertical arrow from the arrow representing l to the arrow representing r. This situation can be denoted by $R : l \Rightarrow r : \underline{n} \rightarrow \underline{1}$, which also states that l and r are arrows from $\underline{n}$ to $\underline{1}$, i.e., that they have at most n variables. Given a term rewriting system, the structure obtained in this way, i.e., the algebraic theory of Σ enriched with one cell for each rule, is called a *computad* [30]; from such a computad, a free construction can generate a (cartesian) 2-category by adding all identity cells, and closing cells with respect to horizontal and vertical composition. The interesting fact is that the resulting 2-category faithfully represents all the possible rewriting sequences of the original system [26, 25, 23, 30]. In fact, horizontal composition of cells generates all the possible instantiations of the rules and, at the same time, places rules in all possible contexts. Vertical composition acts instead as sequential composition.

Cartesian (2-)categories can be defined as *symmetric monoidal (2-)categories* equipped with two *(2-)natural (symmetric monoidal) (2-)transformations* denoted by ∇ (the *duplicator*) and ! (the *discharger*), respectively. The definition of cartesian (2-)categories can be weakened slightly by dropping the requirement of (2-)naturality for ∇ and !: we call the resulting (2-)categories *gs-monoidal*. As anticipated above, the main result of [8] shows that term graphs over a given signature Σ are in one-to-one correspondence with the arrows of the *gs-monoidal theory* of Σ, i.e., a gs-monoidal category freely generated by Σ. In this paper, building on this result, after introducing a definition of term graph rewriting, we show how to generate from a term graph rewriting system a free gs-monoidal 2-category, called its *gs-monoidal 2-theory*. The main result shows that in such a 2-theory there is a cell between two term graphs if and only if they are related by a rewriting sequence where the redexes have to satisfy a mild restriction.

As expected, the categorical framework allows us to relate in a clean way term graph rewriting and term rewriting. In fact, by the above the only difference with terms is that the (2-)naturality of the duplicator and of the discharger does not hold anymore. And this fact has a clear interpretation in terms of "sharing of subterms" and of "garbage collection", in the following sense. Suppose that $g : \underline{n} \rightarrow \underline{1}$ is the arrow corresponding to a n-ary operator $g(x_1, \ldots, x_n)$. Then, using ";" for composition of arrows in diagrammatic order, arrow $a_1 = g \,; \nabla_{\underline{1}} : \underline{n} \rightarrow \underline{2}$ represents, intuitively, a structure having two pointers to $g(x_1, \ldots, x_n)$, while arrow $a_2 = \nabla_{\underline{n}} \,; g \oplus g : \underline{n} \rightarrow \underline{2}$ represents a structure where two copies of g share the same variables (Figure 1 shows the term graphs corresponding to a_1 and a_2, using conventions to be introduced in Section 2; in particular the numbers to the left represent the variables, and those to the right the roots).

Now, regarded as term graphs the two structures must be distinct, because they exhibit a different degree of sharing for some substructure: There are two "pointers" to g in G_1, but only one to each copy of g in G_2. Indeed, arrows a_1 and a_2 are distinct in the free gs-monoidal category concerned. On the contrary, in the free cartesian category, the naturality of ∇ implies that $g \,; \nabla_{\underline{1}} = \nabla_{\underline{n}} \,; g \oplus g$, which means that the two arrows are provably the same. Therefore structures with different degree of sharing are identified, and, by convention, one can take as representative of an equivalence class of such arrows the structure with less sharing, that is, a tuple of trees (in our example, arrow a_1 $(= a_2)$ in the concerned

Fig. 1. Two term graphs that correspond to the same tuple of terms.

algebraic theory represents the tuple of terms $\langle g(x_1, \ldots, x_n), g(x_1, \ldots, x_n)\rangle$.

By similar arguments it can be shown that the categorical counterpart of "implicit garbage collection" is the naturality of the discharger "!". Let us call "garbage" a substructure that is not accessible by any root. Then arrow $g; !_{\underline{1}} : \underline{n} \to \underline{0}$ can be regarded as the structure $g(x_1, \ldots, x_n)$ where g is garbage, because, intuitively, its only root is deleted by the discharger $!_{\underline{1}}$. On the other hand, arrow $!_{\underline{n}} : \underline{n} \to \underline{0}$ represents an empty structure. Now the naturality of ! implies that $g; !_{\underline{1}} = !_{\underline{n}}$, i.e., that any structure with some garbage is equivalent to the same structure with the garbage removed, and we may call this "implicit garbage collection".

The paper is organized as follows. In Section 2 we introduce *ranked term graphs*, i.e., equivalence classes of directed acyclic graphs labeled over a signature Σ, with distinguished lists of *variable* and *root* nodes. Such nodes are used for defining an operation of "composition", which is the counterpart of term substitution, and that provides term graphs with a natural categorical structure. We then introduce our definition of term graph rewriting. Section 3 recalls the basic definitions about *symmetric monoidal 2-categories*, and introduces *gs-monoidal 2-categories*: Symmetric monoidal 2-categories equipped with two transformations, satisfying a few properties. In Section 4 the main result of the paper is presented: Namely, that for each term graph rewriting system, a gs-monoidal 2-category can be freely generated, such that its cells faithfully represent the rewriting sequences of the original system. In Section 5 we discuss the relationship between our notion of term graph rewriting and term rewriting. Finally Section 6 concludes suggesting further investigations, which include a comparative study of the classical definition of term graph rewriting in [2]; the analysis of the properties of the equivalence induced on rewriting sequences by their representation as cells; and lifting the results presented here to *cyclic* term graph rewriting.

2 Term Graphs and Term Graph Rewriting

This section introduces (ranked) term graphs as isomorphism classes of (ranked) directed acyclic graphs, as well as the basic definitions related to Term Graph Rewriting. Our main concern in the next sections will be to stress the underlying categorical structures, hence the following presentation slightly departs from the standard definition However, apart from the restriction to the acyclic case, we can safely see our term graphs as a minor variation of those defined in the seminal paper [2]: A full discussion of their relationship can be found in [8].

Definition 1 (directed acyclic graphs, dags). Let Σ be a signature, i.e., a ranked set of operator symbols, and let *arity* be the function returning the arity of an operator symbol, i.e., $arity(f) = n$ iff $f \in \Sigma_n$. A *labeled graph d (over Σ)* is a triple $d = \langle N, l, s\rangle$, where N is a set of *nodes* (ranged over by $a, b, \ldots$), $l : N \rightharpoonup \Sigma$ is a partial function called the *labeling function*, $s : N \rightharpoonup N^*$ is a partial function called the *successor function*, and such that the following conditions are satisfied:

- $dom(l) = dom(s)$, i.e., labeling and successor functions are defined on the same subset of N; a node $a \in N$ is called *empty* if $a \notin dom(l)$.
- for each node $a \in dom(l)$, $arity(l(a)) = length(s(a))$, i.e., each non-empty node has as many successor nodes as the arity of its label.

If $s(n) = \langle n_1, \ldots, n_k\rangle$, we say that n_i is the *i-th successor* of n and denote it by $s(n)_i$. A labeled graph is *discrete* if all its nodes are empty. A *path* in d is a sequence $\langle a_0, i_0, a_1, \ldots, i_{m-1}, a_m\rangle$, where $m \geq 0$, $a_0, \ldots, a_m \in N$, $i_0, \ldots, i_{m-1} \in \mathbb{N}$ (the natural numbers), and a_k is the i_{k-1}-th successor of a_{k-1} for $k \in \{1, \ldots, m\}$. The *length* of this path is m; if $m = 0$, the path is *empty*. A *cycle* is a path like above where $a_0 = a_m$.

A *directed acyclic graph* or *dag (over Σ)*, is a labeled graph which does not contain any non-empty cycle. If $a \in N$ is a node of a dag $d = \langle N, l, s\rangle$, then by $d|a$ we denote the sub-dag of d rooted at a, defined in the obvious way. For a dag d we shall often denote its components by $N(d)$, l_d and s_d, respectively. Moreover, $N_\emptyset(d)$ and $N_\Sigma(d)$ denote the set of empty and non-empty nodes of d, respectively (thus $N(d) = N_\Sigma(d) \uplus N_\emptyset(d)$, where $\uplus$ denotes disjoint union). □

Definition 2 (dag morphisms, category $\mathbf{Dag}_\Sigma$). Let d and d' be two dags. A *(dag) morphism* $f : d \to d'$ is a function $f : N(d) \to N(d')$ that preserves labeling and successors, i.e., such that for each node $a \in N_\Sigma(d)$, $l_{d'}(f(a)) = l_d(a)$, and $s_{d'}(f(a))_i = f(s_d(a)_i)$ for each $i \in \{1, \ldots, arity(l_d(a))\}$. A dag morphism $f : d \to d'$ is *Σ-injective* if its restriction to $N_\Sigma(d)$ is injective; it is *strongly acyclic* if for all $a_1 \in N_\emptyset(d)$ and $a_2 \in N_\Sigma(d)$ there is no path from $f(a_1)$ to $f(a_2)$ in d'; and it is *plain* if it is both Σ-injective and strongly acyclic. It is easy to verify that Σ-injective morphisms compose, as well as strongly acyclic ones.

Dags over Σ and dag morphisms clearly form a category that will be denoted $\mathbf{Dag}_\Sigma$. □

In the following, for each $n \in \mathbb{N}$ we shall denote by $\underline{n}$ the set $\underline{n} = \{1, \ldots, n\}$ (thus $\underline{0} = \emptyset$).

Definition 3 (ranked dags and term graphs). An *(i, j)-ranked dag* (or also, a *dag of rank (i, j)*) is a triple $g = \langle r, d, v\rangle$, where d is a dag with exactly j empty nodes, $r : \underline{i} \to N(d)$ is a function called the *root mapping*, and $v : \underline{j} \to N_\emptyset(d)$

is a bijection between $\underline{j}$ and the empty nodes of d, called the *variable mapping*. Node $r(k)$ is called the k*-th root* of d, and $v(k)$ is called the k*-th variable* of d, for each admissible k.

Two (i,j)-ranked dags $g = \langle r, d, v \rangle$ and $g' = \langle r', d', v' \rangle$ are *isomorphic* if there exists a *ranked dag isomorphism* $\phi : g \to g'$, i.e., a dag isomorphism $\phi : d \to d'$ such that $\phi \circ r = r'$ and $\phi \circ v = v'$. A (i,j)*-ranked term graph* G (or *with rank* (i,j)) is an isomorphism class of (i,j)-ranked dags. Sometimes we will write G^i_j to recall that G has rank (i,j). □

We introduce now two operations on ranked term graphs. The *composition* of two ranked term graphs is obtained by gluing the variables of the first one with the roots of the second one, and it is defined only if their number is equal. This operation allows us to define a category having ranked term graphs as arrow. Next the *union* of term graphs is introduced: it is always defined, and it is a sort of disjoint union where roots and variables are suitably renumbered. This second operation provides the category of term graphs with a monoidal structure that will be made explicit in the next sections.

Definition 4 (composition of ranked term graphs). Let $G'^j_k = [\langle r', d', v' \rangle]$ and $G^i_j = [\langle r, d, v \rangle]$ be two ranked term graphs. Their *composition* is the ranked term graph $H^i_k = G'^j_k ; G^i_j$ defined as $H^i_k = [\langle in_d \circ r, d'', in_{d'} \circ v' \rangle]$, where $d'', in_d : d \to d''$ and $in_{d'} : d' \to d''$ are obtained as follows. Assuming that $d = \langle N(d), l_d, s_d \rangle$ and $d' = \langle N(d'), l_{d'}, s_{d'} \rangle$, we have $d'' = \langle (N(d) \uplus N(d'))/_{\approx}, l'', s'' \rangle$, where $\approx$ is the least equivalence relation such that $v(i) \approx r'(i)$ for $i \in \underline{j}$, and l'' and s'' are determined by l_d and s_d, respectively, for all $\approx$-equivalence classes containing only nodes of d, and by $l_{d'}$ and $s_{d'}$, respectively, for all other classes. Furthermore, the injections $in_d : d \to d''$ and $in_{d'} : d' \to d''$ map each node to its $\approx$-equivalence class. □

It is easy to check that both morphisms in_d and $in_{d'}$ are plain. It is worth stressing that the composed dag can be characterized elegantly as a pushout, in the sense that $\langle d'', in_d, in_{d'} \rangle$ is a pushout of $\langle v : \underline{j} \to d, r' : \underline{j} \to d' \rangle$ in $\mathbf{Dag}_\Sigma$ (set $\underline{j}$ is regarded as a discrete dag). Thus the well-definedness of composition of term graphs easily follows from the uniqueness of pushouts up to isomorphism.[4]

Definition 5 (the category of term graphs). For a given signature Σ, $\mathbf{TG}_\Sigma$ denotes the category having as objects underlined natural numbers, and as arrows from $\underline{n}$ to $\underline{m}$ all (m,n)-ranked term graphs. Arrow composition is defined as in Definition 4, and the identity on $\underline{n}$ is the term graph $G^{\underline{n}}_{id}$ of rank (n,n) having n nodes, and where the i-th root is also the i-th variable, for all $i \in \underline{n}$. Then it is easy to check that $\mathbf{TG}_\Sigma$ is a well-defined category, because composition is associative, and the identity laws hold. □

[4] Actually, the pushout of two arrows in category $\mathbf{Dag}_\Sigma$ does not always exist: it does exist however in the case we are interested in, since morphism $v : \underline{j} \to d$ is injective and has only empty nodes in the codomain. See [10] for necessary and sufficient conditions for the existence of pushouts in the equivalent category of *jungles*.

Definition 6 (union of ranked term graphs). Let $G^i_j = [\langle r, d, v\rangle]$ and $G'^k_l = [\langle r', d', v'\rangle]$ be two ranked term graphs. Their *union* or *parallel composition* is the term graph of rank $(i+k, j+l)$ $G^i_j \oplus G'^k_l = [\langle r'', d \uplus d', v''\rangle]$, where $r'' : \underline{i+k} \to d \uplus d'$ and $v'' : \underline{j+l} \to d \uplus d'$ are defined as

- $r''(x) = \begin{cases} r(x) & \text{if } x \in \underline{i} \\ r'(x-i) & \text{if } x \in \{i+1, \ldots, i+k\}. \end{cases}$
- $v''(x) = \begin{cases} v(x) & \text{if } x \in \underline{j} \\ v'(x-i) & \text{if } x \in \{j+1, \ldots, j+l\}. \end{cases}$

□

Example 1 (term graphs, composition and union). Four term graphs are shown in Figure 2. Empty nodes are represented by the natural numbers corresponding to their position in the list of variables, and are depicted as a vertical sequence on the left; non-empty nodes are represented by their label, from where the edges pointing to the successors leave; the list of numbers on the right represent pointers to the roots: A dashed arrow from j to a node indicates that it is the j-th root. For example, the first term graph G_1 has rank $(4, 2)$, four nodes (two empty, 1 and 2, and two non-empty, f and g), the successors of g are the variables 2 and 1 (in this order), the successors of f are g and 2, and the four roots are g, f, 2, and f. These graphical conventions make easy the operation of composition, that can be performed by matching the roots of the first graph with the variables of the second one, and then by eliminating them. For example, term graph $G_1 ; G_2$ is the composition of G_1 and of G_2 of rank $(1, 4)$. The last term graph is $G_1 \oplus G_2$, the union of G_1 and G_2, of rank $(5, 6)$. □

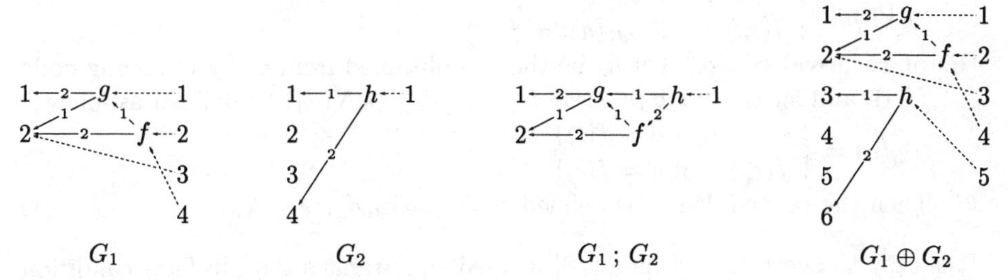

Fig. 2. Two term graphs, their composition and their union.

Definition 7 (rule, redex, term graph rewriting system). A *rule* is a triple $\mathbf{R} = \langle d, r_l, r_r\rangle$, where d is a dag, and r_l, r_r are two distinguished nodes of d, called the *left root* and the *right root*, respectively. Furthermore, we require that (1) node r_l is not empty, (2) all nodes of d are reachable from r_l or r_r, (3) all empty nodes in d are reachable from r_l, and (4) there is not path from r_r to r_l. A *redex* in a dag d_0 is a pair $\Delta = \langle \mathbf{R}, f\rangle$ where $\mathbf{R}$ is a rule as above, and $f : d|r_l \to d_0$ is a dag morphism from $d|r_l$, the sub-dag of d rooted at r_l, to d_0. Node $f(r_l)$ of d_0 is called the *root* of the redex Δ. Redex $\langle \mathbf{R}, f\rangle$ is *plain* if so is f.

A *(term graph) rewriting system* is a pair $\mathcal{R} = \langle \{\mathbf{R}_i\}_{i \in I}, \Sigma \rangle$ such that for all $i \in I$, $\mathbf{R}_i$ is a rule over Σ. □

Note that condition (4) above is not required for rules in [2]. We don't allow for a non-empty path from r_r to r_l to ensure that the result of an application of the rule is an acyclic dag. And we don't allow for an empty path ($r_l = r_r$) even if we could, because such rules would result in trivial identities, making nevertheless some statements more awkward in the rest of the paper.

Definition 8 (ranked dag rewriting). Let $g_0 = \langle r_0, d_0, v_0 \rangle$ be an (i,j)-ranked dag, and let $\Delta = \langle \langle d, r_l, r_r \rangle, f : d|r_l \to d_0 \rangle$ be a redex in it. We say that g_0 *rewrites to* g_3 *via* Δ, denoted as $g_0 \to_\Delta g_3$, if g_3 is the (i,j)-ranked dag obtained through the following procedure:

[**Build phase**] Define d_1 as the dag having as nodes the disjoint union of $N(d_0)$ and $N(d|r_r) - N(d|r_l)$. Functions l_{d_1} and s_{d_1} are defined as

– $l_{d_1}(a) = \begin{cases} l_{d_0}(a) & \text{if } a \in N(d_0) \\ l_d(a) & \text{if } a \in N(d|r_r) - N(d|r_l). \end{cases}$

– $s_{d_1}(a)_i = \begin{cases} s_{d_0}(a)_i & \text{if } a \in N(d_0) \\ s_d(a)_i & \text{if } a, s_d(a)_i \in N(d|r_r) - N(d|r_l) \\ f(s_d(a)_i) & \text{if } a \in N(d|r_r) - N(d|r_l), s_d(a)_i \in N(d|r_l). \end{cases}$

Let $\hat{f} : d \to d_1$ be the obvious extension of morphism $f : d|r_l \to d_0$ such that $\hat{f}(a) = a$ for each $a \in N(d|r_r) - N(d|r_l)$.

[**Redirection phase**] Define d_2 as the dag obtained by replacing all references to the root of the redex $f(r_l)$ by references to the image of the right root r_r. More precisely, $d_2 = \langle N(d_1), l_{d_1}, s_{d_2} \rangle$, where

$s_{d_2}(a)_i = \begin{cases} s_{d_1}(a)_i & \text{if } s_{d_1}(a)_i \neq f(r_l) \\ \hat{f}(r_r) & \text{if } s_{d_1}(a)_i = f(r_l). \end{cases}$

[**Root removal phase**] Let d_3 be the dag obtained from d_2 by removing node $f(r_l)$, and let the *track function* $tr_\Delta : N(d_0) \to N(d_3)$ be defined as follows:

$tr_\Delta(a) = \begin{cases} a & \text{if } a \neq f(r_l) \\ \hat{f}(r_r) & \text{if } a = f(r_l). \end{cases}$

Then the ranked dag g_3 is defined as $\langle tr_\Delta \circ r_0, d_3, tr_\Delta \circ v_0 \rangle$. □

It is easy to check that g_3 is a well-defined (i,j)-ranked dag. In fact, condition (4) of Definition 7 ensures that d_2 is acyclic and that its node $f(r_l)$ is not the successor of any node. Moreover, $tr_\Delta \circ v_0$ is a bijection between the set of empty nodes of d_3 and $\underline{j}$ by condition (3).

The definition of dag rewriting is easily extended to term graphs.

Definition 9 (term graph rewriting). Let G be a term graph, and let $\Delta = \langle \mathbf{R}, f \rangle$ be a redex in one of its elements, say $g = \langle r, d, v \rangle$. Then for each $h \in G$ and ranked dag isomorphism $\phi : g \to h$, Δ induces a redex in h, obtained by composing f with ϕ. In this situation we say that Δ is a redex in G. Furthermore, if $g \to_\Delta g'$ and $h \to_{\langle \mathbf{R}, \phi \circ f \rangle} h'$, then it is easy to check that g' and h' are isomorphic. Thus we safely extend rewriting to term graphs, writing $G \to_\Delta G'$, where $G' = [g']$. A *term graph rewriting sequence* $G_0 \to^*_{\mathcal{R}} G_n$ over a rewriting system $\mathcal{R}$ is a sequence of $n \geq 0$ rewriting steps $G_0 \to_{\Delta_1} G_1 \ldots G_{n-1} \to_{\Delta_n} G_n$, where at each step a rule of $\mathcal{R}$ is used. A rewriting sequence is *plain* if so are all redexes in it. □

3 GS-Monoidal 2-Categories

In this section we briefly present some notions about 2-categories needed in the rest of the paper. We first recall the basic definitions: For an introduction, we refer the reader to the classical work [16]. Then we introduce the original concept of *gs-monoidal 2-categories*, which will be fundamental for our characterization of term graph rewriting.

Roughly, a 2-category simply is a category $\mathbf{C}$ such that, given any two objects a, b, the hom-set $\mathbf{C}[a, b]$, i.e., the class of arrows from a to b, is a category: moreover, these *hom-categories* satisfy suitable composition properties. An arrow in $\mathbf{C}[a, b]$, a *cell*, is denoted as $\alpha : f \Rightarrow g : a \to b$ (where $f, g : a \to b$); or graphically, as

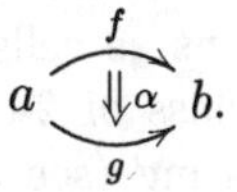

$$a \underset{g}{\overset{f}{\Downarrow\alpha}} b.$$

Definition 10 (2-Categories). Let $\mathbf{C}$ be a category such that each hom $\mathbf{C}[a, b]$ is a category. Moreover, let us assume that for each triple a, b, c of objects there is a (horizontal) composition function $- * -$ such that, given $\alpha : f \Rightarrow h : a \to b$ and $\beta : g \Rightarrow i : b \to c$, then $\gamma = \alpha * \beta : f; g \to h; i \in \mathbf{C}[a, c]$. Graphically,

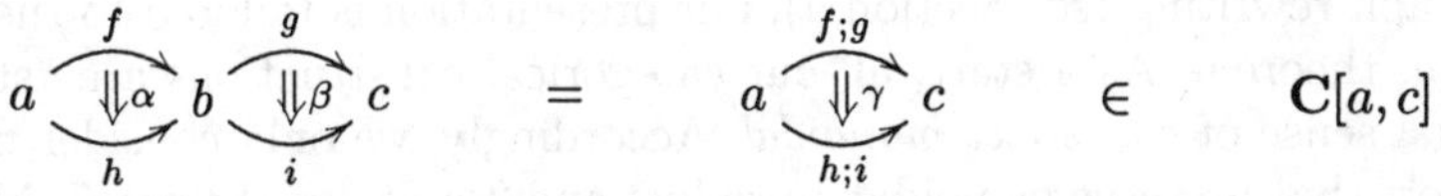

$$a \underset{h}{\overset{f}{\Downarrow\alpha}} b \underset{i}{\overset{g}{\Downarrow\beta}} c \quad = \quad a \underset{h;i}{\overset{f;g}{\Downarrow\gamma}} c \quad \in \quad \mathbf{C}[a, c]$$

where $-;-$ denotes composition inside $\mathbf{C}$. Let us consider the following cells:

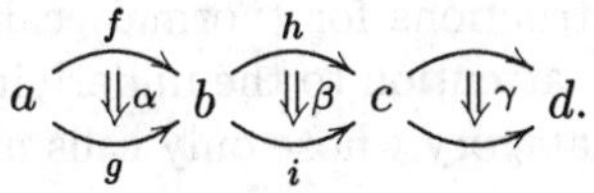

$$a \underset{g}{\overset{f}{\Downarrow\alpha}} b \underset{i}{\overset{h}{\Downarrow\beta}} c \; \Downarrow\gamma \; d.$$

A *2-category* $\underline{\mathbf{C}}$ is a category $\mathbf{C}$ (called the *underlying* category) with the structure above defined, such that the composition functions satisfy the equations:

(1) $id_a * \alpha = \alpha = \alpha * id_b$; (2) $(\alpha * \beta) * \gamma = \alpha * (\beta * \gamma)$;

(3) $f * h = f; h$ (4) $(\alpha * h) \cdot (g * \beta) = \alpha * \beta = (f * \beta) \cdot (\alpha * i)$.

where $- \cdot -$ denotes the (vertical) composition insides hom-categories.[5] □

[5] For sake of readability, since no confusion can arise, we often denote the cell corresponding to the identity of an arrow by the arrow itself. For example, axiom (1) reads as $id_{id_a} * \alpha = \alpha = \alpha * id_{id_b}$; axiom (3) reads as $id_f * id_h = id_{f;h}$; and so on.

Axioms (1) and (2) assure that there exists also an underlying horizontal category, with the same objects as **C** and cells as arrows, and (3)-(4) express the functoriality of each composition in the other. In fact, 2-categories could also be described as suitable *internal categories* in **Cat** (see [3]), that is, categories such that both its classes of arrows and of objects are categories: The remaining definitions are actually biased towards this "internal" approach.

Definition 11 (2-Functors and (2-Natural) 2-Transformations). Let $\underline{\mathbf{C}}$ and $\underline{\mathbf{D}}$ be two 2-categories. A *2-functor* $F : \underline{\mathbf{C}} \to \underline{\mathbf{D}}$ is a triple $\langle F_O, F_A, F_C \rangle$ of functions, mapping objects to objects, arrows to arrows and cells to cells, respectively, preserving identities and compositions of all kinds.

Let $F, G : \underline{\mathbf{C}} \to \underline{\mathbf{D}}$ be two parallel 2-functors: a *2-transformation* $\eta : F \Rightarrow G$ is a family of arrows of **D** indexed by objects of **C**, $\eta = \{\eta_a : F_O(a) \to G_O(a) \mid a \in |\mathbf{C}|\}$; it is *2-natural* if moreover, for every cell $\alpha : f \Rightarrow g : a \to b \in \underline{\mathbf{C}}$, it verifies $F_C(\alpha) * \eta_b = \eta_a * G_C(\alpha)$. □

The paradigmatic example of 2-category—and one whose use is widespread in Computer Science—is $\underline{\mathbf{Cat}}$, having small categories as objects, functors as arrows, and natural transformations as cells. As far as rewriting is concerned, it is well-known that a suitable class of 2-categories, *algebraic 2-theories*, can be used for describing term rewriting (see e.g. [26, 25, 23, 30], and the more recent [29, 9]). As anticipated in the Introduction, we want to provide a similar categorical model for term graph rewriting: To this aim we introduce now *gs-monoidal 2-categories*—roughly, symmetric monoidal 2-categories equipped with two 2-transformations.

In particular, since the focus of the paper is the correspondence result for term graph rewriting (see Section 4), our presentation is tailored to the need of our main theorem. As a start, all our categorical constructions are "strict", in the usual sense of e.g. *strict monoidal*. Accordingly, we only consider monoidal 2-functors that preserve monoidal 2-product and 2-unit "on the nose". Moreover, for the sake of brevity, we present some definitions directly at the 2-categorical level: The analogous constructions for (normal) categories can be obtained in principle by restricting the attention to the underlying category (since each category can be seen as a 2-category whose only cells are identities).

Definition 12 (symmetric monoidal 2-categories). A *monoidal 2-category* $\underline{\mathbf{C}}$ is a triple $\langle \underline{\mathbf{C}}_0, \otimes, e \rangle$, where $\underline{\mathbf{C}}_0$ is a 2-category, $e \in |\mathbf{C}_0|$ is a distinguished object and $\otimes : \underline{\mathbf{C}}_0 \times \underline{\mathbf{C}}_0 \to \underline{\mathbf{C}}_0$ is a 2-functor, satisfying the axioms: $(\alpha \otimes \beta) \otimes \gamma = \alpha \otimes (\beta \otimes \gamma)$ and $\alpha \otimes id_e = id_e \otimes \alpha = \alpha$ for all cells $\alpha, \beta, \gamma \in \underline{\mathbf{C}}_0$.

A *symmetric monoidal 2-category* $\underline{\mathbf{C}}$ is a four-tuple $\langle \underline{\mathbf{C}}_0, \otimes, e, \rho \rangle$ where $\langle \underline{\mathbf{C}}_0, \otimes, e \rangle$ is a monoidal 2-category, and $\rho : \otimes \Rightarrow \otimes \circ \mathrm{X} : \underline{\mathbf{C}}_0 \times \underline{\mathbf{C}}_0 \to \underline{\mathbf{C}}_0$ is a 2-natural 2-transformation (where X is the 2-functor that swaps its two arguments) such that $\rho_{e,a} = id_a$ for all $a \in |\mathbf{C}_0|$, and satisfying:

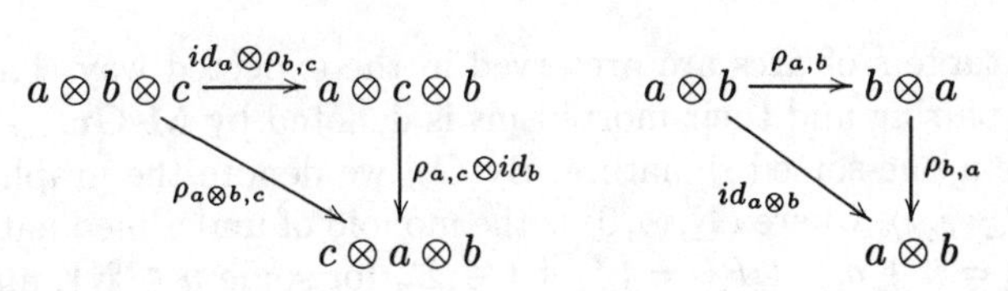

A *monoidal 2-functor* $F : \underline{\mathbf{C}} \to \underline{\mathbf{C}}'$ is a 2-functor $F : \underline{\mathbf{C}}_0 \to \underline{\mathbf{C}}'_0$ such that $F(e) = e'$ and $F(\alpha \otimes \beta) = F(\alpha) \otimes' F(\beta)$ for all $\alpha, \beta \in \underline{\mathbf{C}}_0$. It is *symmetric* if moreover $F(\rho_{a,b}) = \rho'_{F(a),F(b)}$. □

If we restrict our attention from the symmetric monoidal 2-category $\underline{\mathbf{C}} = \langle \underline{\mathbf{C}}_0, \otimes, e, \rho \rangle$ to the underlying category $\mathbf{C}_0$ and the derived operators $\otimes_r, \rho_r$ (defined in the obvious way), it turns out that the category $\mathbf{C} = \langle \mathbf{C}_0, \otimes_r, e, \rho_r \rangle$ is symmetric (strict) monoidal, according to the classical definition (see e.g. [22]).

Definition 13 (gs-monoidal 2-categories). A *gs-monoidal 2-category* $\underline{\mathbf{C}}$ is a six-tuple $\langle \underline{\mathbf{C}}_0, \otimes, e, \rho, \nabla, ! \rangle$, where $\langle \underline{\mathbf{C}}_0, \otimes, e, \rho \rangle$ is a symmetric monoidal 2-category and $! : Id_{\underline{\mathbf{C}}_0} \Rightarrow e : \underline{\mathbf{C}}_0 \to \underline{\mathbf{C}}_0$, $\nabla : Id_{\underline{\mathbf{C}}_0} \Rightarrow \otimes \circ \Delta : \underline{\mathbf{C}}_0 \to \underline{\mathbf{C}}_0$ are two 2-transformations (Δ is the diagonal 2-functor), such that $!_e = \nabla_e = id_e$ and satisfying:

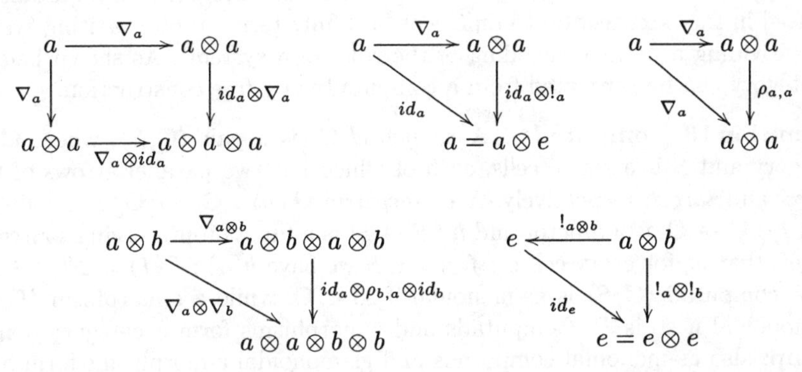

A *gs-monoidal 2-functor* $F : \underline{\mathbf{C}} \to \underline{\mathbf{C}}'$ is a symmetric monoidal 2-functor such that $F(!_a) = !'_{F(a)}$ and $F(\nabla_a) = \nabla'_{F(a)}$. The category of small gs-monoidal 2-categories and gs-monoidal 2-functors is denoted by **GSM-2Cat**. □

Again, if a 2-category $\underline{\mathbf{C}}$ is gs-monoidal, also the underlying category $\mathbf{C}$ has such a structure: We denote by **GSM-Cat** the category of small *gs-monoidal categories* and *gs-monoidal functors*. Next we recall from [8] how it is possible to generate from a signature a free gs-monoidal category, called its *gs-monoidal theory*.

Definition 14 (graphs with pairing and signature graphs). A *graph with pairing* G is a six-tuple $\langle N, \times, e, A, s, t \rangle$, where $\langle N, \times, e \rangle$ is a monoid of nodes, A a set of arcs, and $s, t : A \to N$ are (the source and target) functions. A morphism of graphs with pairing $\langle f_N, f_A \rangle : G \to G'$ is a pair where $f_N : \langle N, \times, e \rangle \to \langle N', \times', e' \rangle$ is a monoid homomorphism and $f_A : A \to A'$ is a function such that sources and targets of arcs are preserved in the expected way. The category of graphs with pairing and their morphisms is denoted by **M-Gr**.

Let Σ be a one-sorted signature. By G_Σ we denote the graph with pairing $\langle \underline{\mathbb{N}}, \otimes, \underline{0}, OP_\Sigma, s, t\rangle$, where $\langle \underline{\mathbb{N}}, \otimes, \underline{0}\rangle$ is the monoid of underlined natural numbers (with $\underline{n} \otimes \underline{m} = \underline{n+m}$), $OP_\Sigma = \{f_\Sigma \mid f \in \Sigma_n \text{ for some } n \in \mathbb{N}\}$, and source and target are determined as $f_\Sigma : \underline{n} \to \underline{1}$ iff $f \in \Sigma_n$, for all $f_\Sigma \in OP_\Sigma$ (without loss of generality, we assume that the sets Σ_n are disjoint). □

Proposition 15 (gs-monoidal theories). *Let* V : **GSM-Cat** $\to$ **M-Gr** *be the forgetful functor mapping a gs-monoidal category to the underlying graph with pairing: It admits a left (free) adjoint* GSM : **M-Gr** $\to$ **GSM-Cat**.

The gs-monoidal theory **GS-Th**(Σ) *of a given (one-sorted) signature* Σ*, is defined as* $GSM(G_\Sigma)$*, i.e., the free gs-monoidal category generated by the graph with pairing associated with* Σ*.* □

The main result of [8], which we recall in Theorem 18, states the existence of a one-to-one correspondence between arrows of the hom-set **GS-Th**$(\Sigma)[n, m]$ and the family of term graphs with n variables and m roots. We introduce now a structure, called *computad* [30], that plays for 2-categories the same role that a graph with pairing plays for a gs-monoidal categories. Computads will be used in the next section to represent faithfully term graph rewriting systems (by providing a finitary encoding of the rules of a system). As shown below, a 2-category can be generated from a computad via a free construction.

Definition 16 (computads). A *computad* $\mathbf{C}_c$ is a pair $\langle \mathbf{C}, S\rangle$, where $\mathbf{C}$ is a category and S is a set of cells, each of which has two parallel arrows of $\mathbf{C}$ as *source* and *target*, respectively. A *c-morphism* $\langle F, h\rangle : \mathbf{C}_c \to \mathbf{C}'_c$ is a pair such that $F : \mathbf{C} \to \mathbf{C}'$ is a functor and $h : S \to S'$ is a function preserving source and target, that is, for every cell $\alpha : f \Rightarrow g \in S$ we have $h(\alpha) : F(f) \Rightarrow F(g) \in S'$.

A computad $\langle \mathbf{C}, S\rangle$ is gs-monoidal if so is $\mathbf{C}$, while a c-morphism $\langle F, h\rangle$ is gs-monoidal if so is F. Computads and c-morphisms form a category denoted **Comp**; also gs-monoidal computads and gs-monoidal c-morphisms form a category, denoted **GSM-Comp**. □

Proposition 17 (free gs-monoidal 2-categories). *Let* V_2 : **GSM-2Cat** $\to$ **GSM-Comp** *be the forgetful functor mapping a gs-monoidal 2-category to the underlying gs-monoidal computad (simply forgetting cell composition): It admits a left (free) adjoint* GSM_2 : **GSM-Comp** $\to$ **GSM-2Cat**. □

Intuitively, the free adjoint GSM_2 composes the cells of a computad in all the possible ways, both horizontally and vertically, imposing further equalities in order to satisfy the axioms of a 2-category and preserving the gs-monoidal structure on the underlying category.

4 Rewriting Sequences as Cells of a 2-Theory

This section presents the main result of the paper. We first show how to generate in a free way a gs-monoidal 2-category from a given term graph rewriting system

$\mathcal{R}$; next we prove that cells of this category faithfully correspond to plain rewriting sequences using the rules of $\mathcal{R}$. Part of the construction of the mentioned gs-monoidal 2-category is based on the main result presented in [8], namely that there is an isomorphism between the category of term graphs over Σ (Definition 5) and the gs-monoidal theory of Σ (Proposition 15).

Theorem 18 (the category of term graphs as gs-monoidal theory). *Let* $\mathbf{TG}_\Sigma$ *be the category introduced in Definition 5. Then the structure* $\langle \mathbf{TG}_\Sigma, \oplus, \underline{0}, G_{\Pi(\_,\_)}, G_{\nabla(\_)}, G_{!}\rangle$ *is a gs-monoidal category isomorphic to* $\mathbf{GS\text{-}Th}(\Sigma)$*, where:*

- *Functor* $\oplus : \mathbf{TG}_\Sigma \times \mathbf{TG}_\Sigma \to \mathbf{TG}_\Sigma$ *is defined on arrows as in Definition 6, and on objects as* $\underline{n} \oplus \underline{m} = \underline{n+m}$*;*
- *For each* $n, m \in \mathbb{N}$*,* $G_{\Pi(\underline{n},\underline{m})}$ *is the discrete term graph of rank* $(n+m, n+m)$ *such that (denoting by* r *and* v *the root and variable functions of one of its representatives)* $r(x) = v(n+x)$ *if* $x \leq m$*, and* $r(x) = v(x-m)$ *if* $m < x \leq n+m$*.*
- *For each* $n \in \mathbb{N}$*,* $G_{\nabla(\underline{n})}$ *is the discrete term graph of rank* $(2*n, n)$ *such that* $r(x) = x$ *if* $x \leq n$*, and* $r(x) = x - n$ *if* $n < x \leq 2*n$*.*
- $G_{!}^{\underline{n}}$ *is the discrete term graph with* n *variables and with no roots.* □

The gs-monoidal 2-category generated by a term graph rewriting system over a signature Σ will have $\mathbf{GS\text{-}Th}(\Sigma)$ as underlying category. The result just presented allows us to use term graphs over Σ to denote the arrows of that category, and to make free use of the equations of gs-monoidal categories when reasoning about them. We first show how to associate (in a quite straightforward way) with each rewriting system $\mathcal{R}$ its *gs-monoidal 2-theory* $\mathbf{GS\text{-}2Th}(\mathcal{R})$, and then relate cells in this 2-category and rewriting sequences over $\mathcal{R}$.

Definition 19 (rule as cell). Let $\mathbf{R} = \langle d, r_l, r_r\rangle$ be a rule, as in Definition 7. Let n be the number of empty nodes of d, and let m be the number of nodes of $d|r_l$. Then we define the term graphs $L_\mathbf{R}$ and $R_\mathbf{R}$ of rank (m, n) as follows:

- $L_\mathbf{R} = [\langle r, d|r_l, v\rangle]$, where $r : \underline{m} \to N(d|r_l)$ is a bijection, and $v : \underline{n} \to N_\emptyset(d|r_l)$ is a bijection between $\underline{n}$ and the set of empty nodes of $d|r_l$. Note that r and v are well-defined by the definition of n and m; moreover they could be fixed in a canonical way, but we don't need this.
- $R_\mathbf{R} = [\langle r', d - \{r_l\}, v\rangle]$, where $d - \{r_l\}$ is the dag obtained by removing from d the node r_l, v is the same as in the previous point, and for all $x \in \underline{m}$,

$$r'(x) = \begin{cases} r(x) & \text{if } r(x) \neq r_l \\ r_r & \text{if } r(x) = r_l \end{cases}$$

Again it is easy to check that $R_\mathbf{R}$ is a well-defined term graph of rank (m, n).

The cell representing rule $\mathbf{R}$ is the cell $\alpha(\mathbf{R}) : L_\mathbf{R} \Rightarrow R_\mathbf{R} : \underline{m} \to \underline{n}$. □

Definition 20 (the gs-monoidal 2-theory of a rewriting system). Let $\mathcal{R} = \langle \{\mathbf{R}_i\}_{i \in I}, \Sigma \rangle$ be a rewriting system, and let $\mathbf{GS\text{-}Th}(\Sigma)$ be the gs-monoidal theory of Σ. The *gs-monoidal computad representing* $\mathcal{R}$ is defined as the computad $\mathbf{C}_{\mathcal{R}} = \langle \mathbf{GS\text{-}Th}(\Sigma), \{\alpha(\mathbf{R}_i)\}_{i \in I} \rangle$, where for each rule $\mathbf{R}_i$ in $\mathcal{R}$ the cell $\alpha(\mathbf{R}_i) : L_{\mathbf{R}_i} \Rightarrow R_{\mathbf{R}_i}$ is as in Definition 19. The *gs-monoidal 2-theory of* $\mathcal{R}$, denoted by $\mathbf{GS\text{-}2Th}(\mathcal{R})$ is the free gs-monoidal 2-category generated by $\mathbf{C}_{\mathcal{R}}$, as described in Proposition 17. □

We need now a few technical definitions and lemmas before presenting the main result relating cells of $\mathbf{GS\text{-}2Th}(\mathcal{R})$ and rewriting sequences over $\mathcal{R}$.

Definition 21 (term graph contexts). A *(term graph) context* $C[i,j]$ is a well-formed, ranked term graph expression containing exactly one (ranked) placeholder $[i,j]$ (for some $i,j \in \mathbb{N}$). If $C[i,j]$ is a context and G is a term graph of rank (i,j), then by $C[G]$ we denote the term graph obtained by evaluating the expression C after replacing $[i,j]$ by G. □

Next lemma provides a "syntactical" characterization of the existence of a plain morphism between two term graphs by using contexts. Due to space limitations, the (quite long) proof will only appear in the full version of the paper.

Lemma 22 (plain morphisms and contexts). *Let G and H be two term graphs such that all nodes of G are roots. Then there is a context $C[m,n]$ (where (m,n) is the rank of G) such that $H = C[G]$ if and only if for every dag $d_G \in G$ and $d_H \in H$ there is a plain dag morphism $f : d_G \to d_H$ (see Definition 2).* □

The last lemma allows us to reformulate the definition of plain term graph rewriting in a more "algebraic" way.

Lemma 23 (term graph rewriting as contextualization). *Let $\mathbf{R} = \langle d, r_l, r_r \rangle$ be a rule and $L_{\mathbf{R}}, R_{\mathbf{R}}$ be the (m,n)-ranked term graphs of Definition 19. Then for every context $C[m,n]$ and for every dag d_0 in $C[L_{\mathbf{R}}]$ there is a plain redex $\Delta = \langle \mathbf{R}, f : d|r_l \to d_0 \rangle$ such that $C[L_{\mathbf{R}}] \to_{\Delta} C[R_{\mathbf{R}}]$.*

Proof. Let $f : d|r_l \to d_0$ be the plain dag morphism whose existence is ensured by Lemma 22. The fact that $C[L_{\mathbf{R}}] \to_{\langle \mathbf{R}, f \rangle} C[R_{\mathbf{R}}]$ can be shown by comparing the construction in Definition 8 with the definition of $L_{\mathbf{R}}$ and $R_{\mathbf{R}}$. In fact the application of rule $\mathbf{R}$ to d_0 has the effect of adding to d_0 a copy of $d|r_r - d|r_l$ and of removing node $f(r_l)$, which is exactly the difference between $R_{\mathbf{R}}$ and $L_{\mathbf{R}}$. Furthermore, all references to $f(r_l)$ are redirected to $\hat{f}(r_r)$. The same happens by replacing $L_{\mathbf{R}}$ by $R_{\mathbf{R}}$ in an arbitrary context, because (by the conditions on rules in Definition 7) the only reference to r_l in $L_{\mathbf{R}}$ is a root, which is "redirected" to r_r in $R_{\mathbf{R}}$, by Definition 19. □

We are now ready to present our main result.

Theorem 24 (rewriting sequences as cells of the gs-monoidal 2-theory). *Let G and H be two term graphs having the same rank (m,n). Then there is a cell $\alpha : G \Rightarrow H : \underline{n} \to \underline{m}$ in $\mathbf{GS\text{-}2Th}(\mathcal{R})$ if and only if there is a plain rewriting sequence $G \to^*_{\mathcal{R}} H$.*

Proof. Only if part. Assume that there is a cell $\alpha : G \Rightarrow H : \underline{n} \rightarrow \underline{m}$ in $\mathbf{GS\text{-}2Th}(\mathcal{R})$. We prove the existence of a plain rewriting sequence from G to H by induction on the number of *2-generators* in α (i.e., the components of α of the form $\alpha(\mathbf{R})$, for $\mathbf{R}$ a rule in $\mathcal{R}$). Note that this is correct, because the axioms for gs-monoidal 2-categories do not identify cells containing a different number of 2-generators. If α does not contain any 2-generator, then it is an identity cell, and the statement trivially holds true. Suppose now that α contains $k+1$ 2-generators. Using essentially the *interchange law*, it is possible to decompose it into an equivalent cell $\alpha_1 \cdot \alpha_2 \cdot \ldots \cdot \alpha_{k+1}$ such that each α_i contains exactly one 2-generator. Let K be the target graph of α_1, i.e., $\alpha_1 : G \Rightarrow K$. Since α_1 contains one 2-generator, it must be of the form $C[\alpha(\mathbf{R})]$ for a suitable context C and a rule $\mathbf{R} \in \mathcal{R}$. By the definition of $\alpha(\mathbf{R})$, we have that $G = C[L_{\mathbf{R}}]$ and $K = C[R_{\mathbf{R}}]$, and by Lemma 23 there is a plain redex Δ such that $G \rightarrow_{\Delta} K$. The statement follows because by induction hypothesis over cell $\alpha_2 \cdot \ldots \cdot \alpha_{k+1} : K \Rightarrow H$ (which contains k 2-generators), there is a plain rewriting sequence from K to H.
If Part. If $G \rightarrow_{\Delta} H$ using a plain redex $\langle \mathbf{R}, f \rangle$ with $\mathbf{R} \in \mathcal{R}$, by Lemma 22 there is a context C_f such that $G = C_f[L_{\mathbf{R}}]$. Moreover, we have that $H = C_f[R_{\mathbf{R}}]$ by Lemma 23. By Definitions 19 and 20 the cell $\alpha(\mathbf{R}) : L_{\mathbf{R}} \Rightarrow R_{\mathbf{R}}$ is a 2-generator of $\mathbf{GS\text{-}2Th}(\mathcal{R})$, and thus (by the generation rules for cells) $C_f[\alpha(\mathbf{R})] : C_f[L_{\mathbf{R}}] \Rightarrow C_f[R_{\mathbf{R}}]$ is a cell. Then the statement follows because concatenation of rewriting steps corresponds faithfully to vertical composition of cells. □

5 Relationship with Term Rewriting

Our 2-categorical presentation of term graph rewriting is directly inspired by that of term rewriting introduced for example in [26, 25, 23]: the goal of this section is to relate the two theories by exploiting the categorical framework.

Firstly, let us briefly recall the relationship between the gs-monoidal theory of a signature Σ, $\mathbf{GS\text{-}Th}(\Sigma)$, and its algebraic theory, $\mathbf{Th}(\Sigma)$, whose definition dates back to Lawvere's thesis [21] (for a more detailed discussion about the various definitions of algebraic theories and the relationship with gs-monoidal theories we refer to [8]).

Definition 25 (algebraic theory over a signature). An *algebraic theory* $\mathbf{C}$ is a category whose objects are underlined natural numbers, and which for each $\underline{n}$ is equipped with an n-tuple of maps $\langle \pi_i^n : \underline{n} \rightarrow \underline{1} \mid i = 1 \ldots n \rangle$, making $\underline{n}$ the n-fold cartesian product of $\underline{1}$, that is, $\underline{n} = \underline{1}^n$.

Given a signature Σ, the *algebraic theory of* Σ is the algebraic theory $\mathbf{Th}(\Sigma)$ such that the hom-set $\mathbf{Th}(\Sigma)[\underline{n}, \underline{m}]$ is the set of m-tuples of terms over variables $\{x_1, \ldots, x_n\}$; for each $\underline{n} \in \mathbb{N}$, the required n-tuple of maps is given by $\langle x_1, \ldots, x_n \rangle$; and arrow composition is defined as term substitution. □

It can be shown that $\mathbf{Th}(\Sigma)$ is indeed an algebraic theory. An equivalent way of introducing the algebraic theory of Σ is via a free construction, as for gs-monoidal theories in Proposition 15. In fact, there is an adjunction between the category $\mathbf{M\text{-}Gr}$ and $\mathbf{C\text{-}Cat}$, a suitable category of cartesian categories, which factorizes through the adjunction mentioned in Proposition 15. This fact allows us to relate the gs-monoidal and the algebraic theories of a signature.

Proposition 26 (relating algebraic and gs-monoidal theories). *Let* **C-Cat** *be the category of small cartesian categories with strictly associative products as objects, and functors preserving products "on the nose" as arrows, and let* **GSM-Cat** *and* **M-Gr** *be the categories of Definition 13 and 14, respectively.*

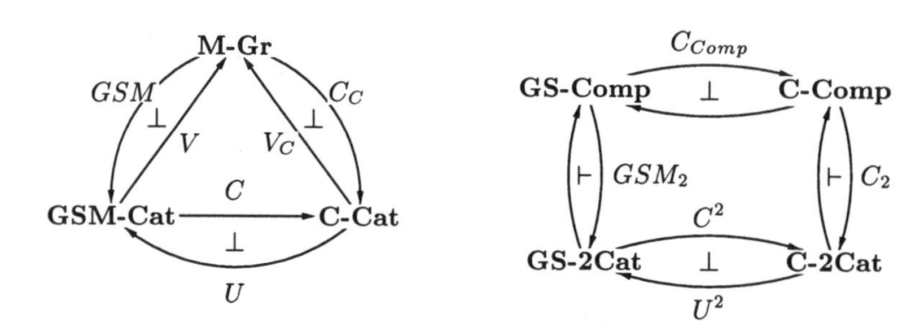

Fig. 3. Two diagrams of adjunctions.

Then the left diagram of Figure 3 commutes, where V and V_C are the obvious forgetful functors that map a gs-monoidal or cartesian[6] *category to the underlying graph with pairing, C_C generates from a graph with pairing a cartesian category, by adding all the missing structure, U is an inclusion (because every cartesian category is gs-monoidal), GSM is as in Proposition 15, and C maps a gs-monoidal category M to the quotient category obtained by imposing the following naturality axioms for all objects a, b and arrows $f : a \to b \in M$:*

$$f; \nabla_b = \nabla_a; (f \otimes f) \qquad f; !_b = !_a$$

Furthermore, we have $\mathbf{Th}(\Sigma) = C_C(G_\Sigma)$, where G_Σ is as in Definition 14, and therefore $\mathbf{Th}(\Sigma) = C(\mathbf{GS\text{-}Th}(\Sigma))$.

Proof outline. To show that a cartesian category is gs-monoidal, define $\nabla_a = \langle id_a, id_a \rangle$ (using pairing), and $!_a$ as the arrow to the terminal object; then it is easy to see that all axioms are satisfied. Concerning functor C, the fact that $C(M)$ is cartesian is based on a result which can be traced back to the early Seventies [6], rediscovered and become folklore in recent years [15, 19]. The other functors in the diagram can be defined by a tedious but straightforward inspection of the object structure of the concerned categories. For a characterization of the algebraic theory as a free construction see for example [12]. Finally, the relationship between algebraic and gs-monoidal theories follows by uniqueness of left adjoints. □

[6] By cartesian we always mean "cartesian with strictly associative products".

The adjunction $C \dashv U$: **GSM-Cat** $\to$ **C-Cat** allows us to relate terms and term graphs over the same signature in a clean way. In fact, the component of the unit of the above adjunction on a gs-monoidal theory, i.e, $\eta_{\text{GS-Th}(\Sigma)}$: **GS-Th**$(\Sigma) \to U(C(\textbf{GS-Th}(\Sigma))) = \textbf{Th}(\Sigma)$, maps each term graph of rank (m, n) to the m-tuple of terms obtained by *unraveling* the term graph from the m roots. Such a mapping can be regarded as a clean categorical version of the unraveling function defined in many papers on term graph rewriting to extract a term from a term graph. Also, it provides a clear interpretation of the relationship between term and term graphs: *a term is an equivalence class of term graphs, modulo the naturality of* ∇ *and* !.

Let us consider now the relationship between the corresponding 2-theories. Let **C-Comp** be the category of computads having objects of **C-Cat** as underlying categories, and **C-2Cat** be the category of small *cartesian 2-categories.*[7] Then the right diagram of Figure 3 commutes, where the vertical adjunctions are as in Proposition 17 and the horizontal ones are based on the adjunction between **GSM-Cat** and **C-Cat** presented above.

These adjunctions allow us to relate term graph rewriting with term rewriting. Given a term graph rewriting system $\mathcal{R}$ over Σ and its computad $\mathbf{C}_{\mathcal{R}}$ (as in Definition 20), the cartesian computad $C_{comp}(\mathbf{C}_{\mathcal{R}})$ is a faithful representation of the term rewriting system $\mathcal{T}_{\mathcal{R}}$ obtained by transforming each rule $\langle d, r_l, r_r\rangle$ of $\mathcal{R}$ into the *term* rule $\eta(d|r_l) \to \eta(d|r_r)$, where $\eta(d')$ is, informally, the term obtained by unraveling dag d' from its root.

By uniqueness of left adjoints, we have $C^2(\textbf{GS-2Th}(\mathcal{R})) = C^2(GSM_2(\mathbf{C}_{\mathcal{R}})) = C_2(C_{comp}(\mathbf{C}_{\mathcal{R}})) = \textbf{2Th}(\mathcal{T}_{\mathcal{R}})$, where the last one is the algebraic 2-theory of system $\mathcal{T}_{\mathcal{R}}$, whose cells faithfully represent term rewriting sequences (see [25, 23, 9]). Therefore the unit of the adjunction $C^2 \dashv U^2$ maps each term graph rewriting sequence over $\mathcal{R}$ to a term rewriting sequence over $\mathcal{T}_{\mathcal{R}}$. Simple considerations about the 2-naturality of ∇ and ! show that this mapping removes all garbage rewritings, and duplicates a rewriting steps if the root of its redex is shared in the starting term graph.

6 Conclusion and Future Work

The main result of this paper shows that plain term graph rewriting sequences (without garbage collection) over a term graph rewriting system $\mathcal{R}$ have a faithful representation as cells of the gs-monoidal 2-theory of $\mathcal{R}$, i.e., a symmetric monoidal 2-category with some additional structure freely generated by the rules of $\mathcal{R}$. We also showed that the categorical framework relates in a clean way term and term graph rewriting. Space limitations prevent us from dealing with subjects like including garbage collection in this framework, weakening the restriction to plain rewriting sequences, and comparing our formalism with other notion of term graph rewriting. While for the first point all our constructions lift

[7] That is, 2-categories with strictly associative *2-products* and a *2-terminal* object, which are defined as the usual products and terminal objects, but where the universality statements are quantified over cells rather than over arrows.

smoothly (basically, we only need to assume the 2-naturality of the discharger !) making the comparison with other formalisms rather straightforward, weakening plain redexes is a key point, linked e.g. to classical problems of un/sharing in functional languages: Any other consideration is deferred to the full paper.

The representation of rewriting sequences as cells induces an obvious equivalence relation on them, relating two sequences if they are represented by the same cell. In the case of term rewriting, such equivalence coincides with the so-called *permutation equivalence* [5, 14], due to the axioms of cartesian 2-categories [20, 11]. The precise characterization of this equivalence for the case of term graph rewriting is left as a topic for further research. A promising direction consists in relating the definition of term graph rewriting presented here with the algebraic approach to graph transformation for which a rich theory concerning aspects of concurrency and parallelism has been developed [7], including various notion of equivalence on graph transformation sequences.

Another interesting topic is the generalization of the results presented in this paper and in [8] to the case of possibly cyclic term graphs, and to relate them to rational terms, i.e., possibly infinite terms having a finite number of distinct subterms. Such a topic is already considered for example in [17], but it would be interesting to see if it could be addressed using categorical techniques along the line of this paper. Some preliminary considerations suggest that *Iteration Theories* [4] could fruitfully be used to this aim, in place of algebraic theories.

References

1. Z.M. Ariola and J.W. Klop. Equational Term Graph Rewriting. *Fundamenta Informaticae*, 26:207–240, 1996.
2. H.P. Barendregt, M.C.J.D. van Eekelen, J.R.W. Glauert, J.R. Kennaway, M.J. Plasmeijer, and M.R. Sleep. Term graph reduction. In *Proceedings PARLE*, volume 259 of *LNCS*, pages 141–158. Springer Verlag, 1987.
3. M. Barr and C. Wells. *Category Theory for Computing Science.* Prentice Hall, 1990.
4. S. Bloom and Z. Ésik. *Iteration Theories.* EATCS Monographs on Theoretical Computer Science. Springer Verlag, 1991.
5. G. Boudol. Computational semantics of term rewriting systems. In M. Nivat and J. Reynolds, editors, *Algebraic Methods in Semantics*, pages 170–235. Cambridge University Press, 1985.
6. L. Budach and H.-J. Hoenke. *Automaten und Functoren.* Akademie-Verlag, 1975.
7. A. Corradini. Concurrent Graph and Term Graph Rewriting. In U. Montanari and V. Sassone, editors, *Proceedings CONCUR'96*, volume 1119 of *LNCS*, pages 438–464. Springer Verlag, 1996.
8. A. Corradini and F. Gadducci. An algebraic presentation of term graphs, via ps-monoidal categories. Submitted for publication. Available at `http://www.di.unipi.it/~gadducci/papers/aptg.ps`, 1997.
9. A. Corradini, F. Gadducci, and U. Montanari. Relating two categorical models of term rewriting. In *Rewriting Tecniques and Applications*, volume 914 of *LNCS*, pages 225–240. Springer Verlag, 1995.

10. A. Corradini and F. Rossi. Hyperedge Replacement Jungle Rewriting for Term Rewriting Systems and Logic Programming. *Theoret. Comput. Sci.*, 109:7–48, 1993.
11. G. Gadducci. *On the Algebraic Approach to Concurrent Term Rewriting.* PhD thesis, University of Pisa - Department of Computer Science, 1996.
12. J.A. Goguen, J.W. Tatcher, E.G. Wagner, and J.R Wright. Some Fundamentals of Order-Algebraic Semantics. In *Mathematical Foundations of Computer Science*, volume 45 of *LNCS*, pages 153–168. Springer Verlag, 1976.
13. B. Hoffmann and D. Plump. Implementing Term Rewriting by Jungle Evaluation. *Informatique théorique et Applications/Theoretical Informatics and Applications*, 25:445–472, 1991.
14. G. Huet and J-J. Lévy. Computations in Orthogonal Rewriting Systems, I. In J.-L. Lassez and G. Plotkin, editors, *Computational Logic: Essays in honour of J. A. Robinson*, pages 395–414. MIT Press, 1991.
15. B. Jacobs. Semantics of Weakening and Contraction. *Annals of Pure and Applied Logic*, 69:73–106, 1994.
16. G.M. Kelly and R.H. Street. Review of the elements of 2-categories. volume 420 of *Lecture Notes in Mathematics*, pages 75–103. Springer Verlag, 1974.
17. J.R. Kennaway, J.W. Klop, M.R. Sleep, and F.J. de Vries. On the Adequacy of Graph Rrewriting for Simulating Term Rewriting. *ACM Trans. Program. Lang. Syst.*, 16:493–523, 1994.
18. A. Kock and G.E. Reyes. Doctrines in Categorical Logic. In J. Bairwise, editor, *Handbook of Mathematical Logic*, pages 283–313. North Holland, 1977.
19. Y. Lafont. Equational Reasoning with 2-dimensional Diagrams. In *Term Rewriting, French Spring School of Theoretical Computer Science*, volume 909 of *LNCS*, pages 170–195. Springer Verlag, 1995.
20. C. Laneve and U. Montanari. Axiomatizing permutation equivalence in the λ-calculus. *Mathematical Structures in Computer Science*, 6:219–249, 1996.
21. F.W. Lawvere. Functorial Semantics of Algebraic Theories. *Proc. National Academy of Science*, 50:869–872, 1963.
22. S. Mac Lane. *Categories for the working mathematician.* Springer Verlag, 1971.
23. J. Meseguer. Conditional rewriting logic as a unified model of concurrency. *Theoret. Comput. Sci.*, 96:73–155, 1992.
24. M.J. Plasmeijer and M.C.J.D. van Eekelen. *Functional Programming and Parallel Graph Rewriting.* Addison Wesley, 1993.
25. A.J. Power. An abstract formulation for rewrite systems. In *Proceedings Category Theory in Computer Science*, volume 389 of *LNCS*, pages 300–312. Springer Verlag, 1989.
26. D.E. Rydehard and E.G. Stell. Foundations of equational deductions: A categorical treatment of equational proofs and unification algorithms. In *Proceedings Category Theory in Computer Science*, volume 283 of *LNCS*, pages 114–139. Springer Verlag, 1987.
27. M.R. Sleep, M.J. Plasmeijer, and M.C. van Eekelen, editors. *Term Graph Rewriting: Theory and Practice.* Wiley, London, 1993.
28. J. Staples. Computation of graph-like expressions. *Theoret. Comput. Sci.*, 10:171–195, 1980.
29. J.G. Stell. *Categorical Aspects of Unification and Rewriting.* PhD thesis, University of Manchester, 1992.
30. R. Street. Categorical structures. In M. Hazewinkel, editor, *Handbook of Algebra*, pages 529–577. Elsevier, 1996.

Presheaf Models for the π-Calculus

Gian Luca Cattani, Ian Stark and Glynn Winskel

BRICS*, Department of Computer Science, University of Aarhus, Denmark

Abstract. Recent work has shown that presheaf categories provide a general model of concurrency, with an inbuilt notion of bisimulation based on open maps. Here it is shown how this approach can also handle systems where the language of actions may change dynamically as a process evolves. The example is the π-calculus, a calculus for 'mobile processes' whose communication topology varies as channels are created and discarded. A denotational semantics is described for the π-calculus within an indexed category of profunctors; the model is fully abstract for bisimilarity, in the sense that bisimulation in the model, obtained from open maps, coincides with the usual bisimulation obtained from the operational semantics of the π-calculus. While attention is concentrated on the 'late' semantics of the π-calculus, it is indicated how the 'early' and other variants can also be captured.

1 Introduction

The gap between domain theory and the theory of concurrency is narrowing. In particular, the π-calculus, which for a long time resisted all but operational semantics, has yielded to a fully abstract denotational semantics; a key idea was to move to domains indexed by a category of name sets (Stark 1996; Fiore, Moggi, and Sangiorgi 1996; Hennessy 1996). Why add yet another model, based this time not on familiar, complete partial orders but instead on presheaf categories?

Several reasons can be given, even at this preliminary stage.

Models for concurrency are best presented as categories where one can take advantage of the universality of constructions (see Winskel and Nielsen (1995)). If domain theory is to meet models for concurrency, it seems that the points of information in a complete partial order need to be replaced by more detailed objects in a category. A problem with the categories of traditional models is that they do not support higher-order constructions. Presheaf categories, on the other hand, not only include important traditional models like synchronisation trees and event structures,[1] but also support function spaces. We see presheaf models as taking us towards a a new domain theory in which presheaf models are analogous to nondeterministic domains (Hennessy and Plotkin 1979).

Another motivation comes in getting a more systematic and algebraic understanding of bisimulation. A traditional way to proceed in giving a theory to

* Basic Research in Computer Science, a centre of the Danish National Research Foundation.

[1] In the sense that these traditional models embed fully, faithfully and densely in particular presheaf models (Joyal, Nielsen, and Winskel 1996).

a process language has been first to endow it with an operational semantics, provide a definition of bisimulation, follow this by the task of verifying that the bisimulation is a congruence (maybe by modifying it a bit), and then establish proof rules. Often the pattern is standard, but sometimes even getting a passable definition of bisimulation can be tricky, as, for example, when higher-order features are involved. An advantage of presenting models for concurrency as categories has been that they then support a general definition of bisimulation based on open maps (Joyal et al. 1996). We can then exploit the universality of various constructions in showing that they preserve bisimulation (Cattani and Winskel 1997). Presheaf categories come along with a concept of open map and bisimulation. They themselves form a category (strictly a bicategory of profunctors), in which the objects are presheaf categories, yielding ways to combine presheaf categories and their bisimulations.

The problem with general definitions is that it's not always so easy to see what their instances amount to. Indeed, a major task of this paper is showing that the bisimulation on processes of the π-calculus obtained from open maps coincides with a traditional definition (following up on earlier work for value-passing processes (Winskel 1996)). However, the presheaf model for the π-calculus contributes more than this. Along the way, the presheaf model casts light on bisimulation for the π-calculus and why operations preserve it; the normal form for processes in the π-calculus (Fiore et al. 1996), which can be read off from the definition of the model (Theorem 5); and suggests smooth translations between variants of the π-calculus (Section 5). We also claim that, compared to the domain model, the "domain equation" of our model is rather simpler, because we seek a category of paths, not processes; the presheaf construction then fills in the necessary nondeterminism. As a result the system is particularly flexible: here we present the full 'late' π-calculus in detail, but we also sketch how the same category holds models of the 'early' π-calculus and other popular variants.

The recent domain models for the π-calculus lie within a functor category $\mathcal{C}po^{\mathcal{I}}$ (Fiore et al. 1996; Stark 1996). Here $\mathcal{I}$ is the category of finite sets of names and injections between them, representing the fact that over time new names may be created and old names relabelled to avoid clashes. Hennessy (1996) has followed the same approach in his model for testing equivalences. The key to capturing the π-calculus is that the categorical requirements of functoriality and naturality give uniformity over varying name sets. Most notably $\mathcal{C}po^{\mathcal{I}}$ is cartesian closed, and the function space correctly handles the fact that old processes must be prepared to receive new names. This paper makes the same step up in the presheaf approach, to give a model of the π-calculus not in $\mathcal{P}rof$ but in $\mathcal{P}rof^{\mathcal{I}}$.

1.1 The π-calculus

The version of the π-calculus we use is entirely standard. We summarise it only very briefly here: for discussion and further detail see the original papers (Milner, Parrow, and Walker 1992a,b; Milner 1991). Processes have the following syntax

$$P ::= \bar{x}y.P \mid x(y).P \mid \nu x\, P \mid [x{=}y]P \mid 0 \mid P+P \mid P|P \mid !P$$

with x and y ranging over some infinite supply of *names*. Note that we include the match operator $[x{=}y]P$, unguarded sum and unguarded replication $!P$. This selection is fairly arbitrary: our model copes equally well with mismatch $[x{\neq}y]P$ and processes defined by recursion, guarded or unguarded. Similarly, it makes no difference if we restrict to one of the popular subsets, such as the asynchronous π-calculus (Boudol 1992).

To simplify presentation we identify processes up to a *structural congruence*, the smallest congruence relation satisfying

$$[x = x]P \equiv P \qquad !P \equiv P \,|\, !P \qquad \begin{array}{ll} x(y).P \equiv x(z).P[z/y] & z \notin fn(P) \\ \nu y\, P \equiv \nu z\, P[z/y] & z \notin fn(P) \end{array}$$

$$\begin{array}{ccc} P + 0 \equiv P & P + Q \equiv Q + P & (P + Q) + R \equiv P + (Q + R) \\ P \,|\, 0 \equiv P & P \,|\, Q \equiv Q \,|\, P & (P \,|\, Q) \,|\, R \equiv P \,|(Q \,|\, R). \end{array}$$

Here $P[z/y]$ denotes capture-avoiding substitution — which may of course require in turn the α-conversion of subexpressions. This equivalence is not as aggressive as the structural congruence of, say, Definition 3.1 in (Milner 1992), which allows name restriction $\nu x(-)$ to change its scope. Nevertheless it cuts down the operational rules we shall need, with none at all for matching and replication. All this is to some degree a matter of taste: if we treat process terms as concrete syntax, with no structural identification, the model is still valid. Indeed full abstraction then allows us to read off the fact that α-conversion, commuting '+' and so forth all respect bisimilarity (replacing, for example, the proofs of Theorems 1 to 9 in (Milner et al. 1992b, §3)).

The operational semantics of processes are given by transitions of four kinds: internal or 'silent' action τ, input $x(y)$, free output $\bar{x}y$ and bound output $\bar{x}(y)$. We denote a general transition by α, and define its free and bound names thus:

$$\begin{array}{lll} fn(\tau) = \emptyset & fn(\bar{x}y) = \{x, y\} & fn(x(y)) = fn(\bar{x}(y)) = \{x\} \\ bn(\tau) = \emptyset & bn(\bar{x}y) = \emptyset & bn(x(y)) = bn(\bar{x}(y)) = \{y\}. \end{array}$$

The transitions that a process may perform are given inductively by the rules in Figure 1. This is a *late* semantics, in that input substitution happens in the (COM) rule when communication actually occurs, rather than at (IN). The chief difference between these rules and Table 2 of (Milner et al. 1992b) is that we let structural congruence do some of the work. Thus there are no symmetric forms for the four right-hand rules, and *sometimes processes must be α-converted before they can interact.* Of course the possible transitions derived are exactly the same as with the original definitions.

A symmetric relation $\mathcal{S}$ between processes is a *bisimulation* if for every $(P, Q) \in \mathcal{S}$ the following conditions hold.

- For $\alpha = \tau, \bar{x}y, \bar{x}(y)$, if $P \xrightarrow{\alpha} P'$ then there is Q' such that $Q \xrightarrow{\alpha} Q'$ and $(P', Q') \in \mathcal{S}$.
- If $P \xrightarrow{x(y)} P'$ then there is Q' such that $Q \xrightarrow{x(y)} Q'$ and for any name z, $(P'[z/y], Q'[z/y]) \in \mathcal{S}$.

$$\text{OUT} \quad \bar{x}y.P \xrightarrow{\bar{x}y} P \qquad \text{SUM} \quad \frac{P \xrightarrow{\alpha} P'}{P + Q \xrightarrow{\alpha} P'}$$

$$\text{IN} \quad x(y).P \xrightarrow{x(y)} P \qquad \text{PAR} \quad \frac{P \xrightarrow{\alpha} P'}{P \,|\, Q \xrightarrow{\alpha} P' \,|\, Q} \; bn(\alpha) \cap fn(Q) = \emptyset$$

$$\text{RES} \quad \frac{P \xrightarrow{\alpha} P'}{\nu x\, P \xrightarrow{\alpha} \nu x\, P'} \; x \notin fn(\alpha) \qquad \text{COM} \quad \frac{P \xrightarrow{x(y)} P' \quad Q \xrightarrow{\bar{x}z} Q'}{P \,|\, Q \xrightarrow{\tau} P'[z/y] \,|\, Q'}$$

$$\text{OPEN} \quad \frac{P \xrightarrow{\bar{x}y} P'}{\nu y\, P \xrightarrow{\bar{x}(y)} P'} \; x \neq y \qquad \text{CLOSE} \quad \frac{P \xrightarrow{x(y)} P' \quad Q \xrightarrow{\bar{x}(y)} Q'}{P \,|\, Q \xrightarrow{\tau} \nu y(P' \,|\, Q')}$$

Fig. 1. Transition rules for π-calculus processes

To check this second condition it is only necessary that z ranges over the free names of P and Q, and one fresh name. This relation is *strong*, in that τ-actions must match, and *late*, in that input actions must match before the transmitted value is known. Two processes are (strong, late) *bisimilar* if there is some bisimulation relating them. We write $P \dot{\sim} Q$ and observe that bisimilarity is itself a bisimulation, and contains all others.

Bisimilarity is preserved by all process constructors except for input prefix $x(y).P$. This is because bisimilarity assumes all names are distinct, while the substitution that happens on input can cause names to become identified. One thus defines processes to be *equivalent* $P \sim Q$ if they are bisimilar under all possible name substitutions. Equivalence is then the smallest congruence containing bisimilarity.

1.2 Presheaf Models and Bisimulation

Let P be a small category. The category of *presheaves over* P, often denoted by $\widehat{\mathsf{P}}$ or by $\mathit{Set}^{\mathsf{P}^{op}}$, is the category whose objects are contravariant functors from P to *Set* (the category of sets and functions) and whose arrows are the natural transformations between such functors.

A category of presheaves, $\widehat{\mathsf{P}}$, is accompanied by the *Yoneda embedding*, a functor $y_{\mathsf{P}} : \mathsf{P} \to \widehat{\mathsf{P}}$, which fully and faithfully embeds P in the category of presheaves.

Via the Yoneda embedding we can regard P essentially as a full subcategory of $\widehat{\mathsf{P}}$. In our applications, the category P is to be thought of as consisting of path objects, or computation-path shapes. The Yoneda Lemma, by providing a natural bijection between $\widehat{\mathsf{P}}(y_{\mathsf{P}}(p), X)$ and $X(p)$, justifies the intuition that a presheaf $X : \mathsf{P}^{op} \to \mathit{Set}$ can be thought of as specifying for a typical path object p the set $X(p)$ of computation paths of shape p. The presheaf X acts on a morphism $m : p \to q$ in P to give a function $X(m)$ saying how q-paths restrict to p-paths. A presheaf being a colimit of path objects can be thought of as a collection of computation paths glued together by identifying subpaths.

Bisimulation on presheaves is derived from notion of open map between presheaves (Joyal and Moerdijk 1994). Recall, a morphism $h : X \to Y$, between presheaves X, Y, is P-*open* iff for all morphisms $m : p \to q$ in P, the square

$$\begin{array}{ccc} X(p) & \xleftarrow{Xm} & X(q) \\ {\scriptstyle h_p}\downarrow & & \downarrow{\scriptstyle h_q} \\ Y(p) & \xleftarrow{Ym} & Y(q) \end{array}$$

is a quasi-pullback, *i.e.* whenever $x \in X(p)$ and $y \in Y(q)$ satisfy $h_p(x) = (Ym)(y)$, then there exists $x' \in X(q)$ such that $(Xm)(x') = x$ and $h_q(x') = y$. (This definition of open map, translates via the Yoneda Lemma to an equivalent path-lifting property of h—see (Joyal et al. 1996).)

We say that presheaves X, Y in $\widehat{\mathsf{P}}$ are P-*bisimilar* iff there is a span of surjective open maps between them. This is equivalent to there being a subobject $R \hookrightarrow X \times Y$ such that the compositions with the projections are surjective open.

The category $\widehat{\mathsf{P}}$, over a small category P, is the free colimit completion of P. In more detail, the Yoneda embedding $y_{\mathsf{P}} : \mathsf{P} \to \widehat{\mathsf{P}}$ satisfies the universal property that for any functor $F : \mathsf{P} \to \mathcal{E}$, where $\mathcal{E}$ is a cocomplete category, there is a colimit-preserving functor $Lan_{y_{\mathsf{P}}}(F) : \widehat{\mathsf{P}} \to \mathcal{E}$, the *left Kan extension of* F *along* y_{P}, unique to within isomorphism, such that $F \cong Lan_{y_{\mathsf{P}}}(F) \circ y_{\mathsf{P}}$. Consequently, if $G : \widehat{\mathsf{P}} \to \mathcal{E}$ is a colimit-preserving functor then, to within isomorphism, G is the left Kan extension $Lan_{y_{\mathsf{P}}}(G \circ y_{\mathsf{P}})$. Recall, in addition, that $Lan_{y_{\mathsf{P}}}(F)$ is left adjoint to the functor taking Y in $\mathcal{E}$ to the presheaf $\mathcal{E}(F(\text{-}), Y)$.

Here we shall only be interested in left Kan extensions where $\mathcal{E}$ is a presheaf category $\widehat{\mathsf{Q}}$. They, and so any colimit-preserving functor between presheaf categories, preserve open maps. Because they preserve colimits they necessarily preserve surjectivity of maps too (and hence bisimulation). Spelt out:

Lemma 1 (Cattani and Winskel 1997). *Let* $G : \widehat{\mathsf{P}} \to \widehat{\mathsf{Q}}$ *be any colimit-preserving functor between presheaf categories. Then* G *preserves (surjective) open maps: If* h *is a (surjective)* P-*open map in* $\widehat{\mathsf{P}}$*, then* $G(h)$ *is a (surjective)* Q-*open map in* $\widehat{\mathsf{Q}}$.

Notation: If $F : \mathsf{P} \nrightarrow \widehat{\mathsf{Q}}$ is a functor, we will often write $F_!$ for the left Kan extension $Lan_{y_{\mathsf{P}}}(F)$ and F^* its right adjoint, mentioned above. This notation is reminiscent of a more usual one (that we shall also employ) that given a functor $F : \mathsf{P} \to \mathsf{Q}$ between small categories writes $F_! \dashv F^*$ for the left Kan extension $Lan_{y_{\mathsf{P}}}(y_{\mathsf{Q}}F)$ and its right adjoint. In this second case a further right adjoint is also present $F^* \dashv F_*$, making F^* into a colimit preserving functor.

Thus we have a variety of general methods to obtain functors $\widehat{\mathsf{P}} \to \widehat{\mathsf{Q}}$, several of them automatically preserving colimits and hence bisimilarity. The following construction gives a relevant example.

A Lifting Construction Given any category P, define $\mathsf{P}_\perp$ as the category isomorphic to P but for a new initial object $\perp$ freely added. Clearly there exists a full embedding $l(ift) : \mathsf{P} \to \mathsf{P}_\perp$. This gives rise to a triple of adjoint functors, $l_! \dashv l^* \dashv l_\star$

$$\widehat{\mathsf{P}} \underset{l_!}{\overset{l_\star}{\rightleftarrows}} \widehat{\mathsf{P}}_\perp \quad (l^* : \widehat{\mathsf{P}}_\perp \to \widehat{\mathsf{P}})$$

where for any presheaf $X \in \widehat{\mathsf{P}}$ and path $p \in \mathsf{P}$ we have $l_\star(X)(p) \cong X(p)$ and $l_\star(X)(\perp) \cong \{\star\}$; from which $l^*(l_\star(X)) \cong X$.

It is not difficult to prove the following proposition analogous to Lemma 2.2 in (Joyal and Moerdijk 1995).

Proposition 2. *$l_\star$ preserves surjective open maps.*

1.3 Profunctors

There are several equivalent ways of presenting profunctors (also called distributors and bimodules, see Borceux (1994)). For us, a *profunctor* $F : \mathsf{P} \nrightarrow \mathsf{Q}$, between small categories P and Q, is a functor $F : \mathsf{P} \to \widehat{\mathsf{Q}}$. The composition $G \circ F$ of profunctors $F : \mathsf{P} \nrightarrow \mathsf{Q}$ and $G : \mathsf{Q} \nrightarrow \mathsf{R}$ is defined to be the composition of functors

$$(Lan_{y_{\mathsf{Q}}}(G)) \circ F \ ,$$

which is only defined to within isomorphism. Thus, we obtain a *bicategory*[2] of profunctors $\mathcal{P}rof$.

It is helpful to view $\mathcal{P}rof$ as a category of "nondeterministic domains" (Hennessy and Plotkin 1979; Hennessy 1996), analogous to those obtained as the Kleisli category of a powerdomain monad. We exhibit an adjunction to back up this claim. Write ω-$\mathcal{A}cc$ for the category of finitely accessible categories, the category analogue of algebraic cpo's; morphisms are functors preserving filtered colimits. The functor

$$(-)^0 : \omega\text{-}\mathcal{A}cc \to \mathcal{P}rof$$

takes a finitely accessible category $\mathcal{C}$ to its "basis" $\mathcal{C}^0$, a choice of skeletal subcategory of its finitely presentable elements; a filtered-colimit preserving functor $H : \mathcal{B} \to \mathcal{C}$ is sent to the profunctor $H^0 : \mathcal{B}^0 \nrightarrow \mathcal{C}^0$ such that $H^0(B) = \mathcal{C}(-, H(B)) : (\mathcal{C}^0)^{op} \to \mathcal{S}et$. The functor

$$\widehat{(-)} : \mathcal{P}rof \to \omega\text{-}\mathcal{A}cc$$

takes a profunctor $F : \mathsf{P} \nrightarrow \mathsf{Q}$ to a choice of left Kan extension $Lan_{y_{\mathsf{P}}}(F)$. There is an equivalence of categories

$$\omega\text{-}\mathcal{A}cc(\mathcal{C}, \widehat{\mathsf{P}}) \simeq \mathcal{P}rof(\mathcal{C}^0, \mathsf{P}) \ ,$$

[2] However, henceforth we won't be pernickety in our category-theoretic terminology, and use terms from traditional category theory, even when constructions, strictly speaking, take place in a bicategory setting.

given by restriction to finitely presentable objects, expressing an adjunction between ω-$\mathcal{A}cc$ and $\mathcal{P}rof$:

$$\omega\text{-}\mathcal{A}cc \underset{\widehat{(-)}}{\overset{(-)^0}{\rightleftarrows}} \mathcal{P}rof \quad (\bot)$$

With bicategorical quibbles, there is good reason to call $\mathcal{P}rof$ the Kleisli category of the monad $\widehat{(-)^0}$, and to think of the monad as analogous to powerdomains in that it introduces nondeterminism.

Turning the monad around we obtain a comonad $!\mathsf{P} = (\widehat{\mathsf{P}})^0$, amounting to the finite-colimit completion of P, whose co-Kleisli category consists of presheaf categories with filtered-colimit-preserving functors as morphisms. In addition, the bicategory $\mathcal{P}rof$ is rich in constructions which enable us to handle higher-order processes as presheaves. Many of these constructions are associated with $\mathcal{P}rof$ being a model of classical linear logic. The tensor $\otimes$ is of particular interest: on objects $\mathsf{P} \otimes \mathsf{Q}$ is the product of categories $\mathsf{P} \times \mathsf{Q}$; while on morphisms if $F : \mathsf{P} \nrightarrow \mathsf{P}'$ and $G : \mathsf{Q} \nrightarrow \mathsf{Q}'$ are profunctors then $F \otimes G : \mathsf{P} \otimes \mathsf{Q} \nrightarrow \mathsf{P}' \otimes \mathsf{Q}'$ acts according to:

$$((F \otimes G)(p,q))(p',q') = ((F(p))(p')) \times ((G(q))(q')) \ ,$$

where $\times$ is the product functor in $\mathcal{S}et$. The unit of $\otimes$ is $\mathbf{1}$, the category with a single object and morphism. Defining the linear function space $\mathsf{P} \nrightarrow \mathsf{Q}$ to be the product of categories $\mathsf{P}^{op} \times \mathsf{Q}$, we see the natural bijection

$$\mathcal{P}rof(\mathsf{P}, [\mathsf{Q} \nrightarrow \mathsf{R}]) \cong \mathcal{P}rof(\mathsf{P} \otimes \mathsf{Q}, \mathsf{R}) \ .$$

It is easy to see that presheaves over $\mathsf{P} \nrightarrow \mathsf{Q}$ correspond to profunctors, and so, to within isomorphism, to colimit-preserving functors from $\widehat{\mathsf{P}}$ to $\widehat{\mathsf{Q}}$. Filtered-colimit-preserving functors are represented, with the help of the exponential !, as presheaves over $!\mathsf{P} \nrightarrow \mathsf{Q}$. Products (&) and coproducts ($+, \Sigma$) in $\mathcal{P}rof$ coincide on objects, where both are given by coproduct of categories; similarly $\top$ and 0 are both the empty category. Linear involution $\mathsf{P}^{\bot}$ is isomorphic to $\mathsf{P} \nrightarrow \mathbf{1}$ and so to P^{op}. $\mathcal{P}rof$ is compact-closed: par ($\parr$) coincides with tensor ($\otimes$), and $\bot$ with $\mathbf{1}$.

A Diagonal Later, in our treatment of replication in the π-calculus, we shall use a diagonal map, which we now construct. Let P and Q be two small categories. Take $w_{\mathsf{P}} : \mathsf{P} \times \mathsf{Q} \to \mathsf{P}$ and $w_{\mathsf{Q}} : \mathsf{P} \times \mathsf{Q} \to \mathsf{Q}$ to be the two projection functors. By composing with the Yoneda embedding they can be promoted to the status of maps in $\mathcal{P}rof$ and therefore give rise to a universal profunctor $w_{\mathsf{P},\mathsf{Q}} = \langle w_{\mathsf{P}}, w_{\mathsf{Q}} \rangle : \mathsf{P} \otimes \mathsf{Q} \nrightarrow \mathsf{P}\&\mathsf{Q}$.[3] This induces an adjoint pair between the associated presheaf categories:

$$\widehat{\mathsf{P} \times \mathsf{Q}} \underset{w^*}{\overset{w_!}{\rightleftarrows}} \widehat{\mathsf{P}} \times \widehat{\mathsf{Q}} \quad (\bot) \ .$$

[3] Given the "linear logic" structure of $\mathcal{P}rof$, the map w does correspond to a form of *weakening*.

Since $\mathcal{P}rof$ has products we also have that for any small category P a diagonal arrow $\Delta_{\mathsf{P}} : \mathsf{P} \nrightarrow \mathsf{P}\&\mathsf{P}$ is definable. By composing the extension $\Delta_{\mathsf{P},!}$ with $w^{\star}_{\mathsf{P},\mathsf{P}}$, we obtain a functor

$$d_{\mathsf{P}} : \widehat{\mathsf{P}} \to \widehat{\mathsf{P} \times \mathsf{P}} .$$

It is not difficult to prove that for any presheaf X over P and for any two objects p_1, p_2 of P:

$$d_{\mathsf{P}} X \langle p_1, p_2 \rangle = X(p_1) \times X(p_2) .$$

Like the lifting of Section 1.2, this is a *structural* arrow, and it makes sense to establish the following preservation property.

Proposition 3. *Let* P *and* Q *be two small categories; then* $w^{\star}_{\mathsf{P},\mathsf{Q}}$ *preserves surjective open maps and, consequently,* d_{P} *does too.*

1.4 The Functor Category $\mathcal{P}rof^{\mathcal{I}}$

As a π-calculus process evolves, the ambient set of channel names may change. To take account of this variability, our path category for the π-calculus will be indexed by $\mathcal{I}$, the category of finite name sets and injective maps between them. Rather than $\mathcal{P}rof$ itself then, we are specifically interested in the functor category $\mathcal{P}rof^{\mathcal{I}}$.

Our first construction is an *object of names* N, the functor $\mathsf{N} : \mathcal{I} \to \mathcal{P}rof$ given by the composition $\mathcal{I} \subseteq \mathcal{S}et \to \mathcal{C}at \to \mathcal{P}rof$, an inclusion followed by two embeddings. This takes a set $s \in \mathcal{I}$ to the corresponding discrete category, which we write also as s.

Several $\mathcal{P}rof$ operations can be extended pointwise to $\mathcal{P}rof^{\mathcal{I}}$: lifting $(-)_\perp$, the involution $(-)^{op}$, tensor '$\otimes$', which is the same as par '$\parr$', and product '&', which is the same as coproduct '+'.

We do not know whether or not $\mathcal{P}rof^{\mathcal{I}}$ is monoidal closed with respect to $\otimes$. There is, however, a Yoneda lemma expressing an isomorphism

$$\widehat{\mathsf{A}(s)} \cong \mathcal{P}rof^{\mathcal{I}}(\mathcal{I}(s,-), \mathsf{A})$$

for s in $\mathcal{I}$ and A in $\mathcal{P}rof^{\mathcal{I}}$. Thus given objects $\mathsf{A}, \mathsf{B} \in \mathcal{P}rof^{\mathcal{I}}$ there is always a candidate for the function space

$$\text{“}\widehat{(\mathsf{A} \nrightarrow \mathsf{B})(s)}\text{”} \cong \mathcal{P}rof^{\mathcal{I}}(\mathcal{I}(s,-) \otimes \mathsf{A}, \mathsf{B}) ,$$

but this is only meaningful if we can exhibit it as an actual presheaf. Conveniently, for the purposes of the π-calculus it is enough to take function spaces $\mathsf{N} \nrightarrow \mathsf{A}$ and it happens that these do exist. On objects they turn out to be given by

$$(\mathsf{N} \nrightarrow \mathsf{A})(s) = s \times \mathsf{A}(s) + \mathsf{A}(s+1) . \tag{1}$$

A presheaf over this comprises a pair $\langle F, Y \rangle$, where $F : s \nrightarrow \mathsf{A}(s)$ is a profunctor arrow from the discrete category s, and $Y \in \widehat{\mathsf{A}(s+1)}$.

Paths of $(\mathsf{N} \twoheadrightarrow \mathsf{A})(s)$ can be seen, rather loosely, as elements of the graph of a function. Thus a path in the $s \times \mathsf{A}(s)$-component of (1) we write as $(x \mapsto p)$ for name $x \in s$ and path $p \in \mathsf{A}(s)$. A path in the $\mathsf{A}(s+1)$-component we write as $(* \mapsto p')$ for $p' \in \mathsf{A}(s+1)$. In a similar spirit we can inject a presheaf $X \in \widehat{\mathsf{A}(s)}$ into the left x-component as $(x \mapsto X)$, and a presheaf $Y \in \widehat{\mathsf{A}(s+1)}$ into the right component as $(* \mapsto Y)$.

Here we have used a convention that we shall follow throughout, that for any set of names s we write '$*$' for the extra element provided in $s+1$, subscripting '$*_s$' when necessary.

This said, we can now describe the action of the profunctor $(\mathsf{N} \twoheadrightarrow \mathsf{A})(i)$ when $i : s \to s'$ is an arrow in $\mathcal{I}$. On the component $s \times \mathsf{A}(s)$

$$(x \mapsto p) \mapsto (i(x) \mapsto \mathsf{A}(i)(p)), \tag{2}$$

while on $\mathsf{A}(s+1)$

$$(*_s \mapsto p') \mapsto \sum_{y \notin Im(i)} (y \mapsto \mathsf{A}[i,y](p')) + (*_{s'} \mapsto \mathsf{A}(i+1)(p')) . \tag{3}$$

Here $[i,y] : s+1 \to s'$ is the injection that extends i over $s+1$ by taking $*_s$ to $y \in (s' \setminus Im(i))$. Note how in this second equation the path $(*_s \mapsto p')$ serves as a 'seed' to set paths for all the fresh names of s' not in the image of i.

To handle name creation we use a construction δ on $\mathcal{P}rof^{\mathcal{I}}$, defined by $\delta \mathsf{A} = \mathsf{A}(\_ + 1)$. This is in fact also a form of function space from N to A, arising from the construction of Day (1970) which lifts the disjoint union '+' of $\mathcal{I}$ to a symmetric monoidal closed structure on $\mathcal{P}rof^{\mathcal{I}}$. Approximately speaking, δA comprises functions that will only accept a fresh name as argument. This allows processes to synchronize on the choice of a local name, as required by the (CLOSE) rule of Figure 1. Stark (1996) expands on this a little.

2 The Equation

We derive a suitable π-calculus path object P in $\mathcal{P}rof^{\mathcal{I}}$ from the following equations:

$$\mathsf{P} \cong \mathsf{P}_\perp + Out + In$$

$$Out = (\mathsf{N} \otimes \mathsf{N} \otimes \mathsf{P}_\perp) + (\mathsf{N} \otimes (\delta \mathsf{P})_\perp) \tag{4}$$

$$In = \mathsf{N} \otimes (\mathsf{N} \twoheadrightarrow \mathsf{P})_\perp \tag{5}$$

Unfolding, the four components of P represent silent action, free output, bound output and input respectively. We give the solution to this equation in two stages: first we describe recursively at each set s the corresponding path category $\mathsf{P}(s)$; and then we specify the profunctor arrow that connnects $\mathsf{P}(s)$ to $\mathsf{P}(s')$ for any injective function $i : s \to s'$.

From the descriptions of the constructors δ and $\twoheadrightarrow$, we can think of the family $\mathsf{P}(-)$ as being recursively described by

$$\mathsf{P}(s) = \mathsf{P}(s)_\perp + s \times s \times \mathsf{P}(s)_\perp + s \times \mathsf{P}(s+1)_\perp + s \times (s \times \mathsf{P}(s) + \mathsf{P}(s+1))_\perp .$$

Our minimal $\mathsf{P}(s)$ is thus a poset, in fact a forest of trees. We have four kinds of root: $\tau.$, $x!y.$, $x!*.$ and $x?$ for any $x, y \in s$. Above these in the order relation we find respectively: $\tau.p$, $x!y.p$, $x!*.p'$, $x?(y{\mapsto}p)$ and $x?(*{\mapsto}p')$, where the last two lie above $x?$ and where p is an object of $\mathsf{P}(s)$ while p' is an object of $\mathsf{P}(s+1)$. The arrows of $\mathsf{P}(s)$ are the prefix order, with for example

$$\begin{aligned} x?(y{\mapsto}p) \leq x?(y{\mapsto}\bar{p}) \quad &\text{iff } p \leq \bar{p} \text{ in } \mathsf{P}(s) \\ x?(*{\mapsto}p') \leq x?(*{\mapsto}\bar{p}') \quad &\text{iff } p' \leq \bar{p}' \text{ in } \mathsf{P}(s+1). \end{aligned}$$

Definition 4. We have already used for the objects of $\mathsf{P}(s)$ a notation suggesting the sequence of "atomic" actions that a path represents. In order to describe the arrow part of the functor P, and later on for the semantics of processes, we now give presheaf analogues of the evident prefix operations on paths.

- For α one of τ or $x!y$ we have a prefixing functor $\alpha : \mathsf{P}(s)_\perp \to \mathsf{P}(s)$. This gives an operation on presheaves $\alpha. = \alpha_! \circ l_\star : \widehat{\mathsf{P}(s)} \to \widehat{\mathsf{P}(s)}$ with

$$\alpha.X(p) = \begin{cases} \{\star\} & \text{if } p = \alpha. \\ X(p') & \text{if } p = \alpha.p' \\ \emptyset & \text{otherwise.} \end{cases}$$

  Similarly with $Y \in \widehat{\mathsf{P}(s+1)}$ we apply $(x!*)_! \circ l_\star$ to obtain $x!*.Y \in \widehat{\mathsf{P}(s)}$.
- Given $X \in \widehat{\mathsf{P}(s)}$ and $x, y \in s$ we define $x?(y{\mapsto}X) \in \widehat{\mathsf{P}(s)}$ by

$$x?(y \mapsto X)\,(p) = \begin{cases} \{\star\} & \text{if } p = x? \\ X(p') & \text{if } p = x?(y{\mapsto}p') \\ \emptyset & \text{otherwise.} \end{cases}$$

  As above we could also have obtained this from a path map $(x?(y{\mapsto}\_))$.
- Suppose that $\langle F, Y\rangle$ is a presheaf over $(\mathsf{N} \nrightarrow \mathsf{P})(s)$, *i.e.* a profunctor $F : s \nrightarrow \mathsf{P}(s)$ and a presheaf $Y \in \widehat{\mathsf{P}(s+1)}$. We define

$$x?\langle F, X\rangle(p) = \begin{cases} \{\star\} & \text{if } p = x? \\ F(y)(p') & \text{if } p = x?(y{\mapsto}p') \\ Y(p') & \text{if } p = x?(*{\mapsto}p') \\ \emptyset & \text{otherwise.} \end{cases}$$

  Once again we could instead have derived this from the operations on paths taking $\langle y, p\rangle \in s \times \mathsf{P}(s)$ to $x?(y{\mapsto}p)$ and $p' \in \mathsf{P}(s+1)$ to $x?(*{\mapsto}p')$.

Lemma 1 and Proposition 2 are sufficient to show that all these functors preserve surjective open maps.

Moving on to the morphism part of P, we need a profunctor $\mathsf{P}(i) : \mathsf{P}(s) \nrightarrow \mathsf{P}(s')$ for every injection $i : s \to s'$. We work by induction on the structure of paths in $\mathsf{P}(s)$. In the base cases minimal paths in $\mathsf{P}(s)$ go to the same in $\mathsf{P}(s')$, regarded via Yoneda as presheaves:

$$\mathsf{P}(i)(\tau.) = \tau. \quad \mathsf{P}(i)(x!y.) = i(x)!i(y). \quad \mathsf{P}(i)(x!*_s.) = i(x)!*_{s'}. \quad \mathsf{P}(i)(x?) = i(x)?$$

The inductive steps are:

$$\begin{array}{rr} \mathsf{P}(i)(\tau.p) = \tau.\mathsf{P}(i)(p) & \mathsf{P}(i)(x!y.p) = i(x)!i(y).\mathsf{P}(i)(p) \\ \mathsf{P}(i)(x!*_s.p') = i(x)!*_{s'}.\mathsf{P}(i+1)(p') & \mathsf{P}(i)(x?(y \mapsto p)) = i(x)?(i(y) \mapsto \mathsf{P}(i)(p)) \\ & \mathsf{P}(i)(x?(* \mapsto p')) = i(x)?(\mathsf{N} \twoheadrightarrow \mathsf{P})(i)(p') \,. \end{array}$$

In the last of these we use the non-trivial action of $(\mathsf{N} \twoheadrightarrow \mathsf{P})(i)$ from (3) to 'fill in' input behaviour on receiving names from $(s' \setminus Im(i))$.

2.1 A Decomposition Result

We will now observe that every presheaf $X \in \widehat{\mathsf{P}(s)}$ decomposes into a sum of disjoint components rooted at one of the minimal path objects $\tau., x!y., x!*., x?$ where $x, y \in s$. The decomposition not only allows us to read off a normal form for presheaves, but also leads to a natural notion of transitions for presheaves.

Let m be a minimal object in $\mathsf{P}(s)$ and $X \in \widehat{\mathsf{P}(s)}$. Any $x \in X(m)$ determines a sub-presheaf C of X as follows:

$$C(p) \;=\; \begin{cases} \{y \in X(p) \mid X(m,p)(y) = x\} & \text{if } m \leq p \\ \emptyset & \text{otherwise} \end{cases}$$

for $p \in \mathsf{P}(s)$, and when $p \leq q$ define the function $C(p,q) : C(q) \to C(p)$ by

$$C(p,q)(z) = X(p,q)(z) \quad \text{for } z \in C(q)$$

— because X is a contravariant functor it follows that

$$X(m,p)(X(p,q)(z)) \;=\; X(m,q)(z) \;=\; x$$

so that $X(p,q)(z) \in C(p)$. It is easily checked that C is a presheaf and indeed a sub-presheaf of X because its action on morphisms (p,q), when $p \leq q$, restricts that of X.

Notation: In this situation, we shall say that C is a *rooted component* of X at x.

Rooted components of X are pairwise disjoint in the sense that if m, m' are minimal objects of $\mathsf{P}(s)$ and C is a rooted component at $x \in X(m)$ and C' is a rooted component at $x' \in C(m')$, then if $C(p) \cap C'(p) \neq \emptyset$ for any $p \in \mathsf{P}(s)$, then $m = m'$ and $x = x'$. Thus, X is isomorphic to a sum of its rooted components:

$$X \cong \sum_m \sum_{x \in X(m)} C_x \tag{6}$$

where m ranges over minimal objects of $\mathsf{P}(s)$ and C_x is the rooted component of X at x.

We analyse further the form of rooted components of $X \in \widehat{\mathsf{P}(s)}$. A rooted component C_i at $i \in X(\tau.)$ is isomorphic to $\tau.X_i$ where $X_i \in \widehat{\mathsf{P}(s)}$ is given by

$$X_i(p) = C_i(\tau.p), \text{ on objects } p \in \mathsf{P}(s), \text{ and}$$
$$X_i(p,q) = C_i(\tau.p, \tau.q) : X_i(q) \to X_i(p), \text{ on morphisms } p \leq q \text{ of } \mathsf{P}(s).$$

We write $X \xrightarrow{\tau} X'$ when there is $i \in X(\tau.)$ such that $X' = X_i$. The assignment $i \mapsto X_i$ is a bijection between the sets $X(\tau.)$ and $\{X' \mid X \xrightarrow{\tau} X'\}$.

A rooted component C_j at $j \in X(x!y.)$, for $x, y \in s$, is isomorphic to $x!y.X_j$, where $X_j \in \widehat{\mathsf{P}(s)}$ is given by

$$X_j(p) = C_j(x!y.p), \text{ on objects } p \in \mathsf{P}(s), \text{ and}$$
$$X_j(p,q) = C_j(x!y.p, x!y.q), \text{ on morphisms } p \leq q \text{ of } \mathsf{P}(s).$$

We write $X \xrightarrow{x!y} X'$ when there is $j \in X(x!y)$ such that $X' = X_j$. The assignment $j \mapsto X_j$ is a bijection between the sets $X(x!y.)$ and $\{X' \mid X \xrightarrow{x!y} X'\}$.

A rooted component C_k at $k \in X(x!*.)$, for $x \in s$, is isomorphic to $x!*.X_k$, where $X_k \in \widehat{\mathsf{P}(s+1)}$ is given by

$$X_k(p') = C_k(x!*.p'), \text{ on objects } p' \in \mathsf{P}(s+1), \text{ and}$$
$$X_k(p',q') = C_k(x!*.p', x!*.q'), \text{ on morphisms } p' \leq q' \text{ of } \mathsf{P}(s+1).$$

We write $X \xrightarrow{x!*} X'$ when there is $k \in X(x!*.)$ such that $X' = X_k$. The assignment $k \mapsto X_k$ is a bijection between the sets $X(x!*.)$ and $\{X' \mid X \xrightarrow{x!*} X'\}$.

Let C_l be a rooted component at $L \in X(x?)$. Define for any $y \in s$, $X_l^y \in \widehat{\mathsf{P}(s)}$ as

$$X_l^y(p) = C_l(x?(y \mapsto p)), \text{ and}$$
$$X_l^y(p,q) = C_l(x?(y \mapsto p), x?(y \mapsto q)) : X_l^y(q) \to X_l^y(p).$$

Define also $X_l^* \in \widehat{\mathsf{P}(s+1)}$ as

$$X_l^*(p') = C_l(x?(* \mapsto p')), \text{ and}$$
$$X_l^*(p',q') = C_l(x?(* \mapsto p'), x?(* \mapsto q')) : X_l^*(q') \to X_l^*(p').$$

Then X_l can be regarded as a pair $\langle \lambda y.X_l^y, X_l^* \rangle$ with $\lambda y.X_l^y : s \nrightarrow \mathsf{P}(s)$. We write $X \xrightarrow{x?} \langle F, Y \rangle$ when there is $l \in X(x?)$ such that $F(y)$ is isomorphic to X_l^y for every $y \in s$ and Y is isomorphic to X_l^*. The assignment $l \mapsto \langle \lambda y.X_l^y, X_l^* \rangle$ is a bijection between the sets $X(x?)$ and $\{\langle F, Y \rangle \mid X \xrightarrow{x?} \langle F, Y \rangle\}$.

This analysis of rooted components transforms (6) into the following result.

Theorem 5 (Decomposition of Presheaves). *Let $X \in \widehat{\mathsf{P}(s)}$. Then*

$$X \cong \sum_{i \in X(\tau.)} \tau.X_i + \sum_{x,y \in s} \sum_{j \in X(x!y.)} x!y.X_j + \sum_{x \in s} \sum_{k \in X(x!*.)} x!*.X_k + \sum_{x \in s} \sum_{l \in X(x?)} x?\langle \lambda y.X_l^y, X_l^* \rangle \,.$$

This gives a decomposition of morphisms, which preserves surjective opens.

Proposition 6. *Let X, Y be two presheaves over $\mathsf{P}(s)$ with $f : X \to Y$ a surjective open map beween them. Then the following "restrictions" of f are surjective open:*

$$\begin{array}{ll} f_i : X_i \to Y_{i'} & \textit{where } i \in X(\tau.) \textit{ and } i' = f_{\tau.}(i) \\ f_j : X_j \to Y_{j'} & \textit{where } j \in X(x!y.) \textit{ and } j' = f_{x!y.}(j) \\ f_k : X_k \to Y_{k'} & \textit{where } k \in X(x!*.) \textit{ and } k' = f_{x!*.}(k) \\ f_l^y : X_l^y \to Y_{l'}^y & \textit{where } l \in X(x?) \textit{ and } l' = f_{x?}(l) \\ f_l^* : X_l^* \to Y_{l'}^* & \textit{where } l \in X(x?) \textit{ and } l' = f_{x?}(l). \end{array}$$

2.2 Indexed Late Bisimilarity for P

The previous section gave a notion of transitions on presheaves; naturally enough, this leads to a form of bisimilarity.

Definition 7. A P-*late bisimulation* is a family $(R_s)_{s \in \mathcal{I}}$ of symmetric binary relations on presheaves in $\widehat{\mathsf{P}(s)}$ such that for any finite name set s and any two presheaves X, Y over $\mathsf{P}(s)$, if $X \; R_s \; Y$ then

$$\begin{aligned} X \xrightarrow{\tau} X' \;&\Rightarrow\; \exists Y'.\, Y \xrightarrow{\tau} Y' \;\&\; X' \; R_s \; Y' \\ X \xrightarrow{x!y} X' \;&\Rightarrow\; \exists Y'.\, Y \xrightarrow{x!y} Y' \;\&\; X' \; R_s \; Y' \\ X \xrightarrow{x!*} X' \;&\Rightarrow\; \exists Y'.\, Y \xrightarrow{x!*} Y' \;\&\; X' \; R_{s+1} \; Y' \\ X \xrightarrow{x?} \langle F, X' \rangle \;&\Rightarrow\; \exists \langle G, Y' \rangle.\, Y \xrightarrow{x?} \langle G, Y' \rangle \;\&\; X' \; R_{s+1} \; Y' \\ &\qquad \& \; \forall y \in s.\; F(y) \; R_s \; G(y) \,. \end{aligned}$$

We say that $X, Y \in \widehat{\mathsf{P}(s)}$ are P-*late bisimilar* iff $X \; R_s \; Y$ for some P-late bisimulation $(R_s)_{s \in \mathcal{I}}$.

Lemma 8. *P-late bisimilarity is an equivalence relation.*

Using Proposition 6 we can show that this P-late bisimilarity corresponds exactly to open map bisimilarity:

Lemma 9. *Suppose X and Y are presheaves over $\mathsf{P}(s)$. Then:*

(i) *If $f : X \to Y$ is a surjective open map then X and Y are P-late bisimilar.*
(ii) *If $X \; R_s \; Y$ for some P-late bisimulation $(R_s)_{s \in \mathcal{I}}$ then X and Y are related by a span of surjective open maps.*

Combining these gives:

Proposition 10. *Two presheaves X and Y over $\mathsf{P}(s)$ are P-late bisimilar if and only if they are connected by a span of surjective open maps.*

Moving to a larger set of free names does not affect P-late bisimilarity.

Proposition 11 (Weakening). *If $X, Y \in \widehat{\mathsf{P}(s)}$ are P-late bisimilar then so are $\mathsf{P}(i)_!(X)$ and $\mathsf{P}(i)_!(Y)$ for any injection $i : s \to s'$.*

Moving to smaller name sets is a little more complicated. For any $i : s \to s'$ in $\mathcal{I}$ define $e_i : \mathsf{P}(s) \to \mathsf{P}(s')$ by induction as follows (omitting the trivial base cases):

$$e_i(\tau.p) = \tau.e_i(p) \qquad e_i(x!y.p) = i(x)!i(y).e_i(p)$$

$$e_i(x!*_s.p) = i(x)!*_{s'}.e_{i+1}(p) \qquad e_i(x?(y \mapsto p)) = i(x)?(i(y) \mapsto e_i(p))$$

$$e_i(x?(*_s \mapsto p')) = i(x)?(*_{s'} \mapsto e_{i+1}(p)) .$$

This differs from $\mathsf{P}(i)$ in having a much simpler action on input of unknowns $x?(*{\mapsto}p')$. Even so e_i^*, which by Proposition 1 preserves open maps, turns out to be a left inverse to $\mathsf{P}(i)_!$. This allows us to prove the following result.

Proposition 12 (Strengthening). *For $X, Y \in \widehat{\mathsf{P}(s)}$ and $i : s \to s'$ in $\mathcal{I}$, if $\mathsf{P}(i)_!(X)$ and $\mathsf{P}(i)_!(Y)$ are P-late bisimilar over $\mathsf{P}(s')$ then so are X and Y over $\mathsf{P}(s)$.*

These results suggest that we could have imposed similar uniformity constraints on the family $(R_s)_{s \in \mathcal{I}}$ in Definition 7: we conjecture that without loss of generality we can require that $\mathsf{R} \rightarrowtail\!\!\!\!\!\!+ (\mathsf{P} \,\&\, \mathsf{P})$ be a cartesian subobject in $\mathcal{P}rof^{\mathcal{I}}$.

3 Constructions

3.1 A Restriction Operator

We define here the operator that will be used to interpret name restriction in π-calculus processes. It arises as a natural family of profunctor arrows indexed by finite sets s and their elements:

$$\nu_{y \in s} : \mathsf{P}(s) \,\,\dashrightarrow\!\!\!\!\!\!\!+\,\, \mathsf{P}(s - \{y\}) .$$

In particular we can observe that the family $(\nu_{* \in s+1})_s$ define a natural transformation $\nu : \delta(\mathsf{P}) \to \mathsf{P}$.

We define the $\nu_{y \in s}$ simultaneously for all s by induction on the structure of the paths. So for each path in $\mathsf{P}(s)$, according to its structure:

$$\nu_{y \in s}(\tau.p) = \tau.\nu_{y \in s}(p)$$

$$\nu_{y \in s}(x!z.p) = \begin{cases} x!z.\nu_{y \in s}(p) & \text{if } x, z \neq y \\ x!*_{s-\{y\}}.\mathsf{P}(b_{s,y})(p) & \text{if } x \neq y \text{ and } z = y \\ \emptyset & \text{otherwise} \end{cases}$$

$$\nu_{y\in s}(x!*_s.p') = \begin{cases} x!*_{s-\{y\}}.\nu_{y\in s+1}(p') & \text{if } x \neq y \\ \emptyset & \text{otherwise} \end{cases}$$

$$\nu_{y\in s}(x?(z \mapsto p)) = \begin{cases} x?(z \mapsto \nu_{y\in s}(p)) & \text{if } x, z \neq y \\ \emptyset & \text{otherwise} \end{cases}$$

$$\nu_{y\in s}(x?(*_s \mapsto p') = \begin{cases} x?(*_{s-\{y\}} \mapsto \nu_{y\in s+1}(p')) & \text{if } x \neq y \\ \emptyset & \text{otherwise} \end{cases}$$

where $b_{s,y} : s \to (s-\{y\}) \cup \{*_{s-\{y\}}\}$ is the bijection that renames y to $*_{s-\{y\}}$. The base cases are the evident instances of these for null p: for example $\nu_{y\in s}(\tau.) = \tau$.

The only complication in this definition is the clauses that ensure restriction correctly turns free output into bound output, as summarised in this result:

Lemma 13. *Let X be a presheaf over $\mathsf{P}(s)$ and let $x, y \in s$. If $X \xrightarrow{x!y} X'$, then $\nu_{y\in s,!}(X) \xrightarrow{x!*_{s-\{y\}}} \mathsf{P}(b_{s,y})_!(X')$.*

Observe that if y, z are two different elements of a set s, then

$$\nu_{z\in s-\{y\}} \circ \nu_{y\in s} = \nu_{y\in s-\{z\}} \circ \nu_{z\in s}.$$

This suggests a definition of ν as a contravariant functor from $\mathcal{I}$ to $\mathcal{P}rof$ with $\nu(s) = \mathsf{P}(s)$ and $\nu(i) : \mathsf{P}(s') \nrightarrow \mathsf{P}(s)$ the restriction, in any order, of the elements of $(s' \setminus Im(i))$.

3.2 Parallel Composition

We give an inductive definition for the parallel composition of two presheaves over $\mathsf{P}(s)$ based on the decomposition result of Section 2.1. We then outline how the same definition arises from an interleaving operation on paths, and so deduce that surjective open maps are preserved.

Definition 14. Let X and Y be two presheaves over $\mathsf{P}(s)$ with the respective decompositions indexed by i, j, k, l and i', j', k', l'. Define $X \|_s Y$, inductively as follows, where $i : s \to s+1$ below is the obvious inclusion function.

$$\begin{aligned}
&\sum_{i\in I} \tau.(X_i \|_s Y) + \sum_{x,y\in s} \sum_{j\in J_{x!y}} x!y.(X_j \|_s Y) + \sum_{x\in s} \sum_{k\in K_x} x!*_s.(X_k \|_{s+1} \mathsf{P}(i)_!(Y)) \\
&\qquad + \sum_{x\in s} \sum_{l\in L_x} x?\langle \lambda y.(X_l^y \|_s Y), X_l^* \|_{s+1} \mathsf{P}(i)_!(Y)\rangle \\
&+ \sum_{i'\in I'} \tau.(X \|_s Y_{i'}) + \sum_{x,y\in s} \sum_{j'\in J'_{x!y}} x!y.(X \|_s Y_{j'}) + \sum_{x\in s} \sum_{k'\in K'_x} x!*_s.(\mathsf{P}(i)_!(X) \|_{s+1} Y_{k'}) \\
&\qquad + \sum_{x\in s} \sum_{l'\in L'_x} x?\langle \lambda y.(X \|_s Y_{l'}^y), \mathsf{P}(i)_!(X) \|_{s+1} Y_{l'}^* \rangle \\
&+ \sum_{x,y\in s} \sum_{j\in J_{x!y}} \sum_{l'\in L'_x} \tau.(X_j \|_s Y_{l'}^y) + \sum_{x\in s} \sum_{k\in K_x} \sum_{l'\in L'_x} \tau.\nu_{*\in s+1,!}(X_k \|_{s+1} Y_{l'}^*) \\
&+ \sum_{x,y\in s} \sum_{j'\in J'_{x!y}} \sum_{l\in L_x} \tau.(X_l^y \|_s Y_{j'}) + \sum_{x\in s} \sum_{k'\in K'_x} \sum_{l\in L_x} \tau.\nu_{*\in s+1,!}(X_l^* \|_{s+1} Y_{k'})
\end{aligned}$$

Much as with the prefixing operations of Definition 4, a more systematic approach begins with the profunctor arrow

$$\|_{s,\perp} : \mathsf{P}(s)_\perp \otimes \mathsf{P}(s)_\perp \nrightarrow \mathsf{P}(s)_\perp$$

that interleaves pairs of paths. Judicious use of Kan extension and the lifting maps of Proposition 2 give the parallel composition of presheaves as a bifunctor

$$\|_s : \widehat{\mathsf{P}(s)} \times \widehat{\mathsf{P}(s)} \longrightarrow \widehat{\mathsf{P}(s)}$$

with the following preservation property:

Proposition 15. *Let X, Y, Z, W be presheaves over $\mathsf{P}(s)$. If maps $f : X \to Z$ and $g : Y \to W$ are surjective open, then so is $f\|_s g : X\|_s Y \to Z\|_s W$.*

3.3 Replication

In the same way that the operational semantics of π-calculus replication vanishes into structural congruence, its presheaf interpretation is built entirely from arrows already at hand. Given any small category P, define the "replicated" category P^∞ by solving the following recursive equation[4]

$$\mathsf{P}^\infty = \mathbf{1} \,\&\, (\mathsf{P} \otimes \mathsf{P}^\infty).$$

This P^∞ is easily calculated to be equivalent to $\sum_{n\in\omega} \mathsf{P}^n$, where $\mathsf{P}^0 = \mathbf{1}$ and $\mathsf{P}^{n+1} = \mathsf{P} \otimes \mathsf{P} \cdots \otimes \mathsf{P}$, $(n+1)$ times. Taking now the diagonal and parallel composition maps at $\mathsf{P}(s)$ we can inductively define replicated versions

$$d_s^\infty : \widehat{\mathsf{P}(s)} \to \widehat{\mathsf{P}^\infty(s)} \quad \text{and} \quad \|_s^\infty : \widehat{\mathsf{P}^\infty(s)} \to \widehat{\mathsf{P}(s)}.$$

Replication itself is simply their composition $!_s = \|_s^\infty \circ d_s^\infty$, and the inductive definition ensures for any presheaf X over $\mathsf{P}(s)$ that $!_s(X) = X \|_s !_s(X)$.

4 The Interpretation

4.1 Semantics

Following (Stark 1996), we give the interpretation to process terms in two steps. First we associate a process P with free names in s to a presheaf $(\!|P|\!)_s \in |\widehat{\mathsf{P}(s)}|$. Then later, in the full interpretation, we take account of all possible name substitutions by giving a process P with free names s a denotation as a natural transformation:

$$[\![P]\!] : N^{|s|} \stackrel{\cdot}{\longrightarrow} \mathsf{P}\,.$$

---

[4] Note the relationship beween the way P^∞ is defined and the way the linear logic ! can be defined in presence of infinite products.

Let s be a set of names. For π-calculus processes whose free names lie in s we inductively define:

$$(\!|0|\!)_s = \emptyset \qquad (\!|P+Q|\!)_s = (\!|P|\!)_s + (\!|Q|\!)_s \qquad (\!|[x=x]P|\!)_s = (\!|P|\!)_s$$
$$(\!|\bar{x}y.P|\!)_s = x!y.(\!|P|\!)_s \qquad (\!|P \mid Q|\!)_s = (\!|P|\!)_s \|_s (\!|Q|\!)_s \qquad (\!|[x=y]P|\!)_s = \emptyset \text{ if } (x \neq y)$$
$$(\!|!P|\!)_s = !_s((\!|P|\!)_s) \qquad (\!|\nu x\, P|\!)_s = \nu_{x \in s+\{x\}}((\!|P|\!)_{s+\{x\}})$$
$$(\!|x(y).P|\!)_s = x?\langle F, Y\rangle \text{ where } F(z) = (\!|P[z/y]|\!)_s \text{ for any } z \in s, \text{ and } Y = (\!|P[*_s/y]|\!)_{s+1}.$$

Lemma 16. *Let $i : s \to s'$ be an injective function between finite sets, with $\mathbf{x} = \langle x_1, x_2, \ldots, x_{|s|}\rangle$ the names in s. Then for any process P using these names,*

$$\mathsf{P}(i)_!((\!|P|\!)_s) \cong (\!|P[i(\mathbf{x})/\mathbf{x}]|\!)_{s'} .$$

The free names of a process may be bound differently in different contexts. To cope with this, we interpret a process P with $|s|$ free names as a natural transformation $[\![P]\!] : \mathsf{N}^{|s|} \stackrel{\cdot}{\longrightarrow} \mathsf{P}$, where

$$[\![P]\!]_{s'} : \overbrace{s' \times s' \cdots \times s'}^{|s|-\text{times}} \longrightarrow \mathsf{P}(s')$$
$$\langle a_1, a_2, \ldots, a_{|s|}\rangle \longmapsto (\!|P[\mathbf{a}/\mathbf{x}]|\!)_{s'}$$

Thus the denotation of a process with free names s carries an environment $N^{|s|}$ as a parameter. The proof that this is indeed a natural transformation depends on Lemma 16, that the $(\!|-|\!)$-interpretation respects name substitution.

4.2 Full Abstraction

We can now show our major result, that bisimulation between processes in the π-calculus coincides with that obtained in the model via open maps.

The first two propositions establish a bisimulation between a process P with free names in s and its denotation $(\!|P|\!)_s$.

Proposition 17. *Let P be a process whose free names lie in s. Then*

- $P \xrightarrow{\bar{x}y} Q$ *implies* $(\!|P|\!)_s \xrightarrow{x!y} (\!|Q|\!)_s$
- $P \xrightarrow{\bar{x}(y)} Q$ *implies* $(\!|P|\!)_s \xrightarrow{x!*_s} (\!|Q[*_s/y]|\!)_{s+1}$
- $P \xrightarrow{x(y)} Q$ *implies* $(\!|P|\!)_s \xrightarrow{x?} \langle F, Y\rangle$ *with* $F(z) \cong (\!|Q[z/y]|\!)_s$ *and* $Y \cong (\!|Q[*_s/y]|\!)_{s+1}$
- $P \xrightarrow{\tau} Q$ *implies* $(\!|P|\!)_s \xrightarrow{\tau} (\!|Q|\!)_s$

Proposition 18. *Let P be a process whose free names lie in s. Then*

- $(\!|P|\!)_s \xrightarrow{x!y} X$ *implies* $\exists Q$ *with* $P \xrightarrow{\bar{x}y} Q$ *and* $(\!|Q|\!)_s \cong X$
- $(\!|P|\!)_s \xrightarrow{x!*_s} X$ *implies* $\exists Q, y$ *with* $P \xrightarrow{\bar{x}(y)} Q$ *and* $(\!|Q[*_s/y]|\!)_{s+1} \cong X$

- $(\!|P|\!)_s \xrightarrow{x?} \langle F, Y\rangle$ *implies* $\exists Q, y$ *with* $P \xrightarrow{x(y)} Q$ *and* $(\!|Q[*_s/y]|\!)_{s+1} \cong Y$ *and* $F(z) \cong (\!|Q[z/y]|\!)_s$
- $(\!|P|\!)_s \xrightarrow{\tau} X$ *implies* $\exists Q$ *with* $P \xrightarrow{\tau} Q$ *and* $(\!|Q|\!)_s \cong X$.

Using these results and Proposition 10 we can deduce the following.

Theorem 19. *Let P and Q be two π-calculus processes with free names in s. Then P is late bisimilar to Q if and only if $(\!|P|\!)_s$ and $(\!|Q|\!)_s$ are connected by a span of surjective open maps.*

Suppose now that P is a π-calculus process with free names s_P. Then for any larger set of names s, an injection $i : s_P \to s$ induces a natural transformation $\pi^{s_P,i} : \mathsf{N}^{|s|} \dot{\to} \mathsf{N}^{|s_P|}$ that projects $|s|$-tuples of names to $|s_P|$-tuples. When i is simply an inclusion and no confusion arises we write this as π^{s_P}.

Theorem 20. *Let P and Q be two π-calculus processes with free names s_P and s_Q respectively. Take $s_{P,Q}$ to be the union $s_P \cup s_Q$. Then P is late equivalent (bisimulation congruent) to Q if and only if for any finite set s and any $|s_{P,Q}|$-tuple $\mathbf{a}$ of elements of s, $[\![P]\!]_s \pi_s^{s_P}(\mathbf{a})$ and $[\![Q]\!]_s \pi_s^{s_Q}(\mathbf{a})$ are connected by a span of surjective open maps.*

Note that it is sufficient here to take s to be exactly the free names $s_{P,Q}$ of the two processes. We can also present this result using the 2-categorical setting of our model:

Corollary 21. *Let P and Q be two π-calculus processes with free names s_P and s_Q respectively. Then P is late equivalent to Q if and only if $[\![P]\!] \circ \pi^{s_P}$ and $[\![Q]\!] \circ \pi^{s_Q}$ are connected by a span of modifications whose components are surjective open maps.*

These results show a precise correspondence between operational and denotational notions of process equivalence; notice though that we do not interpret equivalent process by *equal* elements in the model. This could be arranged, by quotienting the categories $\widehat{\mathsf{P}(s)}$ to make every open map invertible, but there seems little reason to do so. In particular this is not likely to be very well-behaved as open maps do not form a calculus of fractions.

5 Other Results and Future Work

The model we have described is just one drawn from a spectrum of name-passing process calculi that can be described within $\mathcal{P}rof^{\mathcal{I}}$. We outline some possibilities.

5.1 Late *vs.* Early

We have given the π-calculus here in its late version, where a process $x(y).P$ carries out input in two stages: it first synchronizes with another process that is prepared to send on channel x; then, later, the transmitted value is substituted for y in the body of P. There is an alternative *early* semantics where these two steps happen together and processes synchronize on (channel,value) pairs. The operational consequences of this choice are discussed in (Milner et al. 1992b, §2.3). There is a corresponding early bisimulation '$\dot{\sim}_E$' and early equivalence '$\sim_E$', which are both strictly coarser than their late forms.

We can follow these late and early alternatives in our denotational semantics. In presheaf models, synchronization points are marked by lifting $(-)_\perp$ in the equation for the path category. An early version of (5) would be

$$In_E = N \otimes (N \multimap \mathsf{P}_\perp) \tag{7}$$

where instead of paths $x?$, $x?(y{\mapsto}p)$ and $x?(*{\mapsto}p')$ we now have $x?y$, $x?*$, $x?y.p$ and $x?*.p'$. Solving this new equation in $\mathcal{P}rof$ gives an object P_E, and we conjecture that this provides a fully abstract model for the early π-calculus, for suitably adjusted interpretation functions $(\!|-|\!)^E$ and $[\![-]\!]^E$.

More directly, this early model seems to arise as a collapse of the late one. A recursively defined morphism of path categories, $k : \mathsf{P}_E \to \mathsf{P}$, induces an arrow in $\mathcal{P}rof^{\mathcal{I}}$

$$k^\star : \mathsf{P} \multimap\!\!\!\!\multimap \mathsf{P}_E \tag{8}$$

which we conjecture maps the late semantics for processes into the early one:

$$k^\star((\!|P|\!)_s) = (\!|P|\!)_s^E \in \mathsf{P}_E(s) \qquad\qquad k^\star \circ [\![P]\!]_s = [\![P]\!]_s^E : N^s \to \mathsf{P}_E$$

for any process P with free names in s.

We can even move the synchronization point for output: clause (4) has a later variant

$$Out_L = N \otimes (N \otimes \mathsf{P})_\perp . \tag{9}$$

It turns out that this makes no difference to process bisimilarity, but it does correspond closely to the presentation style of (Milner 1991). There processes synchronize on channel names alone, $P \xrightarrow{\bar{x}} C$ or $P \xrightarrow{x} F$, becoming *concretions* C (name-process pairs, $N \otimes \mathsf{P}$) and *abstractions* F (name-to-process functions, $N \multimap \mathsf{P}$) respectively. Actual communication is represented by the application of abstractions to concretions $F \bullet C$.

The domain models of the π-calculus in (Fiore et al. 1996; Stark 1996) do not cover the early version, chiefly because rearrangements like equation (7) are harder to express. There the domain equation for processes uses the Plotkin powerdomain to mark synchronization; while our equation for paths uses the much simpler lifting operation.

5.2 Other π-calculi

The development of our model has been purely denotational, with no operational manipulation of processes through expansion laws or the like. As a consequence, there are no *required* operators in the language, and the model remains valid for any subset of the π-calculus. Even so, particular sublanguages may fit simpler equations. For example, the asynchronous π-calculus of Boudol (1992) constrains output to the form $\bar{x}y.0$, suggesting the clause

$$Out_A = N \otimes N \tag{10}$$

to replace (4). The πI-calculus of Sangiorgi (1995) allows only bound output $\bar{x}(y).P$, equivalence to $\nu y(\bar{x}y.P)$ in the original π-calculus. Every communication now passes a fresh name, and we would replace (5) and (4) with

$$In_I = Out_I = N \otimes \delta \mathsf{P} . \tag{11}$$

Moreover the morphism $\mathsf{P}(i)$ now arises from the category map e_i introduced just before Proposition 12, with the restrictions $\nu(i)$ from the end of Section 3.1 being $e_i^\star$, the left inverse, and now also right adjoint, to $\mathsf{P}(i)_!$. This gives some support to Sangiorgi's claim that the πI-calculus is a simpler, more symmetric version of the π-calculus.

These examples show the flexibility of our approach by drawing on the rich categorical structure of $\mathcal{P}rof^{\mathcal{I}}$. As ever in category theory, this also leads us to look at the maps *between* models: we hope to find further morphisms like (8), from 'late' to 'early', that might tie together the wide selection of customized π-calculi proposed in recent years.

We do not have a general function space in $\mathcal{P}rof^{\mathcal{I}}$ so variability of names together with process-passing in systems like CHOCS (Thomsen 1993) or even the full higher-order π-calculus of Sangiorgi (1992) do not yet fit into our framework. Generally, the problem with treating higher-order processes lies not just in writing down and solving plausible "domain" equations, but also in understanding the operational content of the semantics and bisimilarities that arise.

Bibliography

Francis Borceux. *Handbook of Categorical Algebra, vol. 1*, volume 50 of *Encyclopedia of Mathematics and its Applications.* Cambridge University Press, 1994.

Gérard Boudol. Asynchrony and the π-calculus. Rapport de recherche 1702, INRIA, Sophia Antipolis, 1992.

Gian Luca Cattani and Glynn Winskel. Presheaf models for concurrency. In *Computer Science Logic 1996*, To appear in Lecture Notes in Computer Science. Springer-Verlag, 1997. A preliminary version appeared as BRICS Report RS-96-35.

Brian J. Day. On closed categories of functors. In *Reports of the Midwest Category Seminar IV*, Lecture Notes in Mathematics 137, pages 1–38. Springer-Verlag, 1970.

Marcelo Fiore, Eugenio Moggi, and Davide Sangiorgi. A fully-abstract model for the π-calculus. In (LICS 1996).

Matthew Hennessy. A fully abstract denotational semantics for the π-calculus. Technical Report 96:04, School of Cognitive and Computing Sciences, University of Sussex, 1996.

Matthew Hennessy and Gordon Plotkin. Full abstraction for a simple parallel programming language. In *Mathematical Foundations of Computer Science: Proceedings of the 8th International Symposium*, Lecture Notes in Computer Science 74, pages 108–120. Springer-Verlag, 1979.

André Joyal and Ieke Moerdijk. A completeness theorem for open maps. *Annals of Pure and Applied Logic*, 70:51–86, 1994.

André Joyal and Ieke Moerdijk. *Algebraic Set Theory*, volume 220 of *London Mathematical Society Lecture Note Series*. Cambridge University Press, 1995.

André Joyal, Mogens Nielsen, and Glynn Winskel. Bisimulation from open maps. *Information and Computation*, 127(2):164–185, 1996.

LICS 1996. *Proceedings of the Eleventh Annual IEEE Symposium on Logic in Computer Science, New Brunswick, New Jersey, July 27–30, 1996.* IEEE Computer Society Press, 1996.

Robin Milner. The polyadic π-calculus — a tutorial. Technical Report ECS-LFCS-91-180, Laboratory for Foundations of Computer Science, University of Edinburgh, 1991.

Robin Milner. Functions as processes. *Mathematical Structures in Computer Science*, 2(2):119–141, 1992.

Robin Milner, Joachim Parrow, and David Walker. A calculus of mobile processes, part I. *Information and Computation*, 100:1–40, 1992a.

Robin Milner, Joachim Parrow, and David Walker. A calculus of mobile processes, part II. *Information and Computation*, 100:41–77, 1992b.

Davide Sangiorgi. *Expressing Mobility in Process Algebras: First-Order and Higher-Order Paradigms.* PhD Thesis CST-99-93, Department of Computer Science, University of Edinburgh, 1992.

Davide Sangiorgi. π-calculus, internal mobility, and agent-passing calculi. Rapport de recherche 2539, INRIA, Sophia Antipolis, 1995.

Ian Stark. A fully abstract domain model for the π-calculus. In (LICS 1996), pages 36–42.

Bent Thomsen. Plain CHOCS. *Acta Informatica*, 30:1–59, 1993.

Glynn Winskel. A presheaf semantics of value-passing processes. In *CONCUR '96: Proceedings of the 7th International Conference on Concurrency Theory*, Lecture Notes in Computer Science 1119, pages 98–114. Springer-Verlag, 1996. An extended version appears as BRICS Report RS-96-44.

Glynn Winskel and Mogens Nielsen. Models for concurrency. In *Handbook of Logic in Computer Science*, volume IV. Oxford University Press, 1995.

Categorical Modelling of Structural Operational Rules

Case Studies

Daniele Turi
<dt@dcs.ed.ac.uk>

Department of Computer Science
Laboratory for Foundations of Computer Science
University of Edinburgh, The King's Buildings
Edinburgh EH9 3JZ, Scotland

Abstract. This paper aims at substantiating a recently introduced categorical theory of 'well-behaved' operational semantics. A variety of concrete examples of structural operational rules is modelled categorically illustrating the versatility and modularity of the theory. Further, a novel functorial notion of guardedness is introduced which allows for a general and formal treatment of guarded recursive programs.

Introduction

The predominant approach to operational semantics is Plotkin's *SOS* [13], which is based on structural rules. One finds in the literature various formats of structural rules which guarantee a good behaviour such as having adequate denotational models and behavioural equivalence (eg bisimulation) being a congruence.

In [17], it is shown that the rules in the best known of these formats, namely *GSOS* [5], are in 1-1 correspondence with natural transformations of a suitable type, depending on specific *functorial* notions of syntax and behaviour. This led to studying abstract rules as natural transformations parametric in the syntax and behaviour, and to a corresponding general theory which elegantly explains the good behaviour of the (abstract) rules in terms of *distributive laws* of monads over comonads.

The present study aims at substantiating the generality of the above theory. One way to do this would be to find formats other than GSOS which fit in the theory. But here a different direction is taken in order to illustrate a key feature of the theory: *modularity*. A few elementary examples of rules such as those for sequential and for parallel composition of programs are considered and the parametricity of the theory is exploited in order to distil abstract rules, parametric in the behaviour, which – by retaining only the essential, intrinsic structure – can be instantiated to various situations.

* Research supported by *EuroFOCS* and the European Union *TMR* programme.

For instance, it is shown how, by varying the choice of the behaviour, the same rules can be interpreted both deterministically and non-deterministically, both in uniform (ie with atomic actions) and non-uniform (ie with side-effects) settings, both with respect to bisimulation and to trace equivalence. Moreover the same rules are interpretable in several categories, such as categories of sets, cpos, metric spaces, semi-lattices, etc. All this while retaining the desirable properties typical of GSOS.

Languages without variable binding, but possibly multi-sorted, can be described as Σ-algebras [6]. This gives rise to a well-known functorial notion of syntax, parametric in the language constructs [2]. It is less clear what a general functorial notion of behaviour is. Here some examples are given; interestingly, most of them are *computational monads* in the sense of [12], but only the fact that the functors are strong is relevant.

Further, a novel functorial notion of *guard*, again parametric in syntax and behaviour, is introduced. This strongly generalizes the specific and somewhat informal notion of guardedness as used for defining recursive programs in languages à la *CCS* [10]. Unguarded recursion also fits nicely in this framework, interpreting the rules in Kleisli categories of lifting monads; but for lack of space this is left for a future presentation.

Contents. After a first section of preliminaries, Section 2 recalls the definition of abstract operational rules from [17], while the next four sections illustrate how a wide range of examples of structural rules can be modelled in such an abstract form. Following [17], Section 7 shows how both an operational and a denotational model can be canonically derived from the abstract rules; a simple process algebra is used as example and interpreted with respect to both bisimulation and trace equivalence. The last section is devoted to guarded recursion.

Thanks to: Gordon Plotkin for joint foundational work; Marcelo Fiore and Alex Simpson for discussions. Part of this work is based on the author's thesis; he wishes to thank Jaco de Bakker and Bart Jacobs for their guidance. Diagrams drawn using *XFig*.

1 Preliminaries and Notation

A non-standard notation first: given two maps $f : X \to Y$ and $g : Y \to Z$ in a category, the expression

$$\frac{\begin{array}{l} f : X \longrightarrow Y \\ g : \quad\;\; \longrightarrow Z \end{array}}{h : X \longrightarrow Z}$$

says that the map $h : X \to Z$ is defined to be the composite $g \circ f$. In general, the objects of a category are ranged over by X and Y. In **Set**, X itself is ranged over by x, y, and z.

Initial and final (ie terminal) objects are written as 0 and 1 respectively and the unique arrow from an object X to 1 is denoted by $! : X \to 1$. The **power** X^A denotes the product of $|A|$-copies of X. The i-th injection (projection) of an n-ary coproduct (product) is denoted by ι_i (π_i). For every set A, the **copower** $A \cdot X$ denotes the coproduct of $|A|$-copies of X, hence it comes with an injection $\iota_a : X \to A \cdot X$ for each a in A; if X is a set, $\iota_a(x)$ is more suggestively written as $a \cdot x$. In Set, the expression $a \cdot x$ can also be understood as the pair $\langle a, x \rangle$ since the copower $A \cdot X$ is isomorphic to the product $A \times X$; also, since the power X^A is isomorphic (in Set) to the set $\mathsf{Set}(A, X)$ of functions of type $A \to X$, the 'thick lambda' notation $\boldsymbol{\lambda} a.x$ is used to denote its elements.

The associativity, symmetry, and unit natural isomorphisms of a symmetric monoidal category (**SMC**) $\langle \mathcal{C}, \otimes, I \rangle$ are denoted by α, γ, and λ respectively. (The latter is often omitted.) The symbol $X^{\otimes n}$ denotes the n-fold monoidal power of X.

All categories in this paper are SMCs with finite products and coproducts and with tensors $\otimes$ distributing over coproducts; the distributivity natural isomorphism is denoted by δ. Two models are used, namely **Set** and the category $SL(\mathsf{Set})$ of semi-lattices and join-preserving functions, but other categories used in semantics, such as categories of cpos and of metric spaces, could have been chosen. In **Set** $\otimes$ and $\times$ coincide, while in $SL(\mathsf{Set})$ they are distinct ($\otimes$ is the classifier of bilinear functions). The categories $SL(\mathcal{C})$ considered here are **affine** SMCs, that is, the unit I of the tensor is the final object 1.

In order to ease the notation, natural transformations are written without subscript in this paper. For instance:

$$\lambda : I \otimes X \to X \cong X \quad \text{instead of} \quad \lambda_X : I \otimes X \to X \cong X \ .$$

Recall that a functor F between SMC's is **strong** [9] if there exists a **strength** natural transformation $st : X \otimes FY \to F(X \otimes Y)$ satisfying the unit and the associativity condition of [9]. For simplicity, the same symbol denotes also the map $F\gamma \circ st \circ \gamma : FX \otimes Y \to F(X \otimes Y)$. If F is a monad, there are two canonical **double strength** maps of type $FX \otimes FY \to F(X \otimes Y)$ (which are equal if the monad is commutative); here they are both denoted by dst. Finally, strong functors compose: $\langle F, st \rangle \circ \langle F', st' \rangle = \langle FF', F(st') \circ st \rangle$.

Semi-Lattices. Let $\mathcal{C}$ be a cartesian category. A **semi-lattice** in $\mathcal{C}$ is a pair $\langle X, \vee \rangle$ such that the 'carrier' X is an object of $\mathcal{C}$ and the 'join'

$$\vee : X \times X \to X$$

is a map of $\mathcal{C}$ satisfying the following three axioms.

$$\vee \circ (id \times \vee) = \vee \circ (\vee \times id) \circ \alpha \ , \qquad \vee = \vee \circ \gamma \ , \qquad \vee \circ \langle id, id \rangle = id \ .$$

If $\mathcal{C}$ is Set, then these axioms read as $x \vee (y \vee z) = (x \vee y) \vee z$, $x \vee y = y \vee x$, and $x \vee x = x$ respectively, for all x, y, z in X.

The category $SL(\mathcal{C})$ has semi-lattices in $\mathcal{C}$ as objects; the maps are the evident join-preserving maps of $\mathcal{C}$, also called linear maps.

Semi-lattices are used here to model non-deterministic programs, binary (hence finite, non-empty) joins modelling binary (and finite, non-empty) choices. Several variations can be considered according to the cardinality of the non-determinism available in the language under study. For instance might consider semi-lattices with a neutral element

$$e : 1 \to X$$

such that $id \cong \vee \circ (id \times e)$, which in Set reads as $x \vee \bot = x$. The corresponding category of finite joins semi-lattices in $\mathcal{C}$ is denoted by $SL_*(\mathcal{C})$.

Given a cartesian category $\mathcal{C}$, the corresponding **free semi-lattice monad** $\mathcal{P}_\mathrm{f}$ exists if the forgetful functor $\langle X, \vee\rangle \mapsto X$ mapping a semi-lattice to its carrier has a left adjoint. Similarly, $\mathcal{P}_\mathrm{fi}$ denotes the monad corresponding to $SL_*(\mathcal{C})$. In the sequel, it is assumed that the forgetful functor of semi-lattices is monadic and that both $\mathcal{P}_\mathrm{f}$ and $\mathcal{P}_\mathrm{fi}$ are commutative.

In Set, the monads $\mathcal{P}_\mathrm{f}$ and $\mathcal{P}_\mathrm{fi}$ are the (covariant) non-empty finite power-set and finite power-set respectively; for both, the unit is the function $x \mapsto \{x\}$ and the multiplication is the 'big union' function. Also, if $\mathcal{C}$ is the category of ω-algebraic cppos and continuous functions, then $\mathcal{P}_\mathrm{f}$ is the Plotkin powerdomain. All these are examples of commutative monads. In Set, the double strength $dst : \mathcal{P}_\mathrm{f}X \times \mathcal{P}_\mathrm{f}Y \to \mathcal{P}_\mathrm{f}(X \times Y)$ of $\mathcal{P}_\mathrm{f}$ is $\langle \{x_i\}_I, \{y_j\}_J\rangle \mapsto \{\langle x_i, y_j\rangle\}_{I,J}$.

Algebras and Coalgebras. Given any endofunctor F on $\mathcal{C}$ (ie $F : \mathcal{C} \to \mathcal{C}$), an F-**algebra** is a pair $\langle X, h\rangle$, where the carrier X is an object and the structure $h : FX \to X$ is a map of $\mathcal{C}$. One often identifies an algebra $\langle X, h\rangle$ with its structure h. The F-algebras form a category F-Alg, with homomorphisms $f : \langle X, h\rangle \to \langle X', h'\rangle$ the maps $f : X \to X'$ between the carriers such that $f \circ h = h' \circ (Ff)$. The carrier of the initial algebra (if it exists) is denoted by $\mu X.FX$; by Lambek's Lemma (see, eg, [15, Lemma 1]), its structure is an isomorphism.

Consider the evident forgetful functor $U^F : F\text{-Alg} \to \mathcal{C}$: one can easily prove that if it has a left adjoint, then it is monadic. The corresponding monad T is called the **monad freely generated by** F. If $\mathcal{C}$ has finite coproducts, then $TX = \mu Y.X + FY$ for every object X of $\mathcal{C}$. A clash of notation: the same symbol μ is used for the initial algebra constructor and for the multiplication $\mu : T^2 \to T$ of the monad T.

Dually, one has a category F-Coalg of F-**coalgebras** $k : X \to FX$ with homomorphisms $f : \langle X, k\rangle \to \langle X', k'\rangle$ maps $f : X \to X'$ between the carriers such that $k' \circ f = (Ff) \circ k$. We write $\nu X.FX$ for the carrier of the final coalgebra and $\varphi : \nu X.FX \cong F(\nu X.FX)$ for its structure which, by the dual of Lambek's Lemma, is an isomorphism.

A Functorial Notion of Syntax. Consider the language with signature Σ consisting of a distinguished constant symbol 'nil', a set A of constant symbols ranged over by a, and two binary operators ' ; ' and '∥'. This signature freely generates, for every set X of variables ranged over by x, the set of terms t given by the following abstract grammar:

$$t ::= x \mid \mathsf{nil} \mid a \mid t \,;\, t \mid t \parallel t \ . \qquad (1)$$

One can model the above signature in any monoidal category with coproducts $\langle \mathcal{C}, \otimes, I \rangle$ as an endofunctor with the same name:

$$\Sigma X = I + A \cdot I + X \otimes X + X \otimes X \ . \tag{2}$$

For $\langle \mathcal{C}, \otimes, I \rangle = \langle \mathsf{Set}, \times, 1 \rangle$, one has that $TX = \mu Y.X + \Sigma Y$ is the above set of terms.

In general, for any signature Σ, one can define an endofunctor

$$\Sigma X = \coprod_{\phi \in \Sigma} X^{\otimes arity(\phi)} \tag{3}$$

on $\langle \mathcal{C}, \otimes, I \rangle$. (Cf [2, §10] for the case $\mathcal{C}$ is cartesian.) In the sequel, $\mathcal{C}$ is assumed to be such that every such Σ freely generates a monad T (which $\mathcal{C}$ 'thinks' as the syntax of the language).

For every object X, the structure of the free Σ-algebra over X is denoted by:

$$\psi : \Sigma T X \to T X \ .$$

It is natural in X. Let ϕ by a binary operator of a signature Σ and let f and g be two maps of type $X \to TX$. The mixed syntax/semantics notation $\phi(f, g)$ is used here to denote the map $X \otimes X \xrightarrow{f \otimes g} TX \otimes TX \xrightarrow{\iota_\phi} \Sigma TX \xrightarrow{\psi} TX$. This generalizes to every operator of any arity.

The unit $\eta : X \to TX$ of the monad T is the insertion-of-variables map; in Set, this is the formal operation mapping a variable x in X to a term x in TX. Using the above notation, one can form the maps $\eta \,;\, \eta$ and $\eta \parallel \eta$ of type $X \otimes X \to TX$.

There is a similar functorial notion of syntax for multi-sorted signatures: it suffices to consider endofunctors on power categories $\mathcal{C}^n$, where n is the number of sorts. It remains a major open question whether also variable binding constructs can be modelled functorially.

A Functorial Notion of Behaviour. In *initial algebra semantics* [6] Σ-algebras are regarded as denotational models and the unique homomorphism from the initial algebra of programs to a denotational model is the canonical compositional interpretation relative to that model. Dually, in *final coalgebra semantics* [1,14] one looks for a 'behaviour' endofunctor B such that its coalgebras are operational models. In particular, the final B-coalgebra is a solution of the domain equation $X \cong BX$ – the canonical domain of interpretation for programs with behaviour of type B.

The typical example of a behaviour endofunctor is $BX = \mathcal{P}(A \cdot X)$ on Set. Its coalgebras are in 1-1 correspondence with relations of type $X \times A \times X$, ie with **labelled transition systems** [13]. Indeed, writing $x \overset{a}{\leadsto} x'$ for $\langle x, a, x' \rangle \in \leadsto$, one has:

$$\frac{\leadsto \subseteq X \times A \times X}{\overline{k \ : X \to \mathcal{P}_{fi}(A \cdot X)}} \qquad x \overset{a}{\leadsto} x' \iff a \cdot x' \in k(x) \tag{4}$$

Note that the following isomorphisms hold.

$$\mathcal{P}(A \cdot X) \;\cong\; \mathcal{P}(A \times X) \;\cong\; \mathsf{Set}(A, \mathcal{P}X) \;\cong\; (\mathcal{P}X)^A \;. \tag{5}$$

For cardinality reasons, $\mathcal{P}$ has no final coalgebra, therefore one usually replaces it by the finite power-set functor $\mathcal{P}_{\mathrm{fi}}$;the correspondence in (4) cuts down then to 'finitely branching' transition systems. The final coalgebra is then the set of rooted, finitely branching trees, with branches labelled by actions a in A, quotiented by strong bisimulation. (Cf [4].)

Here more examples are considered, taken from the following class of endofunctor, not only on Set but in any SMC with enough structure to define them; the definition is parametric in a set S of states or actions.

$$B ::= \mathit{Id} \mid 1 + B \mid S \cdot B \mid B^S \mid \mathcal{P}_{\mathrm{f}} B \mid \mathcal{P}_{\mathrm{fi}} B \tag{6}$$

This is by no means intended as an exhaustive list of behaviour endofunctors. For concrete behaviours, β is used to range over BX.

Convention: in the two models Set and $SL(\mathsf{Set})$, the initial algebra and final coalgebra isomorphisms as well as the monad operations η, μ and the coproduct injections ι_1, ι_2 are often omitted.

2 Abstract Operational Rules

In [17], it is shown that if the operational rules for a language with signature Σ and behaviour of type B can be modelled as a map

$$\rho : \Sigma(X \times BX) \longrightarrow BTX \tag{7}$$

natural in X, then the rules enjoy several desirable properties. Further, it is shown that for the specific behaviour endofunctor $BX = (\mathcal{P}_{\mathrm{fi}} X)^A$ on Set there is an (essentially) 1-1 correspondence between natural transformations as in (7) and 'image finite' sets of rules in *GSOS* format [5] (relative to a signature Σ and a finite set of actions A).

In the sequel, various examples of rules involving behaviours other than the one for GSOS are modelled as in (7). The endofunctor Σ is as in (3), therefore it suffices to model the rules for each operator ϕ separately as a map

$$[\![\phi]\!] : \underbrace{(X \times BX) \otimes \cdots \otimes (X \times BX)}_{arity(\phi)} \longrightarrow BTX$$

natural in X, and then define ρ as the coparing of such maps.

3 Exceptions and Side-Effects

As a first example, let us consider the rules for the sequential composition of programs involving a simple example of exceptions, namely explicit termination.

Furthermore, let us consider programs with side-effects, assuming a given set S of states ranged over by σ. The rules are:

$$\frac{\langle x, \sigma\rangle \rightsquigarrow \langle x', \sigma'\rangle}{\langle x\,;y, \sigma\rangle \rightsquigarrow \langle x'\,;y, \sigma'\rangle}\,, \qquad \frac{\langle x, \sigma\rangle \rightsquigarrow \sigma'}{\langle x\,;y, \sigma\rangle \rightsquigarrow \langle y, \sigma'\rangle}\,.$$

Let $\langle \mathcal{C}, \otimes, 1\rangle$ be an affine SMC, with finite products and coproducts, powers, and tensors distributing over coproducts. If the programs are deterministic, then one can use the endofunctor

$$BX = (S\cdot(1+X))^S \tag{8}$$

on $\mathcal{C}$ as behaviour. Note that in Set the isomorphism $S\cdot(1+X)\cong S+S\cdot X$ holds, but the expression on the left hand side can be interpreted also in categories where S is not an object (such as categories of semi-lattices and of pointed partial orders). Also, note that the endofunctor $X\mapsto(S\cdot X)^S$ is strong; in Set the corresponding strength $st:(S\cdot X)^S\times Y\to(S\cdot(X\times Y))^S$ is the map $\langle(\boldsymbol{\lambda}\sigma.(\sigma'\cdot x)), y\rangle\mapsto\boldsymbol{\lambda}\sigma.(\sigma'\cdot\langle x,y\rangle)$. Then sequential composition can be modelled as follows.

$$\begin{array}{rll}
\pi_2\otimes\pi_1 & : (X\times(S\cdot(1+X))^S)\otimes(X\times(S\cdot(1+X))^S) & \\
 & \qquad\longrightarrow (S\cdot(1+X))^S\otimes X & \\
st & : \qquad\longrightarrow (S\cdot((1+X)\otimes X))^S & \\
(S\cdot\delta)^S & : \qquad\cong\ (S\cdot(X+(X\otimes X))^S & \\
(S\cdot[\eta,(\eta\,;\eta)])^S & : \qquad\longrightarrow (S\cdot TX)^S & \\
(S\cdot\iota_2)^S & : \qquad\longrightarrow (S\cdot(1+TX))^S & \\
\hline
[\![\,;]\!] & : (X\times(S\cdot(1+X))^S)\otimes(X\times(S\cdot(1+X))^S) & \\
 & \qquad\longrightarrow (S\cdot(1+TX))^S &
\end{array}$$

Note that this map in natural because all its components are. The same holds for all the rules modelled in this paper.

In Set:

$$(x,\beta)[\![\,;]\!](y,\beta') = \boldsymbol{\lambda}\sigma.\begin{cases}\sigma'\cdot(x'\,;y) & \text{if } \beta(\sigma)=\sigma'\cdot x'\\ \sigma'\cdot y & \text{if } \beta(\sigma)=\sigma'\end{cases}\,.$$

4 Non-Determinism

Consider the following rules for the parallel composition operator of CCS [10], relative to a set A of actions ranged over by a, with a distinguished 'silent step' action τ, and a dual action $\overline{a}$ for every a in A different from τ.

$$\frac{x\overset{a}{\rightsquigarrow}x'}{x\parallel y\overset{a}{\rightsquigarrow}x'\parallel y}\,,\qquad \frac{y\overset{a}{\rightsquigarrow}y'}{x\parallel y\overset{a}{\rightsquigarrow}x\parallel y'}\,,\qquad \frac{x\overset{a}{\rightsquigarrow}x'\quad y\overset{\overline{a}}{\rightsquigarrow}y'}{x\parallel y\overset{\tau}{\rightsquigarrow}x'\parallel y'}\,. \tag{9}$$

One possible behaviour endofunctor for these rules is

$$BX = (\mathcal{P}_{\mathrm{fi}}X)^A \tag{10}$$

on a distributive category $\mathcal{C}$ with powers and a commutative free semi-lattice monad $\mathcal{P}_{fi}$. Note that $BX = (\mathcal{P}_{fi}X)^A$ inherits from $\mathcal{P}_{fi}X$ a finite semi-lattice structure, say $\vee : (\mathcal{P}_{fi}X)^A \times (\mathcal{P}_{fi}X)^A \to (\mathcal{P}_{fi}X)^A$ and $e : 1 \to (\mathcal{P}_{fi}X)^A$, natural in X. Further, B is the composition of strong endofunctors, hence it is strong.

Let us model each rule separately and then join them using the semi-lattice structure of BTX. The first rule can be modelled as:

$$\begin{array}{rl}
\pi_2 \times \pi_1 & : (X \times (\mathcal{P}_{fi}X)^A) \times (X \times (\mathcal{P}_{fi}X)^A) \longrightarrow (\mathcal{P}_{fi}X)^A \times X \\
st & : \qquad\qquad\qquad\qquad\qquad\qquad\qquad \longrightarrow (\mathcal{P}_{fi}(X \times X))^A \\
(\mathcal{P}_{fi}(\eta \parallel \eta))^A & : \qquad\qquad\qquad\qquad\qquad\qquad\qquad \longrightarrow (\mathcal{P}_{fi}TX)^A \\
\hline
\textit{l-merge} & : (X \times (\mathcal{P}_{fi}X)^A) \times (X \times (\mathcal{P}_{fi}X)^A) \longrightarrow (\mathcal{P}_{fi}TX)^A
\end{array}$$

The second rule is symmetric, hence it can be modelled as symply the composite $\textit{l-merge} \circ \gamma$. As for the third rule, let us denote its modelling by *synch*. Since the rule only contributes to the τ-component of $(\mathcal{P}_{fi}TX)^A$, we can take $\pi_a \circ \textit{synch}$ to be the composite map $(X \times (\mathcal{P}_{fi}X)^A) \times (X \times (\mathcal{P}_{fi}X)^A) \xrightarrow{!} 1 \xrightarrow{e} \mathcal{P}_{fi}TX$. Next, the τ-component of *synch* is the join over all actions a different from τ of the following family of maps $\textit{synch}_a$, where *dst* is the double strength of the commutative monad $\mathcal{P}_{fi}$. (Note that a join of cardinality $|A|$ is needed here, hence if A is infinite one needs to replace $\mathcal{P}_{fi}$ by a free monad for semi-lattices with joins of cardinality greater than or equal to $|A|$.)

$$\begin{array}{rl}
\pi_2 \times \pi_2 & : (X \times (\mathcal{P}_{fi}X)^A) \times (X \times (\mathcal{P}_{fi}X)^A) \longrightarrow (\mathcal{P}_{fi}X)^A \times (\mathcal{P}_{fi}X)^A \\
\pi_a \times \pi_{\overline{a}} & : \qquad\qquad\qquad\qquad\qquad\qquad\qquad \longrightarrow \mathcal{P}_{fi}X \times \mathcal{P}_{fi}X \\
dst & : \qquad\qquad\qquad\qquad\qquad\qquad\qquad \longrightarrow \mathcal{P}_{fi}(X \times X) \\
\mathcal{P}_{fi}(\eta \parallel \eta) & : \qquad\qquad\qquad\qquad\qquad\qquad\qquad \longrightarrow \mathcal{P}_{fi}TX \\
\hline
\textit{synch}_a & : (X \times (\mathcal{P}_{fi}X)^A) \times (X \times (\mathcal{P}_{fi}X)^A) \longrightarrow \mathcal{P}_{fi}TX
\end{array}$$

Then:

$$[\![\parallel]\!] = \textit{l-merge} \vee (\textit{l-merge} \circ \gamma) \vee \textit{synch} \ .$$

In Set:

$$\begin{array}{rl}
(x, \beta)[\![\parallel]\!](y, \beta') = & (a \mapsto (\{x' \parallel y \mid x' \in \beta(a)\} \cup \{x \parallel y' \mid y' \in \beta'(a)\})) \cup \\
& \cup (\tau \mapsto \{x' \parallel y' \mid \exists a : x' \in \beta(a),\, y' \in \beta'(\overline{a})\})
\end{array} .$$

Another simple example of an operator modellable in this setting is the usual binary choice operator 'or':

$$\begin{array}{rl}
\pi_2 \times \pi_2 & : (X \times (\mathcal{P}_{fi}X)^A) \times (X \times (\mathcal{P}_{fi}X)^A) \longrightarrow (\mathcal{P}_{fi}X)^A \times (\mathcal{P}_{fi}X)^A \\
\vee & : \qquad\qquad\qquad\qquad\qquad\qquad\qquad \longrightarrow (\mathcal{P}_{fi}X)^A \\
(\mathcal{P}_{fi}\eta)^A & : \qquad\qquad\qquad\qquad\qquad\qquad\qquad \longrightarrow (\mathcal{P}_{fi}TX)^A \\
\hline
[\![\mathsf{or}]\!] & : (X \times (\mathcal{P}_{fi}X)^A) \times (X \times (\mathcal{P}_{fi}X)^A) \longrightarrow (\mathcal{P}_{fi}TX)^A
\end{array}$$

In Set: $(x, \beta)[\![\mathsf{or}]\!](y, \beta') = \boldsymbol{\lambda} a.(\beta(a) \cup \beta'(a))$.

Negative Premises. In order to model rules with negative premises one can use the following variant of the above behaviour endofunctor:

$$BX = (1 + \mathcal{P}_{\mathrm{f}}X)^A \ . \tag{11}$$

(In Set, $1 + \mathcal{P}_{\mathrm{f}}X$ is isomorphic to $\mathcal{P}_{\mathrm{fi}}X$, but this is not true in general.)

As an example consider the following rules describing a unary operator '$\tau \gg$' which gives priority to τ-steps:

$$\frac{x \overset{\tau}{\rightsquigarrow} x'}{(\tau \gg x) \overset{\tau}{\rightsquigarrow} x'} \ , \qquad \frac{x \overset{\tau}{\not\rightsquigarrow} \quad x \overset{a}{\rightsquigarrow} x'}{(\tau \gg x) \overset{a}{\rightsquigarrow} x'} \ a \neq \tau \ . \tag{12}$$

For every action a different from τ, the second rule can be modelled as:

$$\begin{array}{rl} \pi_2 & : X \times (1 + \mathcal{P}_{\mathrm{f}}X)^A \longrightarrow (1 + \mathcal{P}_{\mathrm{f}}X)^A \\ \langle id, id \rangle & : \qquad\qquad\qquad\quad \longrightarrow (1 + \mathcal{P}_{\mathrm{f}}X)^A \times (1 + \mathcal{P}_{\mathrm{f}}X)^A \\ \pi_a \times \pi_\tau & : \qquad\qquad\qquad\quad \longrightarrow (1 + \mathcal{P}_{\mathrm{f}}X) \times (1 + \mathcal{P}_{\mathrm{f}}X) \\ \delta & : \qquad\qquad\qquad\quad \longrightarrow 1 + \mathcal{P}_{\mathrm{f}}X + \mathcal{P}_{\mathrm{f}}X + (\mathcal{P}_{\mathrm{f}}X)^2 \\ id + id + [!, !] & : \qquad\qquad\qquad\quad \longrightarrow 1 + \mathcal{P}_{\mathrm{f}}X + 1 \\ \gamma & : \qquad\qquad\qquad\quad \cong \ 1 + 1 + \mathcal{P}_{\mathrm{f}}X \\ [id, id] + \mathcal{P}_{\mathrm{f}}\eta & : \qquad\qquad\qquad\quad \longrightarrow 1 + \mathcal{P}_{\mathrm{f}}TX \\ \hline \pi_a \circ [\![\tau \gg]\!] & : (X \times (1 + \mathcal{P}_{\mathrm{f}}X)^A) \longrightarrow 1 + \mathcal{P}_{\mathrm{f}}TX \end{array}$$

while the first rule yields: $\pi_\tau \circ [\![\tau \gg]\!] = (1 + \mathcal{P}_{\mathrm{f}}\eta) \circ \pi_\tau \circ \pi_2$.

The examples in this section should convince the reader that GSOS rules can be modelled using the behaviour B as in (11) on any distributive category with powers and a commutative free semi-lattice monad $\mathcal{P}_{\mathrm{f}}$. (Cf [17, Theorem 2.1].)

5 Modularity

One of the advantages of the present approach to opertional semantics is that, by decomposing each rule in a series of natural transformations, one brings to light the amount of structure of the behaviour which is essential for the rule. As a consequence, one can interpret the same rules modularly with respect to different behaviours with the same structure.

Parallel Composition. For modelling the first rule in (9), all one needs is B to be a strong endofunctor. Correspondingly, the map *l-merge* can be abstracted to:

$$\textit{l-merge}_B = B(\eta \parallel \eta) \circ st \circ (\pi_2 \otimes \pi_1) : (X \times BX) \otimes (X \times BX) \to BTX \ ,$$

parametric in B. Therefore, if BX is endowed with a join $\vee : BX \times BX \to BX$, one can abstract the modelling of the first two rules for parallel composition (thus without synchronization!) as:

$$[\![\parallel]\!]_{B,\vee} = \textit{l-merge}_B \vee (\textit{l-merge}_B \circ \gamma) \ . \tag{13}$$

In §6, it is shown how this map $[\![\parallel]\!]_{B,\vee}$ can be instantiated to the behaviour on semi-lattices as used for trace equivalence.

A simple variation of the above $l\text{-}merge_B$ is obtained by considering behaviours of the form $BX = B'(1+X)$:

$$l\text{-}merge_{B'(1+\_)} = B'(\iota_2) \circ B'([\eta, (\eta \parallel \eta)]) \circ B'(\delta) \circ st \circ (\pi_2 \otimes \pi_1) \ .$$

One can then instantiate this rule to $B'X = (\mathcal{P}_f(S \cdot X))^S$. In **Set**:

$$l\text{-}merge((x,\beta),(y,\beta')) = \boldsymbol{\lambda}\sigma.(\{\sigma' \cdot (x' \parallel y) \mid (\sigma' \cdot x') \in \beta(\sigma)\} \cup \{\sigma' \cdot y \mid \sigma' \in \beta(\sigma)\})$$

which corresponds to the usual rules of the left merge operator for non-deterministic programs with side-effects and explicit termination.

Sequential Composition. Also the modelling $[\![\, ; \,]\!]$ of the rules for sequential composition in §3 can be made parametric in $BX = B'(1+X)$, for some strong endofunctor B'. This simply yields the above $l\text{-}merge_{B'(1+\_)}$ with '$\parallel$' replaced by ' ; ':

$$[\![\, ; \,]\!]_{B'(1+\_)} = B'(\iota_2) \circ B'([\eta, (\eta \, ; \, \eta)]) \circ B'(\delta) \circ st \circ (\pi_2 \otimes \pi_1) \ .$$

For $B'X = (\mathcal{P}_f(S \cdot (X)))^S$ on **Set** one has then the non-deterministic version of the map $[\![\, ; \,]\!]$ in §3.

The rules for sequential composition with atomic actions instead of side-effects are slightly different. One can use negative premises as in the previous section or a distinguished transition, say $x \rightsquigarrow *$, to express explicit termination, in which case the rules are:

$$\frac{x \overset{a}{\rightsquigarrow} x'}{x \, ; \, y \overset{a}{\rightsquigarrow} x' \, ; \, y} \ , \qquad \frac{x \rightsquigarrow * \quad y \overset{a}{\rightsquigarrow} y'}{x \, ; \, y \overset{a}{\rightsquigarrow} y'} \ , \qquad \frac{x \rightsquigarrow * \quad y \rightsquigarrow *}{x \, ; \, y \overset{a}{\rightsquigarrow} *} \ . \tag{14}$$

These rules can be modelled parametrically in a strong endofunctor B' on an affine SMC with finite products and coproducts, and tensors distributing over coproducts:

$$\begin{array}{rll}
\pi_2 \otimes id & : (X \times (1+B'X)) \otimes (X \times (1+B'X)) & \\
 & \quad \longrightarrow (1+B'X) \otimes (X \times (1+B'X)) & \\
\delta & : \quad \cong (X \times (1+B'X)) + ((B'X \otimes X) \times (1+B'X)) & \\
\pi_2 + \pi_1 & : \quad \longrightarrow (1+B'X) + (B'X \otimes X) & \\
id + st & : \quad \longrightarrow 1 + B'X + B'(X \otimes X) & \\
id + [B'\eta, B'(\eta \, ; \, \eta)] & : \quad \longrightarrow 1 + B'TX & \\
\hline
[\![\, ; \,]\!]_{1+B'} & : ((X \times (1+B'X)) \otimes (X \times (1+B'X)) \longrightarrow 1 + B'TX &
\end{array}$$

In **Set**, for $B'X = (\mathcal{P}_f X)^A$:

$$(x,\beta)[\![\, ; \,]\!]_{1+B'}(y,\beta') = \begin{cases} * & \text{if } \beta = * = \beta' \\ \boldsymbol{\lambda}a.(\{y' \mid y' \in \beta'(a)\}) & \text{if } \beta = *, \beta' \neq * \\ \boldsymbol{\lambda}a.(\{x' \, ; \, y \mid x' \in \beta(a)\}) & \text{otherwise} \end{cases} \ .$$

Choice. The map $[\![\, \mathsf{or} \,]\!]$ modelling the choice operator of CCS can be abstracted and made parametric in any endofunctor B on an affine SMC (so that the tensor comes with two projections, say $\pi_1^{\otimes} : X \otimes Y \to X$ and $\pi_2^{\otimes} : X \otimes Y \to Y$) such that BX is endowed with a semi-lattice structure $\vee$:

$$[\![\, \mathsf{or} \,]\!]_{B,\vee} = B\eta \circ \vee \circ (\pi_2 \times \pi_2) \circ \langle \pi_1^{\otimes}, \pi_2^{\otimes} \rangle : (X \times BX) \otimes (X \times BX) \longrightarrow BTX \ . \quad (15)$$

Actions as Outputs. An alternative to the behaviour $B_1 X = (\mathcal{P}_{\mathrm{fi}} X)^A$ is the behaviour $B_2 X = \mathcal{P}_{\mathrm{fi}}(A \cdot X)$. In Set, if A is finite, these two behaviours are isomorphic, but this is not true in general, although there is a canonical map *embed* : $\mathcal{P}_{\mathrm{fi}}(A \cdot X) \to (\mathcal{P}_{\mathrm{fi}} X)^A$ (which can be used to interpret the rule for synchronization – ie the third rule in (9) – using B_2 instead of B_1). The endofunctor B_1 with actions as powers corresponds to consider actions as (testable!) inputs, while the endofunctor B_2 with actions as copowers corresponds to actions as outputs. In the presence of negative premises (and exceptions in general) it seems the use of powers is really needed, but otherwise copowers seem more natural. Take, for instance, the axiom $a \overset{a}{\leadsto} \mathsf{nil}$ for atomic actions a. Using B_1 one has to map all actions different from a to the neutral element of $B_1 TX$, while using B_2 one can define $[\![A]\!] : A \cdot 1 \to \mathcal{P}_{\mathrm{fi}}(A \cdot TX)$ to be the composition of the unit of the monad $\mathcal{P}_{\mathrm{fi}}$ with B_2 applied to the insertion of the constant nil into TX (which is denoted here by nil as well). More generally, for every pointed endofunctor $\langle H, \nu \rangle$, one can define:

$$[\![A]\!]_H = \nu \circ (A \cdot \mathsf{nil}) : A \cdot 1 \to H(A \cdot TX) \ . \quad (16)$$

Clearly, also $[\![\ \|\ ]\!]_{B,\vee}$ and $[\![\, \mathsf{or} \,]\!]_{B,\vee}$ can be instantiated to $BX = \mathcal{P}_{\mathrm{fi}}(A \cdot X)$ as well as to $BX = 1 + \mathcal{P}_{\mathrm{f}}(A \cdot X)$. The latter behaviour can be used also for $[\![\ ;\ ]\!]_{1+B'}$.

Replication. Another example of a modular modelling is the one for a unary loop operator. It can be defined parametrically in a strong endofunctor B and a natural transformation $\otimes : X \times Y \to X \otimes Y$ as follows:

$$\frac{\begin{array}{rl} \otimes & : X \times BX \longrightarrow X \otimes BX \\ \gamma & : \qquad\qquad \longrightarrow BX \otimes X \\ st & : \qquad\qquad \longrightarrow B(X \otimes X) \\ B(\eta\ ;\ \mathsf{loop}(\eta)) & : \qquad\qquad \longrightarrow BTX \end{array}}{[\![\mathsf{loop}]\!]_B^{\otimes} \quad : X \times BX \longrightarrow BTX}$$

In Set, for $BX = 1 + \mathcal{P}_{\mathrm{f}}(A \cdot X)$ and for $\otimes$ simply the identity, one has:

$$[\![\mathsf{loop}]\!](x, \beta) = \begin{cases} * & \text{if } \beta = * \\ \{a \cdot (x'\ ;\ \mathsf{loop}(x)) \mid a \cdot x \in \beta\} & \text{otherwise} \end{cases}$$

which models the rules:

$$\frac{x \overset{a}{\leadsto} x'}{\mathsf{loop}(x) \overset{a}{\leadsto} x'\ ;\ (\mathsf{loop}(x))} \ , \qquad \frac{x \leadsto *}{\mathsf{loop}(x) \leadsto *} \ . \quad (17)$$

Note that, by using parallel instead of sequential composition, the operator loop has essentially the same behaviour as the replication operator of the π-calculus [11].

6 Non-Determinism II: Trace Equivalence

In [7], non-determinism with side-effects and explicit termination is treated by simply interpreting the (deterministic) behaviour $BX = (S \cdot (1 + X))^S$ in a category of semi-lattices, instead of using the behaviour $BX = (\mathcal{P}_f(S \cdot (1 + X)))^S$. In terms of behavioural equivalences, this amounts to consider (completed) *trace equivalence* instead of bisimulation.

When modelling rules in categories of semi-lattices one needs the full generality of monoidal rather than cartesian categories. Indeed, although, if $\mathcal{C}$ is cartesian, the category $SL(\mathcal{C})$ inherits from it the cartesian structure (which is useful for applying the structural recursion theorem with accumulators), for modelling the syntax one needs tensors rather than cartesian products: the natural interpretation of a binary operators in a category of semi-lattice is not as linear but as ***bilinear*** maps, ie as maps which are linear in each component separately; the tensor product $\otimes$ is the classifier of bilinear maps. (Cf [7].)

As a first example of trace equivalence semantics, consider the map $[\![\,;\,]\!]$ defined in §3: clearly, it can be interpreted in categories of semi-lattices. For instance, in $SL(\mathsf{Set})$, one has the following, where $x \leq y$ stands for $x \vee y = y$:

$$(x, \beta)[\![\,;\,]\!](y, \beta') = \boldsymbol{\lambda}\sigma.\left(\bigvee_{(\sigma'\cdot x') \leq \beta(\sigma)} \sigma' \cdot (x'\,;\,y) \vee \bigvee_{\sigma' \leq \beta(\sigma)} \sigma' \cdot y\right) .$$

Note that copowers $S \cdot X$ in $SL(\mathcal{C})$ are right-linear, hence in $SL(\mathsf{Set})$:

$$\sigma \cdot (x \vee y) = (\sigma \cdot x) \vee (\sigma \cdot y)$$

for every σ in S, and for all x, y in X. Therefore, one could write $\sigma \cdot x$ as the more familiar $\{\sigma\} \circledast x$ of [7], where $\circledast : X \times Y \to X \otimes Y$ is the universal bilinear natural transformation associated with the tensor.

Other examples are the maps $[\![\ \|\ ]\!]_{B,\vee}$, $[\![\,;\,]\!]_{1+B'}$, $[\![\mathsf{or}]\!]_{B,\vee}$, $[\![A]\!]_H$, and $[\![\mathsf{nil}]\!]$ which can be interpreted in $SL(\mathsf{Set})$ with respect to $BX = 1 + A \cdot X$, $B'X = A \cdot X$, and $HX = 1 + X$ as follows:

$$(x, \beta)[\![\ \|\ ]\!](y, \beta') = \begin{cases} * \vee \bigvee_{a \cdot x' \leq \beta} a \cdot (x' \parallel y) \vee \bigvee_{a \cdot y' \leq \beta'} a \cdot (x \parallel y') & \text{if } \beta \geq * \leq \beta' \\ \bigvee_{a \cdot x' \leq \beta} a \cdot (x' \parallel y) \vee \bigvee_{a \cdot y' \leq \beta'} a \cdot (x \parallel y') & \text{otherwise} \end{cases}$$

$$(x, \beta)[\![\,;\,]\!](y, \beta') = \begin{cases} * \vee \bigvee_{a \cdot y' \leq \beta'} a \cdot y' \vee \bigvee_{a \cdot x' \leq \beta} a \cdot (x'\,;\,y) & \text{if } \beta \geq * \leq \beta' \\ \bigvee_{a \cdot y' \leq \beta'} a \cdot y' \vee \bigvee_{a \cdot x' \leq \beta} a \cdot (x'\,;\,y) & \text{if } \beta \geq * \not\leq \beta' \\ \bigvee_{a \cdot x' \leq \beta} a \cdot (x'\,;\,y) & \text{otherwise} \end{cases}$$

$$(x, \beta)[\![\mathsf{or}]\!](y, \beta') = \beta \vee \beta'$$

$$[\![a]\!] = a \cdot \mathsf{nil} \qquad [\![\mathsf{nil}]\!] = *$$

7 From the Rules to the Models

In this section, it is shown how both an operational and a denotational model can be derived from abstract operational rules $\rho : \Sigma(Id \times B) \longrightarrow BT$. General results from [17] ensure that the denotational model is always adequate with respect to the operational one. All this is exemplified using the language in (1) and the natural transformation

$$\rho = [\,[\![\mathsf{nil}]\!], [\![A]\!]_H, [\![\;\|\;]\!]_{B,\vee}, [\![\;;\;]\!]_{1+B'}\,] : \Sigma(X \times BX) \longrightarrow BTX \qquad (18)$$

interpreted either in Set with respect $BX = 1 + \mathcal{P}_f(A \cdot X)$, $B'X = \mathcal{P}_f(A \cdot X)$, and $HX = 1 + \mathcal{P}_f X$, or in $SL(\mathsf{Set})$ with respect to $BX = 1 + A \cdot X$, $B'X = A \cdot X$, and $HX = 1 + X$. Note the isomorphism $1 + \mathcal{P}_f(A \cdot X) \cong \mathcal{P}_{fi}(A \times X)$ which holds in Set; this is used extensively in the sequel without further mention.

The main tool in this section is the following 'folklore' structural recursion theorem with accumulators, ie with terms as extra parameters of the recursion. (Cf [17].)

Theorem 1 (Structural Recursion with Accumulators). Let T be the monad freely generated by an endofunctor F on a cartesian category $\mathcal{C}$ and let $\psi : FTX \to TX$ be the structure of the free F-algebra over an object X of $\mathcal{C}$. For all maps $f : X \to Y$ and $h : F(TX \times Y) \to Y$ in $\mathcal{C}$ there exists a unique map $f^\sharp : TX \to Y$ in $\mathcal{C}$ such that

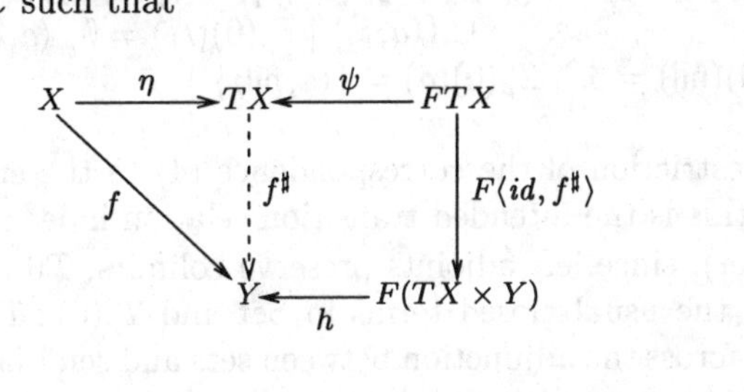

commutes. □

A special case is when X is the initial object 0. Then the left triangle is vacuous, ψ is an isomorphism, and the unique map from $T0$ to Y is called the **iterator** (or the inductive extension) of h:

$$\begin{array}{ccc} T0 & \xleftarrow{\;\psi\;} & FT0 \\ {\scriptstyle it(h)}\big\downarrow & & \big\downarrow{\scriptstyle F\langle id,\, it(h)\rangle} \\ Y & \xleftarrow[\;h\;]{} & F(T0 \times Y) \end{array}$$

In this case it suffices that $\langle T0, \psi\rangle$ is the initial F-algebra, rather than the stronger hypothesis that F freely generates T.

7.1 The Induced Operational Model

Notation. [17] For Σ, B, and T as above and for every natural transformation $\rho : \Sigma(Id \times B) \stackrel{\cdot}{\longrightarrow} BT$, put:

$$\varrho = B\mu \circ \rho_T : \Sigma(T \times BT) \stackrel{\cdot}{\longrightarrow} BT \ . \tag{19}$$

Note that the mapping $\rho \mapsto \varrho$ is a 1-1 correspondence between natural transformations of type $\Sigma(Id \times B) \stackrel{\cdot}{\longrightarrow} BT$ and those of type $\Sigma(T \times BT) \stackrel{\cdot}{\longrightarrow} BT$.

The **canonical operational model** $T_\rho(0)$ induced by some abstract operational rules $\rho : \Sigma(Id \times B) \stackrel{\cdot}{\longrightarrow} BT$ is the iterator of ϱ at 0:

$$\begin{array}{ccc} T0 & \xleftarrow{\ \psi\ } & \Sigma T0 \\ {\scriptstyle T_\rho(0)} \Big\downarrow & & \Big\downarrow {\scriptstyle \Sigma\langle id, T_\rho(0)\rangle} \\ BT0 & \xleftarrow[\ \varrho\ ]{} & \Sigma(T0 \times BT0) \end{array}$$

In Set, chasing the above diagram one obtains:

$$\begin{aligned}
T_\rho(0)(t_1 \parallel t_2) &= \{\langle a, t_1' \parallel t_2\rangle \mid \langle a, t_1'\rangle \in T_\rho(0)(t_1)\} \cup \\
&\quad \cup \{\langle a, t_1 \parallel t_2'\rangle \mid \langle a, t_2'\rangle \in T_\rho(0)(t_2)\} \\
T_\rho(0)(t_1 \,;\, t_2) &= \{\langle a, t_1' \,;\, t_2\rangle \mid \langle a, t_1'\rangle \in T_\rho(0)(t_1)\} \cup \\
&\quad \cup \{\langle a, t_2'\rangle \mid T_\rho(0)(t_1) = \emptyset,\ \langle a, t_2'\rangle \in T_\rho(0)(t_2)\} \\
T_\rho(0)(\mathsf{nil}) = \emptyset & \quad T_\rho(0)(a) = \{\langle a, \mathsf{nil}\rangle\}
\end{aligned}$$

Using the restriction of the correspondence (4) to the endofunctor $\mathcal{P}_{\mathrm{fi}}$, one can check that this is the intended transition relation indeed.

In $SL(\mathsf{Set})$, since left adjoints preserve colimits, $T0$ is simply the free semi-lattice over the usual closed terms in Set and $T_\rho(0) : T0 \to 1 + A \cdot T0$ can be transposed across the adjunction between sets and semi-lattices, yielding a transition system (with explicit termination):

$$t \stackrel{a}{\leadsto} t' \iff a \cdot t' \leq T_\rho(0) \quad \text{and} \quad t \leadsto * \iff * \leq T_\rho(0) \ ,$$

for all closed terms t and t'. This should help in understanding the following: $T_\rho(0)(\mathsf{nil}) = *$, $T_\rho(0)(a) = a \cdot \mathsf{nil}$,

$$T_\rho(0)(t_1 \parallel t_2) = \begin{cases} * \vee \bigvee_{a \cdot t_1' \leq T_\rho(0)(t_1)} a \cdot (t_1' \parallel t_2) \vee \\ \quad \vee \bigvee_{a \cdot t_2' \leq T_\rho(0)(t_2)} a \cdot (t_1 \parallel t_2') & \text{if } T_\rho(0)(t_1) \geq * \leq T_\rho(0)(t_2) \\ \bigvee_{a \cdot t_1' \leq T_\rho(0)(t_1)} a \cdot (t_1' \parallel t_2) \vee \\ \quad \vee \bigvee_{a \cdot t_2' \leq T_\rho(0)(t_2)} a \cdot (t_1 \parallel t_2') & \text{otherwise} \end{cases}$$

and a similar clause for $T_\rho(0)(t_1 \,;\, t_2)$.

7.2 The Coinduced Denotational Model

The construction of the model $T_\rho(0)$ extends to a functor $T_\rho : B\text{-}\mathsf{Coalg} \to B\text{-}\mathsf{Coalg}$ as follows. For every coalgebra $k : X \to BX$, the coalgebra $T_\rho(k) : TX \to BTX$ is the map:

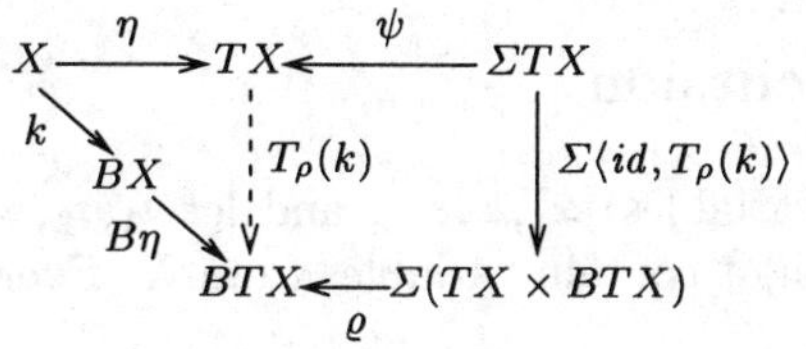

$$\begin{array}{ccccc} X & \xrightarrow{\eta} & TX & \xleftarrow{\psi} & \Sigma TX \\ {\scriptstyle k}\searrow & & \vdots\, {\scriptstyle T_\rho(k)} & & \downarrow {\scriptstyle \Sigma\langle id, T_\rho(k)\rangle} \\ & BX \xrightarrow{B\eta} & BTX & \xleftarrow{\varrho} & \Sigma(TX \times BTX) \end{array}$$

given by the structural recursion theorem. The naturality of ρ ensures this is a functor. (Moreover, T_ρ is a monad lifting the monad T to the B-coalgebras [17].) In particular, T_ρ can be applied to the final B-coalgebra

$$\varphi : \nu X.BX \cong B(\nu X.BX)$$

yielding the coalgebra $T_\rho(\varphi) : T(\nu X.BX) \to BT(\nu X.BX)$. The coalgebra $T_\rho(\varphi)$ is a 'conservative extension' of both the final coalgebra φ and the canonical operational model $T_\rho(0)$, because of the commutativity of the triangle in the above diagram and of the functoriality of T_ρ. In Set, this means that:

$$\langle a, p'\rangle \in \varphi(p) \iff \langle a, p'\rangle \in T_\rho(\varphi)(p)\ ,\ \langle a, t'\rangle \in T_\rho(0)(t) \iff \langle a, t'\rangle \in T_\rho(\varphi)(t)$$

for all elements p, p' of $\nu X.BX$ and for all closed terms t, t'.

Most importantly, the final coalgebra can be endowed with a canonical T-algebra structure, namely the **final coalgebra semantics** [1,14] $\mathcal{M}_\varphi$ of the operational model $T_\rho(\varphi)$:

$$\begin{array}{ccc} T(\nu X.BX) & \xrightarrow{\mathcal{M}_\varphi} & \nu X.BX \\ {\scriptstyle T_\rho(\varphi)}\downarrow & & \downarrow {\scriptstyle \varphi}\ \cong \\ BT(\nu X.BX) & \xrightarrow[B\mathcal{M}_\varphi]{} & B(\nu X.BX) \end{array}$$

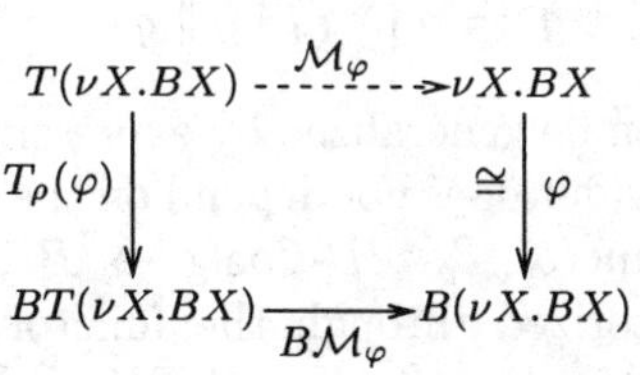

Thus $\mathcal{M}_\varphi$ is the unique homomorphism from $T_\rho(\varphi)$ to the final coalgebra φ.

The theory in [17] shows that $\mathcal{M}_\varphi$, which is written there as $D_\rho(1)$, qualifies as the **canonical denotational model** for ρ and that it is dual, in a suitable sense, to the operational model $T_\rho(0)$. One can derive from it a denotation for every n-ary operator ϕ of the language, ie a map

$$\widehat{\phi} : \underbrace{\nu X.BX \otimes \cdots \otimes \nu X.BX}_{n} \longrightarrow \nu X.BX$$

by simply precomposing $\mathcal{M}_\varphi$ with the insertion of $(\nu X.BX)^{\otimes n}$ in $T(\nu X.BX)$, ie $\widehat{\phi}(p_1, \ldots, p_n) = \mathcal{M}_\varphi(\phi(p_1, \ldots, p_n))$ for all elements $p_1, \ldots, p_n : 1 \to \nu X.BX$.

For instance, in Set:

$$\begin{aligned} p_1 \mathbin{\widehat{\|}} p_2 &= \mathcal{M}_\varphi(p_1 \parallel p_2) \\ &= \{\langle a, \mathcal{M}_\varphi(p_1' \parallel p_2)\rangle \mid \langle a, p_1'\rangle \in \varphi(p_1)\} \cup \\ &\qquad \cup \{\langle a, \mathcal{M}_\varphi(p_1 \parallel p_2')\rangle \mid \langle a, p_2'\rangle \in \varphi(p_2)\} \end{aligned} .$$

8 Guarded Recursion

Let X be a set (of variables) $x_1, x_2, \ldots$ and let $t_1, t_2, \ldots$ be elements of TX, hence terms which might contain variables from X. Every system

$$\Theta = \begin{cases} x_1 = t_1 \\ x_2 = t_2 \\ \vdots \end{cases} \tag{20}$$

of mutually recursive term-equations in X can be seen as a T-coalgebra having as carrier the set X of variables appearing in the system and as structure the function

$$k : X \to TX \qquad\qquad x_i \mapsto t_i .$$

Conversely, every T-coalgebra defines a system of term-equations.

In structural operational semantics, the usual way to deal with recursive programs in Set is to use 'guarded terms'. This is a rather informal notion and the aim of this section is to use the categorical modelling of the rules to make it formal and general.

Typically, a term is called guarded if it is of the form $a\,;t$, that is, if it starts with an action. (Cf [5,3].) For instance, following [3], the guarded terms corresponding to the language in (1) are:

$$g ::= a \mid g\,;t \mid g \parallel g . \tag{21}$$

The idea is that this can be generalized by expressing, in a formal way, that a term is guarded if its behaviour does not depend on the behaviour of its variables.

First note that the functor T_ρ : B-Coalg $\to$ B-Coalg defined in the previous section can be factorized through the functor mapping a B-coalgebra $k : X \to BX$ to the coalgebra $B\eta \circ k : X \to BTX$ of the composite endofunctor BT. The other component of the factorization is denoted, with a slight abuse of notation, also as T_ρ : BT-Coalg $\to$ B-Coalg. It is defined as:

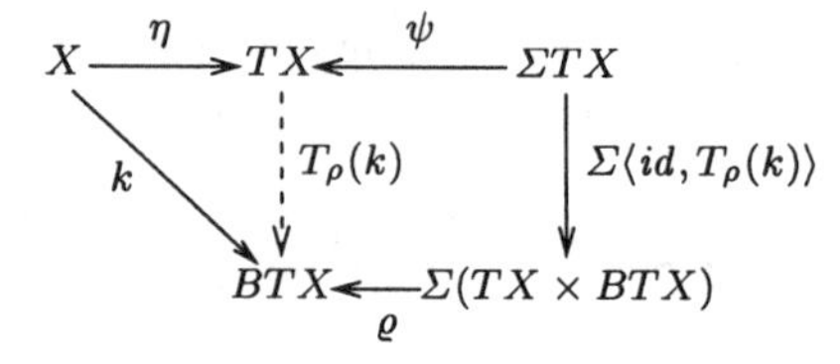

Next, let U : BT-Coalg $\to \mathcal{C}$ be the forgetful functor from the BT-coalgebras to their carriers. Following [8], one can prove there is a 1-1 correspondence between 'liftings of functors' and natural transformations of a suitable type which allows one to regard the functor $T_\rho : BT\text{-}\mathsf{Coalg} \to B\text{-}\mathsf{Coalg}$ as a natural transformation $T_\rho : TU \stackrel{\cdot}{\longrightarrow} BTU$. (Cf 'Liftings as Coactions' in [16, §4].)

Definition 2. A **guard** for some abstract operational rules ρ as in (7) is a 'U-cone' over the induced natural transformation $T_\rho : TU \stackrel{\cdot}{\longrightarrow} BTU$, ie a span

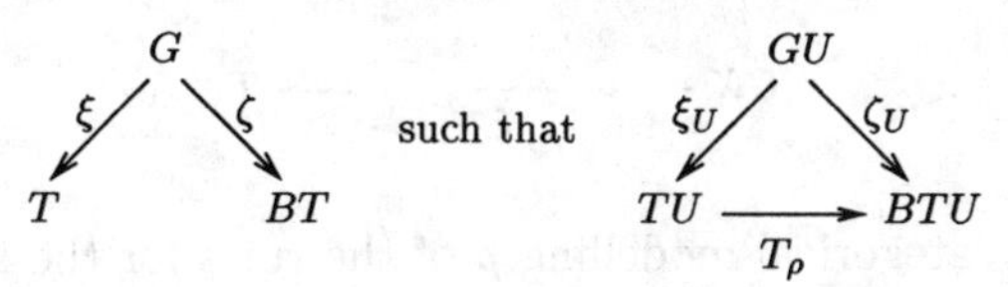

commutes. □

Given a guard $\langle G, \xi, \zeta \rangle$, the coalgebras $\Theta : X \to GX$ can be understood as systems of guarded term-equations. Indeed The second leg ζ of the span can be used to turn Θ into a BT-coalgebra and then the operational model $T_\rho(\zeta \circ \Theta)$ can be defined:

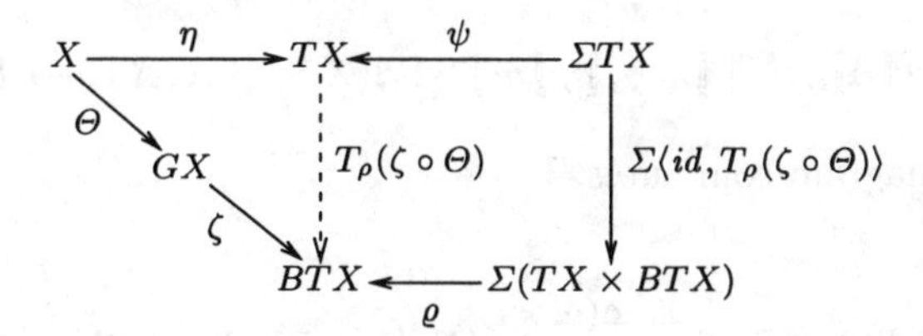

The corresponding final coalgebra semantics

$$\begin{array}{ccc} TX & \stackrel{\mathcal{M}_\Theta}{\dashrightarrow} & \nu X.BX \\ {\scriptstyle T_\rho(\zeta \circ \Theta)}\downarrow & & \downarrow{\scriptstyle \cong\ \varphi} \\ BTX & \xrightarrow[B\mathcal{M}_\Theta]{} & B(\nu X.BX) \end{array}$$

gives the desired recursive interpretation of the system Θ.

How does one prove that this notion specialises to the above example? Firstly, note that the functorial notion of syntax corresponding to the grammar in (21) is the functor

$$\Sigma_G : \mathcal{C} \times \mathcal{C} \to \mathcal{C} \qquad (X_1, X_2) \mapsto A \cdot I + X_1 \otimes X_2 + X_1 \otimes X_1 \ .$$

Indeed, for every X, the initial algebra

$$GX = \mu Y.\Sigma_G(Y, TX)$$

specialises in Set to the set of terms given by the grammar in (21). Further, note that composing Σ_G with the diagonal $\Delta : \mathcal{C} \to \mathcal{C} \times \mathcal{C}$ one obtains a subfunctor of the endofunctor in (2) freely generating the monad $TX = \mu Y.X + \Sigma Y$ corresponding to the grammar in (1); let $i : \Sigma_G \circ \Delta \stackrel{\cdot}{\longrightarrow} \Sigma$ be the corresponding insertion map. The first leg $\xi : G \stackrel{\cdot}{\longrightarrow} T$ of the span can then be defined recursively using the initiality of GX with respect to the endofunctor $\Sigma_G(\_\,, TX)$:

$$\begin{array}{ccc}
\Sigma_G(GX,TX) & \xrightarrow{\Sigma_G(\xi,TX)} & \Sigma_G(TX,TX) \\
 & & \downarrow \\
\cong \downarrow & & \Sigma TX \\
 & & \downarrow \psi \\
GX & \dashrightarrow_{\xi} & TX
\end{array}$$

Next, let the categorical modelling ρ of the rules for the language (1) be as in (18) with respect to an arbitrary B such that $B = 1 + B'$ and both B and B' are endowed with a semi-lattice structure. Consider the following restricted version of sequential composition:

$$[\![\, ; \,]\!]_{B'} = B'(\eta \,;\, \eta) \circ st \circ (\pi_2 \otimes id) : (X \times B'X) \otimes X \to B'TX \ .$$

Then, for:

$$\rho' = [\, [\![A]\!]_H, [\![\, \| \,]\!]_{B',\vee}, [\![\, ; \,]\!]_{B'} \,] : \Sigma_G(X \times B'X, X) \longrightarrow B'TX$$

the following diagram commutes.

$$\begin{array}{ccccc}
\Sigma_G(X \times B'X, X \times B'X) & \xrightarrow{\Sigma_G \Delta(id \times \iota_2)} & \Sigma_G(X \times BX, X \times BX) & \xrightarrow{i} & \Sigma(X \times BX) \\
\Sigma_G(id,\pi_1) \downarrow & & & & \downarrow \rho \\
\Sigma_G(X \times B'X, X) & \xrightarrow[\rho']{} & B'TX & \xrightarrow[\iota_2]{} & 1 + B'TX = BTX
\end{array}$$

Since the variables in GX are those stemming from the second component of Σ_G, the fact that the restriction of the rules ρ to $\Sigma_G(X \times B'X, X \times B'X)$ factorizes through $\Sigma_G(id, \pi_1)$ means that the behaviour of the variables in a guarded term is irrelevant for its behaviour. In particular, this allows us to define the second leg of the span as $\zeta = \iota_2 \circ \zeta' : G \stackrel{\cdot}{\longrightarrow} BT$, where ζ' is the iterator of the composite structure $\varrho' \circ \Sigma_G(\xi \times B'TX, TX)$ with respect to the initial algebra GX:

$$\begin{array}{ccc}
\Sigma_G(GX,TX) & \xrightarrow{\Sigma_G(\langle id,\zeta' \rangle,TX)} & \Sigma_G(GX \times B'TX, TX) \\
 & & \downarrow \Sigma_G(\xi \times B'TX, TX) \\
\cong \downarrow & & \Sigma_G(TX \times B'TX, TX) \\
 & & \downarrow \varrho' \\
GX & \dashrightarrow_{\zeta'} & B'TX
\end{array}$$

Proposition 3. The above $\langle G, \xi, \zeta \rangle$ is a guard for the abstract operational rules ρ as in (18).

Proof. Some diagram chasing shows that, for every BT-coalgebra k, that both $T_\rho(k) \circ \xi$ and ζ fit as the unique iterator of the composite structure

$$\varrho \circ i \circ \Sigma_G(\xi \times id, \langle id, T_\rho(k)\rangle) : \Sigma_G(GX \times BTX, TX) \longrightarrow BTX$$

with respect to the initial algebra $GX = \mu Y.\Sigma_G(Y, TX)$. □

Thus, for instance, the following system Θ of term-equations in $X = \{x, y\}$ is guarded with respect to the above guard and $BX = 1 + \mathcal{P}_f(A \cdot X)$ on Set:

$$x = a\,;x\ , \qquad y = (a\,;y) \parallel (b\,;x)\ .$$

Omitting, as usual, the insertion-of-variables η and the final coalgebra isomorphism, one has:

$$\mathcal{M}_\Theta(x) = \{\langle a, \mathcal{M}_\Theta(x)\rangle\}\ , \quad \mathcal{M}_\Theta(y) = \{\langle a, \mathcal{M}_\Theta(y)\rangle, \langle b, \mathcal{M}_\Theta(x)\rangle\}$$

which is the intended interpretation of Θ as a set of recursive processes. Similarly, for $BX = 1 + A \cdot X$ on SL(Set):

$$\mathcal{M}_\Theta(x) = a \cdot \mathcal{M}_\Theta(x)\ , \quad \mathcal{M}_\Theta(y) = (a \cdot \mathcal{M}_\Theta(y)) \vee (b \cdot \mathcal{M}_\Theta(x))\ .$$

References

1. P. Aczel. *Non-well-founded sets.* Number 14 in Lecture Notes. CSLI, 1988.
2. M.A. Arbib and E.G. Manes. *Arrows, Structures, and Functors.* Academic Press, 1975.
3. J.W. de Bakker and J.-J.Ch. Meyer. Metric semantics for concurrency. *BIT*, 28:504–529, 1988.
4. M. Barr. Terminal coalgebras in well-founded set theory. *Theoretical Computer Science*, 144(2):299–315, 1993.
5. B. Bloom, S. Istrail, and A.R. Meyer. Bisimulation can't be traced. *Journal of the ACM*, 42(1):232–268, jan 1995. A preliminary report appeared in *Proc. 3rd LICS*, pages 229-239, 1988.
6. J.A. Goguen, J.W. Thatcher, and E.G. Wagner. An initial algebra approach to the specification, correctness and implementation of abstract data types. In R.T. Yeh, editor, *Current Trends in Programming Methodology*, volume IV, pages 80–149. Prentice Hall, 1978.
7. M.C.B. Hennessy and G.D. Plotkin. Full abstraction for a simple parallel programming language. In J. Bečvář, editor, *Proc. 8th MFCS*, volume 74 of *LNCS*, pages 108–120. Springer-Verlag, 1979.
8. P.T. Johnstone. Adjoint lifting theorems for categories of algebras. *Bull. London Math. Soc.*, 7:294–297, 1975.
9. A. Kock. Strong functors and monoidal monads. *Arch. Math. (Basel)*, 23:113–120, 1972.
10. R. Milner. *A Calculus of Communicating Systems*, volume 92 of *LNCS*. Springer-Verlag, 1980.
11. R. Milner. Functions as processes. In M.S. Paterson, editor, *Proc. of 17th ICALP*, 1990.

12. E. Moggi. Notions of computation and monads. *Information and Computation*, 93:55–92, 1991.
13. G.D. Plotkin. A structural approach to operational semantics. Technical Report DAIMI FN-19, Computer Science Department, Aarhus University, 1981.
14. J. Rutten and D. Turi. On the foundations of final semantics: non-standard sets, metric spaces, partial orders. In J. de Bakker et al., editors, *Proc. of the REX workshop Semantics – Foundations and Applications*, volume 666 of *LNCS*, pages 477–530. Springer-Verlag, 1993.
15. M. Smyth and G. Plotkin. The category-theoretic solution of recursive domain equations. *SIAM J. Comput.*, 11:761–783, 1982.
16. D. Turi. *Functorial Operational Semantics and its Denotational Dual.* PhD thesis, Free University, Amsterdam, June 1996. Accessible from `<http://www.dcs.ed.ac.uk/home/dt/>`.
17. D. Turi and G.D. Plotkin. Towards a mathematical operational semantics. In *Proc. 12th LICS Conf.* IEEE, Computer Society Press, 1997.

Specifying Interaction Categories

D. Pavlović*[1] and S. Abramsky[2]

[1] COGS, University of Sussex, Brighton BN1 9QH
[2] Department of Computer Science, University of Edinburgh, Edinburgh EH9 3JZ

Abstract. We analyse two complementary methods for obtaining models of typed process calculi, in the form of interaction categories. These two methods allow adding new features to previously captured notions of process and of type, respectively. By combining them, all familiar examples of interaction categories, as well as some new ones, can be built starting from some familiar categories.
Using the presented constructions, interaction categories can be analysed without fixing a set of axioms, merely in terms of the way in which they are specified — just like algebras are analysed in terms of equations and relations, independently on abstract characterisations of their varieties.

1 Introduction

Interaction categories [1] are proposed as a general, yet practical tool for reasoning about functional and concurrent computation. They are not meant to be a definitive formal system, but an open list of features that should be considered and corelated. The paradigm of processes as relations extended in time is proposed as a conceptual basis for integrating type theory and process calculus, in a categorical framework. The interaction of processes is captured by the composition of arrows; the identity arrows interpret buffers.

We make a step towards determining the structure of interaction categories by analysing the ways in which they come about. It turns out that all of the existing examples, and several new ones, can obtained by alternating two *specification methods*, which respectively determine a notion a of process and a notion of a type.

The notion of specification is understood as a method of building a structure from some given material. For instance, universal algebra is the method of specifying structures by operations and equations; forcing is a method of specifying new models of set theory over the old ones. Note that the Birkhoff variety theorem, which axiomatises categories arising in universal algebra, as well as the Giraud theorem, providing axioms for those which may arise from forcing, came only *after* the corresponding specification methods have been extensively developed and thoroughly analysed. Studying the constructions leading to a class of structures is usually prcedes its abstract axiomatic characterisation.

* This work was partially supported under ONR grant N00014-92-J-1974.

2 Specifications and categories derived from them

Both specification methods that we are about to describe start from an arbitrary interaction category $\mathcal{R}$. It can be a degenerate example: specifications will then add some relevant structure in it.

The first of our specification methods leads a category with the same objects as $\mathcal{R}$, but with the morphisms capturing a richer notion of process. The second one refines the type structure, leaving the morphisms essentially unchanged.

2.1 Specifying processes

Definition 1. A functor $\mathsf{h} : \mathcal{R} \to \mathcal{Q}$ between monoidal categories [16, sec. 1.1.] is said to be *lax monoidal* if it is given with a natural family

$$\mu_{AB} : \mathsf{h}A \otimes \mathsf{h}B \longrightarrow \mathsf{h}(A \otimes B) \text{ and an arrow}$$
$$\eta : \top \longrightarrow \mathsf{h}\top,$$

which are coherent in the sense that for all A, B, C the following diagrams commute

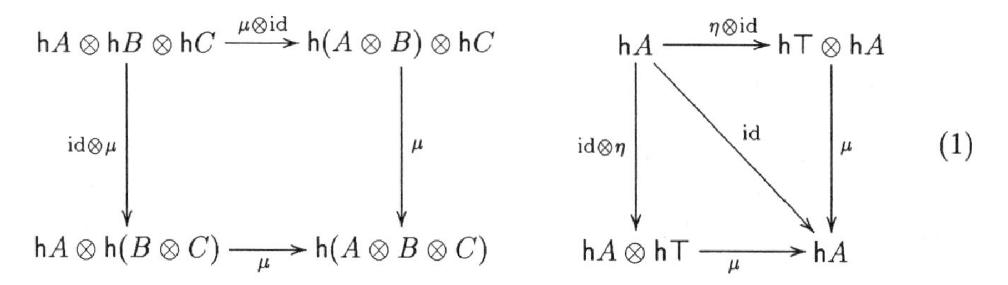

— with monidal isomorphisms $(A \otimes B) \otimes C \cong A \otimes (B \otimes C)$ and $A \otimes \top \cong A \cong \top \otimes A$ omitted for simplicity. h is said to be *(strong) monoidal* when μ and η are isomorphisms.

Remark. Recall that a monoidal category $\mathcal{R}$ is $\star$-autonomous [5] if and only if it is self-dual, and the duality[1] $(-)^\star : \mathcal{R}^{op} \to \mathcal{R}$, $A \cong A^{\star\star}$, induces the cotensor $B \multimap C$ in the form $(B \otimes C^\star)^\star$, thus making $\mathcal{R}$ autonomous (i.e. closed symmetric monoidal [16, sec. 1.5]). Now a $\star$-autonomous $\mathcal{R}$ satisfies the MIX rule [13] if and only if the duality functor $(-)^\star$ is lax monoidal; and it is compact closed [17] if and only if the duality is monoidal.

Process specifications. All these concepts readily generalise to enriched categories [16]. *Process specifications* are meant to be $\mathsf{Pos}_\perp$-enriched lax monoidal functors $\mathsf{h} : \mathcal{R} \to \mathsf{Pos}_\perp$, where $\mathcal{R}$ is at least autonomous and $\mathsf{Pos}_\perp$ is the category of posets with the bottom; morphisms are the bottom preserving monotone

[1] As a small contribution to the notational confusion in this context, we denote the two monoidal structures in $\star$-autonomous categories by $(\otimes, \top)$ and $(\invamp, \perp)$. The usual notations $(\times, 1)$ and $(+, 0)$ for products and coproducts are often replaced respectively with $(\&, 1)$ and $(\oplus, 0)$.

maps, and the monoidal structure is induced by the cartesian products. The base $\mathcal{R}$ is to be thought of as a category of abstract sets and relations. The functor h defines a notion of process by specifying for each set the processes typed by it. They are partially ordered by computational power, i.e. their ability to *simulate* one another.

For simplicity, we presently leave the enriched aspects aside, and formalize process specifications as functors to Set. Formally, though, any monoidal category would do.

Construction. The category of processes $\mathcal{R}_h$ induced by a specification $\mathsf{h} : \mathcal{R} \to \mathsf{Set}$ will have the same objects as $\mathcal{R}$. To define the morphisms, we compose the internal hom-functor $\multimap : \mathcal{R}^{op} \times \mathcal{R} \longrightarrow \mathcal{R}$ with $\mathsf{h} : \mathcal{R} \longrightarrow \mathsf{Set}$. Hence

$$\mathcal{R}_{\mathsf{h}}(A, B) = \mathsf{h}(A \multimap B). \tag{2}$$

The structure of $\mathcal{R}$ then readily lifts to $\mathcal{R}_{\mathsf{h}}$. The identity on A is obtained using the transposition $\ulcorner \mathrm{id}_A \urcorner : \top \longrightarrow (A \multimap A)$ of the corresponding identity in $\mathcal{R}$:

$$\mathrm{id}_A = \mathsf{h}\ulcorner \mathrm{id}_A \urcorner(\eta). \tag{3}$$

The composite of $f \in \mathcal{R}_{\mathsf{h}}(A, B)$ and $g \in \mathcal{R}_{\mathsf{h}}(B, C)$, i.e. of $\langle f, g\rangle \in \mathsf{h}(A \multimap B) \times \mathsf{h}(B \multimap C)$, becomes

$$f\,; g = \mathsf{h}m(\mu\langle f, g\rangle), \tag{4}$$

where $m : (A \multimap B) \otimes (B \multimap C) \longrightarrow (A \multimap C)$ is the internal composition in $\mathcal{R}$. The autonomous structure on the objects is inherited directly, while the arrow part is first internalised. For instance, the functor $X \otimes (-)$ on $\mathcal{R}$ induces a family of arrows from $A \multimap B$ to $(X \otimes A) \multimap (X \otimes B)$ in $\mathcal{R}$, the h-image of which is a family of functions from $\mathcal{R}_{\mathsf{h}}(A, B)$ to $\mathcal{R}_{\mathsf{h}}(X \otimes A, X \otimes B)$.

Note that $\mathcal{R}_{\mathsf{h}}$ comes with a functor $J = J_{\mathsf{h}} : \mathcal{R} \to \mathcal{R}_{\mathsf{h}}$. It is identity on the objects, and it maps $f \in \mathcal{R}(A, B)$ to $Jf = \mathsf{h}\ulcorner f \urcorner(\eta)$, where $\ulcorner f \urcorner \in \mathcal{R}(\top, A \multimap B)$ is the transpose of f. It is autonomous by the very definition of the autonomous structure in $\mathcal{R}_{\mathsf{h}}$.

The other way around, any functor $F : \mathcal{R} \to \mathcal{Q}$ induces a representation $\mathsf{h} = \mathsf{h}_F : \mathcal{R} \to \mathsf{Set}$, with $\mathsf{h}A = \mathcal{Q}(\top, FA)$. If F is monoidal, h is lax monoidal, with $\eta = \mathrm{id}_\top$ and μ_{AB} induced by tensoring the arrows $\top \to FA$ and $\top \to FB$ to get $\top \longrightarrow FA \otimes FB \cong F(A \otimes B)$. If F is autonomous, we can construct $\mathcal{R}_{\mathsf{h}}$ as above, and it will be isomorphic with $\mathcal{Q}$ if and only if F is bijective on objects. In fact, any essentially surjective F induces a weak equivalence $F' : \mathcal{R}_{\mathsf{h}} \to \mathcal{Q}$, with $F = (J\,; F')$. We spell out just the 1-dimensional part of the underlying 2-adjunction. Note that it extends to $\mathcal{V}$-enriched categories for any monoidal $\mathcal{V}$ in place of Set.

Fix an autonomous $\mathcal{R}$ and consider the category $\mathcal{R}/\mathsf{Bij}$ of bijective on objects, autonomous functors out of it. A morphism from such an $F : \mathcal{R} \to \mathcal{Q}$ to $G : \mathcal{R} \to \mathcal{P}$ will be an autonomous functor $M : \mathcal{Q} \to \mathcal{P}$, satisfying $(F\,; M) = G$ (and necesssarily bijective on objects too).

On the other hand, let $[\mathcal{R}, \mathsf{Set}]_{lax\otimes}$ be the category of lax monoidal functors and lax monoidal transformations. A natural transformation $\varphi : \mathsf{h} \to \mathsf{h}'$ is said to be lax monoidal if $\eta\,; \varphi_\top = \eta'$ and $(\mu_{AB}\,; \varphi_{A \otimes B}) = ((\varphi_A \times \varphi_B)\,; \mu'_{AB})$.

Proposition 2. $\mathcal{R}/\mathsf{Bij} \simeq [\mathcal{R}, \mathsf{Set}]_{lax\otimes}$

2.2 Specifying types

Definition 3. Let $\mathcal{R}$ be a category and $\mathcal{B}$ a bicategory [7]. A *lax functor* $\mathsf{P} : \mathcal{R} \to \mathcal{B}$ is an assignment for each object A of $\mathcal{R}$ of an object $\mathsf{P}A$ in $\mathcal{B}$ and for each arrow $f : A \to B$ of a 1-cell $\mathsf{P}f : \mathsf{P}A \to \mathsf{P}B$ in $\mathcal{B}$. Furthermore, P comes equipped with the 2-cells

$$\mu_{fg} : \mathsf{P}f \,;\mathsf{P}g \longrightarrow \mathsf{P}(f\,;g) \quad \text{for every composable } f \text{ and } g, \text{ and}$$
$$\eta_A : id_{\mathsf{P}A} \longrightarrow \mathsf{P}(\mathrm{id}_A) \quad \text{for every object } A,$$

satisfying coherence conditions similar to (1).

The lax monoidal functors from 1 are just lax functors between monoidal categories, regarded as bicategories with one object.

Type specifications. To refine the type structure of an interaction category $\mathcal{R}$, we assign for each type $A \in \mathcal{R}$ a set $\mathsf{P}A$ of new "properties", or "predicates" over it. Putting them all together, we construct a new interaction category $\mathcal{R}^{\mathsf{P}}$. No new processes are added: an $\mathcal{R}^{\mathsf{P}}$-morphism from $\alpha \in \mathsf{P}A$ to $\beta \in \mathsf{P}B$ will be just an $\mathcal{R}$-morphism $f : A \to B$, mapping the elements that satisfy α to those that satisfy β. Which αs and βs does f connect in this way will be specified by a relation $\mathsf{P}f \subseteq \mathsf{P}A \times \mathsf{P}B$. Clearly, such relations will usually not satisfy more than

$$\alpha\{\mathsf{P}f\}\beta \wedge \beta\{\mathsf{P}g\}\gamma \Longrightarrow \alpha\{\mathsf{P}(f\,;g)\}\gamma \tag{5}$$
$$\alpha = \alpha' \Longrightarrow \alpha\{\mathsf{P}(\mathrm{id})\}\alpha' \tag{6}$$

where $\alpha\{\mathsf{P}f\}\beta$ abbreviates $\langle\alpha, \beta\rangle \in \mathsf{P}f$. A type specification thus turns out to be a lax functor P from an interaction category $\mathcal{R}$ to the $\mathsf{Pos}_\perp$-category Rel of sets and relations.

Extracting from such a specification $\mathsf{P} : \mathcal{R} \to \mathsf{Rel}$ an interaction category $\mathcal{R}^{\mathsf{P}}$ is not essentially more complicated than extracting $\mathcal{R}_{\mathsf{h}}$ in 2.1, but it has very general background and deep conceptual roots.

Comprehension for categories. Consider the bicategory Span: its objects are sets, and a morphism from A to B is a pair of functions $A \leftarrow M \to B$. A 2-cell to another such pair $A \leftarrow M' \to B$ is just a function $\varphi : M \to M'$, commuting with the pairs. Given a span $B \leftarrow N \to C$, the composite $A \leftarrow (M\,;N) \to C$ is obtained by calculating a pullback of $M \to B$ and $B \leftarrow N$. Identities will clearly be in the form $A \xleftarrow{\mathrm{id}} A \xrightarrow{\mathrm{id}} A$. A span $A \xleftarrow{a} M \xrightarrow{b} B$ can also be viewed as an $A \times B$-matrix of sets, with $\langle a, b\rangle^{-1}(i, j)$ as the (i, j)-th entry. The 2-cells are obviously just entry-wise families of functions. The described composition then corresponds the usual matrix multiplication, using the set-theoretical sums and products.

Now any lax functor $\mathsf{P} : \mathcal{R} \to \mathsf{Span}$ induces the *total category* $\int_{\mathcal{R}} \mathsf{P}$, defined:

$$|\int_{\mathcal{R}} \mathsf{P}| = \sum_{X \in |\mathcal{R}|} \mathsf{P}X \tag{7}$$

$$\int_{\mathcal{R}} \mathsf{P}(\langle A, \alpha \rangle, \langle B, \beta \rangle) = \sum_{f \in \mathcal{R}(A,B)} \alpha\{\mathsf{P}f\}\beta \tag{8}$$

where $\alpha\{\mathsf{P}f\}\beta$ is the (α, β)-th entry of the matrix $\mathsf{P}f$. The composite of $\langle f, \varphi \rangle : \langle A, \alpha \rangle \longrightarrow \langle B, \beta \rangle$ and $\langle g, \psi \rangle : \langle B, \beta \rangle \longrightarrow \langle C, \gamma \rangle$ in $\int_{\mathcal{R}} \mathsf{P}$, is $\Big\langle (f\,; g),\, \mu^{\alpha\gamma}_{fg}(\beta, \varphi, \psi) \Big\rangle$, where

$$\mu^{\alpha\gamma}_{fg} : \sum_{\beta \in \mathsf{P}B} \alpha\{\mathsf{P}f\}\beta \times \beta\{\mathsf{P}g\}\gamma \longrightarrow \alpha\,\{\mathsf{P}(f\,; g)\}\,\gamma \tag{9}$$

is the (α, γ)-th component of the 2-cell μ_{fg}. The identity on $\langle A, \alpha \rangle$ is $\langle \mathrm{id}_A, \eta_A(\alpha) \rangle$.

While the total category comes with the obvious projection $I : \int_{\mathcal{R}} \mathsf{P} \longrightarrow \mathcal{R}$, any functor $F : \mathcal{Q} \longrightarrow \mathcal{R}$ (say, between small categories) induces a lax functor $\mathsf{P}_F : \mathcal{R} \to \mathsf{Span}$, with an isomorphism $F' : \mathcal{Q} \to \int_{\mathcal{R}} \mathsf{P}_F$ satisfying $F = (F'\,; I)$. The lax functor P_F sends each $A \in \mathcal{R}$ to the set $\mathsf{P}A = \{\alpha \in \mathcal{Q} | F\alpha = A\}$, and each arrow $f : A \to B$ to the $\mathsf{P}A \times \mathsf{P}B$-matrix of sets

$$\alpha\{\mathsf{P}f\}\beta = \{\varphi \in \mathcal{Q}(\alpha, \beta) | F\varphi = f\}. \tag{10}$$

The described correspondence extends to the equivalence

Proposition 4.[2] $\mathsf{Cat}/\mathcal{R} \;\simeq\; [\mathcal{R}, \mathsf{Span}]_{lax}$

between the category of functors to $\mathcal{R}$, with commutative triangles as morphisms, and the category of lax functors $\mathcal{R} \to \mathsf{Span}$ and the functional lax transformations. A lax transformation $\varphi : \mathsf{P} \to \mathsf{Q} : \mathcal{R} \to \mathsf{Span}$ is a family of matrices $\varphi_A : \mathsf{P}A \nrightarrow \mathsf{Q}A$ with a coherent 2-cell $(\mathsf{P}f\,; \varphi_B) \longrightarrow (\varphi_A\,; \mathsf{Q}f)$ for every $f : A \to B$. It is said to be *functional* if all components φ_A are functions.

The establisned equivalence extends in various directions. By dropping the functionality requirement, and varying the notion of lax transformation on the right-hand side, one gets various interesting classes of morphisms on the left-hand side: indexed profunctors and anafunctors [18], and a categorical form of simulations. On the other hand, it restricts to the Conduché correspondence [23], to the Grothendieck construction [15], and so on, until it boils down to the familiar correspondence $\mathsf{Set}/R \simeq [R, \mathsf{Set}]$ of the functions to a set R and the R-indexed sets — and, finally, to the *comprehension scheme* $\mathsf{Sub}/R \cong [R, \Omega]$, connecting the subobjects of R with the predicates over it. Indeed, just as the

[2] Although this correspondence, at least on the level of objects, seems too basic to be unknown, we remain unable to find any reference to it, in literature or folklore. A more complicated one, relating $\mathsf{Cat}/\mathcal{R}$ with the *normalised* lax functors from $\mathcal{R}$ to *categories and profunctors* is often mentioned and has been known for long [8], yet even there, the arrow part may still deserve attention, even from experts.

extension $\{x \in R | p(x)\} \hookrightarrow R$ can be obtained as a pullback of the truth $t : 1 \to \Omega$ along the predicate $p : R \to \Omega$, the construct $\int_{\mathcal{R}} \mathsf{P} \longrightarrow \mathcal{R}$ can be obtained as a pullback along $\mathsf{P} : \mathcal{R} \to \mathsf{Span}$ of the obvious projection $\mathsf{t} : \mathsf{Span}^\bullet \longrightarrow \mathsf{Span}$, where $\mathsf{Span}^\bullet$ is the total category of the identity on Span.

To restrict to the lax functors $\mathsf{P} : \mathcal{R} \to \mathsf{Rel}$, note that a relation $R \hookrightarrow A \times B$ is a jointly monic span $A \leftarrow R \to B$, i.e. a matrix of 0s and 1s. The canonical functor $\mathsf{Span} \to \mathsf{Rel}$ is thus obtained by taking monic images of spans, or, in terms of matrices of sets, by reducing each nonempty entry to 1. The category $[\mathcal{R}, \mathsf{Rel}]_{lax}$ is thus a reflective subcategory of $[\mathcal{R}, \mathsf{Span}]_{lax}$. Restricting to it the right-hand side in proposition 4, we get on the left-hand side the category $\mathsf{Fait}/\mathcal{R}$ spanned by the faithful functors to $\mathcal{R}$. Indeed, according to (8), $\int_{\mathcal{R}} \mathsf{P} \longrightarrow \mathcal{R}$ is faithful if and only if each $\alpha\{\mathsf{P}f\}\beta$ is just 0 or 1.

The interaction category specified by $\mathsf{P} : \mathcal{R} \to \mathsf{Rel}$ will be denoted $\mathcal{R}^{\mathsf{P}} = \int_{\mathcal{R}} \mathsf{P}$.

2.3 Lifting the structure

In principle, the signature of an interaction category combines linear logic with delay monads, *in* an enriched setting. In the present paper, we can only comment on the first of these three aspects.

The linear structure of $\mathcal{R}$ lifts to $\mathcal{R}_{\mathsf{h}}$ in a fairly straightforward way. First of all, $\mathcal{R}_{\mathsf{h}}$ is $\star$-autonomous (resp. compact closed) if and only if $\mathcal{R}$ is. Namely, any endofunctor D on $\mathcal{R}$ lifts to an endofunctor on D_{h} on $\mathcal{R}_{\mathsf{h}}$: the arrow part is again the h-image of the obvious family $(A \multimap B) \longrightarrow (DA \multimap DB)$. In this way, the duality lifts from $\mathcal{R}$ to $\mathcal{R}_{\mathsf{h}}$.

Moreover, any natural transformation between lifted endofunctors lifts too — along the functor $\mathcal{R} \to \mathcal{R}_{\mathsf{h}}$. Monads and comonads thus induce monads and comonads. Recall that a *bang* is a monoidal comonad $! : \mathcal{R} \to \mathcal{R}$ the coalgebras of which are $\otimes$-comonoids. This can be expressed by natural transformations $e_A : !A \to \top$ and $d_A : !A \to !A \otimes !A$, imposing the required structure [9]. A bang thus lifts from $\mathcal{R}$ to $\mathcal{R}_{\mathsf{h}}$. However, the couniversal bang, sending each object to the corresponding cofree $\otimes$-comonoid, may lose its property in lifting.

Finally, using just definition (2), one easily shows that the (weak) products *and* coproducts are preserved and thus created by the functor $\mathcal{R} \to \mathcal{R}_{\mathsf{h}}$ as soon as the specification $\mathsf{h} : \mathcal{R} \to \mathsf{Set}$ preserves the (weak) products. However, we shall see that it usually does not. Process specifications thus yield categories with few limits and colimits. Adding more types corrects this.

Lifting structures along type specifications is less straightforward, although quite uniform. Looking at the correspondence from proposition 4, one sees that any, say, binary functorial operation $\diamond$, preserved by $\int_{\mathcal{R}} \mathsf{P} \longrightarrow \mathcal{R}$, corresponds to a functional lax transformation $\mathsf{P}A \times \mathsf{P}B \xrightarrow{\diamond} \mathsf{P}(A \diamond B)$, with $\langle A, \alpha \rangle \diamond \langle B, \beta \rangle = \langle A \diamond B, \alpha \diamond \beta \rangle$. In order to lift $\diamond$ from $\mathcal{R}$ to $\mathcal{R}^{\mathsf{P}}$, we must thus specify the corresponding transformations. This is where we depart from the degeneracies of $\mathcal{R}$.

3 Examples

The idea is to start from a simple model $\mathcal{R}$, and successively refine it by specifying

$$\mathcal{R} \longrightarrow \mathcal{R}_{\mathsf{h}_1} \longleftarrow (\mathcal{R}_{\mathsf{h}_1})^{\mathsf{P}_1} \longrightarrow ((\mathcal{R}_{\mathsf{h}_1})^{\mathsf{P}_1})_{\mathsf{h}_2} \longleftarrow (((\mathcal{R}_{\mathsf{h}_1})^{\mathsf{P}_1})_{\mathsf{h}_2})^{\mathsf{P}_2} \longrightarrow \cdots$$

The view of processes as relations in time suggests that any category of relations could be taken as the base $\mathcal{R}$. Namely, the calculus of relations as jointly monic spans can be developed not just over sets but over more general categories $\mathcal{C}$ [12]. The obtained category $\mathsf{Rel}(\mathcal{C})$ is always compact closed, but varying $\mathcal{C}$ allows additional structure on *actions*.

3.1 Synchrony

The simplest case is of course $\mathsf{Rel} = \mathsf{Rel}(\mathsf{Set})$. Let the process specification $\mathsf{s} : \mathsf{Rel} \to \mathsf{Set}$ assign to every set A the poset $\mathsf{s}A$ of nonempty, prefix-closed sets of finite strings from A. These strings are to be thought of as "the elements of A extended in time", so that the elements of $\mathsf{s}A$ become "the subsets of A extended in time". Algebraically, they can be presented as one-sided multiplicative systems of the free monoid A^*, i.e., the complements of the one-sided ideals of A^*.

The arrow part of s will map a relation $A \leftarrow R \to B$ to the function $\mathsf{s}R : \mathsf{s}A \to \mathsf{s}B$, defined

$$\mathsf{s}R(S) = \{t \in B^* | \exists s \in S.sR^*t\}, \tag{11}$$

where $A^* \leftarrow R^* \to B^*$ is the componentwise extension of R to strings. The lax monoidal structure consists of the function $\mu_{AB} : \mathsf{s}A \times \mathsf{s}B \longrightarrow \mathsf{s}(A \otimes B)$, where

$$\mu_{AB}(S,T) = \{u \in (A \otimes B)^* | \pi_A^*(u) \in S \wedge \pi_B^*(u) \in T\}, \tag{12}$$

and $\eta \in \mathsf{s}1$ consisting of all finite strings of $\bullet \in 1$.

The category $\mathsf{sproc} = \mathsf{Rel}_{\mathsf{s}}$, obtained by the construction from 2.1, is a rudimentary interaction category of synchronous processes, modulo the trace equivalence. Finer notions of behaviour are obtained by taking as the elements of $\mathsf{s}A$ transition systems, or A-labelled trees, rather than just the traces $S \subseteq A^*$. Definitions (11) and (12) readily extend. Working modulo bisimilarity complicates matters [19, 20], but everything goes through.

The synchronous interaction category SProc [1] is obtained by a further type specification $\mathsf{S} : \mathsf{sproc} \to \mathsf{Rel}$. Its object part will actually be the same as for the above process specification. Its arrow part should take the process $U \in \mathsf{sproc}(A, B)$ to the relation $\mathsf{S}A \leftarrow \mathsf{S}U \to \mathsf{S}B$ defined

$$S\{\mathsf{S}U\}T \iff \forall u \in U.\ \pi_A^*(u) \in S \wedge \pi_B^*(u) \in T \tag{13}$$

If sproc is taken modulo bisimilarity, the process U in this definition should be replaced by the corresponding set of traces.

The category SProc is thus $(\mathsf{Rel}_{\mathsf{s}})^{\mathsf{S}}$. This order of specifying can be changed, as one can easily see by constructing the pullback of the functors $\mathsf{Rel} \longrightarrow \mathsf{sproc} \longleftarrow$

SProc, obtained from specifications. Although it is intuitively simpler to first specify the notion of process, the advantage of first specfying the types is that the biproducts and the cofree comonoids of SProc — neither of which are present in sproc — can then be traced back to Rel.

3.2 Asynchrony

To capture the asynchrony, one can start from the calculus of relations $\mathsf{Rel}^\bullet$ developed over the category $\mathsf{Set}^\bullet$ of *pointed* sets, all containing a fixed element $\bullet$, which all functions must preserve. $\bullet$ represents the *idle* action, which allows processes to wait. $\mathsf{Set}^\bullet$ is the Kleisli category for the monad $1+(-) : \mathsf{Set} \to \mathsf{Set}$, but it is sometimes useful to view it as the category of sets and *partial* functions. $\mathsf{Rel}^\bullet = \mathsf{Rel}(\mathsf{Set}^\bullet)$ can thus be presented either as the full subcategory of Rel spanned by the objects in the form $1+A$, or as the category of sets and partial relations. A partial relation $A \leftharpoondown R \rightharpoonup B$ actually boils down to a triple $\langle R_A, R_\times, R_B\rangle$, where $R_\times \hookrightarrow A\times B$ is an ordinary binary relation, while $R_A \hookrightarrow A$ and $R_B \hookrightarrow B$ the parts where R is undefined. The tensor and the cotensor are $A \otimes B = A+B+A\times B$, and the embedding $1+(-) : \mathsf{Rel}^\bullet \to \mathsf{Rel}$ preserves them. The weak biproducts $A+B$ are also preserved, and note that $\mathsf{Rel}^\bullet$ does not have the strong ones, which is reflected in the asynchronous interaction categories.

To specify $\mathsf{as}^\bullet : \mathsf{Rel}^\bullet \to \mathsf{Set}$, identify $\mathsf{Rel}^\bullet$ with its image in Rel and note that $\bullet$ must be the unit of any monoid in $\mathsf{Set}^\bullet$. Rather than $(1+A)^*$, the free monoid over $1+A$ is thus $1+A^+$, where A^+ consists of all *nonempty* strings from A.

The object part of $\mathsf{as}^\bullet$ thus takes $1+A$ to the set of prefix-closed subsets of $1+A^+$, each containing $\bullet$. The arrow part is defined using the monoid homomorphism $\widetilde{(-)} : (1+A)^* \longrightarrow 1+A^+$, which removes $\bullet$ from all nontrivial strings, and induces the weak equivalence $s \approx t \iff \widetilde{s} = \widetilde{t}$. A relation $1+A \leftarrow R \rightarrow 1+B$ now goes to the function $\mathsf{as}^\bullet R : \mathsf{as}^\bullet A \longrightarrow \mathsf{as}^\bullet B$, defined

$$\mathsf{as}^\bullet R(S) = \{t \in 1+B^+ \,|\, \exists s \in S.\ s \approx R^* \approx t\}. \tag{14}$$

In words, a string t belongs to $\mathsf{as}^\bullet R(S)$ if there is a string s in S such that s and t can be filled up with sequences of $\bullet$ in such a way that they become componentwise R-related.

By a similar trick, the function $\mu_{AB} : \mathsf{as}^\bullet A \times \mathsf{as}^\bullet B \longrightarrow \mathsf{as}^\bullet(A \otimes B)$ shuffles the strings:

$$\mu_{AB}(S,T) = \left\{u \in ((1+A)\times(1+B))^+ \,|\, \widetilde{\pi_A^*}(u) \in S \wedge \widetilde{\pi_B^*}(u) \in T\right\} \tag{15}$$

An element of $\mu_{AB}(S,T)$ is obtained by taking some $s \in S$ and $t \in T$, possibly of different length, interpolating $\bullet$ in them at will, to get $s' = \alpha_1 \ldots \alpha_n$ and $t' = \beta_1 \ldots \beta_n$, and then forming $u = \langle\alpha_1, \beta_1\rangle \ldots \langle\alpha_n, \beta_n\rangle$. The unit is $\eta = \{\bullet\}$.

The asynchronous interaction category $\mathsf{as}^\bullet\mathsf{proc} = \mathsf{Rel}^\bullet_{\mathsf{as}^\bullet}$ is obtained as before. A version depicting a finer notion of behaviou can again obtained using $(1+A)$-labelled trees or transition systems, this time modulo weak or branching bisimilarity. A full fledged asynchronous category $\mathsf{AS^\bullet Proc}$, with *weak* biproducts and a *weakly* couniversal bang, is obtained by adding more types along a specification $\mathsf{AS^\bullet} : \mathsf{as^\bullet proc} \longrightarrow \mathsf{Rel}$, similar to S from section 3.1, but relaxed modulo $\approx$.

The original asynchronous category ASProc [1, sec. 5] is obtained in the same way, but using relations in place of partial functions, i.e. starting from $\mathsf{Req} = \mathsf{Rel}(\mathsf{Rel})$ rather than $\mathsf{Rel^\bullet} = \mathsf{Rel}(\mathsf{Set^\bullet})$. Req is the category of sets and the *partial equivalence* relations on $A+B$ as the morphisms from A to B. Namely, a relation $A \nleftrightarrow R \nleftrightarrow B$ in Rel boils down to a jointly surjective pair $A \rightarrow R \leftarrow B$ in $\mathsf{Set^\bullet}$. Alternatively, Req can be viewed as the full subcategory of Rel spanned by the power sets $\wp A$. The tensor preservation along the embedding $\wp : \mathsf{Req} \rightarrow \mathsf{Rel}$ boils down to the exponential laws $\wp(A+B) \cong \wp A \times \wp B$ and $\wp 0 = 1$.

The specification $\mathsf{as} : \mathsf{Req} \rightarrow \mathsf{Set}$ assigns to each A the set of nonempty prefix closed sets of sequences from $\wp^+ A = \wp A - \emptyset$. The empty set is deleted because it plays the role of $\bullet$. The resulting category $\mathsf{asproc} = \mathsf{Req_{as}}$ compares to $\mathsf{as^\bullet proc}$ just as Req compares to $\mathsf{Rel^\bullet}$. For instance, bang comonads are precluded by the fact that any functor $! : \mathsf{Req} \rightarrow \mathsf{Req}$ with a natural family $e_A : !A \rightarrow 0$ must be trivial.

The structure of actions can be further enriched using other monads on Set. E.g., consider the one sending A to $1+A+A$. (If its unit is chosen to include A in $1+A+A$ as the first copy, then the multiplication should send the first two As from $1+(1+A+A)+(1+A+A)$ to $1+A+A$ in order, and twist the last two of them.) Besides the idling $\bullet$, this monad captures the input/output distinction — between the elements of the two copies of A. The Kleisli category $\mathsf{Set^{\overline{\bullet}}}$ for this monad can now be viewed as the category of sets, with pairs $\langle f, F\rangle$ as morphisms from A to B, where f is a partial function $A \rightharpoonup B$ and F is a subset of A. The composite of $\langle f, F\rangle : A \rightarrow B$ and $\langle g, G\rangle : B \rightarrow C$ consists of the usual composite of partial function $(f\,; g)$, accompanied with the set $\big(F \cap \varphi^{-1}(G)\big) \cup \big(\overline{F} \cap \varphi^{-1}(\overline{G})\big)$, where $\overline{F}, \overline{G}$ denote the complements. The free monoid $A^{\overline{*}}$ over A in $\mathsf{Set^{\overline{\bullet}}}$ will be the quotient of $1+(A+A)^+$ satisfying $\alpha\overline{\alpha} = \bullet$ for all $\alpha \in A$, with $\overline{(-)} : A+A \longrightarrow A+A$ denoting the twist map. All monoids in $\mathsf{Set^{\overline{\bullet}}}$ are thus groups — which means that any computation can be "consumed" and "internalized" as $\bullet$. One is thus led to consider the *infix closed* sets $S \subseteq A^{\overline{*}}$, i.e. such that

$$\alpha s \overline{\alpha} \in S \Longrightarrow s \in S \tag{16}$$

for any $\alpha \in A$. They correspond to normal subgroups of $A^{\overline{*}}$ roughly like the prefix closed sets correspond to the ideals of A^*, the underlying idea being that, in reversible time, computations develop in two directions.

The specification $\mathsf{as^{\overline{\bullet}}} : \mathsf{Rel^{\overline{\bullet}}} \rightarrow \mathsf{Set}$, where $\mathsf{Rel^{\overline{\bullet}}} = \mathsf{Rel}(\mathsf{Set^{\overline{\bullet}}})$, will now assign to each A the set of the infix closed subsets of $A^{\overline{*}}$. The arrow part can be formally defined just as for $\mathsf{as^\bullet}$ — but the kernel of $\widetilde{(-)} : (1+A+A)^* \longrightarrow A^{\overline{*}}$ will be much larger and instead of sequences of $\bullet$, we shall be interpolating more general strings of the input and output actions, that reduce to $\bullet$.

For finer notions of behaviour, instead of infix closed sets, one could use transition systems without a distinguished initial state, or labelled acyclic graphs, modulo the corresponding notion of bisimilarity...

3.3 Coherence

Each of categories constructed so far can be refined by first extending Rel, say, by the notion of *coherence*. It can be introduced by lax functor $\mathsf{C} : \mathsf{Rel} \to \mathsf{Rel}$, assigning to each set A the set of all symmetric, irreflexive binary relations on it. A relation $A \leftarrow R \rightarrow B$ now induces $\mathsf{C}A \leftarrow \mathsf{C}R \rightarrow \mathsf{C}B$, defined

$$\Phi\{\mathsf{C}R\}\Psi \iff \forall \alpha\alpha' \in A\beta\beta' \in B.\ (\alpha R\beta \wedge \alpha' R\beta') \Rightarrow (\alpha\Phi\alpha' \Rightarrow \beta\Psi\beta'). \tag{17}$$

The total category $\mathsf{Rel}^{\mathsf{C}}$ will be the familiar category Coh of coherence spaces [14]. By imposing on each set of traces $S \subseteq A^*$ (or on labelled trees, or transition systems) the coherence requirement

$$s\alpha, s\alpha' \in S \Longrightarrow \alpha\Phi\alpha' \tag{18}$$

for all $\alpha \neq \alpha' \in A$, all previously described specifications lift to Coh, and yield interaction categories with a grain of true concurrency. It is interesting to notice that already the synchronous ones can be specified in many different, meaningful ways.

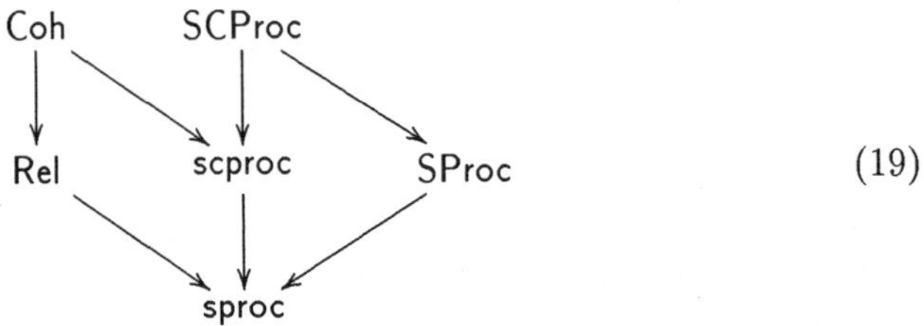

3.4 Games

Categories of games are specifed starting from *signed* sets $\mathsf{Set}_{\pm}$. A signed set A is a pair $\langle A_-, A_+\rangle$, of ordinary sets; a signed function $f : A \rightarrow B$ is an ordinary partial function $f : A_+ + B_- \rightharpoonup B_+ + A_-$. To compose it with $g : B \rightarrow C$, i.e. $g : B_+ + C_- \rightharpoonup C_+ + B_-$, follow the images of each $x \in A_+$ along the tower $A_+ \xrightarrow{f} B_+ \xrightarrow{g} B_- \xrightarrow{f} B_+ \xrightarrow{g} \cdots$. If an image ever leaves B and lands in A_- or in C_+, it will be the value of $(f\,;g)$ at x. Otherwise, $(f\,;g)$ remains undefined at x. To define $(f\,;g)$ on C_-, follow the tower $C_- \xrightarrow{g} B_- \xrightarrow{f} B_+ \xrightarrow{g} B_- \cdots$. The identity on $A = \langle A_-, A_+\rangle$ is obviously the identity on $A_+ + A_-$. The obtained category is compact closed, with the structure

$$A \otimes B = \langle A_- + B_-, A_+ + B_+\rangle, \tag{20}$$
$$A^* = \langle A_+, A_-\rangle. \tag{21}$$

The free compact closed categories are effectively described in terms of $\mathsf{Set}_\pm$ [17, sec. 3] — in fact, by a type specification over it. A further, somewhat more complicated refinement yields the free $\star$-autonomous categories.

A basic category of games is obtained by a type specification $\mathsf{G} : \mathsf{Set}_\pm \to \mathsf{Rel}$, sending each A to the set of all nonempty prefix closed subsets of $(A_- \times A_+)^*$. The functions $\mathsf{G}A \times \mathsf{G}B \xrightarrow{\otimes} \mathsf{G}(A \otimes B)$ and $\mathsf{G}A \times \mathsf{G}B \xrightarrow{\multimap} \mathsf{G}(A^\star \otimes B)$ shuffle these sets, the latter in the subtle way described in [3]. A signed function $f : A \to B$ then induces $\mathsf{G}A \leftarrow \mathsf{G}f \to \mathsf{G}B$, relating Ψ and Φ if and only if f yields a *history free strategy* for $\Phi \multimap \Psi$ [3].

The history sensitive strategies can now be introduced in a process specification over the total category $\mathsf{Set}_\pm^\mathsf{G}$. Furthermore, the winning positions can be added in a further type specification, or just extending the specification G. Clearly, G can also be extended to include the equivalence relations on positions, essential for [4]. The relations $\mathsf{G}f$ will be supplied with the requirement that f preserves the equivalences on the games being related. The category from [4] will follow from an additional process specification, identifying the equivalent strategies.

References

1. S. Abramsky, S.J. Gay and R. Nagarajan, Interaction categories and the foundations of the typed concurrent programming. In: *Deductive Program Design: Proceedings of the 1994 Marktoberdorf International Summer School*, (Springer 1996)
2. S. Abramsky, S.J. Gay and R. Nagarajan, Specification structures and propositions-as-types for concurrency. In: *Logics for Concurrency: Structure vs. Automata. Proceedings of the VIIIth Banff Workshop*, Lecture Notes in Computer Science(Springer 1996)
3. S. Abramsky and R. Jagadeesan, Games and full completeness for multiplicative linear logic, *J. Symbolic Logic*
4. S. Abramsky, R. Jagadeesan and P. Malacaria, Full abstraction for PCF, *submitted* (1996)
5. M. Barr, *$\star$-Autonomous Categories*, Lecture Notes in Mathematics 752 (Springer 1979)
6. M. Barr, $\star$-Autonomous categories and linear logic, *Math. Structures Comput. Sci.* 1/2(1991), 159–178
7. J. Bénabou, Introduction to bicategories, in: *Reports of the Midwest Category Seminar I*, Lecture Notes in Mathematics 47 (Springer, 1967) 1–77
8. J. Bénabou, 2-dimensional limits and colimits of distributors, abstract of a talk given in Oberwolfach (1972)
9. G.M. Bierman, What is a categorical model of intuitionistic linear logic?, in: *Proceedings of Conference on Typed Lambda Calculus and Applications*, M. Dezani-Ciancaglini and G. Plotkin, eds., Lecture Notes in Computer Science 902 (Springer 1995)
10. V. Danos and L. Regnier, The structure of multiplicatives, *Archive form Math. Logic* 28(1989) 181–203
11. T. Fox, Coalgebras and cartesian categories, *Comm. Algebra*, 4/7(1976) 665–667
12. P.J. Freyd and A. Scedrov, *Categories, Allegories*, North-Holland Mathematical Library 39 (North-Holland, 1990)

13. A. Fleury and C. Retoré, The MIX rule, Unpublished note, 1990
14. J.-Y. Girard et al., *Proofs and Types*, Cambridge Tracts in Theoretical Computer Science 7 (Cambridge Univ. Press 1989)
15. A. Grothendieck, Catégories fibrées et descente, Exposé VI, *Revêtements Etales et Groupe Fondamental (SGA1)*, Lecture Notes in Mathematics 224 (Springer, 1971) 145–194
16. G.M. Kelly, *Basic Concepts of Enriched Category Theory*, L.M.S. Lecture Notes 64 (Cambridge Univ. Press 1982)
17. G.M. Kelly and M.L. Laplaza, Coherence for compact closed categories, *J. Pure Appl. Algebra* 19(1980) 193–213
18. M. Makkai, Avoiding the axiom of choice in general category theory, to appear in *J. Pure Appl. Algebra*
19. D. Pavlović, Categorical logic of concurrency and interaction I. Synchronous processes, in: *Theory and Formal Methods of Computing 1994*, C.L. Henkin et al., eds. (World Scientific 1995), 105–141
20. D. Pavlović, Convenient categories of processes and simulations I: modulo strong bisimilarity, *Category Theory and Computer Science '95*, D.H. Pitt et al., eds., Lect. Notes in Comp. Science 953 (Springer, 1995), 3–24
21. D. Pavlović, Maps I: relative to a factorisation system, *J. Pure Appl. Algebra* 99(1995), 9–34; Maps II: Chasing diagrams in categorical proof theory, *J. of the IGPL* 2/4(1996), 159–194
22. R.A.G. Seely, Linear logic, $\star$-autonomous categories and cofree coalgebras, in: J. Gray and A. Scedrov (eds.), *Categories in Computer Science and Logic*, *Contemp. Math.* 92 (Amer. Math. Soc., 1989), 371-382
23. R. Street, Conduché functors, *a hand written note*, dated 15 October 1986, 4 pp.

Shedding New Light in the World of Logical Systems

Uwe Wolter Alfio Martini*

Technische Universität Berlin, FB Informatik, Sekr. 6-1
Franklinstr. 28/29, D-10587 Berlin, Germany
{wolter,alfio}@cs.tu-berlin.de

Abstract

The notion of an *Institution* [5] is here taken as the precise formulation for the notion of a logical system. By using elementary tools from the core of category theory, we are able to reveal the underlying mathematical structures lying "behind" the logical formulation of the satisfaction condition, and hence to acquire a both suitable and deeper understanding of the institution concept. This allows us to systematically approach the problem of describing and analyzing relations between logical systems. Theorem 2.10 redesigns the notion of an institution to a purely categorical level, so that the satisfaction condition becomes a functorial (and natural) transformation from specifications to (subcategories of) models and vice versa. This systematic procedure is also applied to discuss and give a natural description for the notions of institution morphism and institution map. The last technical discussion is a careful and detailed analysis of two examples, which tries to outline how the new categorical insights could help in guiding the development of a unifying theory for relations between logical systems.

1 Introduction

The need to live within a multitude of logical systems in order to tackle the different problems arising in the field of formal software specification and development is well-recognized. Such a necessity is essentially rooted in the practical requirements to capture various aspects of the system's specification, adequacy of the underlying logic (with respect to the problem to be specified), and also in the availability and reusability of supporting tools, as for instance, theorem provers [3, 2].

*Research supported in part by a CNPq-grant 200529/94-3.

In this context, a "logical system" is usually understood as any (algebraic) framework consisting of a collection of admissible languages (signatures), a collection of admissible structures on that language (models), and for every language, a set of sentences, which are used to axiomatize collections of models.

In order to formally capture the above informal notion of logical system and to provide a basis for doing much of the work on software specification independently of the underlying logical system chosen, Goguen and Burstall [5] introduced the notion of an *institution.* The major requirement here is that there is a satisfaction relation between models and sentences which is consistent under change of notation. The formal definition goes as follows:

Definition 1.1 (Institution) An *institution* $\mathcal{I} = (SIGN, Sen, Mod, \models)$ consists of a category $SIGN$, whose objects are called *signatures*; a functor $Sen : SIGN \to SET$, giving for each signature a set whose elements are called *sentences* over that signature; a functor $Mod : SIGN^{op} \to CAT$, giving for each signature Σ a category whose objects are called Σ-models, and whose arrows are called Σ-morphisms; and a function $\models$ associating to each signature Σ a relation $\models_\Sigma \subseteq |Mod(\Sigma)| \times Sen(\Sigma)$, called ($\Sigma$-)*satisfaction relation*, such that for each arrow $\phi : \Sigma_1 \to \Sigma_2$ in $SIGN$ the *satisfaction condition*

$$\textbf{(SC)} \quad M_2 \models_{\Sigma_2} Sen(\phi)(\varphi_1) \iff Mod(\phi)(M_2) \models_{\Sigma_1} \varphi_1,$$

holds for any $M_2 \in |Mod(\Sigma_2|$ and any $\varphi_1 \in Sen(\Sigma_1)$.

However, working in an arbitrary but fixed "institution" is not always adequate, since realistic software specification often requires not only a span of different logical systems but also a possibility of migrating from one logical system to another. Therefore, a first proposal for a formalization of a relation between two logical systems was introduced in [5] under the name of *institution morphism.* Since then, in order to capture different intuitions, requirements, and achieve more flexibility as well (translation of proofs and even proof calculi), several other concepts for relating logical systems were developed (for a survey, see, e.g., [3, 9]).

Such an explosion of possibilities called inevitably to an attempt to discuss about the possibility of sketching some notion of a taxonomy in order to classify such maps. The proposal in [9] was based on the directions of the arrows (accounting for translation of signatures, sentences, and models) involved in the different formulations. It is additionally mentioned that the satisfaction condition, and the various forms in which it appears in these mappings, may contribute to add some extra complexity to this problem as well. In [8], attention is also brought to the need of having a better understanding of the component implications of the satisfaction condition.

In this paper we use elementary tools from the core of category theory in order to get a deeper understanding of the institution concept, i.e., by revealing the underlying mathematical structures lying "behind" the logical formulation of the satisfaction condition. In doing so, we are able to approach systematically the problem of describing and analyzing relations between logical systems, as long as they can be formally reflected by the concept of an institution (or a weakening of it).

The main ideas introduced in this paper are as follows: we first reveal some (additional) underlying structures that can be extracted from the triple $(Sen, Mod, \models)$ in definition 1.1. For instance, given $\phi : \Sigma_1 \to \Sigma_2$ in $SIGN$ and $\Sigma \in |SIGN|$, we can by using elementary powerset constructions, verify straightforwardly the existence of the adjunctions $Mod(\phi) \dashv Mod(\phi)^{-1}$ and $Sen(\phi)^{-1} \dashv Sen(\phi)$. Besides, from the satisfaction relation, and using again the powerset lifting, one gets the so-called adjunction between specifications and models $th(\Sigma) \dashv mod(\Sigma)$. Using these adjuctions, it can be shown that the satisfaction condition is equivalent to the assumption of a functorial (and natural) transformation from specifications to (subcategories of) models and vice versa. This allow us to give a constructive and (natural) categorical description for an institution (theorem 2.10 - Institutional (Co-)Frame). We also examine the component implications of this condition and show that they can be given a suitable categorical interpretation by using lax indexed functors.

The above procedure gives rise to some sort of two-level methodology for investigating relations between logical systems.

- Firstly, we analyze on the level of institutional frames the following questions: What relations are there? What are the categorical properties of these relations? Which compatibility conditions are satisfied?
- secondly, we describe how the relations on the abstract level arise possibly from more fine grained transitions on the level of sentences and models, and then verify how the compatibility conditions on the abstract level can be proved possibly by checking certain properties of the fine grained transitions.

This systematic procedure is then applied to discuss and give a natural description for the notions of institution morphism and institution map. The paper also explores the systematic use of Grothendieck constructions which naturally provide other concepts used in the literature for describing and relating logical systems. The last technical discussion is a careful and detailed analysis of two examples, which tries to outline how the new categorical insights could help in guiding the development of a unifying theory for relations between logical systems.

2 Institutions "structurally"

In this section we systematically reveal some categorical structures that are intrinsic to logical systems as long as they can be formally reflected by the concept of an institution. Firstly, these investigations provide new insights into the conceptual nature of logical systems so that we can, for instance, give a very simple, although enlightening categorical description of the *satisfaction condition.* Secondly, the derived structures offer an appropriate frame for describing and analyzing relations between logical systems, as presented in the sequel.

Assuming an institution $\mathcal{I} = (SIGN, Sen, Mod, \models)$ as in 1.1 and given a set of Σ-sentences, i.e., a *specification* $\Gamma \subseteq Sen(\Sigma)$, we define the category $mod(\Sigma)(\Gamma)$ as the full subcategory of $Mod(\Sigma)$ determined by those models $M \in |Mod(\Sigma)|$ that satisfy all the sentences in Γ, i.e., we have

2.1 $|mod(\Sigma)(\Gamma)| = \{M \in |Mod(\Sigma)| \mid \forall \varphi \in \Gamma : M \models_\Sigma \varphi\}$.

Analogously, we define for a subcategory $\mathcal{M} \subseteq Mod(\Sigma)$ of Σ-models the specification $th(\Sigma)(\mathcal{M}) \subseteq Sen(\Sigma)$ given by those sentences $\varphi \in Sen(\Sigma)$ which are satisfied by all models in $\mathcal{M}$, i.e., we set

2.2 $th(\Sigma)(\mathcal{M}) = \{\varphi \in Sen(\Sigma) \mid \forall M \in |\mathcal{M}| : M \models_\Sigma \varphi\}$.

Based on these definitions, [5] presents an equation which describes the satisfaction condition more abstractly and compactly on the conceptual level of specifications and of subcategories of models. The idea here is to carry out a thorough analysis of these kinds of relations in order to find out the categorical structure "behind" the satisfaction condition.

Obviously, $mod(\Sigma)$ and $th(\Sigma)$ induce mappings between Σ-specifications and subcategories of Σ-models and vice versa. By applying power set and power category constructions respectively, we obtain the correct domains and codomains of these mappings.

Definition 2.3 (Specifications and Subcategories) Given a signature $\Sigma \in |SIGN|$, the category $Spec(\Sigma)$ of all Σ-specifications has as objects all specifications $\Gamma \subseteq Sen(\Sigma)$ and as arrows all inverse inclusions $\Gamma_1 \supseteq \Gamma_2$. Analogously, the category $Sub(\Sigma)$ of all subcategories of Σ-models has as objects all subcategories $\mathcal{M} \subseteq Mod(\Sigma)$ and as arrows all inclusion functors $\mathcal{M}_1 \subseteq \mathcal{M}_2$.

By definition of $Spec(\Sigma)$ and $Sub(\Sigma)$, we can formulate the usual categorical presentation of the Galois correspondence arising from any (satisfaction) relation as an adjunction $th(\Sigma) \dashv mod(\Sigma)$.

In [4], the description of a Galois correspondence in the language of adjoint functors is given as a typical example for *categorical overkill*. However, such a formulation will turn out to be a suitable way to get a better insight into the structure of logical systems.

Proposition 2.4 (Galois Correspondence) *Given a signature $\Sigma \in |SIGN|$ the equations (2.1) and (2.2) define functors*

$$mod(\Sigma) : Spec(\Sigma) \to Sub(\Sigma) \quad \text{and} \quad th(\Sigma) : Sub(\Sigma) \to Spec(\Sigma)$$

such that $th(\Sigma)$ is left-adjoint to $mod(\Sigma)$.

The power set and power category construction lift also a function or a functor respectively to a Galois correspondence. That is, for any signature morphism $\phi : \Sigma_1 \to \Sigma_2$ we obtain functors

2.5 $Sen(\phi)^{-1} : Spec(\Sigma_2) \to Spec(\Sigma_1)$ and $Sen(\phi) : Spec(\Sigma_1) \to Spec(\Sigma_2)$

such that $\Gamma_2 \supseteq Sen(\phi)(\Gamma_1)$ iff $Sen(\phi)^{-1}(\Gamma_2) \supseteq \Gamma_1$ for any specifications $\Gamma_1 \in |Spec(\Sigma_1)|$ and $\Gamma_2 \in |Spec(\Sigma_2)|$. This means that $Sen(\phi)^{-1} \dashv Sen(\phi)$ where the unit is given by the relation $\Gamma_2 \supseteq Sen(\phi)\,(Sen(\phi)^{-1}(\Gamma_2))$ and the co-unit by $Sen(\phi)^{-1}\,(Sen(\phi)(\Gamma_1)) \supseteq \Gamma_1$. In the same way we get functors

2.6 $\quad Mod(\phi) : Sub(\Sigma_2) \to Sub(\Sigma_1) \quad$ and $\quad Mod(\phi)^{-1} : Sub(\Sigma_1) \to Sub(\Sigma_2)$

with $Mod(\phi) \dashv Mod(\phi)^{-1}$, i.e., with $\mathcal{M}_2 \subseteq Mod(\phi)^{-1}(\mathcal{M}_1)$ iff $Mod(\phi)(\mathcal{M}_2) \subseteq \mathcal{M}_1$ for any $\mathcal{M}_1 \in |Sub(\Sigma_1)|$, $\mathcal{M}_2 \in |Sub(\Sigma_2)|$, and with the unit $\mathcal{M}_2 \subseteq Mod(\phi)^{-1}\,(Mod(\phi)(\mathcal{M}_2))$ and the co-unit $Mod(\phi)\,(Mod(\phi)^{-1}(\mathcal{M}_1)) \subseteq \mathcal{M}_1$.

Note that the subcategory $Mod(\phi)^{-1}(\mathcal{M}_1)$ of $Mod(\Sigma_2)$ is given by the set-theoretical pre-image of $\mathcal{M}_1$ w.r.t. $Mod(\phi)$, whereas $Mod(\phi)(\mathcal{M}_2)$ denotes the smallest subcategory of $Mod(\Sigma_1)$ containing the set-theoretical image of $\mathcal{M}_2$ with respect to (short w.r.t.) $Mod(\phi)$.

So far we have just presented a simple abstraction step from the conceptual level of single sentences and single models to the level of specifications and subcategories of models. The abstraction step produces a diagram of adjunctions that keeps the complete information provided by the triple $(Sen,\ Mod,\ \models)$ for any signature morphism $\phi : \Sigma_1 \to \Sigma_2$, as the following diagram shows:

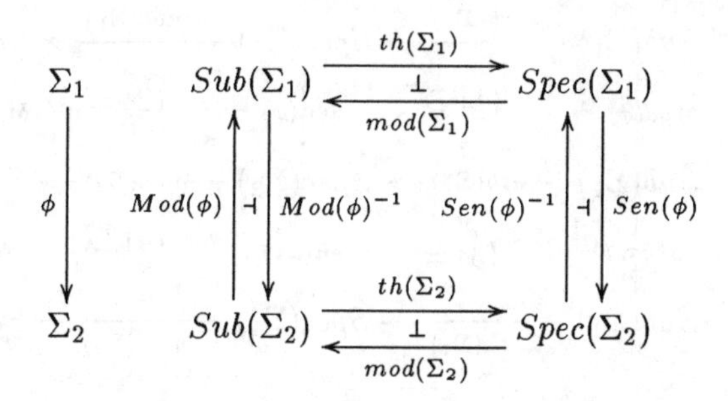

The nice observation will be that the satisfaction condition fits properly into the perspective of this diagram or, more correctly, into the perspective of the four single diagrams obtained by unfolding the above one. However, before delving into this question, we execute now a careful analysis of this condition by first breaking it into its component implications:

$$(\mathbf{SC}_{Sen\Rightarrow Mod}) \quad M_2 \models_{\Sigma_2} Sen(\phi)(\varphi_1) \;\Rightarrow\; Mod(\phi)(M_2) \models_{\Sigma_1} \varphi_1$$

$$(\mathbf{SC}_{Mod\Rightarrow Sen}) \quad M_2 \models_{\Sigma_2} Sen(\phi)(\varphi_1) \;\Leftarrow\; Mod(\phi)(M_2) \models_{\Sigma_1} \varphi_1$$

The set-theoretical meaning of these implications is clarified in the following

Proposition 2.7 (Satisfaction condition) *For each signature morphism $\phi : \Sigma_1 \to \Sigma_2$ any of the following conditions is equivalent to condition* $(\mathbf{SC}_{Sen\Rightarrow Mod})$*:*

1. $Mod(\phi)\,(mod(\Sigma_2)(\Gamma_2)) \subseteq mod(\Sigma_1)\,(Sen(\phi)^{-1}(\Gamma_2))$ *for all* $\Gamma_2 \in |Spec(\Sigma_2)|$

2. $mod(\Sigma_2)\,(Sen(\phi)(\Gamma_1)) \subseteq Mod(\phi)^{-1}\,(mod(\Sigma_1)(\Gamma_1))$ *for all* $\Gamma_1 \in |Spec(\Sigma_1)|$

3. $th(\Sigma_1)\,(Mod(\phi)(\mathcal{M}_2)) \supseteq Sen(\phi)^{-1}\,(th(\Sigma_2)(\mathcal{M}_2))$ *for all* $\mathcal{M}_2 \in |Sub(\Sigma_2)|$

Moreover any of the following conditions is equivalent to condition $(\mathbf{SC}_{Mod\Rightarrow Sen})$*:*

4. $th(\Sigma_2)\,\left(Mod(\phi)^{-1}(\mathcal{M}_1)\right) \supseteq Sen(\phi)\,(th(\Sigma_1)(\mathcal{M}_1))$ *for all* $\mathcal{M}_1 \in |Sub(\Sigma_1)|$

5. $Mod(\phi)^{-1}\,(mod(\Sigma_1)(\Gamma_1)) \subseteq mod(\Sigma_2)\,(Sen(\phi)(\Gamma_1))$ *for all* $\Gamma_1 \in |Spec(\Sigma_1)|$

6. $Sen(\phi)^{-1}\,(th(\Sigma_2)(\mathcal{M}_2)) \supseteq th(\Sigma_1)\,(Mod(\phi)(\mathcal{M}_2))$ *for all* $\mathcal{M}_2 \in |Sub(\Sigma_2)|$

Concerning 2.7.2 and 2.7.5, the reader has to bear in mind that the set-theoretical pre-image of a full subcategory becomes a full subcategory as well.

Since the categories $Spec(\Sigma)$ and $Sub(\Sigma)$ are partial order categories, i.e., only the existence of arrows matters, proposition 2.7 shows that the satisfaction condition is equivalent to each of the following two equations: $Mod(\phi)^{-1} \circ mod(\Sigma_1) = mod(\Sigma_2) \circ Sen(\phi)$ and $Sen(\phi)^{-1} \circ th(\Sigma_2) = th(\Sigma_1) \circ Mod(\phi)$.

Now, the above inequalities can be visualized straightforwardly in the "unfolded" version of our previous diagram of "adjoint situations", as shown below:

$$\begin{array}{cccccc}
\Sigma_1 & Sub(\Sigma_1) & \xrightarrow{th(\Sigma_1)} & Spec(\Sigma_1) & \xrightarrow{mod(\Sigma_1)} & Sub(\Sigma_1) \\
\downarrow \phi & \downarrow Mod(\phi)^{-1} & (1) \supseteq & \downarrow Sen(\phi) & (2) = & \downarrow Mod(\phi)^{-1} \\
\Sigma_2 & Sub(\Sigma_2) & \xrightarrow{th(\Sigma_2)} & Spec(\Sigma_2) & \xrightarrow{mod(\Sigma_2)} & Sub(\Sigma_2) \\
\uparrow \phi & \downarrow Mod(\phi) & (3) = & \downarrow Sen(\phi)^{-1} & (4) \supseteq & \downarrow Mod(\phi) \\
\Sigma_1 & Sub(\Sigma_1) & \xrightarrow[th(\Sigma_1)]{} & Spec(\Sigma_1) & \xrightarrow[mod(\Sigma_1)]{} & Sub(\Sigma_1)
\end{array}$$

Now we put $Sub(\phi) =_{def} Mod(\phi)^{-1}$, and $Spec(\phi) =_{def} Sen(\phi)$ for all $\phi : \Sigma_1 \to \Sigma_2$ in $SIGN$. These definitions induce two (covariant) indexed categories $Spec, Sub : SIGN \to CAT$, once one notes that $Spec =_{def} \wp_{set} \circ Sen$ and $Sub = \overline{\wp}_{cat} \circ Mod^{op}$ with $\wp_{set} : SET \to CAT$ the covariant power set functor and $\overline{\wp}_{cat} : CAT^{op} \to CAT$ the contravariant power category functor. Hence, the commutativity of (2) says that $mod(\Sigma_1)$ is the component at Σ_1 of an indexed functor $mod : Spec \Rightarrow Sub : SIGN \to CAT$.

By the same token, we obtain two op-indexed categories $coSub, coSpec : SIGN^{op} \to CAT$ with $coSub(\phi) =_{def} Mod(\phi)$ and $coSpec(\phi) =_{def} Sen(\phi)^{-1}$ for all $\phi : \Sigma_1 \to \Sigma_2$, i.e., with $coSpec = \overline{\wp}_{set} \circ Sen^{op}$ and $coSub = \wp_{cat} \circ Mod$, where $\overline{\wp}_{set} : SET^{op} \to CAT$ is the contravariant power set functor and $\wp_{cat} : CAT \to CAT$ the covariant power category functor. Now, the commutativity of (3) says that $th(\Sigma_2)$ is the component at Σ_2 of an op-indexed functor $th : coSub \Rightarrow coSpec : SIGN^{op} \to CAT$.

Moreover, the commutativity in (2) and (3), says that the satisfaction condition is equivalent to the assumption of a natural (and functorial) transformation from (set of) sentences to (subcategories of) models and vice versa.

Much of the categorical argumentation behind the proof of 2.7 and the above informal discussion is supported by the observation that, e.g., the family of inclusions in 2.7.1 constitute a natural transformation from $Mod(\phi) \circ mod(\Sigma_2)$ to $mod(\Sigma_1) \circ Sen(\phi)^{-1}$ and by the next

Lemma 2.8 *Let be given functors* $H : C_1 \to C_2, K : D_1 \to D_2$ *and functors* $F_i : C_i \to D_i, G_i : D_i \to C_i$ *for* $i = 1,2$ *such that* $F_i \dashv G_i$ *with unit* $\eta_i : Id_{C_i} \Rightarrow G_i \circ F_i$ *and counit* $\varepsilon_i : F_i \circ G_i \Rightarrow Id_{D_i}$. *Then we can define for any natural transformation* $\alpha : F_2 \circ H \Rightarrow K \circ F_1 : C_1 \to D_2$ *a natural transformation* $\alpha^* : H \circ G_1 \Rightarrow G_2 \circ K : D_1 \to C_2$ *by*

$$\alpha^* =_{def} (G_2 \circ K \circ \varepsilon_1) \bullet (G_2 \circ \alpha \circ G_1) \bullet (\eta_2 \circ H \circ G_1).$$

For any natural transformation $\beta : H \circ G_1 \Rightarrow G_2 \circ K : D_1 \to C_2$ *we obtain a natural transformation* $\beta_* : F_2 \circ H \Rightarrow K \circ F_1 : C_1 \to D_2$ *by*

$$\beta_* =_{def} (\varepsilon_2 \circ K \circ F_1) \bullet (F_2 \circ \beta \circ F_1) \bullet (F_2 \circ H \circ \eta_1).$$

Moreover we have $(\alpha^*)_* = \alpha$ *and* $(\beta_*)^* = \beta$.

The remaining problem is to find a concept that describes the global structures arising from the diagrams (1) or (4), respectively. This is the task of our next

Definition 2.9 (Lax indexed functor) A *lax indexed functor* $\alpha : C \overset{\leq}{\Rightarrow} D : IND \to CAT$ from an indexed category $C : IND \to CAT$ to an indexed category $D : IND \to CAT$ assigns to each $i \in |IND|$ a functor $\alpha(i) : C(i) \to D(i)$ and to each $\sigma : i \to j$ in IND a natural transformation $\alpha(\sigma) : \alpha(j) \circ C(\sigma) \Rightarrow D(\sigma) \circ \alpha(i) : C(i) \to D(j)$ such that for any $\sigma : i \to j$ and $\tau : j \to k$ in IND the following *compositionality condition* is satisfied

$$\textbf{(CC)} \quad \alpha(\tau \circ \sigma) = (D(\tau) \circ \alpha(\sigma)) \bullet (\alpha(\tau) \circ C(\sigma)).$$

A lax indexed functor $\alpha : C \overset{\leq}{\Rightarrow} D$ from an op-indexed category $C : IND^{op} \to CAT$ to an op-indexed category $D : IND^{op} \to CAT$ will be also refered to as a *lax op-indexed functor*. The concept of *op-lax indexed functor* $\alpha : C \overset{\geq}{\Rightarrow} D : IND \to CAT$ is defined in the same way. Only the natural transformations for $\sigma : i \to j$ in IND are assumed to go into the opposite direction $\alpha(\sigma) : D(\sigma) \circ \alpha(i) \Rightarrow \alpha(j) \circ C(\sigma) : C(i) \to D(j)$, with the compositionality condition holding analogously[1].

Now we are able to summarize diagrams (1) and (4). First note that the compositionality condition follows again directly from the fact that the categories $Spec(\Sigma)$ and $Sub(\Sigma)$ are partial order categories. Now, the inequality in (1) says that $(\mathbf{SC}_{Mod \Rightarrow Sen})$ is equivalent to the existence of a lax indexed functor $th(\sigma)$:

[1] This definition and corresponding terminology was based and inspired by a similar notion in [6].

$Sub \overset{\leq}{\Rightarrow} Spec$. Similarly, the inequality in (4) points out that $(\mathbf{SC}_{Sen \Rightarrow Mod})$ is equivalent to the existence of an op-lax op-indexed functor $mod(\phi) : coSpec \overset{\geq}{\Rightarrow} coSub$. In addition, it's not hard to see the concept of lax indexed functor as a generalization of the concept of an indexed functor, where the family of natural transformations indexed by the arrows of the base category are restricted to be identities. Moreover, once the indexed "collection" in the above definition is constrained to be formed by partial order categories, an indexed functor, as provided, e.g., by (2) and (3), reflects essentially the simultaneous existence of both a lax and an op-lax indexed functor (compare proposition 2.7).

The categorical structures intrinsic to institutions are summarized in the following

Theorem 2.10 (Institutional (co-)frame) *For any institution* $\mathcal{I} = (SIGN, Sen, Mod, \models)$ *we can construct the* institutional frame $\mathcal{F}(\mathcal{I}) = (SIGN, Spec, Sub, mod, th)$ associated to $\mathcal{I}$ *such that:*

- *Spec* : $SIGN \to CAT$ *and* $Sub : SIGN \to CAT$ *are indexed categories defined by* $Spec = \wp_{set} \circ Sen$ *and* $Sub = \overline{\wp}_{cat} \circ Mod^{op}$.
- $\models$ *provides for any* Σ *in* $SIGN$ *functors* $mod(\Sigma) : Spec(\Sigma) \to Sub(\Sigma)$ *and* $th(\Sigma) : Sub(\Sigma) \to Spec(\Sigma)$ *such that* $th(\Sigma) \dashv mod(\Sigma)$.
- *The satisfaction condition is a necessary and sufficient condition that the functors* $mod(\Sigma)$ *and* $th(\Sigma)$ *constitute an indexed functor* $mod : Spec \Rightarrow Sub$ *and a lax indexed functor* $th : Sub \overset{\leq}{\Rightarrow} Spec$*, respectively.*

In the same way, we can construct the institutional co-frame $\overline{\mathcal{F}}(\mathcal{I}) = (SIGN, coSpec, coSub, mod, th)$ associated to $\mathcal{I}$ *with an op-indexed functor* $th : coSub \Rightarrow coSpec : SIGN^{op} \to CAT$ *and an op-lax op-indexed functor* $mod : coSpec \overset{\geq}{\Rightarrow} coSub : SIGN^{op} \to CAT$ *locally right adjoint to* th.

The situation in which for a given indexed functor, say, $mod : Spec \Rightarrow Sub$, there exists a left adjoint $th(\Sigma)$ for each $mod(\Sigma)$, is summarized in [10] under the slogan "mod has a left adjoint locally". In general, it is possible to show that, under these assumptions, the family $th(\Sigma), \Sigma \in |SIGN|$, constitutes a lax indexed functor $th : Sub \overset{\leq}{\Rightarrow} Spec$. This fact holds, as suggested by lemma 2.8, even if mod is an op-lax indexed functor $mod : Spec \overset{\geq}{\Rightarrow} Sub$.

In the sequel we demonstrate how concepts used in the literature for describing logical systems arise from institutional frames through systematic, canonical steps, using the Grothendieck construction.

One structure that arises naturally from an institution and that is often used to describe relations between logical systems is the category TH_0 of *theories* (see [7]). TH_0 has as objects pairs (Σ, Γ) with Σ a signature and $\Gamma \in |Spec(\Sigma)|$, and as arrows $\phi : (\Sigma_1, \Gamma_1) \to (\Sigma_2, \Gamma_2)$ signature morphisms $\phi : \Sigma_1 \to \Sigma_2$ such that $\Gamma_2 \supseteq Sen(\phi)(\Gamma_1)$. That is, from our viewpoint a theory morphism from (Σ_1, Γ_1) to (Σ_2, Γ_2) is actually given by an arrow $\phi : \Sigma_1 \to \Sigma_2$ in $SIGN$ and an arrow $i : \Gamma_2 \to Spec(\phi)(\Gamma_1)$ in $Spec(\Sigma_2)$. From the perspective of the pairs

(Σ_1, Γ_1) and (Σ_2, Γ_2) the arrows ϕ and i are going into opposite directions. The Grothendieck construction, as it is usually presented in the literature, can also handle this situation, once one uses the opposite category in the corresponding fibration functor. However, to avoid dealing with "op-manipulations" in more involving definitions and constructions, we will explicit give here the corresponding alternative formulation for the (covariant) Grothendieck construction. The contravariant case can be defined along the same lines.

Definition 2.11 (Grothendieck construction) Given an indexed category $C : IND \to CAT$ we define the category $Fl(C)$ as follows:

- *objects:* are pairs (i, a) where $i \in |IND|$ and $a \in |C(i)|$.
- *arrows:* from (i, a) to (j, b) are pairs (σ, f) where $\sigma : i \to j$ is an arrow in IND and $f : b \to C(\sigma)(a)$ is an arrow in $C(j)$.
- *composition:* Given arrows $(\sigma, f) : (i, a) \to (j, b)$ and $(\tau, g) : (j, b) \to (k, c)$ in $Fl(C)$, let $(\tau, g) \circ (\sigma, f) =_{def} (\tau \circ \sigma, C(\tau)(f) \circ g)$.

  Moreover we obtain a *projection* functor $Proj_C : Fl(C) \to IND$ with $Proj_C(i, a) = i$ and $Proj_C(\sigma, f) = \sigma$ for any (i, a) and any (σ, f) in $Fl(C)$.

It is straightforward to see that $Fl(Spec)$ delivers TH_0. If we apply the (contravariant) Grothendieck construction to $coSpec$, we get $Fl(coSpec)$ with the same objects (Σ, Γ) as $Fl(Spec)$ and with arrows $(\phi, i) : (\Sigma_1, \Gamma_1) \to (\Sigma_2, \Gamma_2)$ such that $\phi : \Sigma_1 \to \Sigma_2$ in $SIGN$ and $Sen(\phi)^{-1}(\Gamma_2) \supseteq \Gamma_1$ in $coSpec(\Sigma_1)$. $Fl(Spec)$ and $Fl(coSpec)$ become isomorphic since $Sen(\phi)^{-1} \dashv Sen(\phi)$, i.e., actually $Proj_{Spec}$ and $Proj_{coSpec}$ become isomorphic in $(CAT \downarrow SIGN)$.

For the same reason, i.e., $Mod(\phi) \dashv Mod(\phi)^{-1}$, $Fl(Sub)$ and $Fl(coSub)$ become isomorphic as well. $Fl(Sub)$ has as objects pairs $(\Sigma, \mathcal{M})$ with $\mathcal{M} \in |Sub(\Sigma)|$ and as arrows $(\phi, i) : (\Sigma_1, \mathcal{M}_1) \to (\Sigma_2, \mathcal{M}_2)$ pairs of a signature morphism $\phi : \Sigma_1 \to \Sigma_2$ and an inclusion $\mathcal{M}_2 \subseteq Sub(\phi)(\mathcal{M}_1)$ in $Sub(\Sigma_2)$.

Before flattening the remaining components of $\mathcal{F}(\mathcal{I})$ and $\overline{\mathcal{F}}(\mathcal{I})$, we first introduce our next

Definition 2.12 (Flattened lax functor) Given a lax indexed functor $\alpha : C \stackrel{\Rightarrow}{\leq} D : IND \to CAT$ the *flatten functor* $Fl(\alpha) : Fl(C) \to Fl(D)$ with $Proj_D \circ Fl(\alpha) = Proj_C$ is defined as follows:

- *objects:* $Fl(C)(i, a) = (i, \alpha(i)(a))$ for any $(i, a) \in |Fl(C)|$.
- *arrows:* $Fl(C)(\sigma, f) = (\sigma, \alpha(\sigma)(a) \circ \alpha(j)(f))$ for any arrow $(\sigma, f) : (i, a) \to (j, b)$ in $Fl(C)$.

Example 2.13 For $th : Sub \stackrel{\Rightarrow}{\leq} Spec : SIGN \to CAT$ we obtain:

- $Fl(th)(\Sigma, \mathcal{M}) = (\Sigma, th(\Sigma)(\mathcal{M}))$
- $Fl(th)(\phi, \mathcal{M}_2 \subseteq Sub(\phi)(\mathcal{M}_1)) = (\phi, th(\Sigma_2)(\mathcal{M}_2) \supseteq th(\Sigma_2)(Sub(\phi)(\mathcal{M}_1)) \supseteq Spec(\phi)(th(\Sigma_1)(\mathcal{M}_1))$.

Since $mod : Spec \Rightarrow Sub : SIGN \rightarrow CAT$ is an indexed functor, i.e., the natural transformations $mod(\phi) : mod(\Sigma_2) \circ Spec(\phi) \Rightarrow Sub(\phi) \circ mod(\Sigma_1)$ are considered to be identities, $Fl(mod)$ looks a little bit simpler:

- $Fl(mod)(\Sigma, \Gamma) = (\Sigma, mod(\Sigma)(\Gamma))$
- $Fl(mod)(\phi, \Gamma_2 \supseteq Spec(\phi)(\Gamma_1)) = (\phi, mod(\Sigma_2)(\Gamma_2) \subseteq mod(\Sigma_2)(Spec(\phi)(\Gamma_1))$ $= Sub(\phi)(mod(\Sigma_1)(\Gamma_1))$.

The flattening of $th : coSub \Rightarrow coSpec$ and $mod : coSpec \overset{\geq}{\Rightarrow} coSub$ produces isomorphic results so that we can state the next

Proposition 2.14 *The flattening of an institutional frame $\mathcal{F}(\mathcal{I})$ and an institutional co-frame $\overline{\mathcal{F}}(\mathcal{I})$ provides (up to isomorphism) the following commutative diagram in CAT*

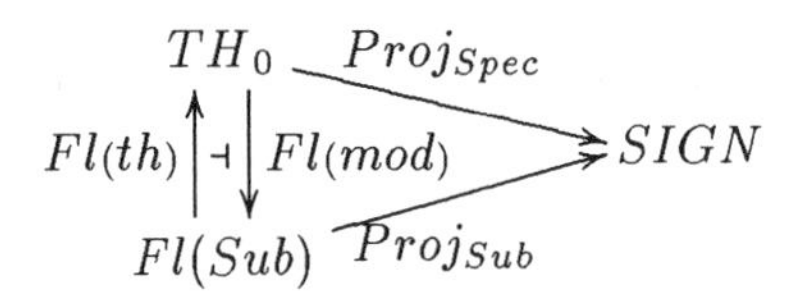

with $Fl(th) \dashv Fl(mod)$.

It can be shown that both $Proj_{Spec}$ and $Proj_{Sub}$ are split fibrations as well as split op-fibrations so that $Spec$ and Sub as well as $coSpec$ and $coSub$ could be reconstructed from the above diagram. Moreover, the commutative diagram carries completely the same information as the institutional frame $\mathcal{F}(\mathcal{I})$, as shown by the following

Theorem 2.15 (Flat versus Lax) *For each functor $G : Fl(C) \rightarrow Fl(D)$ with $Proj_D \circ G = Proj_C$, there exists a lax indexed functor $Dec(G) : C \overset{\leq}{\Rightarrow} D : IND \rightarrow CAT$ with $Fl(Dec(G)) = G$. Moreover, $Dec(Fl(F)) = F$ for any lax indexed functor $F : C \overset{\leq}{\Rightarrow} D : IND \rightarrow CAT$.*

After flattening, the next abstraction step, namely abstracting from indexing, can be done by forgetting the projections. However, the approaches presented so far in the literature are even more rigid. They have not taken into consideration $Fl(Sub)$ and $Fl(th)$. They just present TH_0 and the functor $emb \circ Fl(mod)^{op} : TH_0{}^{op} \rightarrow CAT$ under the name *generalized model functor* [7], where $emb : Fl(Sub)^{op} \rightarrow CAT$ is the obvious embedding functor assigning to $(\Sigma, \mathcal{M})$ the category $\mathcal{M}$ and assigning to $(\phi : \Sigma_1 \rightarrow \Sigma_2, \mathcal{M}_2 \subseteq Sub(\phi)(\mathcal{M}_1))$ the corresponding restriction $Mod(\phi) \downarrow : \mathcal{M}_2 \rightarrow \mathcal{M}_1$ of the functor $Mod(\phi) : Mod(\Sigma_2) \rightarrow Mod(\Sigma_1)$. Please bear in mind that $Sub(\phi)(\mathcal{M}_1) =_{def} Mod(\phi)^{-1}(\mathcal{M}_1)$. From our viewpoint, other approaches have even thrown away $Fl(Sub)$ and especially $Fl(th)$ and this suggests that they have probably lost some power for analyzing and describing relations between logical systems. Nevertheless, only a future analysis of more examples will show if this really matters.

3 Institution morphisms

After the technical preparations in the previous section, it becomes straightforward to analyze structurally the concept of institution morphism which was the original proposal in [5] for a formalization of relations between institutions.

Definition 3.1 (Institution morphism) An *institution morphism* $(\Phi, \alpha, \beta) : \mathcal{I} \to \mathcal{I}'$ between two institutions $\mathcal{I} = (SIGN, Sen, Mod, \models)$ and $\mathcal{I}' =(SIGN'$, Sen', Mod', $\models')$ is given by a functor $\Phi : SIGN \to SIGN'$, a natural transformation $\alpha : Sen' \circ \Phi \Rightarrow Sen : SIGN \to SET$, and a natural transformation $\beta : Mod \Rightarrow Mod' \circ \Phi^{op} : SIGN^{op} \to CAT$ such that for each $\Sigma \in |SIGN|$ the *institution morphism condition*

$$\textbf{(IMC)} \quad M \models_\Sigma \alpha(\Sigma)(\varphi') \iff \beta(\Sigma)(M) \models'_{\Phi(\Sigma)} \varphi',$$

holds for any $M \in |Mod(\Sigma)|$ and any $\varphi' \in Sen'(\Phi(\Sigma))$.

A superficial characterization of the concept of institution morphism could be that the (pointwise) transition of models goes in the same direction as the translation of signatures and that the (pointwise) translation of sentences goes into the opposite direction. However, on one hand, we have in many examples also transitions in the other directions even if they are not reflected by the concept of institution morphism (see section 5). On the other hand, we have already seen that on the abstract level the formation of images and the formation of pre-images are conceptually equivalent. Thus, as we will also see in theorem 3.3, the direction of α and β in definition 3.1 is not relevant for an essential understanding of this concept.

We apply again power set and power category constructions and obtain for any signature Σ in $SIGN$ functors $\beta(\Sigma) : Sub(\Sigma) \to Sub'(\Phi(\Sigma))$ and $\beta(\Sigma)^{-1} : Sub'(\Phi(\Sigma)) \to Sub(\Sigma)$ with $\beta(\Sigma) \dashv \beta(\Sigma)^{-1}$ and functors $\alpha(\Sigma)^{-1} : Spec(\Sigma)) \to Spec'(\Phi(\Sigma))$ and $\alpha(\Sigma) : Spec'(\Phi(\Sigma)) \to Spec(\Sigma)$ with $\alpha(\Sigma)^{-1} \dashv \alpha(\Sigma)$.

Adopting an analogous procedure as in the previous section, allows to state the following equivalent formulations for the institution morphism condition.

Proposition 3.2 (Institution morphism condition) *For each* $\Sigma \in |SIGN|$ *any of the following conditions is equivalent to condition* **(IMC)**

- $mod(\Sigma)\,(\alpha(\Sigma)(\Gamma')) = \beta(\Sigma)^{-1}\,(mod'(\Phi(\Sigma))(\Gamma'))$ *for all* $\Gamma' \in Spec'(\Phi(\Sigma))$
- $\alpha(\Sigma)^{-1}\,(th(\Sigma)(\mathcal{M})) = th'(\Phi(\Sigma))\,(\beta(\Sigma)(\mathcal{M}))$ *for all* $\mathcal{M} \in Sub(\Sigma)$.

Now, the categorical formulation of the notion of an institution morphism, taking into account the above equalities is stated in the next theorem, below. Please keep in mind that the opposite of a natural transformation $\eta : F \Rightarrow G : C \to D$ goes into the opposite direction, i.e., $\eta^{op} : G^{op} \Rightarrow F^{op} : C^{op} \to D^{op}$.

Theorem 3.3 (Institution morphism) *A triple* $(\Phi, \alpha, \beta) : \mathcal{I} \to \mathcal{I}'$ *as in definition 3.1 satisfies condition* **(IMC)** *iff for the natural transformations* $\wp_{set} \circ \alpha :$

Spec$' \circ \Phi \Rightarrow$ *Spec and* $\overline{\wp}_{cat} \circ \beta^{op} : Sub' \circ \Phi \Rightarrow Sub$ *between the corresponding institutional frames* $\mathcal{F}(\mathcal{I}')$ *and* $\mathcal{F}(\mathcal{I})$ *the following equation is satisfied*

$$mod \bullet (\wp_{set} \circ \alpha) = (\overline{\wp}_{cat} \circ \beta^{op}) \bullet (mod' \circ \Phi).$$

Moreover condition **(IMC)** *is equivalent to the equation*

$$(\overline{\wp}_{set} \circ \alpha^{op}) \bullet th = (th' \circ \Phi^{op}) \bullet (\wp_{cat} \circ \beta)$$

for the natural transformations $\overline{\wp}_{set} \circ \alpha^{op} : coSpec \Rightarrow coSpec' \circ \Phi^{op}$ *and* $\wp_{cat} \circ \beta : coSub \Rightarrow coSub' \circ \Phi^{op}$ *between the institutional co-frames* $\overline{\mathcal{F}}(\mathcal{I})$ *and* $\overline{\mathcal{F}}(\mathcal{I}')$.

Theorem 3.3 suggest the following two-level methodology for investigating relations between institutions.

Firstly, we analyze on the level of institutional frames the following questions: What relations are there? What are the categorical properties of these relations? Which compatibility conditions are satisfied? Secondly, we describe how the relations on the abstract level arise possibly from more fine grained transitions on the level of sentences and models; and finally, how the compatibility conditions on the abstract level can be proved possibly by checking certain properties of the fine grained transitions.

Following this methodological line, we can characterize institution morphism as follows: We have as well a morphism γ from $\mathcal{F}(\mathcal{I}')$ to $\mathcal{F}(\mathcal{I})$ as a morphism δ into the opposite direction from $\overline{\mathcal{F}}(\mathcal{I})$ to $\overline{\mathcal{F}}(\mathcal{I}')$. All the components of γ and δ are indexed and op-indexed functors, respectively. γ is compatible with *mod*, i.e., with the assignment of subcategories of models to specifications, and δ is compatible with *th*. Both γ and δ arise by abstraction from a triple (Φ, α, β) as described in definition 3.1, and thus γ and δ can be seen as inherently related to each other. Further, the compatibility of γ and β, respectively, could be proved by showing that condition **(IMC)** is satisfied by the triple (Φ, α, β).

It is worth mentioning that our abstract viewpoint makes explicit that the composition of institution morphisms is trivially well-defined. Note that the characterization of an institution morphism by γ and δ offers two divergent directions of possible generalizations independent from the possibility to relax condition **(IMC)**.

4 Institution maps

The definition in [7] of the concept *institution map* is rather technical and complex. To break down this complexity and be prepared for the discussion in the next section we firstly introduce the concept of *plain institution map* which owns the same granularity as the concept of institution map. Thereafter we outline a revised description of the general concept of institution map.

Definition 4.1 (Plain institution map) A *plain institution map* $(\Phi, \alpha, \beta) : \mathcal{I} \to \mathcal{I}'$ between two institutions $\mathcal{I} = (SIGN, Sen, Mod, \models)$ and $\mathcal{I}' = (SIGN',$

Sen', Mod', $\models'$) is given by a functor $\Phi : SIGN \to SIGN'$, a natural transformation $\alpha : Sen \Rightarrow Sen' \circ \Phi : SIGN \to SET$, and a natural transformation $\beta : Mod' \circ \Phi^{op} \Rightarrow Mod : SIGN^{op} \to CAT$ such that for each $\Sigma \in |SIGN|$ the *plain institution map condition*

$$\textbf{(IPC)} \quad M' \models'_{\Phi(\Sigma)} \alpha(\Sigma)(\varphi) \iff \beta(\Sigma)(M') \models_\Sigma \varphi,$$

holds for any $M' \in |Mod(\Phi(\Sigma))|$ and any $\varphi \in Sen(\Sigma)$.

Following essentially the same steps as in the previous section, we are able to reformulate the above definition and state the expected

Theorem 4.2 (Plain institution map) *A triple $(\Phi, \alpha, \beta) : \mathcal{I} \to \mathcal{I}'$ as in definition 4.1 satisfies condition* **(IPC)** *iff for the natural transformations $\wp_{set} \circ \alpha : Spec \Rightarrow Spec' \circ \Phi$, and $\overline{\wp}_{cat} \circ \beta^{op} : Sub \Rightarrow Sub' \circ \Phi$, between the corresponding institutional frames $\mathcal{F}(\mathcal{I})$ and $\mathcal{F}(\mathcal{I}')$, the following equation is satisfied:*

$$(\overline{\wp}_{cat} \circ \beta^{op}) \bullet mod = (mod' \circ \Phi) \bullet (\wp_{set} \circ \alpha).$$

Although the direction of the arrows comprising the equation stated in theorem 3.3 have here an opposite direction, the concepts institution morphism and plain institution map turn out to be even descriptionally equivalent in many (plain) situations, as the discussion in the next section will try to clarify.

Concernig the concepts institution morphism and plain institution map we have seen that the abstract level of institutional frames is appropriate to give a concise categorical presentation of these concepts. In contrast, the general concept of institution map is even already based on relations between abstract structures associated to institutions. In such a way it remains to show that institutional frames offer a more adequate way for presenting the general concept.

The concept of institution map is mainly motivated by the observation that the transition between logical systems often requires to transform a signature into a specification to gain semantic compatibility. A motivating example in [7] was the simulation of predicates by internal boolean functions where we have to introduce those axioms needed to force the interpretation of the internal sort *bool* to be a two-element boolean algebra. A further typical example could be the embedding of an institution of total algebras into an institution of partial algebras where we have to introduce definedness axioms. That is, the additional axioms are devoted to simulate the semantic restriction "total algebra" on the level of logic.

Let be given two institutions $\mathcal{I} = (SIGN, Sen, Mod, \models)$ and $\mathcal{I}' =(SIGN'$, Sen', Mod', $\models'$). To model the transformation of signatures into specifications, [7] introduces a functor $\Upsilon : TH_0 \to TH_0{}'$ between the corresponding categories $TH_0 = Fl(Spec)$ and $TH_0{}' = Fl(Spec')$ of *theories*. Thereby, the theory $\Upsilon(\Sigma, \emptyset)$ represents the transformed signature Σ. Further, it is required that the transformation of theories goes along with an underlying transformation of signatures, i.e., a functor $\Phi : SIGN \to SIGN'$ is assumed such that $\Phi \circ Proj = Proj' \circ \Upsilon$, where $Proj : Fl(Spec) \to SIGN$ and $Proj' : Fl(Spec') \to SIGN'$.

An equivalent description of these assumptions on the level of institutional frames can be obtained by theorem 2.15 and the following

Lemma 4.3 *For any indexed category $C : IND' \to CAT$ and any functor $G : IND \to IND'$ the assignments*

- $G^\diamond(i,a) = (G(i), a)$ *for any (i,a) in $Fl(C \circ G)$ and*
- $G^\diamond(\sigma, f) = (G(\sigma), f)$ *for any (σ, f) in $Fl(C \circ G)$*

define a functor $G^\diamond : Fl(C \circ G) \to Fl(C)$ such that the following diagram is a pullback diagram in CAT

$$\begin{array}{ccc} Fl(C \circ G) & \xrightarrow{Proj_{C \circ G}} & IND \\ {\scriptstyle G^\diamond}\downarrow & & \downarrow{\scriptstyle G} \\ Fl(C) & \xrightarrow{Proj_C} & IND' \end{array}$$

Corollar 4.4 *Let be given functors $\Upsilon : Fl(Spec) \to Fl(Spec')$ and $\Phi : SIGN \to SIGN'$. Then the equation $\Phi \circ Proj = Proj' \circ \Upsilon$ holds iff there exists a lax indexed functor $\alpha_\Upsilon : Spec \stackrel{\leq}{\Rightarrow} Spec' \circ \Phi$ with $\Upsilon = \Phi^\diamond \circ Fl(\alpha_\Upsilon)$, i.e.. with $\Upsilon(\Sigma, \Gamma) = (\Phi(\Sigma), \alpha_\Upsilon(\Sigma)(\Gamma))$ for any $\Sigma \in |SIGN|$ and any $\Gamma \in |Spec(\Sigma)|$.*

Contrary to syntax, the semantical transitions are given by a pointwise translation of models. That is, [7] assumes a natural transformation $\beta : Mod'_{ex} \circ \Upsilon^{op} \Rightarrow Mod_{ex} : TH_0{}^{op} \to CAT$ between the corresponding generalized model functors $Mod_{ex} = emb \circ Fl(mod)^{op}$ and $Mod'_{ex} = emb' \circ Fl(mod')^{op}$ as described in section 2. However, $\Upsilon(\Sigma, \Gamma) = (\Phi(\Sigma), \alpha_\Upsilon(\Sigma)(\Gamma))$ ensures that only the components $\beta(\Sigma, \emptyset) : Mod'_{ex}(\Upsilon(\Sigma, \emptyset)) \to Mod_{ex}(\Sigma, \emptyset)$ of β really matter since all other functors $\beta(\Sigma, \Gamma)$ become restrictions of $\beta(\Sigma, \emptyset)$. This observation enables us to define for any $\Sigma \in |SIGN|$ a functor $\beta_\Upsilon(\Sigma) : Sub(\Sigma) \to Sub'(\Phi(\Sigma))$ by setting

$$\beta_\Upsilon(\Sigma)(\mathcal{M}) = \beta(\Sigma, \emptyset)^{-1}(\mathcal{M})$$

for all $\mathcal{M} \in |Sub(\Sigma)| = |Mod_{ex}(\Sigma, \emptyset)|$. It can be shown that this defines a lax indexed functor $\beta_\Upsilon : Sub \stackrel{\leq}{\Rightarrow} Sub' \circ \Phi$. Finally the *general institution map condition* can be described by the following equation

$$\beta_\Upsilon \bullet mod = (mod' \circ \Phi) \bullet \alpha_\Upsilon$$

The crucial point is that [7] can not formulate this condition directly. He has to require additionally that α_Υ can be approximated by a transformation of signatures into specifications plus a pointwise translation of sentences. That is, an additional natural transformation $\alpha : Sen \Rightarrow Sen' \circ \Phi$ is assumed that provides a lax indexed functor $\alpha_{ex} : Spec \stackrel{\leq}{\Rightarrow} Spec' \circ \Phi$ by setting $\alpha_{ex} = \alpha_\Upsilon(\Sigma)(\emptyset) \cup \alpha(\Sigma)(\Gamma)$. Further it is required that α_Υ and α_{ex} are semantically equivalent, i.e., $(mod' \circ \Phi) \bullet \alpha_\Upsilon = (mod' \circ \Phi) \bullet \alpha_{ex}$. Using these additional ingredients, [7] is able to present, in the tradition of the satisfaction condition, the following condition which is equivalent to the abstract equation above

$$M' \models_{\Phi(\Sigma)} \alpha(\Sigma)(\varphi) \quad \text{iff} \quad \beta(\Sigma, \emptyset)(M') \models_{\Sigma} \varphi$$

for each $\Sigma \in |SIGN|$, $\varphi \in Sen(\Sigma)$, and $M' \in Mod'_{ex}(\Phi(\Sigma, \emptyset))$.

Note that the complex construction of α_{ex} and the corresponding requirements were actually also motivated by examples, especially by the unfailing Knuth-Bendix completion. However, following our methodology, we can characterize institution maps independent from the question how the abstract relations can be approximated by internal constructions. Taking into account other internal constructions we could also define naturally different variations and generalizations of the concept of institution map all of them keeping the abstract characterization.

[7] calls an institution map *α-simple* if $\alpha_{\Upsilon} = \alpha_{ex}$. Further, it is easy to see that an institution map is *plain* in the sense of 4.1 if $\alpha_{\Upsilon} = \alpha_{ex} = \wp_{set} \circ \alpha$, i.e., if especially $\alpha_{\Upsilon}(\Sigma)(\emptyset) = \emptyset$.

5 Analysis of examples

A development of abstract concepts should be based on a careful and detailed analysis of examples and special cases. This section tries to exemplify in what extend our structural approach could be helpful in analyzing relations between logical systems and in developing a unifying theory for those relations.

In [5], the relation between the institution $\mathcal{I}_{\mathcal{FO}}$ of many-sorted first order logic with equality and the institution $\mathcal{I}_{\mathcal{EQ}}$ of many-sorted equational logic was used as an example to explain the concept of an institution morphism.

First of all, any equational signature $\Sigma = (S, OP)$ can be extended trivially to a first order signature $\Phi(\Sigma) = (S, OP, \emptyset)$ where $\emptyset$ denotes the empty set of predicate symbols. This provides a functor $\Phi : SIGN_{\mathcal{EQ}} \to SIGN_{\mathcal{FO}}$. Note that we assume for $\mathcal{I}_{\mathcal{FO}}$ (and analogously to $\mathcal{I}_{\mathcal{EQ}}$) that the equality is "built in", i.e., it is not to be seen as an interpretation of signature symbols, and thus can be used by the functor $Sen_{\mathcal{FO}} : SIGN_{\mathcal{FO}} \to SET$ to build up sentences. This corresponds to the semantical assumption that equality is to be interpreted as an actual identity in first order models. Secondly, any equation $\varphi \in Sen_{\mathcal{EQ}}(S, OP)$ can be regarded as a first order sentence $\varphi \in Sen_{\mathcal{FO}}(S, OP, \emptyset)$, i.e., we actually have a natural transformation $\alpha : Sen_{\mathcal{EQ}} \Rightarrow Sen_{\mathcal{FO}} \circ \Phi$ where the components $\alpha(\Sigma) : Sen_{\mathcal{EQ}}(\Sigma) \to Sen_{\mathcal{FO}}(\Phi(\Sigma))$ can be regarded as inclusions. Thirdly, any algebra $A \in Mod_{\mathcal{EQ}}(S, OP)$ can be viewed as a first order model $A \in Mod_{\mathcal{FO}}(S, OP, \emptyset)$, and vice versa, any first order model $M \in Mod_{\mathcal{FO}}(S, OP, \emptyset)$ can be regarded as an algebra $M \in Mod_{\mathcal{EQ}}(S, OP)$. That is, we have a natural isomorphism $\beta : Mod_{\mathcal{FO}} \circ \Phi^{op} \Rightarrow Mod_{\mathcal{EQ}}$ with an inverse $\beta^{-1} : Mod_{\mathcal{EQ}} \Rightarrow Mod_{\mathcal{FO}} \circ \Phi^{op}$. Finally, the following equation holds obviously for any equational signature Σ and any specification $\Gamma \in |Spec_{\mathcal{EQ}}|$:

$$\beta(\Sigma)^{-1}(mod_{\mathcal{EQ}}(\Sigma)(\Gamma)) = mod_{\mathcal{FO}}(\Phi(\Sigma))(\alpha(\Sigma)(\Gamma)).$$

Note that $\beta(\Sigma)^{-1}(\mathcal{M}) = \beta^{-1}(\Sigma)(\mathcal{M})$ for each $\mathcal{M} \in |Sub_{\mathcal{EQ}}|$, i.e., since $\beta(\Sigma)$ is an isomorphism, the pre-image of $\mathcal{M}$ with respect to $\beta(\Sigma)$ coincides with the image of $\mathcal{M}$ with respect to the inverse $\beta^{-1}(\Sigma)$. In such a way, we obtain on the abstract level the equation

$$(\overline{\wp}_{cat} \circ \beta^{op}) \bullet mod_{\mathcal{EQ}} = (mod_{\mathcal{FO}} \circ \Phi) \bullet (\wp_{set} \circ \alpha).$$

However, this proves the existence of a plain institution map from $\mathcal{I}_{\mathcal{EQ}}$ to $\mathcal{I}_{\mathcal{FO}}$ and not the promised institution morphism from $\mathcal{I}_{\mathcal{FO}}$ to $\mathcal{I}_{\mathcal{EQ}}$.

The crucial point is that, forgetting predicate symbols, defines a further functor $\Psi : SIGN_{\mathcal{FO}} \to SIGN_{\mathcal{EQ}}$ with $\Psi(S, OP, P) = (S, OP)$ for any first order signature (S, OP, P). This functor was actually presented in [5], and it turns out to be right adjoint to Φ. The unit is the identity, i.e., $\Psi \circ \Phi = Id_{SIGN_{\mathcal{EQ}}}$, and the co-unit $\varepsilon : \Phi \circ \Psi \Rightarrow Id_{SIGN_{\mathcal{FO}}}$ describes the inclusion of $\Phi(\Psi(S, OP, P)) = (S, OP, \emptyset)$ into (S, OP, P).

This adjoint situation causes that the described plain institution map from $\mathcal{I}_{\mathcal{EQ}}$ to $\mathcal{I}_{\mathcal{FO}}$ can be also interpreted as an institution morphism from $\mathcal{I}_{\mathcal{FO}}$ to $\mathcal{I}_{\mathcal{EQ}}$. The equivalence of the concepts plain institution map and institution morphism in situations where we have adjoint functors between the categories of signatures was already observed in [11] and is formulated as a theorem in [1]. Our abstract characterization of the concepts institution morphism and institution map enables us to develop this result by a straightforward categorical argumentation. To see this, let be given a plain institution map

$$(\Phi : SIGN \to SIGN', \alpha : Sen \Rightarrow Sen' \circ \Phi, \beta : Mod' \circ \Phi^{op} \Rightarrow Mod)$$

from an institution $\mathcal{I}$ to an institution $\mathcal{I}'$. For any functor $\Psi : SIGN' \to SIGN$ such that a natural transformation $\varepsilon : \Psi \circ \Phi \Rightarrow Id_{SIGN'}$ exists we can construct natural transformations $\alpha_\varepsilon : Sen \circ \Psi \Rightarrow Sen'$ and $\beta_\varepsilon : Mod' \Rightarrow Mod \circ \Psi^{op}$ by

$$\alpha_\varepsilon =_{def} (Sen' \circ \varepsilon) \bullet (\alpha \circ \Psi) \quad \text{and} \quad \beta_\varepsilon =_{def} (\beta \circ \Psi^{op}) \bullet (Mod' \circ \varepsilon^{op}).$$

By applying the interchange law, it can be shown that $(\overline{\wp}_{cat} \circ \beta^{op}) \bullet mod = (mod' \circ \Phi) \bullet (\wp_{set} \circ \alpha)$ implies $mod \bullet (\wp_{set} \circ \alpha_\varepsilon) = (\overline{\wp}_{cat} \circ \beta_\varepsilon^{op}) \bullet (mod' \circ \Psi)$, i.e., we have actually constructed an institution morphism $(\Psi, \alpha_\varepsilon, \beta_\varepsilon)$ out of the plain institution map (Φ, α, β).

The application of this construction to our example delivers the promised institution morphism $(\Psi, \alpha_\varepsilon, \beta_\varepsilon)$ from $\mathcal{I}_{\mathcal{FO}}$ to $\mathcal{I}_{\mathcal{EQ}}$, where we have, for any first order signature (S, OP, P) an inclusion functor $\alpha_\varepsilon(S, OP, P) : Sen_{\mathcal{EQ}}(S, OP) \to Sen_{\mathcal{FO}}(S, OP, P)$ and a forgetful functor $\beta_\varepsilon(S, OP, P) : Mod_{\mathcal{FO}}(S, OP, P) \to Mod_{\mathcal{EQ}}(S, OP)$.

Analogously, we can also construct for an institution morphism

$$(\Psi : SIGN' \to SIGN, \alpha' : Sen \circ \Psi \Rightarrow Sen', \beta : Mod' \Rightarrow Mod \circ \Psi^{op})$$

from $\mathcal{I}'$ to $\mathcal{I}$ a plain institution map $(\Phi, \alpha'_\eta, \beta'_\eta)$ from $\mathcal{I}$ to $\mathcal{I}'$ if we have a functor $\Phi : SIGN \to SIGN'$ and a natural transformation $\eta : Id_{SIGN} \to \Psi \circ \Phi$.

Moreover, we have $\alpha_{\varepsilon\eta} = \alpha$ and $\beta_{\varepsilon\eta} = \beta$ as well as $\alpha'_{\eta\varepsilon} = \alpha'$ and $\beta'_{\eta\varepsilon} = \beta'$ if η is the unit and ε the co-unit of an adjunction $\Phi \dashv \Psi$. Again this can be shown by a simple application of the interchange law using the additional assumptions $(\varepsilon \circ \Phi) \bullet (\eta \circ \Phi) = Id_\Phi$ and $(\Psi \circ \varepsilon) \bullet (\eta \circ \Psi) = Id_\Psi$.

Therefore, we see that $\mathcal{I_{FO}}$ and $\mathcal{I_{EQ}}$ are related in a plain and very special way. Besides, it seems that those kinds of relations appear at many places. In [11] we have outlined that $\mathcal{I_{EQ}}$ is related in a similar way to the institution $\mathcal{I_{UEQ}}$ of unsorted equational logic. [1] presents a more relevant example – the relationship between linear and branching temporal logic. Moreover, we have seen that the concepts institution morphism and institution map are not really helpful for classifying such simple relations. Firstly, they provide no differentiation, since the situation can be characterized by both concepts equivalently. Secondly, both concepts take into account only some of the existing mappings between components of the institutions in consideration. Finally, they interpret one and the same relation as arrows in opposite directions.

Our methodological proposal in situations as above will be to ignore both concepts and to describe the existing relationship directly using categorical means as they were developed in this paper. Concerning our example, we might propose the following characterization: $\mathcal{I_{FO}}$ is a *plain logical extension* of $\mathcal{I_{EQ}}$: $SIGN_{\mathcal{EQ}}$ can be seen as a full *co-reflective* subcategory of $SIGN_{\mathcal{FO}}$ and there is a natural inclusion $\alpha : Sen_{\mathcal{EQ}} \Rightarrow Sen_{\mathcal{FO}} \circ \subseteq$. The semantics of $\mathcal{I_{EQ}}$ is *preserved*, i.e., we have a natural isomorphism $\beta : Mod_{\mathcal{FO}} \circ \subseteq^{op} \Rightarrow Mod_{\mathcal{EQ}}$ and the equation $(\overline{\wp}_{cat} \circ \beta^{op}) \bullet mod_{\mathcal{EQ}} = (mod_{\mathcal{FO}} \circ \subseteq) \bullet (\wp_{set} \circ \alpha)$ is satisfied.

The relationship between $\mathcal{I_{EQ}}$ and $\mathcal{I_{UEQ}}$ was used in [7] to explain the concept of institution map. A proper characterization might be: $\mathcal{I_{EQ}}$ is a *signature refinement* of $\mathcal{I_{UEQ}}$: $SIGN_{\mathcal{UEQ}}$ can be seen as a full *reflective* subcategory of $SIGN_{\mathcal{EQ}}$, i.e., any unsorted signature can be seen as a one-sorted signature and omitting sorts provides a functor $\Psi : SIGN_{\mathcal{EQ}} \to SIGN_{\mathcal{UEQ}}$ left-adjoint to $\subseteq$. There is a natural isomorphism $\alpha : Sen_{\mathcal{UEQ}} \Rightarrow Sen_{\mathcal{EQ}} \circ \subseteq$ and the semantics of $\mathcal{I_{UEQ}}$ is preserved. Finally, we can compose both characterizations and state that $\mathcal{I_{FO}}$ is a plain logical extension of a signature refinement of $\mathcal{I_{UEQ}}$.

Note that we could also describe $\mathcal{I_{FO}}$ as a signature refinement of a plain logical extension of $\mathcal{I_{UEQ}}$ if we take into account the institution $\mathcal{I_{UFO}}$ of unsorted first order logic. In this way a lot of concepts could be introduced and a lot of questions could be put into consideration as, for instance, under which conditions can an extension of a refinement be transformed into an equivalent refinement of an extension? Which kinds of relations between institutions can be sequentialized into a refinement of an extension? And so on.

6 Concluding Remarks

The prototypical analysis in section 5 shows that the methodology presented in this paper provides a promising basis for developing a unifying theory for relations between logical systems. To approach such a difficult enterprise, we will work in two directions: firstly, a thorough analysis of examples is needed, mainly in order to see how an appropriate and well-formed classification of relations between logical systems should or has to look like. Secondly, as done in section 3 and 4, we will revisit other concepts which were already developed to relate institutions as, for instance, simulation of institutions [3] and pre-institution transformation [8]. This revision is expected to present these concepts as entities of a hierarchy (or net) of concepts that arises from our categorical methodology.

A further line will probably be the incorporation of other components of logical systems, like entailment systems and proof calculus, as done, for instance, in [7]. In [11], it is shown how this can be properly done for the notion of an entailment system, using some of the tools presented in this paper.

References

[1] M. Arrais and J.L. Fiadeiro. Unifying theories in different institutions. In *Proc. WADT11, Oslo*, pages 81–101. Springer, LNCS 1130, 1996.

[2] M. Cerioli and J. Meseguer. May i borrow your logic? (transporting logical structures along maps). *TCS*, 173(2):311–347, 1997.

[3] Maura Cerioli. *Relationships between Logical Formalisms.* PhD thesis, Università di Pisa–Genova–Udine, 1993. TD-4/93.

[4] J. Goguen. A categorical manifesto. *Mathematical Structures in Computer Science*, 1(1):49–67, 1991.

[5] J. A. Goguen and R. M. Burstall. Institutions: Abstract Model Theory for Specification and Programming. *Journals of the ACM*, 39(1):95–146, January 1992.

[6] C. B. Jay. Extending properties to categories of partial maps. Technical Report LFCS 90–107, University of Edinburgh, LFCS, 1990.

[7] J. Meseguer. General logics. In H.-D. Ebbinghaus et. al., editor, *Logic colloquium '87*, pages 275–329. Elsevier Science Publishers B. V.,North Holland, 1989.

[8] S. Salibra and G. Scollo. A soft stairway to institutions. In *Recent Trends in Data Type Specification*, pages 310–329. Springer, 1992. LNCS 655.

[9] A. Tarlecki. Moving between logical systems. In *Recent Trends in Data Type Specification*, pages 478–502. Springer, LNCS 1130, 1996.

[10] A. Tarlecki, R.M. Burstall, and J.A. Goguen. Some fundamental algebraic tools for the semantics of computation. Part III: Indexed categories. *TCS*, 91:239–264, 1991.

[11] U. Wolter. Institutional frames. In *Recent Trends in Data Type Specification*, pages 469–482. Springer, LNCS 906, 1995.

Combining and Representing Logical Systems

Till Mossakowski[1], Andrzej Tarlecki[2] and Wiesław Pawłowski[3]

[1] Department of Computer Science, University of Bremen, P.O.Box 33 04 40, D-28334 Bremen, Germany, E-mail till@informatik.uni-bremen.de, (Corresponding author)
[2] Institute of Informatics, Warsaw University and Institute of Computer Science, Polish Academy of Sciences, Warsaw.
[3] Institute of Computer Science, Polish Academy of Sciences, Gdańsk

Abstract. The paper addresses important problems of building complex logical systems and their representations in universal logics in a systematic way. Following Goguen and Burstall, we adopt the model-theoretic view of logic as captured in the notion of institution and of parchment (a certain algebraic way of presenting institutions).
We propose a modified notion of parchment together with a notion of parchment morphism and representation, respectively. We lift formal properties of the categories of institutions and their representations to this level: the category of parchments is complete, and parchment representations may be put together using categorical limits as well. However, parchments provide a more adequate framework for systematic combination of logical systems than institutions. We indicate how the necessary invention for proper combination of various logical features may be introduced either on an ad hoc basis (when putting parchments together using limits in the category of parchments) or via representations in a universal logic (when parchment combination is driven by their representations).

1 Introduction

There is a wealth of various logical systems necessary for different purposes. As it is unlikely that there ever will be a single universal logical system serving all purposes equally well, the question arises how to handle the needed variety. One key idea is that logical systems should be built and used in a structured way. Two important practically-motivated issues then are:

1. Combining logical systems, which opens the possibility for introducing and adding new concepts to logical systems in a step-by-step manner.
2. Representing logical systems in other logical system(s) with well-known model theory and proof theory, which opens the possibility to re-use tools like theorem provers.

Several different meta-notions of logical system have been introduced. One of them, institutions [8], comes with two different types of arrow:

1. Institution morphisms, which capture how one logical system is built upon another and thus provide a rudimentary framework for combination of logics via category-theoretic limits;

2. Institution representations, which encode one logical system in another and thus provide a basis for re-use of theorem provers (under an extra technical condition).

In [20] the role of these notions and the interplay between them have been studied. A suitable notion of map between institution representations, consisting of an institution morphism and an extra "fitting" component has been introduced. It captures the situation that not only an institution is built over another one, but also the representation of the former in some "universal" institution is built over the representation of the latter. The main new theorem of [20] states that representations can be combined via category-theoretic limits, so that combined institutions can be represented provided that each component of the combination can be represented.

One deficiency of combinations of institutions via limits is that since sentences in institutions have no inner structure, they are simply united in the combination rather then being properly combined, and the features of the combined logics do not really interact in the result. The solution is to move to parchments [7], certain algebraic presentations of institutions providing an abstract syntax and evaluator-based semantics for sentences and therefore more useful for logic combination [13].

In the present work we deal with a slightly redefined notion of parchment, shifting some foundational problems of the original parchment definition (see Sect. 3), preceded by the usual preliminaries in Sect. 1.1 and a brief summary on institutions in Sect. 2. These *λ-parchments* come equipped with the natural notion of *λ-parchment morphism* (Sect. 3.1). We check that the category of λ-parchments is complete (Sect. 4), just as the category of parchments [13], thus providing another framework for systematic combination of logical systems, more adequate for this purpose than the category of institutions. We then propose a notion of *λ-parchment representation* (Sect. 5) and show that λ-parchment representations interact with λ-parchment morphisms quite similarly as institution representations interact with institution morphisms. In particular, we lift the theorem about combination of institution representations to the level of λ-parchment representations (Sect. 6). We sketch an example illustrating intrinsic features of the combinations of λ-parchment representations and the potential usefulness of the proposed framework in Sect. 7. We also address the question how far the combination of logical systems can be done automatically by a straightforward application of our framework and how far purpose-driven human invention is necessary when combining logical systems. The latter is indeed necessary to stay within the realm of "logical" λ-parchments and their logical morphisms when λ-parchments are put together using limits in their respective categories. An important contribution of the present work is that we show how this additional human invention to "massage" the resulting λ-parchment may be replaced by the limit construction on representations in a given universal λ-parchment (Sect. 6).

1.1 Algebraic preliminaries

In the following we will rely on the standard notions and facts concerning relational structures; we recall some of them here, mostly to fix notation and terminology.

We work with relational algebraic signatures $\Sigma = (S, OP, REL)$ consisting of a set of sort symbols $s \in S$, an $S^* \times S$-indexed set OP of total operation symbols and an S^*-indexed set REL of relation symbols. Signature morphisms can be defined in a straightforward way. This gives a category **AlgSig**.

Let $Logic \in |\mathbf{AlgSig}|$ be the relational algebraic signature consisting of a sort $*$ and a relation symbol $D : *$. Let $\mathbf{AlgSig_*}$ be the category of many-sorted relational signatures from **AlgSig** having $Logic$ as a subsignature and signature morphisms being the identity on $Logic$.

For each signature $\Sigma \in \mathbf{AlgSig}$, we then have a category $\mathbf{Str}(\Sigma)$ of Σ-structures and Σ-homomorphisms defined in the standard way, where homomorphisms only preserve, but not necessarily reflect the relations, see [15]. For each signature morphism $\sigma: \Sigma \longrightarrow \Sigma' \in \mathbf{AlgSig}$, there is a forgetful functor $\mathbf{Str}(\sigma): \mathbf{Str}(\Sigma') \longrightarrow \mathbf{Str}(\Sigma)$ which intuitively renames all components of Σ'-models and Σ'-homomorphisms along σ.

The usual term algebra T_Σ (with empty relations) is an initial object in $\mathbf{Str}(\Sigma)$. For each signature morphism $\sigma: \Sigma \longrightarrow \Sigma'$, the forgetful functor $\mathbf{Str}(\sigma): \mathbf{Str}(\Sigma') \longrightarrow \mathbf{Str}(\Sigma)$ has a left adjoint $F_\sigma: \mathbf{Str}(\Sigma) \longrightarrow \mathbf{Str}(\Sigma')$ defined by $F_\sigma(A) = T_{\Sigma'}(|A|)$. The unit of the adjunction is denoted by η^σ, the counit by ϵ^σ. For a Σ-homomorphism $h: X \longrightarrow A|_\sigma$, $h^\#: F_\sigma(X) \longrightarrow A$ denotes the adjoint arrow, and for a Σ'-homomorphism $k: F_\sigma(X) \longrightarrow A$, $k^\natural: X \longrightarrow A|_\sigma$ denotes the co-adjoint arrow.

A Σ-homomorphism $h: A \longrightarrow B$ is called *full*, if for all $R : w \in REL$, $h_w[R_A] = R_B \cap h_w[A_w]$. It is called *closed*, if for all $R : w \in REL$, $h_w^{-1}[R_B] = R_A$. Closedness implies fullness, and full surjections = quotients = regular epis, see [3], Section 2.6.

We will later on need the following properties, which are formulated in [3] for partial algebras, but directly carry over to relational structures (relations behave like domains of partial operations):

Proposition 1 *[3], 2.4.5(d). If $g \circ f$ is a full surjection, then so is g.*

Proposition 2 *[3], 2.4.6. A full surjection which also is a monomorphism (= injection) is an isomorphism.*

Proposition 3 *[3], 2.7.1 and 2.7.2. Let $f: A \longrightarrow B$ a full surjective homomorphism and $g: A \longrightarrow C$ be an arbitrary homomorphism. Then*

1. *There exists a homomorphism $h: B \longrightarrow C$ with $g = h \circ f$ iff $ker(f) \subseteq ker(g)$, and*
2. *h is injective iff $ker(f) = ker(g)$.*

2 Meta-frameworks for logic combination

There are a number of meta-frameworks formalizing the notion of logical system, even if we limit our attention to approaches taking a model-theoretic view of logic. *Specification frames* [5] just consider specifications and models. *Institutions* [8] split specifications into signatures and sentences, and a satisfaction relation (between models and sentences) is provided. *Parchments* [7] further specify an abstract syntax of sentences, along which satisfaction can be defined inductively using an algebra of term evaluators. *Context institutions* [16] allow to deal with variable contexts and substitutions. In the sequel, we first recall the well-known institutions, which provide our basic view of logical systems. Then we introduce λ-parchments, a certain variant of parchments pushing a foundationally and perhaps methodologically dubious aspect of the original parchment definition aside. λ-parchments are presentations of institutions which are more appropriate as a framework for logic combination.

2.1 Institutions and institution morphisms

Any specification formalism is usually based on some notion of signature, model, sentence and satisfaction. These are the usual ingredients of abstract model theory [2] and are the essence of Goguen and Burstall's notion of institution [8]:

An *institution* $I = (\mathbf{Sign}, \mathbf{Sen}, \mathbf{Mod}, \models)$ consists of

1. a category **Sign** of *signatures*,
2. a functor $\mathbf{Sen}: \mathbf{Sign} \longrightarrow \mathbf{Set}$ giving the set of *sentences* $\mathbf{Sen}(\Sigma)$ over a signature Σ, and giving for each signature morphism $\sigma: \Sigma \longrightarrow \Sigma'$ the sentence translation map $\mathbf{Sen}(\sigma): \mathbf{Sen}(\Sigma) \longrightarrow \mathbf{Sen}(\Sigma')$,
3. a functor $\mathbf{Mod}: (\mathbf{Sign})^{op} \longrightarrow \mathbf{Class}$ giving the class[4] of *models* over a given signature, and giving for each signature morphism $\sigma: \Sigma \longrightarrow \Sigma'$ the *reduct functor* $\mathbf{Mod}(\sigma): \mathbf{Mod}(\Sigma') \longrightarrow \mathbf{Mod}(\Sigma)$,
4. a satisfaction relation $\models_\Sigma \subseteq \mathbf{Mod}(\Sigma) \times \mathbf{Sen}(\Sigma)$ for each $\Sigma \in \mathbf{Sign}$

such that for each $\sigma: \Sigma \longrightarrow \Sigma'$ in **Sign** the *Satisfaction Condition*, stating that *truth is invariant under change of notation*, holds:

$$M' \models_{\Sigma'} \mathbf{Sen}(\sigma)(\varphi) \iff \mathbf{Mod}(\sigma)(M') \models_\Sigma \varphi$$

for each $M' \in \mathbf{Mod}(\Sigma')$ and $\varphi \in \mathbf{Sen}(\Sigma)$.

We write $M'|_\sigma$ (the *reduct* of M' under σ) for $\mathbf{Mod}(\sigma)(M')$. In concrete examples, if $\sigma: \Sigma \longrightarrow \Sigma'$ is an inclusion, we also write $M'|_\Sigma$.

As said in the introduction, there are different notions of arrow between institutions, serving basically two different purposes (see [4, 20] for an overview):

1. The first purpose is that of building a more complex institution I over a simpler institution I', which is captured by an *institution morphism* [8] from I to I'. The idea is that I' may contain some "basic" concepts, and I may add some further concepts to I'. The technical definition is given below.
2. The second purpose is that of *representing* an institution I within a sufficiently rich institution I'. This is addressed in Sect. 5.

Given two institutions $I = (\mathbf{Sign}, \mathbf{Sen}, \mathbf{Mod}, \models)$ and $I' = (\mathbf{Sign}', \mathbf{Sen}', \mathbf{Mod}', \models')$, an *institution morphism* $\mu = (\Phi, \alpha, \beta): I \longrightarrow I'$ consists of

- a functor $\Phi: \mathbf{Sign} \longrightarrow \mathbf{Sign}'$,
- a natural transformation $\alpha: \mathbf{Sen}' \circ \Phi \longrightarrow \mathbf{Sen}$ and
- a natural transformation $\beta: \mathbf{Mod} \longrightarrow \mathbf{Mod}' \circ \Phi^{op}$

such that the following satisfaction invariant holds:

$$M \models_\Sigma \alpha_\Sigma(\varphi) \iff \beta_\Sigma(M) \models'_{\Phi(\Sigma)} \varphi'$$

for each $\Sigma \in \mathbf{Sign}$, $M \in \mathbf{Mod}(\Sigma)$ and $\varphi' \in \mathbf{Sen}'(\Phi(\Sigma))$.

Composition of institution morphisms is done in a straightforward way.

[4] There is no problem in allowing *categories* of models here, but model morphisms are entirely orthogonal to the subject of this paper, so we leave them out here.

Theorem 4 [19]. The category **Ins** *of institutions and institution morphisms is complete.* □

Thus we can apply the idea of combining things via colimits [6] to institutions as objects in **Ins**. However, since we write institution morphisms from a richer institution to poorer one, limits rather than colimits provide an appropriate tool for combination of institutions. Most roughly, limits in **Ins** are constructed by taking limits of the categories of signatures and of classes of models, and colimits of sets of sentences (due to the contravariant direction of the translation of sentences). Taking limits in **CAT** and **Class** results in categories of "amalgamated objects": this is used to put together signatures and models, respectively, at the level of single objects. In contrast, individual sentences are not combined, but rather sets of sentences are combined using colimits in **Set**. We refer to [20] for some simple examples. To show how this works and to indicate some problems here, as well as to prepare an easy to follow background for the further developments in the paper, let us introduce some well-known institutions and morphisms between them. We deliberately choose here the most trivial examples of logics and then their most trivial versions in a hope that this will make the problem and our solutions more visible and the presentation easier to follow.

Example 1. The institution ALG of many-sorted algebras without sentences. □

Example 2. The institution $ALG(=)$: many-sorted algebras with equalities between ground terms. Detailed definition and the proof of satisfaction condition may be extracted for instance from [8]. □

Example 3. The institution $PALG$ of partial many-sorted algebras (without sentences) has signatures of form $\Sigma = (S, OP, POP)$ consisting of sort symbols $s \in S$, total operation symbols $op\colon w \longrightarrow s \in OP$ and partial operation symbols $pop\colon w \longrightarrow s \in POP$, with signature morphisms defined as expected.

A Σ-model A consists of a (S, OP)-model in ALG plus a family of partial operations $(pop_A\colon A_w \multimap A_s)_{pop:w\longrightarrow s \in POP}$. Reducts are defined as in ALG. □

Example 4. The institution morphism $\mu^{=} = (Id, \alpha^{=}, Id)\colon ALG(=) \longrightarrow ALG$ is the identity on signatures and models, and for sentences, $\alpha^{=}_{\Sigma}$ is the inclusion. □

Example 5. The institution morphism $\mu^{\mathrm{P}} = (\Phi^{\mathrm{P}}, Id, \beta^{\mathrm{P}})\colon PALG \longrightarrow ALG$ maps a signature $\Sigma = (S, OP, POP)$ to $\Phi^{\mathrm{P}}(\Sigma) = (S, OP)$. Now a Σ-model consists of a $\Phi^{\mathrm{P}}(\Sigma)$-model plus some family of partial operations, and $\beta^{\mathrm{P}}_{\Sigma}$ just forgets this family of partial operations. □

We are now prepared to consider the combination of $ALG(=)$ and $PALG$ via the pullback

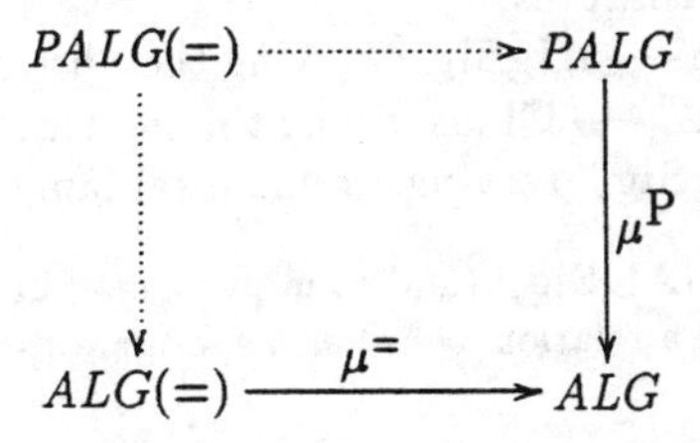

It turns out that signatures and models in $PALG(=)$ are those of $PALG$, while (S, OP, POP)-sentences are (S, OP)-sentences in $ALG(=)$. Thus all equations in $PALG(=)$ are built up from total operation symbols, while a true interaction of equality (as added to ALG in $ALG(=)$) and partiality (as added to ALG in $PALG$) should allow us to write down equations containing partial operation symbols as well.

Notational convention. In the following, we will always use symbols **Sign**, **Sen**, **Mod** and $\models$ for the corresponding components of an institution. We will omit obvious expansions of institutions to such components if some natural decorations are used to make the relationship evident. So, for instance, **Sign**$'$ and $\mathbf{Sign}_1$ will denote the categories of signatures of institutions I' and I_1, respectively. We will tacitly adopt a similar convention for other concepts introduced in this paper (institution morphisms, λ-parchments and their morphisms and representations, etc.) hoping that this will never lead to any real confusion.

3 λ-Parchments

The lack of interaction of concepts when taking limits of institutions shows that it is not enough to have a colimit (for example, a disjoint union) of sentence sets. We rather would like to unite *constructions*, or *operators* on sentence sets.

The idea behind the notion of parchment introduced by Goguen and Burstall [7] is to present sentences as terms over a signature $\mathbf{Lang}(\Sigma)$ giving their abstract syntax, and interpret them semantically using initiality of the term algebra. A combination of abstract syntaxes then causes a true interaction of different syntactical operators in the abstract syntax terms, which also has to be reflected in the semantics. The semantics is given by a special signature acting as a semantical universe (such that models are signature morphisms into it) and a "universal" semantic structure of this signature specifying semantical evaluation.

Thus parchments have a rather syntactical flavour, mixing up signatures and models in a signature category with a large signature. This works fine as technical means to present institutions. However, one may have doubts from the conceptual point of view if putting signatures and models into one category is a good idea. Technically, this causes troubles for instance when presenting institution morphisms and representations, where signatures and models may be mapped differently, even in the opposite directions. Therefore, we now introduce a variant of the notion of parchment, which has a cleaner separation between signatures and models:

Definition 5. A λ-parchment $P = (\mathbf{Sign}, \mathbf{Lang}, \mathbf{Mod}, L, \lambda, \mathcal{G})$ consists of

- a category **Sign** of signatures,
- a functor $\mathbf{Lang}: \mathbf{Sign} \longrightarrow \mathbf{AlgSig}_*$[5], giving the abstract syntax of sentences,
- a functor $\mathbf{Mod}: \mathbf{Sign}^{op} \longrightarrow \mathbf{Class}$ giving the models,
- a signature L in $|\mathbf{AlgSig}_*|$ serving as universal language,

[5] Recall that signatures in $|\mathbf{AlgSig}_*|$ (and so in particular $\mathbf{Lang}(\Sigma)$) always contain *Logic* (with a special sort $*$ and a relation $D : *$) as a subsignature preserved by the morphisms considered.

- a natural transformation λ: **Mod** $\longrightarrow$ *hom*(**Lang**(_), L) selecting appropriate parts of the universal language, and
- a semantical structure $\mathcal{G}$ in **Str**(L), determining the semantical evaluation of the syntactical constructs of P.

$\mathcal{G}_*$ is the *space of truth values* of P. In many standard examples, we have $\mathcal{G}_* =$ **Bool** (with **Bool** $= \{true;\ false\}$), and this was originally required in [7]. To be able to capture multi-valued logics as well as the two-valued ones, we drop this restriction. However, to recover the ultimatively two-valued notion of logical satisfaction, $D_{\mathcal{G}} \subseteq \mathcal{G}_*$ singles out the *designated truth values* in P, which are interpreted as truth, while the other values are interpreted as falsity (if we pass over to a two-valued satisfaction). For example, in Kleene's three-valued logic [9], only *true* is designated, while a paraconsistent three-valued logic [17] would also have the third truth value designated.

Definition 6. A λ-parchment $P = (\mathbf{Sign}, \mathbf{Lang}, \mathbf{Mod}, L, \lambda, \mathcal{G})$ induces an institution $\mathbf{I}(P) = (\mathbf{Sign}, \mathbf{Sen}, \mathbf{Mod}, \models)$ by putting

- $\mathbf{Sen}(\Sigma) = (T_{\mathbf{Lang}(\Sigma)})_*$, where $T_{\mathbf{Lang}(\Sigma)}$ is the initial **Lang**(Σ)-structure,
- $\mathbf{Sen}(\sigma: \Sigma \longrightarrow \Sigma') = (T_{\mathbf{Lang}(\sigma)})_*$ where $T_{\mathbf{Lang}(\sigma)}: T_{\mathbf{Lang}(\Sigma)} \longrightarrow T_{\mathbf{Lang}(\Sigma')}|_{\mathbf{Lang}(\sigma)}$ is the initial homomorphism,
- $M \models_\Sigma \varphi$ iff $M^\heartsuit_*(\varphi) \in D_{\mathcal{G}}$ where $M^\heartsuit: T_{\mathbf{Lang}(\Sigma)} \longrightarrow \mathcal{G}|_{\lambda_\Sigma(M)}$ is the initial homomorphism.

The satisfaction condition follows easily, using initiality of $T_{\mathbf{Lang}(\Sigma)}$ (see [7]).

There is a trivial inverse construction, mapping an institution to a λ-parchment with an algebraic constant in the abstract syntax for each sentence of a given signature, which at least in principle shows that every institution can be presented by a λ-parchment.

Example 6. The institution *ALG* of many-sorted algebras can be generated by a λ-parchment $(\mathbf{Sign}, \mathbf{Lang}, \mathbf{Mod}, L, \lambda, \mathcal{G})$, also denoted by *ALG*:

- **Sign** and **Mod** are taken from the institution *ALG*,
- **Lang** takes an *ALG*-signature (S, OP) to the $\mathbf{AlgSig}_*$-signature *Logic* $+(S, OP)$
- **Lang** extends a signature morphism $\sigma: \Sigma \longrightarrow \Sigma'$ to the signature morphism $\mathbf{Lang}(\sigma): \mathbf{Lang}(\Sigma) \rightarrow \mathbf{Lang}(\Sigma')$ that is the identity on *Logic*,
- $L =$ *Logic* $+$
  sorts s, for each set s[6]
  opns $op: s_1 \times \cdots \times s_n \longrightarrow s$, for each function $op: s_1 \times \cdots \times s_n \longrightarrow s$,
- For a signature Σ and a Σ-model M, $\lambda_\Sigma(M)$ is the identity on *Logic*, maps s to M_s and op to op_M,
- $\mathcal{G}_* =$ **Bool** , $D_{\mathcal{G}} = \{true\}$,
- $\mathcal{G}_s = s$ for each set s, $op_{\mathcal{G}} = op$ for each function $op: s_1 \times \cdots \times s_n \longrightarrow s$. □

[6] There is a foundational problem with this huge L, which can be solved by taking $\mathbf{AlgSig}_*$ to live in a higher Grothendieck universe.

Example 7. $ALG(=)$ is $(\mathbf{Sign}, \mathbf{Lang}, \mathbf{Mod}, L, \lambda, \mathfrak{C})$ with

- $\mathbf{Sign} = \mathbf{Sign}^{ALG}$, and $\mathbf{Mod} = \mathbf{Mod}^{ALG}$ are taken from ALG,
- $\mathbf{Lang}(S, OP) = \mathbf{Lang}^{ALG}(S, OP) +$
  opns $=: s \times s \longrightarrow *$ $(s \in S)$,
- $L = L^{ALG} +$
  opns $=: s \times s \longrightarrow *$ $(s$ a set$)$,
- λ is the straightforward extension of λ^{ALG},
- $\mathfrak{C}|_{L^{ALG}} = \mathfrak{C}^{ALG}$,
- $=_{\mathfrak{C}} (a, b) = \begin{cases} true, & \text{if } a = b \\ false, & \text{if } a \neq b \end{cases}$, □

Example 8. The partial many sorted λ-parchment $PALG = (\mathbf{Sign}, \mathbf{Lang}, \mathbf{Mod}, L, \lambda, \mathfrak{C})$:

- **Sign** is taken from the institution $PALG$,
- $\mathbf{Lang}(S, OP, POP) = \mathbf{Lang}^{ALG}(S, OP \cup POP)$
- $L = Logic +$
  sorts s, for each set s
  opns $pop: s_1 \times \cdots \times s_n \longrightarrow s$, for each partial function $pop: s_1 \times \cdots \times s_n \,-\!\!\circ\!\!\!\rightarrow s$
  (notice that this contains the signature L^{ALG}),
- λ is defined in the same way as λ^{ALG},
- $\mathfrak{C}_* = \mathbf{Bool}$, $\quad D_{\mathfrak{C}} = \{ true \}$,
- $\mathfrak{C}_s = s \uplus \{\bot\}$, where $\bot$ is a special new element,
- $pop_{\mathfrak{C}}(a_1, \ldots, a_n) = \begin{cases} pop(a_1, \ldots, a_n), & \text{if } a_i \neq \bot \text{ and } pop(a_1, \ldots, a_n) \text{ def.} \\ \bot, & \text{otherwise} \end{cases}$

□

As expected, ALG, $ALG(=)$ and $PALG$ are mapped by **I** to ALG, $ALG(=)$ and $PALG$, respectively. This justifies the overloading of names.

Note that the operations in $\mathbf{Lang}^{PALG}(S, OP, POP)$ are not necessary to generate the institution $PALG$, since the latter has no sentences. But the λ-parchment $PALG$ is defined in such a way that we expect to have total and partial function symbols in any kind of term which will be introduced by combining $PALG$ with other λ-parchments.

3.1 λ-Parchment morphisms

How can we express the fact that a λ-parchment P is built over another λ-parchment P'? We have to lift the concept of institution morphism to the level of λ-parchments.

Definition 7. Given two λ-parchments $P = (\mathbf{Sign}, \mathbf{Lang}, \mathbf{Mod}, L, \lambda, \mathfrak{C})$ and $P' = (\mathbf{Sign}', \mathbf{Lang}', \mathbf{Mod}', L', \lambda', \mathfrak{C}')$, a *$\lambda$-parchment morphism* $(\Phi, \alpha, \beta, l, g): P \longrightarrow P'$ consists of

- a functor $\Phi: \mathbf{Sign} \longrightarrow \mathbf{Sign}'$,
- a natural transformation $\alpha: \mathbf{Lang}' \circ \Phi \longrightarrow \mathbf{Lang}$,
- a natural transformation $\beta: \mathbf{Mod} \longrightarrow \mathbf{Mod}' \circ \Phi^{op}$,
- a signature morphism $l: L' \longrightarrow L$ and

– an L'-homomorphism $g\colon \mathfrak{G}' \longrightarrow \mathfrak{G}|_l$

such that for each $\Sigma \in \mathbf{Sign}$ and $M \in \mathbf{Mod}(\Sigma)$

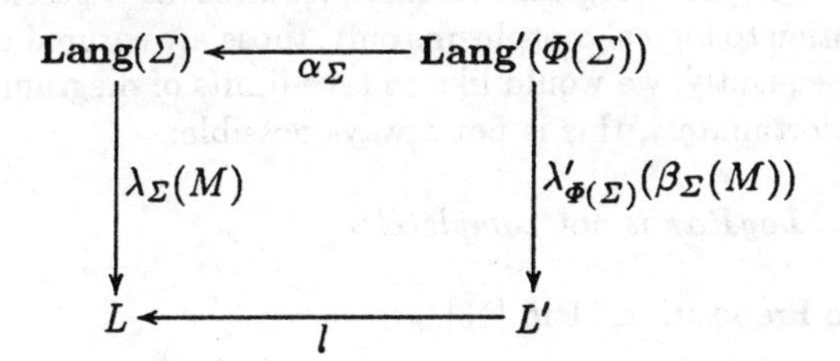

commutes.

Together with a straightforward composition, this gives a category $\lambda\mathbf{Par}$ of λ-parchments and λ-parchment morphisms.

If $\mathfrak{G}|_{Logic} = Id$ (that means, $g_* = Id$ and $D_{\mathfrak{G}} = D_{\mathfrak{G}'}$), we say that $(\Phi, \alpha, \beta, l, g)$ is *logical*. (We always have that $g_*[D_{\mathfrak{G}'}] \subseteq D_{\mathfrak{G}}$, but non-designation of truth values need not be preserved.) Let $\lambda\mathbf{LogPar}$ be the category of λ-parchments and logical λ-parchment morphisms. □

Proposition 8. *I can be extended to a functor* $\mathbf{I}\colon \lambda\boldsymbol{LogPar} \longrightarrow \boldsymbol{Ins}$.

Proof. Put $\mathbf{I}((\mathbf{Sign}, \mathbf{Lang}, \mathbf{Mod}, L, \lambda, \mathfrak{G}) \xrightarrow{(\Phi,\alpha,\beta,l,g)} (\mathbf{Sign}', \mathbf{Lang}', \mathbf{Mod}', L', \lambda', \mathfrak{G}')) =$

$$\mathbf{I}(\mathbf{Sign}, \mathbf{Lang}, \mathbf{Mod}, L, \lambda, \mathfrak{G}) \xrightarrow{(\Phi,\bar{\alpha},\beta)} \mathbf{I}(\mathbf{Sign}', \mathbf{Lang}', \mathbf{Mod}', L', \lambda', \mathfrak{G}')$$

with $\bar{\alpha}_\Sigma = (init_\Sigma)_*$, where $T_{\mathbf{Lang}'(\Phi(\Sigma))} \xrightarrow{init_\Sigma} T_{\mathbf{Lang}(\Sigma)}|_{\alpha_\Sigma}$ is the initial homomorphism.

The satisfaction condition follows by the initiality of $T_{\mathbf{Lang}(\Sigma)}$ from the fact that $g|_{Logic}$ preserves and *reflects* designation of truth values. □

Now the institution morphisms $\mu^=$ and μ^{P} from Examples 4 and 5, respectively, can be lifted to logical λ-parchment morphisms:

Example 9. Let $\mu^=\colon ALG(=) \longrightarrow ALG$ be $(Id, \alpha^=, Id, l^=, Id)$, where $l^=$ and $\alpha^=_\Sigma$ are inclusions. Then $\mathbf{I}(\mu^=)$ is the institution morphism presented in Example 4. □

Example 10. Let $\mu^{\mathrm{P}}\colon PALG \longrightarrow ALG$ be $(\Phi^{\mathrm{P}}, \alpha^{\mathrm{P}}, Id, l^{\mathrm{P}}, g^{\mathrm{P}})$ with

– $\Phi^{\mathrm{P}}(S, OP, POP) = (S, OP)$,
– $\alpha^{\mathrm{P}}_\Sigma$ and l^{P} are inclusions,
– g^{P}_s is the inclusion of s into $s \uplus \{\bot\}$, and g^{P}_* is the identity.

Then $\mathbf{I}(\mu^{\mathrm{P}})$ is the institution morphism presented in Example 5. □

To combine $ALG(=)$ and $PALG$, we would like to take a pullback like at the end of Section 2.1. But first, we have to prove its existence, or more generally, completeness of the category of λ-parchments.

4 Putting λ-parchments together using limits

To adequately put together logical systems presented as λ-parchments, we should restrict our attention to logical morphisms only: those are ensured to yield institution morphisms. Consequently, we would like to take limits of diagrams of λ-parchments in **λLogPar**. Unfortunately, this is not always possible:

Proposition 9. *λLogPar is* not *complete!*

Proof. Similar to Proposition 11 in [13]. □

Therefore, we have to take limits in **λPar**, at the risk of getting λ-parchment morphisms that are *not* logical, even when starting with a diagram in **λLogPar**.

Theorem 10. *The category **λPar** of λ-parchments and their morphisms is complete.*

Proof. (Sketch) Given a diagram of λ-parchments $\langle P_n \rangle_{n\in N}$ and their morphisms $\langle \mu_e : P_{s(e)} \rightarrow P_{t(e)} \rangle_{e\in E}$, where the graph of the diagram has nodes $n \in N$ and edges $e \in E$ with source $s(e) \in N$ and target $t(e) \in N$, proceed as follows:

- Construct the signature category of the limiting λ-parchment by taking a limit (in **CAT**) of the induced diagram of signature categories.
- For each signature $\Sigma \in |\mathbf{Sign}|$ in the resulting category:
  - Define the language of sentences $\mathbf{Lang}(\Sigma)$ as a colimit (in $\mathbf{AlgSig}_*$) of the induced diagram of $\mathbf{AlgSig}_*$-signatures.
  - Define a model category $\mathbf{Mod}(\Sigma)$ as a limit (in **Class**) of the induced diagram of model classes.
  - Define the universal language L as colimit in $\mathbf{AlgSig}_*$ of the induced diagram of $\mathbf{AlgSig}_*$-signatures $\langle L_n \rangle_{n\in N}$.
  - For each Σ-model M, $\lambda_\Sigma(M) : \mathbf{Lang}(\Sigma) \longrightarrow L$ is constructed using the colimiting property of $\mathbf{Lang}(\Sigma)$.
  - Since $l_{s(e)} \circ l_e = l_{t(e)}$, we have an isomorphism between the corresponding free functors induced by these $\mathbf{AlgSig}_*$ signature morphisms: $F_{l_{s(e)}} \circ F_{l_e} \cong F_{l_{t(e)}}$. Define the L-structure $\mathcal{G}$ to be the colimit (in $\mathbf{Str}(L)$) of the diagram of L-structures $\langle F_{l_n}(\mathcal{G}_n) \rangle_{n\in N}$ with L-homomorphisms

$$\langle F_{l_{t(e)}}(\mathcal{G}_{t(e)}) \cong F_{l_{s(e)}}(F_{l_e}(\mathcal{G}_{t(e)})) \xrightarrow{F_{l_{s(e)}}(g_e^{\#})} F_{l_{s(e)}}(\mathcal{G}_{s(e)}) \rangle_{e\in E}$$

  Let the colimit injections be $\overline{g_n} : F_{l_n}(\mathcal{G}_n) \rightarrow \mathcal{G}$, $n \in N$. Then

$$g_n := \overline{g_n}^{\natural} : \mathcal{G}_n \rightarrow \mathcal{G}|_{l_n} \quad (n \in N)$$

  are the ground structure translations of the limit projections constructed in **λPar**.

The above defines a limiting λ-parchment $P = (\mathbf{Sign}, \mathbf{Lang}, \mathbf{Mod}, L, \lambda, \mathcal{G})$ and limit projections $\mu_n = (\Phi_n, \alpha_n, \beta_n, l_n, g_n) : P \longrightarrow P_n$, $n \in N$. □

An example of a pullback of parchments combining $ALG(=)$ and $PALG$ is given in [13]. This can be easily translated to the level of λ-parchments. But the resulting pullback generates a whole bunch of new truth values. In [13], a congruence on $\mathcal{G}$ is defined, identifying these new truth values with either *true* or *false* and thus assigning a meaning to equations between possibly undefined terms. The same can be done here: in the framework of λ-parchments the resulting λ-parchment can be massaged similarly by quotienting $\mathcal{G}$ by an appropriate congruence to obtain the desired result.

However, the need for such additional transformation may be avoided if we deal with λ-parchments encoded in some "universal λ-parchment" – this will be presented in the next section.

5 Representations

While institution and λ-parchment morphisms capture the intuition that a logical system is built over another one (and the morphism is the corresponding projection), representations serve to *encode* a logical system into another one. We begin with institution representations.

Definition 11. Given institutions $I = (\mathbf{Sign}, \mathbf{Sen}, \mathbf{Mod}, \models)$ and $I' = (\mathbf{Sign}', \mathbf{Sen}', \mathbf{Mod}', \models')$, an institution ***representation*** [20] (***plain map of institutions*** in [12]) $\mu = (\Phi, \alpha, \beta): I \longrightarrow I'$ consists of

- a functor $\Phi: \mathbf{Sign} \longrightarrow \mathbf{Sign}'$,
- a natural transformation $\alpha: \mathbf{Sen} \longrightarrow \mathbf{Sen}' \circ \Phi$ and
- a natural transformation $\beta: \mathbf{Mod}' \circ \Phi^{op} \longrightarrow \mathbf{Mod}$.

such that the following ***representation condition*** is satisfied for $\Sigma \in \mathbf{Sign}$, $M' \in \mathbf{Mod}'(\Phi(\Sigma))$ and $\varphi \in \mathbf{Sen}(\Sigma)$:

$$M' \models'_{\Phi(\Sigma)} \alpha_\Sigma(\varphi) \iff \beta_\Sigma(M') \models_\Sigma \varphi$$

Together with a straightforward notion of composition, this gives us a category **InsRep** of institutions and representations.

One important application of representations is the re-use of theorem provers. It is based on the following well-known theorem [1]:

Theorem 12. *Let $\mu: I \longrightarrow I'$ be an institution representation with surjective model components. Then semantical entailment is preserved:*

$$\Gamma \models_\Sigma \varphi \textit{ iff } \alpha_\Sigma(\Gamma) \models'_{\Phi(\Sigma)} \alpha_\Sigma(\varphi)$$

where as usual, $\Gamma \models_\Sigma \varphi$ means that for all Σ-models M, $M \models_\Sigma \Gamma$ implies $M \models_\Sigma \varphi$.

Next, we consider representations of λ-parchments:

Definition 13. Given two λ-parchments $P = (\mathbf{Sign}, \mathbf{Lang}, \mathbf{Mod}, L, \lambda, \mathcal{G})$ and $P' = (\mathbf{Sign}', \mathbf{Lang}', \mathbf{Mod}', L', \lambda', \mathcal{G}')$, a ***$\lambda$-parchment representation*** $(\Phi, \alpha, \beta, l, g): P \longrightarrow P'$ consists of

- a functor $\Phi\colon \mathbf{Sign} \longrightarrow \mathbf{Sign}'$
- a natural transformation $\alpha\colon \mathbf{Lang} \longrightarrow \mathbf{Lang}' \circ \Phi$
- a natural transformation $\beta\colon \mathbf{Mod}' \circ \Phi^{op} \longrightarrow \mathbf{Mod}$
- a signature morphism $l\colon L \longrightarrow L'$
- an L-homomorphism $g\colon \mathcal{G} \longrightarrow \mathcal{G}'|_l$ (note that by the homomorphism condition, g has to preserve, but not necessarily reflect, the set of designated truth values).

such that for each $\Sigma \in \mathbf{Sign}$ and $M' \in \mathbf{Mod}'(\Phi(\Sigma))$

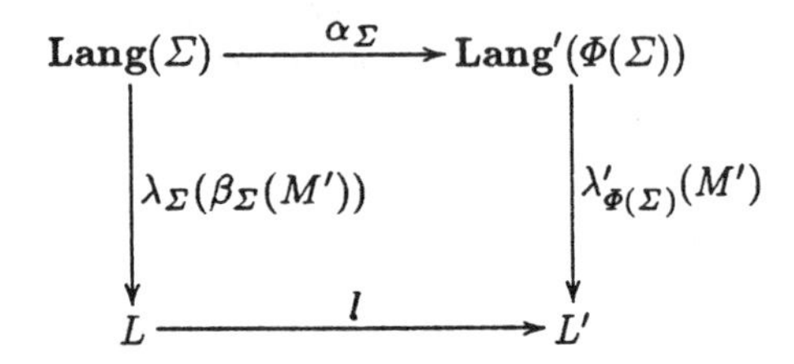

Together with a straightforward composition, this gives us a category λ**ParRep**.

Completely analogously to the case of λ-parchment morphisms, call a λ-parchment representation $\mu = (\Phi, \alpha, \beta, l, g)\colon P \to P'$ *logical*, if $g\,|_{Logic}$ is the identity (which also implies that the set of designated truth values is the same).

This gives us a subcategory λ**LogParRep** of λ**ParRep**, consisting of λ-parchments and logical representations.

Proposition 14. *The construction of institutions out of λ-parchments from Definition 6 can be extended to a functor **I**: λ**LogParRep** $\to$ **InsRep**.*

Proof. Very similar to that of Proposition 8. □

6 Combinations of representations

One way to use the notion of representations is to develop some sufficiently rich "universal" logic, equipped with a powerful proof theory, support tools, etc, and then reuse its facilities for other logics that can be represented in it. The question then arises whether we can systematically build logic representations together with the construction of logics themselves.

In our framework, let us consider a λ-parchment $UP = (\mathbf{USign}, \mathbf{ULang}, \mathbf{UMod}, UL, U\lambda, U\mathcal{G})$, which we will very informally view as such a rich "universal" logic. To study the problems mentioned above, we introduce a *representation map* between λ-parchments represented in *UP*. The idea is that one λ-parchment is built over another one, and moreover, that the representation in *UP* of the former also is built over the representation of the latter.

Definition 15. Given two λ-parchments $P = (\mathbf{Sign}, \mathbf{Lang}, \mathbf{Mod}, L, \lambda, \mathcal{G})$ and $P' = (\mathbf{Sign}', \mathbf{Lang}', \mathbf{Mod}', L', \lambda', \mathcal{G}')$, let $\rho\colon P \longrightarrow UP$ and $\rho'\colon P' \longrightarrow UP$ be their representations in *UP*. A *representation map* from ρ to ρ' consists of

- a λ-parchment morphism $\tilde{\mu} = (\tilde{\Phi}, \tilde{\alpha}, \tilde{\beta}, \tilde{l}, \tilde{g})\colon P' \longrightarrow P$.

– a natural transformation $\theta\colon \Phi \circ \tilde{\Phi} \longrightarrow \Phi'$

such that

– $\alpha' \circ \tilde{\alpha} = (\mathbf{ULang} * \theta) \circ (\alpha * \tilde{\Phi})$, i. e. for each signature $\Sigma' \in |\mathbf{Sign}'|$, $\alpha'_{\Sigma'} \circ \tilde{\alpha}_{\Sigma'} = \mathbf{ULang}(\theta_{\Sigma'}) \circ \alpha_{\tilde{\Phi}(\Sigma')}$,
– $\tilde{\beta} \circ \beta' = (\beta * \tilde{\Phi}^{op}) \circ (\mathbf{UMod} * \theta^{op})$, i. e. for each signature $\Sigma' \in |\mathbf{Sign}'|$, $\tilde{\beta}_{\Sigma'} \circ \beta'_{\Sigma'} = \beta_{\tilde{\Phi}(\Sigma')} \circ \mathbf{UMod}(\theta_{\Sigma'})$,
– $l = l' \circ \tilde{l}$ and
– $g'|_{\tilde{l}} \circ \tilde{g} = g$.

Proposition 16. The λ-parchment morphism in a map between logical λ-parchment representations is logical as well.

Proof. Just notice that under the notation of Def. 15, $g'|_{Logic} \circ \tilde{g}|_{Logic} = g|_{Logic}$, and so $\tilde{g}|_{Logic}$ is identity whenever $g'|_{Logic}$ and $g|_{Logic}$ are identities. □

Definition 17. Consider three λ-parchments $P = (\mathbf{Sign}, \mathbf{Lang}, \mathbf{Mod}, L, \lambda, ⊈)$, $P' = (\mathbf{Sign}', \mathbf{Lang}', \mathbf{Mod}', L', \lambda', ⊈')$ and $P'' = (\mathbf{Sign}'', \mathbf{Lang}'', \mathbf{Mod}'', L'', \lambda'', ⊈'')$, and let $\rho\colon P \longrightarrow UP$, $\rho'\colon P' \longrightarrow UP$ and $\rho''\colon P'' \longrightarrow UP$ be their representations in UP. Given two representation maps $\langle \tilde{\mu}_1, \theta_1 \rangle\colon \rho \longrightarrow \rho'$ and $\langle \tilde{\mu}_2, \theta_2 \rangle\colon \rho' \longrightarrow \rho''$, their *composition* is defined component-wise as:

$$\langle \tilde{\mu}_1 \circ \tilde{\mu}_2, \theta_2 \circ (\theta_1 * \Phi_2) \rangle$$

This yields a category $\lambda\mathbf{ParRep}_{UP}$ of λ-parchment representations in UP and their maps, with the obvious (contravariant) projection functor $\Pi\colon \lambda\mathbf{ParRep}^{op}_{UP} \longrightarrow \lambda\mathbf{Par}$.

*Theorem 18. Suppose that the category of signatures **USign** of the "universal" λ-parchment UP is cocomplete. Then the category $\lambda\mathbf{ParRep}_{UP}$ of λ-parchment representations in UP and their maps is cocomplete, and moreover, the projection functor $\Pi\colon \lambda\mathbf{ParRep}^{op}_{UP} \longrightarrow \lambda\mathbf{Par}$ is continuous.*

Proof. Consider a diagram Δ of λ-parchment representations $\langle \rho_n\colon P_n \longrightarrow UP \rangle_{n \in N}$ and their maps $\langle \langle \tilde{\mu}_e, \theta_e \rangle\colon \rho_{s(e)} \to \rho_{t(e)} \rangle_{e \in E}$, where the graph of the diagram Δ has nodes $n \in N$ and edges $e \in E$ with source $s(e) \in N$ and target $t(e) \in N$.

Recall from the proof of Theorem 10 the construction in $\lambda\mathbf{Par}$ of a limit $P = (\mathbf{Sign}, \mathbf{Lang}, \mathbf{Mod}, L, \lambda, ⊈)$ with projections $\tilde{\mu}_n\colon P \longrightarrow P_n$, $n \in N$, of the diagram $\Pi(\Delta)$. In the following, we construct a representation $\rho\colon P \longrightarrow UP$ and with natural transformations $\theta_n\colon \Phi_n \circ \tilde{\Phi}_n \longrightarrow \Phi$, $n \in N$, which together with the $\tilde{\mu}_n$ will be the colimit injections.

– To define $\Phi\colon \mathbf{Sign} \to \mathbf{USign}$, recall that $\mathbf{Sign}$ with functors $\tilde{\Phi}_n\colon \mathbf{Sign} \to \mathbf{Sign}_n$, $n \in N$, is a limit in **CAT** of the diagram of signature categories $\langle \mathbf{Sign}_n \rangle_{n \in N}$ with functors $\langle \tilde{\Phi}_e\colon \mathbf{Sign}_{t(e)} \to \mathbf{Sign}_{s(e)} \rangle_{e \in E}$ (note the contravariancy).
For each signature $\Sigma \in |\mathbf{Sign}|$, put $\Sigma_n = \tilde{\Phi}_n(\Sigma) \in |\mathbf{Sign}_n|$, $n \in N$, and let $\Phi(\Sigma) \in |\mathbf{USign}|$ with signature morphisms $(\theta_n)_\Sigma\colon \Phi_n(\Sigma_n) \to \Phi(\Sigma)$ be a

colimit in **USign** of the diagram of signatures $\langle \Phi_n(\Sigma_n)\rangle_{n\in N}$ with morphisms $\langle (\theta_e)_{\Sigma_{t(e)}}: \Phi_{s(e)}(\Sigma_{s(e)}) \rightarrow \Phi_{t(e)}(\Sigma_{t(e)})\rangle_{e\in E}$. Then, for each signature morphism $\sigma: \Sigma \rightarrow \Sigma'$, $\Phi(\sigma): \Phi(\Sigma) \rightarrow \Phi(\Sigma')$ is given by the colimiting property of $\Phi(\Sigma)$ for the cocone $\langle (\theta_n)_{\Sigma'} \circ \Phi_n(\tilde{\Phi}_n(\sigma)): \Phi_n(\Sigma_n) \rightarrow \Phi(\Sigma')\rangle_{n\in N}$.

- Consider now a signature $\Sigma \in |\mathbf{Sign}|$, and let $\Sigma_n = \tilde{\Phi}_n(\Sigma) \in |\mathbf{Sign}_n|$, for $n \in N$. Then:
  - To define a signature morphism $\alpha_\Sigma: \mathbf{Lang}(\Sigma) \rightarrow \mathbf{ULang}(\Phi(\Sigma))$, recall that $\mathbf{Lang}(\Sigma)$ with signature morphisms $(\tilde{\alpha}_n)_\Sigma: \mathbf{Lang}(\Sigma_n) \rightarrow \mathbf{Lang}(\Sigma)$, $n \in N$, is a colimit of the diagram Δ^α_Σ of signatures $\langle \mathbf{Lang}(\Sigma_n)\rangle_{n\in N}$ with signature morphisms $\langle (\tilde{\alpha}_e)_{\Sigma_{t(e)}}: \mathbf{Lang}_{s(e)}(\Sigma_{s(e)}) \rightarrow \mathbf{Lang}_{t(e)}(\Sigma_{t(e)})\rangle_{e\in E}$. (Note that, in comparison with Theorem 10, the direction of $\tilde{\alpha}_e$ has changed here since $\tilde{\mu}_e$ goes from $P_{t(e)}$ to $P_{s(e)}$.) Then $\langle \mathbf{ULang}((\theta_n)_\Sigma) \circ (\alpha_n)_{\Sigma_n}\rangle_{n\in N}$ can be shown to be a cocone over Δ^α_Σ. Now by the colimiting property of $\langle \mathbf{Lang}(\Sigma), \langle (\tilde{\alpha}_n)_\Sigma\rangle_{n\in N}\rangle$, there is a signature morphism $\alpha_\Sigma: \mathbf{Lang}(\Sigma) \longrightarrow \mathbf{ULang}(\Phi(\Sigma))$ with $\alpha_\Sigma \circ (\tilde{\alpha}_n)_\Sigma = \mathbf{ULang}((\theta_n)_\Sigma) \circ (\alpha_n)_{\Sigma_n}$.
  - To define a functor $\beta_\Sigma: \mathbf{UMod}(\Phi(\Sigma)) \rightarrow \mathbf{Mod}(\Sigma)$, recall that $\mathbf{Mod}(\Sigma)$ with functors $(\tilde{\beta}_n)_\Sigma: \mathbf{Mod}(\Sigma) \rightarrow \mathbf{Mod}_n(\Sigma_n)$, $n \in N$, is a limit of the diagram of categories $\langle \mathbf{Mod}_n(\Sigma_n)\rangle_{n\in N}$ with functors

    $$\langle (\tilde{\beta}_e)_{\Sigma_{t(e)}}: \mathbf{Mod}_{t(e)}(\Sigma_{t(e)}) \rightarrow \mathbf{Mod}_{s(e)}(\Sigma_{s(e)})\rangle_{e\in E}.$$

    For $UM \in |\mathbf{UMod}(\Phi(\Sigma))|$, define $\beta_\Sigma(UM) \in |\mathbf{Mod}(\Sigma)|$ to be the unique object in $|\mathbf{Mod}(\Sigma)|$ with $(\tilde{\beta}_n)_\Sigma(\beta_\Sigma(UM)) = (\beta_n)_{\Sigma_n}(\mathbf{UMod}((\theta_n)_\Sigma)(UM))$ for $n \in N$; this is well-defined by the construction of $\mathbf{Mod}(\Sigma)$ and of θ_n, $n \in N$, and the appropriate commutativity condition for maps of representations.
  - To define a signature morphism $l: L \longrightarrow UL$, recall that $\langle L, \langle L_n \xrightarrow{\tilde{l}_n} L\rangle_{n\in N}\rangle$ is a colimit of signatures L_n with signature morphisms $\langle \tilde{l}_e: L_{s(e)} \longrightarrow L_{t_e}\rangle_{e\in E}$. By the representation map conditions, $\langle UL, \langle L_n \xrightarrow{l_n} UL\rangle\rangle$ is a cocone over this diagram, so we get an $l: L \longrightarrow UL$ with $l \circ \tilde{l}_n = l_n$, which is the required commutativity condition for the universal language translation.
  - Recall that $\mathfrak{G} \in \mathbf{Str}(L)$ was defined as the colimit of the diagram $\Delta^{\mathfrak{G}}$ of L-structures $\langle F_{(\tilde{l}_n)}(\mathfrak{G}_n)\rangle_{n\in N}$ with L-homomorphisms

    $$\langle F_{\tilde{l}_{s(e)}}(\mathfrak{G}_{s(e)}) \cong F_{\tilde{l}_{t(e)}}(F_{\tilde{l}_e}(\mathfrak{G}_{s(e)})) \xrightarrow{F_{\tilde{l}_{t(e)}}(g_e^\#)} F_{\tilde{l}_{t(e)}}(\mathfrak{G}_{t(e)})\rangle_{e\in E}.$$

    We have to construct $g: \mathfrak{G} \longrightarrow U\mathfrak{G}|_l$. This can be done by finding a cocone over $\Delta^{\mathfrak{G}}$ with tip $U\mathfrak{G}|_l$. For $n \in N$, consider the following L_n-homomorphism:

    $$\mathfrak{G}_n \xrightarrow{g_n} U\mathfrak{G}|_{l_n} = U\mathfrak{G}|_{l\circ\tilde{l}_n}.$$

    When passing over to the adjoint arrows

    $$F_{\tilde{l}_n}(\mathfrak{G}) \xrightarrow{\quad g_n^\# \quad} U\mathfrak{G}|_l$$

$(n \in N)$, by the compositionality theorem for adjoint situations (see [10], Section IV.8, Theorem 1), we get the commutativity of

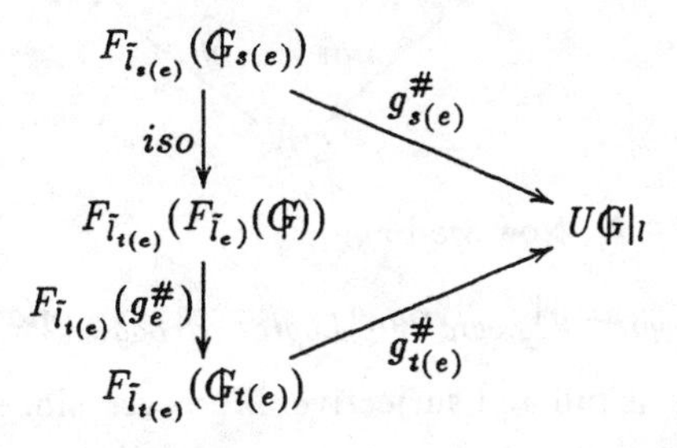

$(e \in E)$. This implies that the $g_n^\#$, $n \in N$, form a cocone over $\Delta^{\mathcal{G}}$. By the colimiting property of $\mathcal{G}$, we get a homomorphism $g: \mathcal{G} \longrightarrow U\mathcal{G}|_l$ with the property that

$$g \circ \overline{g_n} = g_n^\#$$

(where $\overline{g_n}$ is the colimit injection from the proof of Theorem 10) which implies

$$g|_{\bar{l}_n} \circ \tilde{g}_n = g_n$$

which is the ground structure part of the commutativity conditions making $\langle \tilde{\mu}_n, \theta_n \rangle$ into a representation map from ρ_n to ρ.

It is relatively straightforward (although tedious and notationally involved) to check that we have indeed defined a λ-parchment representation $\rho: P \rightarrow UP$. Then, one can also check that natural transformations $\theta_n: \Phi_n \circ \tilde{\Phi}_n \rightarrow \Phi$, $n \in N$, with the corresponding λ-parchment morphisms form maps of representations. Finally, one can check that the representation $\rho: P \rightarrow UP$ with maps $\langle \tilde{\mu}_n, \theta_n \rangle: \rho_n \rightarrow \rho$, $n \in N$, form a colimit of the diagram Δ in $\mathbf{\lambda ParRep}_{UP}$. □

Definition 19. $\mathbf{\lambda LogParRep}_{UP}$ is the full subcategory of $\mathbf{\lambda ParRep}_{UP}$ determined by the λ-parchment representations in UP that are logical.

Notice that Prop. 16 immediately implies that all maps between λ-parchment representations in $\mathbf{\lambda LogParRep}_{UP}$ contain logical λ-parchment morphisms.

Theorem 20. *Suppose that the category of signatures **USign** of the "universal" λ-parchment UP is cocomplete. Then $\boldsymbol{\lambda LogParRep}_{UP}$ has colimits of non-empty diagrams.*

Proof. Given a non-empty diagram Δ of λ-parchment representations $\langle \rho_n: P_n \longrightarrow UP \rangle_{n \in N}$ and their representation maps $\langle \langle \tilde{\mu}_e, \theta_e \rangle: \rho_{s(e)} \rightarrow \rho_{t(e)} \rangle_{e \in E}$, let $\langle \rho = (\Phi, \alpha, \beta, l, g): P \rightarrow UP, \langle \langle \tilde{\mu}_n, \theta_n \rangle: \rho_n \longrightarrow \rho \rangle_{n \in N} \rangle$ be the colimit of Δ in $\mathbf{\lambda ParRep}_{UP}$ due to Theorem 18.

Consider the least congruence $\sim$ on $\mathcal{G}$ such that $t \sim_* t'$ if $g_*(t) = g_*(t')$. The factorization of $\mathcal{G}$ by this congruence leads to a natural homomorphism *nat* which is full and surjective. Clearly, $\sim\ \subseteq ker(g)$. By Proposition 3 (1.), there is a factorization

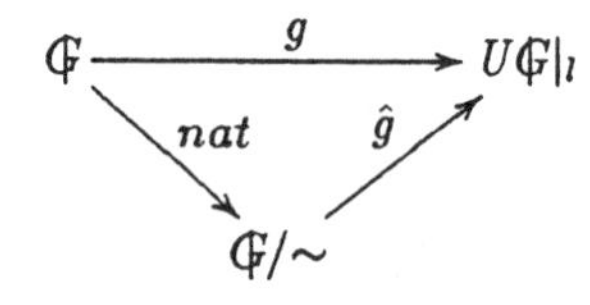

Fix an arbitrary $n \in N$. Now we have

$$Id = g_n\,|_{Logic} = g\,|_{Logic} \circ \tilde{g}_n\,|_{Logic} = \hat{g}\,|_{Logic} \circ nat\,|_{Logic} \circ \tilde{g}_n\,|_{Logic}\,.$$

By Prop. 1, $\hat{g}\,|_{Logic}$ is full and surjective. Moreover, since $ker(nat)\,|_{Logic} = \sim|_{Logic} = ker(g)\,|_{Logic}$, by Prop. 3 (2.), it is also injective, thus an isomorphism. Without loss of generality, we can assume that $\hat{g}\,|_{Logic}$ is the identity. Then the λ-parchment $\hat{P} = (\mathbf{Sign}, \mathbf{Lang}, \mathbf{Mod}, L, \lambda, \mathcal{G}/\sim)$ has a logical representation $\hat{\rho} = (\Phi, \alpha, \beta, l, \hat{g}) : \hat{P} \longrightarrow UP$. On the other hand, $(Id, Id, Id, Id, nat) : \hat{P} \longrightarrow P$ is a λ-parchment morphism, which we also denote by nat. For and for each $n \in N$, $\langle \tilde{\mu}_n \circ nat, \theta_n \rangle : \rho_n \longrightarrow \hat{\rho}$ is a logical representation map due to Prop. 16. These representation maps together form a cocone, since they are obtained from a cocone by composing with nat.

To prove the colimiting property, let $\langle \langle \mu'_n, \theta'_n \rangle : \rho_n \longrightarrow \rho' \rangle_{n \in N}$ be a cocone over Δ in $\lambda\mathbf{LogParRep}_{UP}$. By the colimiting property of $\rho : P \longrightarrow UP$ in $\lambda\mathbf{ParRep}_{UP}$, there is a representation map $\langle \mu_1, \theta_1 \rangle : \rho \longrightarrow \rho'$. Since $g'\,|_{l_1} \circ g_1 = g$ and ρ' is logical, $ker(g_1) \subseteq \sim$. By Prop. 3, there is a homomorphism $\hat{g_1} : \mathcal{G}/\sim \longrightarrow \mathcal{G}'\,|_{l_1}$ with $\hat{g_1} \circ nat = g_1$. Thus $g'\,|_{l_1} \circ \hat{g_1} \circ nat = g'\,|_{l_1} \circ g_1 = g = \hat{g} \circ nat$. Since nat is an epi, this implies $g'\,|_{l_1} \circ \hat{g_1} = \hat{g}$, which is the desired condition making $\langle (\Phi_1, \alpha_1, \beta_1, l_1, \hat{g_1}), \theta_1 \rangle$ a representation map from $\hat{\rho}$ to ρ'. □

For empty diagrams, the above construction yields an initial representation of the terminal λ-parchment into UP. However, this is logical only if $|U\mathcal{G}|_* = \emptyset$.

7 An Example

To illustrate the above construction without cluttering it with the details of any more practical "universal λ-parchment", let us consider an extremely simplified example. Consider as a universal parchment the λ-parchment $PFOL$ for partial first-order logic, with both strong and existential equalities, as e.g. in the logic of the CASL specification formalism [14, 11]. A rather obvious definition of such a parchment follows the lines of the parchment for the usual first-order logic as presented in [18]. In particular:

- $\mathbf{Sign}^{PFOL}$ includes $\mathbf{Sign}^{PALG}$ with the same models,
- $\mathbf{Lang}^{PFOL}(S, OP, POP)$ includes sorts S and operations $OP \cup POP$, as well as at least two operations $\stackrel{e}{=}, \stackrel{s}{=} : s \times s \longrightarrow *$ (augmented with propositional connectives and operations to form terms with variables, open formulae, etc.),
- L^{PFOL} contains L^{PALG} as well as $\stackrel{e}{=}$, $\stackrel{s}{=}$, propositional connectives, etc.,
- $\mathcal{G}^{PFOL}|_{L^{PALG}} = \mathcal{G}^{PALG}$

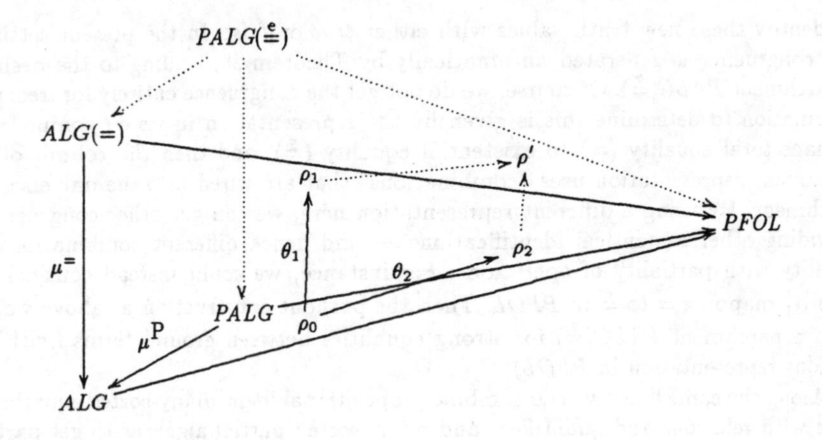

Fig. 1. A pushout in λ**LogParRep**$_{PFOL}$.

- $\stackrel{e}{=}_{\mathbb{C}} (a,b) = \begin{cases} true, & \text{if } a \neq \perp, b \neq \perp, a = b \\ false, & \text{otherwise} \end{cases}$
- $\stackrel{s}{=}_{\mathbb{C}} (a,b) = \begin{cases} true, & \text{if } a = b \\ false, & \text{otherwise} \end{cases}$

Then we can consider the following morphisms and representations of λ-parchments (see diagram in Fig. 1):

- Let $\mu^=: ALG(=) \longrightarrow ALG$ and $\mu^P: PALG \longrightarrow ALG$ be defined as in Examples 9 and 10.
- Let $\rho_0 = (\Phi_0, \alpha_0, Id, l_0, g_0): ALG \longrightarrow PFOL$ be defined using obvious inclusions Φ_0, α_0, l_0 and g_0, where g_s is the inclusion of s into $s \uplus \{\perp\}$, and g is the identity for the remaining sorts.
- Let $\rho_1 = (\Phi_1, \alpha_1, Id, l_1, g_1): ALG(=) \longrightarrow PFOL$ be defined using obvious inclusions Φ_1, l_1, g_1 being g_0, and α_1 mapping everything identical except that $=$ is mapped to $\stackrel{e}{=}$.
- Let $\rho_2 = (\Phi_2, \alpha_2, Id, Id, Id): PALG \longrightarrow PFOL$ be defined using obvious inclusions α_2 and Φ_2.
- Let $\theta_1: \Phi_0 \circ \Phi^= \longrightarrow \Phi_1$ be the identity and $\theta_2: \Phi_0 \circ \Phi^P \longrightarrow \Phi_2$ be the natural inclusion.

Then the pushout in the category λ**LogParRep**$_{PFOL}$ yields a new parchment $PALG(\stackrel{e}{=})$ with its obvious representation into $PFOL$. This parchment now has two truth values and generates the institution of partial algebras with ground existential equalities. Note that, as indicated at the end of Sect. 4, the combination of $ALG(=)$ and $PALG$ via a pullback of λ-parchments leads to a λ-parchment $PALG(\stackrel{?}{=})$ which has many new truth values: since it is not clear which truth value to assign to $=_{\mathbb{C}} (a, \perp)$ and $=_{\mathbb{C}} (\perp, a)$, these applications lead to new freely generated truth values. In [13], a congruence on the $\mathbb{C}$-component of $PALG(\stackrel{?}{=})$ is defined in order to identify these new truth values with either *true* or *false*. In the present setting, this congruence is generated automatically by Theorem 20, leading to the desired λ-parchment $PALG(\stackrel{e}{=})$. Of course, we do not get the congruence entirely for free: the information to determine this is given by the representation in its α_1 component: it maps total equality $(=)$ to existential equality $(\stackrel{e}{=})$, and then the colimit of λ-parchment representation uses "combinations" that are wired into the universal λ-parchment. By using a different representation here, we can get other congruences providing other semantical identifications — and hence different combinations of equality with partiality of operations. For instance, we could instead consider ρ_1' with α_1' mapping $=$ to $\stackrel{s}{=}$ in *PFOL*. Then the pushout construction as above would yield a parchment $PALG(\stackrel{s}{=})$ for strong equalities between ground terms (with its obvious representation in *PFOL*).

Along the same lines, we can combine propositional logic, many-sorted equational logic with relations and quantifiers and many-sorted partial algebras to get partial first-order logic.

8 Conclusion and future work

We have modified Goguen and Burstall's notion of parchment [7] and introduced λ-parchments (Def. 5), their morphisms (Def. 7), representations (Def. 13) and representation maps (Def. 15). We propose these as a framework for combining various features of logical systems presented as λ-parchments. Our main theorems show that it is indeed possible to combine λ-parchments using categorical limits, and that their representations can be combined in a similar manner. Theorem 10 states that the category of λ-parchments is complete (although the limit parchment may have to be further modified to really provide combination of the features in the underlying parchments). Theorem 18 states the cocompleteness of the category of λ-parchment representations, with the colimit of representations providing a representation for the limit of the represented parchments. Finally, Theorem 20 extends this to logical representations, which preserve the truth-space, and so leads to representations between the institutions presented. The colimits given by this last theorem render any modification of the limit λ-parchments unnecessary — the features of the underlying parchments are combined according to what is built into the universal parchment they are represented in.

We have illustrated these general ideas in Sect. 7 by combining many-sorted equational logic $ALG(=)$ with partial algebras *PALG* getting a logic of many-sorted partial algebras with existence equations $PALG(\stackrel{e}{=})$. As "universal" logic, which serves for representing the various logics needed in the combination, we have used partial first-order logic *PFOL*, close to the logic of the CASL specification formalism. It would of course be more interesting to have examples where the concepts that appear in the combination are not already directly present in the universal logic — such examples would need a more flexible notion of representation between λ-parchments.

Thus future work should examine this point in more detail. For example, we have experimented with so-called model-theoretic parchments, where the Procrustean semantical structure $\mathcal{G}$ and its translations in parchment morphisms and representations are defined in a more local way, referring to individual signatures and models.

To carry over the completeness and cocompleteness results to this setting, probably the condition of being logical has to be relaxed. Indeed, the requirement that the truth space component of parchment morphisms and representations is an identity is very strong. A first step would be to work with closed monomorphisms here, which still leads to institution morphisms and representations (and for which our technical results hold as well).

But this is perhaps still too strong in order to get enough representations within higher-order logics. See [20] for a brief discussion and for some hints for relaxing the representation condition for institutions as well. An open question is what this would mean at the level of λ-parchments.

Furthermore, representations should be made more flexible in another direction, analogous to so-called *simple* representations [12]. This notion can be generalized to λ-parchments via a monad on **λParRep** which adds all theories as signatures. To gain even more flexibility, it should also add all derived operations as operation symbols, which would, for example, allow to choose between existential or strong equality even if there is no corresponding operation in the universal logic available.

References

[1] E. Astesiano, M. Cerioli. Relationships between logical frameworks. In M. Bidoit, C. Choppy, eds., *Proc. 8th ADT workshop*, *Lecture Notes in Computer Science* **655**, 126–143. Springer Verlag, 1992.

[2] Jon Barwise. Axioms for abstract model theory. *Annals of Mathematical Logic* **7**, 221–265, 1974.

[3] P. Burmeister. *A model theoretic approach to partial algebras*. Akademie Verlag, Berlin, 1986.

[4] M. Cerioli. *Relationships between Logical Formalisms*. PhD thesis, TD-4/93, Università di Pisa-Genova-Udine, 1993.

[5] H. Ehrig, P. Pepper, F. Orejas. On recent trends in algebraic specification. In *Proc. ICALP'89*, *Lecture Notes in Computer Science* **372**, 263–288. Springer Verlag, 1989.

[6] J. A. Goguen. A categorical manifesto. *Mathematical Structures in Computer Science* **1**, 49–67, 1991.

[7] J. A. Goguen, R. M. Burstall. A study in the foundations of programming methodology: Specifications, institutions, charters and parchments. In D. Pitt et al., ed., *Category Theory and Computer Programming*, *Lecture Notes in Computer Science* **240**, 313–333. Springer Verlag, 1985.

[8] J. A. Goguen, R. M. Burstall. Institutions: Abstract model theory for specification and programming. *Journal of the Association for Computing Machinery* **39**, 95–146, 1992. Predecessor in: LNCS 164, 221–256, 1984.

[9] S.C. Kleene. *Introduction to Metamathematics*. North Holland, 1952.

[10] S. Mac Lane. *Categories for the working mathematician*. Springer, 1972.

[11] CoFI Task Group on Language Design. *CASL* – The *CoFI* Algebraic Specification Language – Summary. *CoFI* Document: CASL/Summary. WWW[7], FTP[8], May 1997.

[12] J. Meseguer. General logics. In *Logic Colloquium 87*, 275–329. North Holland, 1989.

[13] T. Mossakowski. Using limits of parchments to systematically construct institutions of partial algebras. In M. Haveraaen, O. Owe, O.-J. Dahl, eds., *Recent Trends in Data*

[7] http://www.brics.dk/Projects/CoFI/Documents/CASL/Summary/

[8] ftp://ftp.brics.dk/Projects/CoFI/Documents/CASL/Summary/

Type Specifications. 11th Workshop on Specification of Abstract Data Types, Lecture Notes in Computer Science **1130**, 379–393. Springer Verlag, 1996.

[14] Peter D. Mosses. CoFI: The Common Framework Initiative for Algebraic Specification and Development. In M. Bidoit, M. Dauchet, eds., *TAPSOFT 97, Lecture Notes in Computer Science* **1214**, 115–137. Springer Verlag, 1997.

[15] P. Padawitz. *Computing in Horn Clause Theories*. Springer Verlag, Heidelberg, 1988.

[16] W. Pawłowski. Context institutions. In M. Haveraaen, O. Owe, O.-J. Dahl, eds., *Recent Trends in Data Type Specifications. 11th Workshop on Specification of Abstract Data Types, Lecture Notes in Computer Science* **1130**, 436–457. Springer Verlag, 1996.

[17] G. Priest. Inconsistent arithmetics and non-Euclidean geometries. Invited talk at Logic Colloquium, San Sebasitan, 1996.

[18] P. Stefaneas. The first order parchment. Report PRG-TR-16-92, Oxford University Computing Laboratory, 1992.

[19] A. Tarlecki. Bits and pieces of the theory of institutions. In D. Pitt, S. Abramsky, A. Poigné, D. Rydeheard, eds., *Proc. Intl. Workshop on Category Theory and Computer Programming, Guildford 1985, Lecture Notes in Computer Science* **240**, 334–363. Springer-Verlag, 1986.

[20] A. Tarlecki. Moving between logical systems. In M. Haveraaen, O. Owe, O.-J. Dahl, eds., *Recent Trends in Data Type Specifications. 11th Workshop on Specification of Abstract Data Types, Lecture Notes in Computer Science* **1130**, 478–502. Springer Verlag, 1996.

A Deciding Algorithm for Linear Isomorphism of Types with Complexity $O(nlog^2(n))^\star$.

A.Andreev[1] S. Soloviev[2]

[1] Department of Mechanics and Mathematics,
Moscow State University,
Moscow, 119899, Russia,
e-mail: andreev@matis.math.msu.su

[2] Computer Science Department,
Durham University, U.K.,
e-mail:Sergei.Soloviev@durham.ac.uk

Abstract. It is known, that ordinary isomorphisms (associativity and commutativity of "times", isomorphisms for "times" unit and currying) provide a complete axiomatisation of isomorphism of types in multiplicative linear lambda calculus (isomorphism of objects in a free symmetric monoidal closed category). One of the reasons to consider linear isomorphism of types instead of ordinary isomorphism was that better complexity could be expected. Meanwhile, no upper bounds reasonnably close to linear were obtained. We describe an algorithm deciding if two types are linearly isomorphic with complexity $O(nlog^2(n))$.

1 Introduction

The problem of characterisation of isomorphism of types that holds in all models of certain system of typed lambda calculus is closely connected with mathematical semantics of datatypes [1], [2]. The problem allows many equivalent re-formulations, for example, a description of isomorphic objects in free closed categories of different classes: Cartesian Closed (CC) Categories , Symmetric Monoidal Closed (SMC)Categories, Biclosed Categories etc. A presentation of a free Closed Category is given by certain system of propositional calculus (with deductions as morphisms). Another (maybe, more familiar to computer scientists) is provided by lambda calculus.

To define the notion of isomorphism of types in a system of typed lambda calculus one needs a)the presence of functional types $A \to B$; b)a composition, for two terms of types $A \to B$ and $B \to C$ their composition is to be a term of

* The main part of this research was done while both authors were visiting Computer Science Department of Aarhus University; the visits being funded by BRICS, a Centre of the Danish National research Foundation, and the european CLICS grant (for the second author). Final version of this paper was done by S.Soloviev while employed by Durham University and funded by British ESPRC grant (on leave from S.Petersburg Institute for Informatics RAN).

the type $A \to C$; c)the terms $id : A \to A$ representing identity maps for every type A (usually, $\lambda x : A.x$); d)an equivalence relation on terms (which respects the composition).

Of course, these conditions look too abstract for the familiar systems of lambda calculus, but we just use an opportunity to present the problem in more general setting. These conditions satisfied, one defines two types A, B to be isomorphic iff there are terms $t : A \to B$ and $s : B \to A$ such that the composite $t \cdot s$ is equivalent to $id : B \to B$ and the composite $s \cdot t$ to $id : A \to A$.

Suppose two types A, B are given and one would decide, if they are isomorphic. The direct attempt based on the definition above will lead, in general, to the consideration of an infinite set of lambda terms living in the types $A \to B$ and $B \to A$. Usually, one is looking for the algorithms based on transformations of types without any explicite reference to lambda terms.

A complete axiomatization and decision algorithm for isomorphism of objects in free CC Categories was obtained in [3]. That result provided automatically a complete axiomatization and decision algorithm for the isomorphicm of types in the First Order Typed Lambda Calculus with terminal object and surjective pairing. The precise formulation is the following:

Let $\sim$ be the equivalence relation on types generated by the following axioms

1. $A \sim A$;
2. $A \wedge B \sim B \wedge A$;
3. $A \wedge (B \wedge C) \sim (A \wedge B) \wedge C$;
4. $A \wedge I \sim A$;
5. $I \to A \sim A$;
6. $A \wedge B \to C \sim A \to (B \to C)$;
7. $A \to I \sim I$
8. $A \to B \wedge C \sim (A \to B) \wedge (A \to C)$

and rules

$$\frac{A \sim B}{C[A/X] \sim D[B/X]}(subst) \quad \frac{A \sim B}{B \sim A}(sym) \quad \frac{A \sim B \quad B \sim C}{A \sim C}(trans)$$

(here $[A/X]$ denotes substitution of A for type variable X).

The types A and B are isomorphic (in above-mentioned calculus) iff $A \sim B$.

(In [3] a model-theoretic proof was given. A direct proof for typed lambda calculus was published in [4], where the applications to Computer Science were also discussed. Di Cosmo developed the method from [4] to obtain similar results for the Second Order Typed Lambda Calculus.)

Some weaker variants of isomorphism of types are of practical interest, the linear isomorphism of types in particular, because one can expect that the complexity will be reasonnably low.

The linear isomorphism of types corresponds to the isomorphism of objects in free SMC category, and can be also described as the isomorphism of types in the system of lambda calculus which corresponds to intuitionistic multiplicative linear logic. (A description of this system can be found in [5], [6], [7], [8].)

In [7] it was shown that the subsystem of the axiom system above, consisting of the axioms 1)-6) (where $\wedge$ is understood as "times" and $\rightarrow$ as linear implication) with the same rules, defines an equivalence relation on types that coincides with the relation of linear isomorphism of types.

Of course, the use of linear logic or linear terms does not imply linear complexity of corresponding algorithms.

The deciding algorithms for the isomorphism in the First Order Typed Lambda Calculus with terminal object and surjective pairing used a reduction to some normal forms, which were, in general, subexponetially longer than the original types. In case of linear isomorphism there is no growth of the length. The main problem is (recursive) ordering of factors in the subformulas of the form $A_1 \wedge .... \wedge A_n$. More or less obvious algorithms have quadratic complexity. We propose an algorithm with complexity $C{\cdot}nlog^2(n)$.

2 The Algorithm.

In this section we shall denote by $\sim$ the eqivalence relation on types defined by axioms (1)-(5) and the rules above. $\wedge$ can be understood as "times", $\rightarrow$ as linear implication, and I is the unit of $\wedge$. The algorithm is based on the fact [7] that two types A, B are linearly isomorphic iff $A \sim B$ for this relation.

2.1 Regular Formulas.

When low complexity bounds are considered the form of presentation of information is quite important. Usually the types are presented by formulas of intuitionistic linear logic, i.e., by propositional formulas with two binary connectives $\wedge$ and $\rightarrow$ (and constant I).We shall use a kind of prefix notation (not exactly polish notation, because we prefer to have "$\wedge$" with varying number of arguments; that does no harm because of associativity axiom 3)). So, the formulas are defined inductively in the following way:

1. the symbols $X_1, ..., X_n, ...$(type variables) and the constant I are formulas;
2. if A, B are formulas then $(\rightarrow AB)$ is formula;
3. if the $A_1...A_n$ are formulas $(n > 1)$, then the $(\wedge A_1...A_n)$ is formula.

Below we shall use also list notation in formulas with the agreement that the expressions like $(\wedge\Gamma), (\rightarrow\Gamma)$ when Γ contains just one member should be understood as that member itself($\wedge$ and $\rightarrow$ should be omitted)

The syntactic axioms 1) - 6) above that characterize the linear isomorphism of types can be replaced for this presentation by the following axioms:

(i) $A \sim A$ $(refl)$;

(ii) $(\wedge A_1...A_n) \sim (\wedge A_{\sigma(1)}...A_{\sigma(n)})$ (com) where σ is a permutation of the set {1,...,n} $(n > 1)$;

(iii) $(\wedge\Gamma(\wedge\Delta)\Sigma) \sim (\wedge\Gamma\Delta\Sigma)$ (as) (with Γ, Δ, Σ being lists of formulas of appropriate length);
(iv) $(\wedge\Gamma I\Delta) \sim (\wedge\Gamma\Delta)$ (un) (with $\Gamma\Delta$ non-empty);
(v) $(\rightarrow IB) \sim B$ (un')
(vi) $\rightarrow A(\rightarrow BC) \sim\rightarrow (\wedge AB)C (cur)$

The rules for $\sim$ in this syntax are still $(subst), (sym), (trans)$.

We shall write $\Rightarrow_k$ iff $A{\sim}B$ is derivable from an instance of the axiom labelled by k by single application of $(subst)$, i.e., by replacement of an occurrence of a left side of this axiom in A by the right side of the same axiom.

We shall write $A{\sim_k}B$ iff $A{\sim}B$ is derivable from the axiom labelled by k only (obviously, $A{\sim_k}B$ iff there exists a chain $A = A_1, A_2, ..., A_n = B$ such that $A_i{\Rightarrow_k}A_{i+1}$ or $A_{i+1}{\Rightarrow_k}A_i$).

We shall write $A{\sim_\rightarrow}B$ if it is derivable from reflexivity $(refl)$ and axioms $(un), (un')$ and (cur) and $A{\sim_{com}}B$, if only $(refl), (com)$ were used.

Now *regular formulas, regular $\wedge$-formulas and regular $\rightarrow$-formulas* are defined in the following way.

Definition 1. 1. the symbols $X_1, ..., X_n, ...$(type variables) and the constant I are regular formulas;
2. if A is regular formula different from I and B is regular $\wedge$-formula, variable or I then $\rightarrow AB$ is regular $\rightarrow$-formula;
3. if each of$A_1, ..., A_n$ $(n > 1)$ is variable or regular $\rightarrow$-formula, then $(\wedge A_1 ... A_n)$ is regular $\wedge$-formula;
4. regular $\wedge$-formulas and regular $\rightarrow$-formulas are regular formulas.

Let A be a (sub)formula. We shall call its 1-extension any (sub)formula of the form $(\rightarrow \Gamma A\Delta)$, or $(\wedge\Gamma A\Delta)$ with Γ and/or Δ non-empty.

Remark 2. (i) A regular formula is $\wedge$- or $\rightarrow$-regular, except the case when it is a variable or the constant I. (ii)All subformulas of a regular formula are regular formulas. (iii) All regular $\wedge$- subformulas of a regular formula A are maximal in the following sense: their 1-extensions (in A) are $\rightarrow$-subformulas (not $\wedge$-subformulas). (iv)All regular $\rightarrow$- subformulas of a regular formula A are maximal in the following sense: their 1-extensions (in A) are $\rightarrow$-subformulas (not $\rightarrow$-subformulas)

Lemma 3. *For every formula A, there is unique regular formula B such that $A{\sim_\rightarrow}B$.*

Proof. Consider the system of formula-reductions $\Rightarrow_{as}, \Rightarrow_{un}, \Rightarrow_{un'}, \Rightarrow_{cur}$ (an occurrence of the left side of corresponding axiom is to be replaced by the right side). By straightforward induction on the structure of formulas one shows that a formula is in normal form iff it is regular. Now (trivial check) a)the system is Church-Rosser's and b)terminating (since each step decreases the number of $\rightarrow, \wedge$ or I).

Denote by $R(A)$ the regular formula corresponding to A.

Lemma 4. *If* $A \Rightarrow_k B$ *where* k *is* $(as), (un), (un'), (cur)$, *then* $R(A) = R(B)$. *If* $A \Rightarrow_k B$ *where* k *is* (com), *then* $R(A) \sim_{com} R(B)$.

Proof. By induction on the length of reduction sequence from A to $R(A)$ in the system of reductions described above.

As an immediate consequence of the two lemmas, we have

Lemma 5. $A \sim B$ *iff* $R(A) \sim_{com} R(B)$.

Let us call the length $l(A)$ of the formula A the number of occurrences of variables, I, $\wedge$ and $\rightarrow$ in A.

Lemma 6. *If* $A \sim_{com} B$ *then* $l(A) = l(B)$

2.2 Computational Model

Our computational model is random access mashine. The cell size is $O(log(n))$, where n is the length of the formulas that are considered. The addresses of cells are $1, 2, \ldots$.

For our goals we have to use a special presentation of formulas. As it is easily seen, with every occurrence of $\wedge, \rightarrow$, variable or I in a formula A a unique subformula of A is connected. This occurrence will be called the main symbol of a subformula.

Initially, let us represent each occurence of $\wedge, \rightarrow$, variable or I (and corresponding subformula) in more standard way, using the quadriples, described below. (We do not need to represent brackets.)It is supposed that each such quadriple is encoded in certain way and occupies one cell of RAM The address of a subformula is understood as the address in RAM-memory of its main symbol.

The quadriple, representing a symbol, is defined in the following way:

Definition 7. – the first member is the symbol itself (one may think, that it is represented numerically in some way);
- the second member is 0 for I or a variable and the address of the leftmost subformula, if the symbol is $\wedge$ or $\rightarrow$;
- the third member is 0 for I or a variable and the address of the last subformula, if the symbol is $\wedge$ or $\rightarrow$;
- the fourth member is the address of the next subformula in the 1-extension of the subformula, corresponding to the considered symbol. If there is no next subformula, it is 0.

The formula is presented in RAM-memory by collection of quadriples representing its symbols (except brackets). Let $[v]$ denote the address of the cell corresponding to an occurrence v.

Example 8. Consider the formula $\rightarrow_1 a \rightarrow_2 bc$ (i.e., $a \rightarrow_1 (b \rightarrow_2 c)$ in standard notation). Its presentation is

$$\{\rightarrow, [a], [\rightarrow_2], 0, \}; \{a, 0, 0, [\rightarrow_2]\}; \{\rightarrow, [b], [c], 0\}; \{b, 0, 0, [c]\}; \{c, 0, 0, 0\}.$$

If we put the quadriples in the nodes of the syntactic tree of our formula (direction of pointers shown by arrows), we'll have

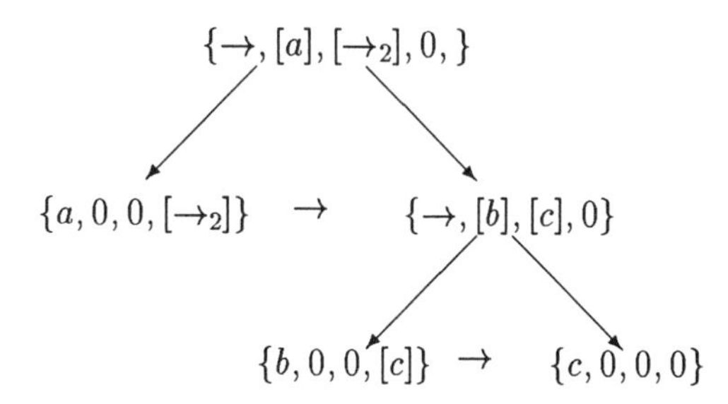

Here $[\rightarrow_1]$ would be the address of first, and $[\rightarrow_2]$ is the address of third quadriple.

The length $l(A)$ of a (sub)formula is the number of cells in this presentation.

It is supposed, that the cells, occupied by different formulas always are disjoint.

Elementary operations are comparisons of quadriples and their members and standard arithmetical and control operations of RAM. Input information of the algorithms considered below consists of the addresses of the main symbols of processed formulas.

The following lemma is obvious.

Lemma 9. *Syntactical identity of two formulas A, B can be checked in time $O(min(l(A), l(B))$.*

Lemma 10. *There is an algorithm transforming the presentation of an arbitrary formula A in the presentation of $R(A)$ with an upper time bound $C(l(A))$ for some constant C.*

Proof. It proceeds by induction on the structure of A and is standard, with one modification due to our use of multiple $\wedge$.

When the formula $\wedge A_1 ... A_n$ is considered, the algorithm should check if $n = 2$ or $n > 2$ (obviously, in constant number of steps).

If $n = 2$ then induction proceeds in ordinary way, if $n > 2$ then the inductive hypothesis (and the algorithm) should be applied to A_1 and $\wedge A_2 ... A_n$, and then in constant number of steps one can "assemble" the presentation of the regular form of the formula $\wedge A_1 ... A_n$. (One does not change registers, only pointers.)

Definition 11. Let us call by extended presentation of a formula it presentation where each quadriple is extended to the quintiple, with the fifth symbol being the length of the subformula, whose main symbol is represented by the quadriple.

Lemma 12. *There is an algorithm transforming the presentation of a formula into its extended presentation in less than $C(l(A))$ steps for some constant C.*

Proof. Consider the following trip. Start in the root of the tree, and go from node to node according the following procedure: a)if there is leftmost subformula (node) not yet visited, go there; b)if it was already visited, go to next subformula in 1-extension; c)if in case b) the node corresponds to the last subformula, go to the node, representing the main symbol of 1-extension (one can keep "return address"), or stop, if we come to the root.

Obviously, each node is visited exactly twice, and (with constant number of extra-operations in each node) one can keep count the number of symbols visited before coming to each node.

Then, if in the first visit this number is written in the 5-th position, one can replace it by the difference at the second visit, which will be the length of the subtree corresponding to subformula.

2.3 Strongly Regular Formulas

Let us define inductively a relation $\leq$ on regular formulas. ($A < B$ is understood as $A \leq B$ and not $A {\sim_{com}} B$.) Remember, that if $A \sim_{com} B$, they have the same length.

Definition 13. – If $l(A) < l(B)$ then $A < B$. If $l(A) = l(B)$, then:
- For variables and constant I,

$$I < a_1 < ... < a_n < ....$$

- If A is a regular $\wedge$-formula and B is a regular $\rightarrow$-formula, then $A < B$;
- If A and B are the regular $\rightarrow$-formulas,

$$A = (\rightarrow A_1 A_2), \; B = (\rightarrow B_1 B_2),$$

  and

$$A_1 < B_1$$

  or

$$A_1 \sim_{com} B_1, \; A_2 \leq B_2$$

  then $A \leq B$;
- If A and B are regular $\wedge$-formulas,

$$A = (\wedge A_1 A_2 ... A_n), \; B = (\wedge B_1 B_2 ... B_n),$$

  and there exist bijections φ and ψ

$$\varphi\{1, 2, ..., m\} \rightarrow \{1, 2, ..., m\},$$

$$\psi\{1, 2, ..., r\} \rightarrow \{1, 2, ..., r\}$$

and the number k,

$$1 \leq k \leq min(m, r),$$

such that

$$A_{\varphi_1} \leq A_{\varphi_2} \leq ... \leq A_{\varphi_m}, \quad B_{\psi_1} \leq B_{\psi_2} \leq ... \leq B_{\psi_r}$$

$$A_{\varphi_1} \sim_{com} B_{\psi_1} \quad A_{\varphi_2} \sim_{com} B_{\psi_2}, ..., A_{\varphi_{k-1}} \sim_{com} B_{\psi_{k-1}}, \quad A_{\varphi_k} < B_{\psi_k}$$

then $A < B$

Lemma 14. *For every regular formulas A, B always $A \leq B$ or $B \leq A$. If $A \leq B$ and $B \leq A$, then $A \sim_{com} B$. (That is, the relation $\leq$ is linear order on $\sim_{com}$-equivalence classes of regular formulas.)*

Proof. By induction on $l(A)$, taking into account, that both $A \leq B$ and $B \leq A$ can hold only if A and B are both of the same type (variables, constants, $\wedge$- or $\rightarrow$-formulas), and have equal length.

Now we are in the position to define strongly regular formulas.

Definition 15. A regular formula is strongly regular iff in each its subformula of the form $\wedge A_1 ... A_n$ there is $A_1 \leq ... \leq A_n$.

Lemma 16. *For every regular formula A there exists exactly one strongly regular formula $S(A)$ in the $\sim_{com}$-equivalence class of A.*

Proof. An $S(A)$ can be obtained by ordering $\wedge$-subformulas of A, that is, using $\wedge_{com}$. By induction on the structure of A (using definition of $\leq$) we show uniqueness.

As an immediate consequence of theis lemma and lemma 5, we have the following theorem:

Theorem 17. $A \sim B \iff S(R(A)) = S(R(B))$

3 Complexity

Lemma 18. *There is an algorithm deciding if $A \leq B$ or $B \leq A$ for strongly regular formulas A, B in time bound by Cn, where $n = min(l(A), l(B))$.*

Proof. By induction on n, taking into account that syntactic identity of two regular formulas can be checked in time proportional to the (minimum) of their lengths (lemma 9)

Theorem 19. *The complexity of construction of the strongly regular form for any regular formula A is at most*

$$O(1) l(A) log^2(l(A)).$$

Proof. The formula is taken in its extended presentation. The addresses of the tuples, representing symbols, do not change (only pointers do).

We prove our theorem by induction on the formula length. If the formula is a variable of I, then the bound is, evidently, true.

Assume, that there exist some constants C_0, C_1 such that for any formula with $l(A) \leq T$ one can construct (the extended presentation of) its strongly regular form in time at most

$$C_0 T log^2 T - C_1.$$

Consider A with the length equal to $T+1$. Of course, $1 \leq T$. Let $L(A)$ denote complexity of constructiong the strongly regular form of A.

Case. $A = \rightarrow BD$.

We have to convert to strongly regular forms the subformulas B, D. The process has the following steps:

- conversion of B to the strongly regular form;
- computing of the address of the first symbol of D;
- conversion of D to the strongly regular form.

As we immediately see, there exists a constant C_2, such that

$$L(A) \leq L(B) + L(D) + C_2.$$

By induction hypothesis,

$$L(A) \leq C_0 l(B) log^2(l(B)) - C_1 + C_0 l(D) log^2(l(D)) - C_1 + C_2 \leq$$

$$\leq C_0 l(A) log^2(l(A)) - 2C_1 + C_2.$$

If

$$C_1 > C_2, \tag{1}$$

then we have

$$L(A) \leq C_0 l(A) log^2(l(A)) - C_1.$$

Case. $A = (\wedge A_1 ... A_m)$.

Let $Q(i)$ denote the following triple: (num_i, len_i, adr_i), where

- $num_i = i$;
- $len_i = l(A_i)$;
- adr_i is the address of the main symbol of the formula A_i.

At the first step we compute the number m and the array

$$\mathbf{Q} = Q(1)Q(2)...Q(m),$$

at the same time converting each A_i to the strongly regular form. The complexity of this process is at most

$$C_3 + \sum_{i=1}^{m}(C_0 l(A_i) log^2(l(A_i)) - C_1 + C_3). \tag{2}$$

At the second step we make sorting of the array $\mathbf{Q}$. After this sorting we have on the i-th place the sequence $Q(\phi(i))$, where ϕ is a bijection

$$\phi : \{1, ..., m\} \rightarrow \{1, ..., m\},$$

such that

$$i \leq j \Rightarrow len_{Q(\phi(i))} \leq len_{Q(\phi(j))}.$$

Complexity of such sorting is at most

$$C_4 m log(m). \tag{3}$$

At the third step we compute the number r and the sequence of integers

$$s_0, ..., s_r$$

such that

$$1 = s_0 < s_1 < ... < s_r = m,$$

$$len_{Q(\phi(s_i))} = len_{Q(\phi(s_i)+1)} = ... = len_{Q(\phi(s_{i+1}-1))},$$

$$len_{Q(\phi(s_{i+1}-1))} < len_{Q(\phi(s_{i+1}))},$$

$$i = 0, 1, ..., r-1.$$

Complexity of this step is at most

$$C_5 m, \tag{4}$$

for some constant C_5.

At the fourth step we are sorting each array

$$Q(\phi(s_i)Q(\phi(s_i) + 1)...Q(\phi(s_{i+1} - 1),$$

if it is nontrivial, i.e.,

$$s_{i+1} - s_i > 1.$$

The order of the new array should correspond to our ordering of formulas of the same length. Complexity of this sorting is at most

$$C_6 len_{Q(\phi(s_i))} log(s_{i+1} - s_i) \tag{5}$$

for some constant C_6, because the comparison of two strongly regular formulas of equal length has complexity $O(n)$(lemma 18).

After this step we have an array

$$Q(\psi(1))...Q(\psi(m)),$$

such that

$$i \leq j \Rightarrow A_{num_\psi(i)} \leq A_{num_\psi(j)}.$$

At the fifth step we have to make permutation of the subformulas in the new order. The complexity of this step is at most

$$C_7 m \tag{6}$$

for some constant C_7. At this step we change only pointers (address labels) in the tuples, representing the main symbols of $A_1, ..., A_m$, but do not move any tuples. (If we would move the subformulas, the bound would not be true.)

Let $l^*(A) = \sum_{i, l(A_i) \leq (1/2) l(A)} l(A_i)$. We have

$$\sum_{i=0}^{r-1} len_{Q(\phi(s_i))} log(s_{i+1} - s_i) \leq l^*(A) log(m),$$

since in our formula it may be at most one subformula with more than half of the whole formula length. This fact implies also

$$mlog(m) \leq l^*(A) log(m).$$

In accordance with the bounds 1 - 6 we have, that

$$L(A) \leq C_3 + \sum_{i=1}^{m} (C_0 l(A_i) log^2(l(A_i)) - C_1 + C_3) + C_8 l^*(A) log(m),$$

for some constant C_8.

Suppose that

$$2C_3 \leq C_1. \tag{7}$$

Then we have

$$L(A) \leq \sum_{i=1}^{m} C_0 l(A_i) log^2 l(A_i) - C_1 + C_8 l^*(A) log(m) \leq$$

$$C_0 (l^*(A) log^2 \frac{l(A)}{2} + (l(A) - l^*(A)) log^2 l(A)) -$$
$$-C_1 + C_8 l^*(A) log(m) \leq$$
$$\leq C_0 (l^*(A)(log^2 l(A) - log l(A)) + (l(A) - l^*(A)) log^2 l(A)) -$$
$$-C_1 + C_8 l^*(A) log(m) =$$
$$C_0 l(A) log^2 l(A) - C_0 l^*(A) log(l(A)) - C_1 + C_8 l^*(A) log(m).$$

Now, we can suppose that

$$C_8 \leq C_0, \tag{8}$$

and then

$$L(A) \leq C_0 l(A) log^2 l(A) - C_1,$$

so that, with the choice of constants as above, our theorem is true. In other words, C_0, C_1 should be chosen in such way, that they make true the conditions 1, 7, 8.

4 Final Remarks

In the main part of this paper we concentrated our efforts on the description of an efficient decision algorithm for the isomorphism of types. In order to obtain better upper complexity bound, we had to look very accurately after the form of presentation and movement of data. (If we tried, as it would seem natural for a pure category theorist, to construct an algorithm not controlling explicitely associativity and commutativity of tensor, i.e., taking input data up to corresponding equivalence, then the complexity would be just hidden in iterated comparison of unordered sets, more difficult to control.) This could give an impression, that there is not so much room for category thery in the paper.

To our mind, the thing should be rather played the other way. That is, to point out, that it is not yet enough attention paid to algorithmic and complexity questions of category theory, as well as optimal presentation of categorical data for computerised applications.

In connection with this remark, let's note also, that in fact the scope of the algorithms of the type described above, is often much larger than it seems. For example, let us consider the case of a free SMC category $C(K)$ generated by some category K, which is not discrete (let's call its objects atoms). All atoms in K are divided into isomorphism classes in obvious way. Assume that in each isomorphism class a canonical representative is fixed. Let us also consider the discrete category K_0, corresponding to K, and corresponding free SMC category $C(K_0)$. The following theorem was proved by S. Soloviev:

Theorem 20. *Two objects A, B are isomorphic in $C(K)$ iff the objects A_0, B_0, obtained by replacement of all atoms by the canonical representatives of their isomorphism classes are isomorphic in $C(K_0)$.*

The proof uses the properties of the Eilenberg-Kelly-Mac Lane naturality graphs and will be published elsewhere.

Thus, the algorithm, described in this paper, provides also an algorithm for $C(K)$ modulo another algorithm, producing canonical representatives of the isomorphism classes in K (rather "user-defined"). Overall complexity will depend in a "modular" way on the complexity of these algorithms.

In case of general SMC categories, the positive result of the algorithm will be sufficient condition of isomorphism of two objects (obtained via tensor and *hom* from some basic objects).

Acknowledgements

We would like to thank Michael Rittri and Roberto Di Cosmo for many stimulating discussions of the matter, and Glynn Winskel and Uffe Engberg for helpful attention to our work while at BRICS.

References

1. M. Rittri. Retrieving library functions by unifying types modulo linear isomorphism. *Proceedings of Conference on Lisp and Functional Programming*, 1992.

2. R. DiCosmo. Isomorphism of Types: from λ-calculus to information retrieval and language design.- Birkhauser, 1995.
3. S.V. Soloviev. The category of finite sets and cartesian closed categories. *Zapiski Nauchnych Seminarov Leningradskogo Otdelenya Matematicheskogo Instituta im. V.A.Steklova AN SSSR*, **105** (1981), 174-194. (English translation in: *Journal of Soviet Mathematics*, **22(3)**(1983), 1387-1400.
4. K. Bruce, R. Di Cosmo and G. Longo. Provable isomorphism of types. *Preprint LIENS-90-14*, Ecole Normale Superieure, Paris (1990).
5. G.E. Mints. Closed categories and Proof Theory.*J.Soviet Math.*, **15** (1981), 45-62.
6. G.-Y. Girard, Y. Lafont. Linear logic and lazy computation. In: Proc.TAPSOFT 87 (Pisa), v.2, p.52-66, *Lecture Notes in Comp.Sci.* **250** (1987).
7. S.V. Soloviev. A complete axiom system for isomorphism of types in closed categories. - *Lecture Notes in Artificial Intelligence*, **698** (1993), 380-392.
8. P.H. Benton, G.M. Bierman, V.C.V. de Paiva and J.M.E. Hyland. A term calculus for Intuitionistic Linear Logic. - In *Proceedings of Typed Lambda calculus and Applications*, *Lecture Notes in Comp.Sci.*, **664** (1992), 75-90.

Effectiveness of the Global Modulus of Continuity on Metric Spaces

Klaus Weihrauch and Xizhong Zheng

Theoretische Informatik
FernUniversität Hagen
58084-Hagen, Germany
klaus.weihrauch@fernuni-hagen.de
xizhong.zheng@fernuni-hagen.de

Abstract. *Let* (X, d_X) *and* (Y, d_Y) *be metric spaces. By definition, there is a function* $h : (f, x, \epsilon) \mapsto \delta$, $(\delta > 0)$, *such that for all continuous function* $f : X \to Y$, $x \in X$ *and* $\epsilon > 0$: $\forall x' \in X(d_X(x, x') < \delta \Longrightarrow d_Y(f(x), f(x')) < \epsilon)$. *By a recent result of Repovš and Semenov [8], there is a function h continuous in f, x and* ϵ *with this property, if* (X, d_X) *is locally compact. Based on Weihrauch's frameworks on computable metric space ([13]), we effectivize this result by showing that there is a computable function of this type. The proof is a direct construction not depending on [8].*

Key words: Modulus of Continuity; Metric Space; Effective Analysis.

1 Introduction

Let (X, d_X) and (Y, d_Y) be metric spaces. A function $f : X \to Y$ is continuous if, for any $x \in X$ and any $\epsilon > 0$, there exists a $\delta > 0$ such that

$$\forall x' \in X \ (d_X(x, x') < \delta \Longrightarrow d_Y(f(x), f(x')) < \epsilon). \tag{1}$$

In other words, f is continuous if and only if there is a (total) function $\tilde{\delta} : X \times \mathbb{R}^+ \to \mathbb{R}^+$ such that, for any $(x, \epsilon) \in X \times \mathbb{R}^+$,

$$\forall x' \in X \ (d_X(x, x') < \tilde{\delta}(x, \epsilon) \Longrightarrow d_Y(f(x), f(x')) < \epsilon). \tag{2}$$

The function $\tilde{\delta}$ is called a *modulus of continuity* of f.

The discussion about modulus of continuity is an interesting and important topic both in classical and effective analysis (see e.g, [2, 5, 10, 12]). For example,

Ko [2] shown that, if $f : [a; b] \to \mathbb{R}^+$ is a computable (hence continuous) real function, then there is a recursive function $m : \omega \to \omega$ such that the function $\tilde{\delta}$ defined by $\tilde{\delta}(x, \epsilon) = 2^{-\lfloor 1/\epsilon \rfloor}$ is a modulus of (uniform) continuity of f. For the classically locally compact metric space X, Repovš and Semenov proved in [8] that, every continuous function $f : X \to Y$ possesses a continuous modulus of continuity. In fact, they have proved even more that, it is possible to determine an appropriate $\delta > 0$ satisfying (1) as a continuous function of the triple (f, x, ϵ) under proper topology. More precisely, let $C(X, Y)$ be the set of all continuous functions from X into Y, endowed with the topology of uniform convergence, i.e., the ϵ-neighbourhood of $f \in C(X, Y)$ is the set $B(f, \epsilon) := \{g \in C(X, Y) : \forall x \in X\ (d_Y(f(x), g(x)) < \epsilon\}$. Then the following has been proved in [8].

Theorem (Repovš and Semenov [8]). *Let (X, d_X) and (Y, d_Y) be metric spaces and suppose that X is locally compact. Then there exists a continuous function $\hat{\delta} : C(X, Y) \times X \times \mathbb{R}^+ \to \mathbb{R}^+$ such that, for every triple $(f, x, \epsilon) \in C(X, Y) \times X \times \mathbb{R}^+$,*

$$\forall x' \in X\ (d_X(x, x') < \hat{\delta}(f, x, \epsilon) \Longrightarrow d_Y(f(x), f(x')) < \epsilon). \tag{3}$$

The function $\hat{\delta}$ in above theorem is called *global modulus of continuity* of function space $C(X, Y)$. The proof of Theorem applies Michael's Selection Theorem (cf. [6]). That is, the function $\hat{\delta}$ exists as a selection function of some lower semi continuous set valued funtion. Thus it is quite ineffective.

Our main purpose of this paper is to discuss the effectiveness about the global modulus of continuity of $C(X, Y)$ when X is (effectively) locally compact. It is, of course, only possible after a computability framework about metric spaces is well established. In [13], Weihrauch introduced computability on metric space by the representation theory. In this theory, computability on finite and infinite sequences over some finite alphabet of symbols are defined explicitly, e.g., by Turing machines, and computability on other sets are introduced by representations, i.e., naming systems, where infinite sequences of symbols are used as names (cf, e.g. [3, 11, 12]). In this paper we will work in this framework and prove finally a computationally effective version of the above theorem, i.e. there is a computable global modulus of continuity of $C(X, Y)$. Thus, we can choose effectively a positive real δ satisfying (1) from any continuous function $f : X \to Y$, element $x \in X$ and positive real ϵ.

We fix Σ to be a finite alphabet containing all symbols we need. Σ^* and Σ^ω are the sets of all finite words and infinite sequences over Σ, respectively. The computability theory on Σ^* and Σ^ω have been well established ([9, 12]). To discuss the computability on other sets, we need their representations by the elements of Σ^* or Σ^ω. For any set A with the cardinality of at most continuum, a representation of A is simply a surjective function $\delta : \Sigma^a \to A$ for $a \in \{*, \omega\}$. For example, let $\Sigma = \{0, 1, \sharp, ¢\}$, the functions $\nu_N : \Sigma^* \to \omega$, $\nu_Q : \Sigma^* \to \mathbb{Q}$ and $\rho_R : \Sigma^\omega \to \mathbb{R}$ be defined respectively by

$$\nu_N(w) = n \iff w = 1^n;$$

$$\nu_Q(w) = r \iff w = 0x0y0z0 \ \& \ r = \frac{\nu_N(x) - \nu_N(y)}{\nu_N(z) + 1};$$

$$\rho_r(p) = a \iff p = \sharp r_0 \sharp r_1 \sharp r_2 \sharp \ \dots \ \& \ \lim_{n\to\infty} \nu_Q(r_n) = a$$
$$\forall n \forall m \geq n(|\nu_Q(r_n) - \nu_Q(r_m)| < 2^{-n}).$$

Then ν_N, ν_Q and ρ_R are the standard representations of the natural numbers ω, rational numbers$\mathbb{Q}$ and real numbers $\mathbb{R}$. By the representations, the fundemental concepts of classical computability theory can be translated directly. Suppose that δ_i are the representations of sets A_i for $i = 0, 1, \dots, n$. An n-ary function $f : A_1 \times \dots \times A_n \to A_0$ is called $(\delta_1, \dots, \delta_n, \delta_0)$-computable if there is a computable function $g : (\Sigma^\omega)^n \to \Sigma^\omega$ such that $\forall p_1, \dots, p_n(f(\delta_1(p_1), \dots, \delta_n(p_n)) = \delta_0 \circ g(p_1, \dots p_n))$.

Suppose that ρ_R, ρ_X and δ_{XY} be reasonable representations of $\mathbb{R}$, X and $C(X,Y)$, respectively. A computable metric space X is effectively locally compact, if we can determine effectively, for any $x \in X$, a neighbourhood B_x of x such that $\overline{B_x}$, the closure of B_x, is compact. Then our main result says that if X and Y are computable metric spaces (see definition of next section) and X is effectively locally compact, then there exists a $(\delta_{XY}, \rho_X, \rho_R, \rho_R)$-computable function $\hat{\delta} : C(X,Y) \times X \times \mathbb{R}^+ \to \mathbb{R}^+$ which satisfies (3) for every triple $(f, x, \epsilon) \in C(X,Y) \times X \times \mathbb{R}^+$.

Our notations are almost same as in [15, 17]. For the notational simplicity, we often do not distinguish explicitly a natural number $i := \nu_N(\bar{i})$ from its name $\bar{i} \in \Sigma^*$ under the standard representation ν_N. In this paper, $B(x,r)$ and $\overline{B}(x,r)$ denote always the open and closed r-ball with center x, respectively. As usual, $\langle \cdot, \cdot \rangle$ is the standard computable pairing function of natural numbers and π_1, π_2 are corresponding first and second reverse functions.

2 Computable Metric Spaces

In this section we review at first the notions of computable metric space and the corresponding representations of Weihrauch [13]. Then the representations of open and compact subsets of a computable metric space and the notion of effectively local compacteness will also be introduced.

Definition (Weihrauch[13]). Let (X,d) be a metric space, $A \subseteq X$ a countable dense subset of X and $\alpha : \omega \to A$ a (bijective) notation of A. If the set D defined by

$$D := \{\langle i,j,k,l \rangle \in \omega : \nu_Q(k) < d(\alpha(i), \alpha(j)) < \nu_Q(l)\}$$

is recursive enumerable, then we call (X, d, A, α), or simply X, a *computable metric space*. The open subset $U_k \subseteq X$ for $k \in \omega$ defined by:

$$U_{\langle i,j \rangle} := \{x \in X : d(\alpha(i), x) < \nu_Q(j)\} \tag{4}$$

is called *basic open set* of X and $\nu_Q(j)$ is its *radius* and denoted by $\mathrm{rd}(U_{\langle i,j \rangle})$ or simply $\mathrm{rd}(\langle i,j \rangle)$.

For example, suppose that d_R is the standard metric on $\mathbb{R}$ and $\alpha_Q : \omega \to \mathbb{Q}$ defined by $\alpha_Q(\langle i,j,k\rangle) = \frac{i-j}{k+1}$ is an enumeration of $\mathbb{Q}$. Then $(\mathbb{R}, d_R, \mathbb{Q}, \alpha_Q)$ is a computable metric space.

It is easy to see that the class of all basic open sets $\{U_i\}_{i\in\omega}$ is a topological basis of the topology deduced by the metric d. In our discussion, the basic open sets play an important role which is very similar to that of the rational intervals in effective analysis. Especially, the elements of the computable metric space and the continuous functions between the computable metric spaces can be represented by some infinite sequences of basic open sets.

Definition (Weihrauch[13]). Let (X, d_X, A_X, α_X) and (Y, d_Y, A_Y, α_Y) be the computable metric spaces. $En(p) := \{(i_0, i_1, \ldots i_m) : \sharp i_0 \natural i_1 \natural \ldots \natural i_m \sharp \sqsubset p\}$ the set enumerated by sequence $p \in \Sigma^\omega$. We define the representations $\rho_X, \rho'_X : \Sigma^\omega \to X$ and $\delta_{XY}, \delta'_{XY} : \Sigma^\omega \to C(X,Y)$ respectively by

$$\rho_X(p) = x \iff p = \sharp i_0 \sharp i_1 \sharp i_2 \sharp \cdots \;\; \& \lim_{n\to\infty} \alpha_X(i_n) = x \;\&$$
$$\forall n \geq m \; (d_X(\alpha_X(i_n), \alpha_X(i_m)) < 2^{-m});$$

$$\rho'_X(p) = x \iff En(p) = \{i \in \omega : x \in U_i\};$$

$$\delta_{XY}(p) = f \iff p = \sharp i_0 \natural j_0 \sharp i_1 \natural j_1 \sharp i_2 \natural j_2 \sharp \cdots \;\; \& \; \forall n \; (f(\overline{U}_{i_n}) \subseteq U_{j_n}) \;\&$$
$$\forall x \in X \; \forall \epsilon > 0 \; \exists \langle i,j\rangle \; (\sharp i \natural j \sharp \sqsubset p \;\&\; x \in U_i \;\&\; \mathrm{rd}(j) < \epsilon);$$

$$\delta'_{XY}(p) = f \iff \forall q \in \mathrm{dom}(f \circ \rho_X) \; (f(\rho_X(q)) = \rho_Y(M_U(p,q))),$$

where M_U is the universal type-2 Turing machine.

As shown in [13], ρ_X is equvelent to ρ'_X and δ_{XY} is equivelt to δ'_{XY} with respect to the effective reduction. So we can use them exchangably. Because any open subset is a union of some basic open sets and any compact subset has a finite cover of basic open sets (we call it a basic finite open cover), we can define the representations of open and compact subsets through basic open sets similar to the case of $\mathbb{R}$ (cf [15, 17]).

Definition 1. Let (X, d, A, α) be a computable metric space. O_X and K_X be the sets of all open and compact subsets of X. Then we can define the representations $\theta_X : \Sigma^\omega \to O_X$ and $\kappa_X : \Sigma^\omega \to K_X$ by

$$\theta_X(p) = O \iff p = \sharp i_0 \sharp i_1 \sharp i_2 \sharp \cdots \;\; \& \; O = \bigcup \{U_{i_n} : n \in \omega\};$$

$$\kappa_X(p) = G \iff En(p) = \{i_0 \natural \cdots \natural i_n : n \in \omega \;\&\; G \subseteq \bigcup\nolimits_{s\leq n} U_{i_s}$$
$$\forall s \leq n \; (U_{i_s} \bigcap G \neq \emptyset)\}.$$

Note that, $\theta_X(p) = O$ means that p enumerates a sequence of basic open sets which exhaust set O. If $\kappa_X(p) = G$ then p enumerates all finite basic open covers of G which intersects A. Next proposition shows that, for a κ-name p of G, it is equivalent to enumerate a Cauchy sequence of the finite basic open covers of G which intersrects G and has limit G in the sense of Hausdorff metric.

Proposition 2. *Let (X, d, A, α) be a computable metric space and d_H the Hausdorff metric defined by*

$$d_H(U,V) = \max\{\sup_{x\in U}\inf_{y\in V} d(x,y),\ \sup_{x\in V}\inf_{y\in U} d(x,y)\}$$

for any sets $U, V \subseteq X$, (cf. e.g.,[1]). If the representation κ'_X is defined by

$$\begin{aligned}\kappa'_X(p) = G \iff & p = \sharp i_0^0 \natural \cdots \natural i_{m_0}^0 \sharp i_0^1 \natural \cdots \natural i_{m_1}^1 \sharp \cdots \ \& \\ & \forall n\ (G \subseteq \bigcup\nolimits_{j\le m_n} U_{i_j^n}\ \&\ \forall s \le m_n (G \bigcap U_{i_s^n} \neq \emptyset))\ \& \\ & \forall t\ \forall s \ge t\ (d_H(\bigcup\nolimits_{j\le m_s} U_{i_j^s}\ ,\ \bigcup\nolimits_{j\le m_t} U_{i_j^t}) < 2^{-t})\end{aligned}$$

Then we have $\kappa_X \equiv \kappa_X$, i.e., there are computable functions $f, g : \Sigma^\omega \to \Sigma^\omega$ such that $\kappa = \kappa' \circ f$ and $\kappa' = \kappa \circ g$.

Proof. Let $\kappa_X(p) = G$ with $p = \sharp i_0^0 \natural \cdots \natural i_{m_0}^0 \sharp i_0^1 \natural \cdots \natural i_{m_1}^1 \sharp \cdots$, then p enumerates all finite basic open covers of G which intersects G. It is easy to see that we can construct a Turing machine M which enumerates, from the input p, a sequence $q := \sharp j_0^0 \natural \cdots \natural j_{m_0}^0 \sharp j_0^1 \natural \cdots \natural j_{m_1}^1 \sharp \cdots$ such that

$$\forall n\ (\sharp j_0^n \natural \cdots \natural j_{m_n}^n \sharp\ \sqsubset p\ \&\ \forall s \le m_n(\mathrm{rd}(j_s^n) < 2^{-n})).$$

Then, $\kappa'_X(q) = G$, hence $\kappa(p) = \kappa'(f_M(p))$. This means that $\kappa \le \kappa'$.

Let $\kappa'(p) = G$. Then p enumerates a Cauchy sequence of finite basic open covers of G which intersects G and has the limit G in the sense of Hausdorff metric. Suppose, without lost of generality, that $p = \sharp i_0^0 \natural \cdots \natural i_{m_0}^0 \sharp i_0^1 \natural \cdots \natural i_{m_1}^1 \sharp \cdots$ and $\forall n\ (d_H(\bigcup_{s\le m_n} U_{i_s^n}\ ,\ G) < 2^{-n})$. Construct a Turing machine M such that, for any input p, M will enumerate all $j_0 \natural j_1 \natural \cdots \natural j_m$ for $m \in \omega$ which satisfy that,

$$\exists n \in \omega \Big(\bigcup\nolimits_{s\le m_n} U_{i_s^n} \subseteq \bigcup\nolimits_{t\le m} U_{j_t}\ \&\ \forall t \le m\ (\bigcup\nolimits_{s\le m_n} \tilde{U}^{(n)}_{i_s^n} \bigcap U_{j_t} \neq \emptyset)\Big),$$

where $\tilde{U}^{(n)}_{\langle i,j\rangle} = \{x \in X : d(\alpha(i), x) < \nu_Q(j) - 2^{-n}\}$. Note that $\tilde{U}^{(n)}_{i_s^n} \subseteq G$ for any $s \le m_n$, because $d_H(\bigcup_{s\le m_n} U_{i_s^n}\ ,\ G) < 2^{-n}$ and G is compact. Then it is not difficult to see that $M(p)$ enumerates all finite basic open covers of G which intersects G, hence $\kappa(f_M(p)) = G$. Therefore, f_M witenesses the reduction that $\kappa' \le \kappa$. □

Classically, a metric space X is called locally compact if, for any $x \in X$, there is a neighbourhood B_x of x such that its closure $\overline{B}_x$ is compact. If such neighbourhood can be obtained effectively in x, then we will call it effectively locally compact. Now we define this concept precisely.

Definition 3. A computable metric space (M, d, A, α) is called *effectively locally compact* if there is a (ρ_X, ρ_R)-computable function $\gamma_X : X \to \mathbb{R}$ such that $\overline{B}(x, \gamma_X(x))$ is compact for any $x \in X$.

Obviously, the space $(\mathbb{R}, d_R, \mathbb{Q}, \alpha_Q)$ is effectively locally compact.

As end of this section, we show a simple property about computable metric space that, the metric d of any computable metric space is a computable function.

Lemma 4. *Let (X, d_X, A, α) be a computable metric space. Then the metric d is a (ρ_X, ρ_X, ρ_R)-computable function.*

Proof Given $x, y \in X$ with $\rho_X(p) = x$, $\rho_X(q) = y$. By the basic properties of the metric, we have $|d_X(x,y) - d_X(\alpha p(n), \alpha q(n))| \leq d_X(x, \alpha p(n)) + d_X(y, \alpha q(n)) < 2^{-n+1}$, and hence

$$\lim_{n\to\infty} d_X(\alpha p(n), \alpha q(n)) = d_X(x,y), \text{ and}$$
$$d_X(\alpha p(n), \alpha q(n)) - 2^{-n+1} < d_X(x,y) < d_X(\alpha p(n), \alpha q(n)) + 2^{-n+1}.$$

Define now inductively two sequences $\{a_n\}$ and $\{b_n\}$ of natural numbers by,

$$\begin{aligned} a_0 &= \mu m(\nu_Q(m) < d_X(\alpha p(0), \alpha q(0)) - 2); \\ b_0 &= \mu m(\nu_Q(m) > d_X(\alpha p(0), \alpha q(0)) + 2); \\ a_{n+1} &= \mu m(\nu_Q(a_n) < \nu_N(m) < d_X(\alpha p(n), \alpha q(n)) - 2^{-n+1}); \\ b_{n+1} &= \mu m(\nu_Q(b_n) > \nu_N(m) > d_X(\alpha p(n), \alpha q(n)) + 2^{-n+1}). \end{aligned}$$

By the recursive enumerability of D in the definition , it is easy to see that the sequences a_n and b_n are recursive in p and q. That is, there is a Turing machine M such that $f_M(p,q) = \sharp a_0 \cent b_0 \sharp a_1 \cent b_1 \sharp \cdots$. Let $r_n^1 = \nu_Q(a_n)$ and $r_n^2 = \nu_Q(b_n)$. Then the sequences r_n^1 and r_n^2 satisfy

$$\forall n(r_n^1 < r_{n+1}^1 < d_X(x,y) < r_{n+1}^2 < r_n^2) \ \& \ \lim_{n\to\infty}(r_n^2 - r_n^1) = 0.$$

This means that $d_X(x,y) = \rho_R(f_M(p,q))$. Hence the metric function d_X is (ρ_X, ρ_X, ρ_R)-computable. □

3 Global Modulus of Continuity

In this section we will prove our main theorem. The proof is a direct construction of an algorithm which determines an appropriate value of δ, for any triple (f, x, ϵ), such that (1) is satisfied. The crucial idea is as follows: given any continuous function $f : X \to Y$, and element $x \in X$, consider at first the function $h : \mathbb{R}^+ \to \mathbb{R}^+$ defined for any $\delta \in \mathbb{R}^+$ by

$$\begin{aligned} h(\delta) :&= \min\{\epsilon \in \mathbb{R}^+ : f(\overline{B}(x,\delta)) \subseteq \overline{B}(f(x), \epsilon)\} \\ &= \max\{d_X(f(x), f(x')) : d_Y(x,x') \leq \delta \ \& \ x' \in X\}. \end{aligned}$$

Here we use the closed ball $\overline{B}$ instead of open ball B because a computable function f maps compact sets to compact sets and this does not work for open sets. h is nondecreasing and it is easy to see that, if $\epsilon = h(\delta_0)$, then δ_0 satisfies (1). Thus it suffices to define $\hat{\delta}$ as the reverse function of h. Unforturnately, h^{-1} does not necessarily exist, if h is not strictly increasing. To avoid this bad situation, let simply $g(\delta) = h(\delta) + \delta$ and $\delta_1 = g^{-1}(\epsilon)$. Then δ_1 satisfies (1) too. The following technical details make sure that the function g^{-1} is computable and, furthermore, depends effectively on the function $f \in C(X, Y)$ and the element $x \in X$.

Next lemma shows that the computable function maps effectively and uniformly any compact set of a computable metric space to a compact set.

Lemma 5. *Let X, Y be computable metric spaces and K_X the set of all compact subsets of X. Then the evaluation function $F : C(X,Y) \times K_X \to K_X$ defined by $F(f, B) := f(B)$ is $(\delta_{XY}, \kappa_X, \kappa_Y)$-computable.*

Proof Suppose that $f = \delta_{XY}(p)$ and $B = \kappa_X(q)$. Then p enumerates a sequence of $i¢j$'s which satisfy $f(\overline{U_i}) \subseteq U_j$ and such that

$$\forall x \in X \; \forall \epsilon > 0 \; \exists \langle i, j \rangle \; (\sharp i¢j\sharp \sqsubset p \; \& \; x \in U_i \; \& \; \mathrm{rd}(j) < \epsilon)$$

and q enumerates all strings $i_0¢i_1¢ \cdots ¢i_n$ for $n \in \omega$ such that $\{U_{i_0}, U_{i_1}, \ldots, U_{i_n}\}$ is a finite basic open covers of B which intersects B.

Construct a Turing machine M as following: input $(p, q) \in \Sigma^\omega \times \Sigma^\omega$, $M(p, q)$ will enumerate all strings $j_0¢j_1¢ \cdots ¢j_n$ for $n \in \omega$ which satisfy that: there are $m \in \omega$, $i_0, i_1, \ldots i_m$ and $k_0, k_1, \ldots k_m$ such that

1. $\sharp i_0¢i_1¢ \cdots ¢i_m\sharp \sqsubset q$;
2. $\forall s \leq m \; (\sharp i_s¢k_s\sharp \sqsubset p)$;
3. $\forall s \leq m \; \exists t \leq n \; (U_{k_s} \subset U_{j_t})$; and
4. $\forall t \leq n \; \exists s \leq m \; (U_{k_s} \subset U_{j_t})$.

Now, if $\{j_0, j_1, \ldots, j_n\}$ satisfies (1 – 4) above for $m \in \omega$ and the finite sets $\{i_0, i_1, \ldots i_m\}$ and $\{k_0, k_1, \ldots k_m\}$, then $\{U_{i_t} : t \leq n\}$ is a finite basic open cover of the compact set B which intersects B, hence $\{U_{j_s} : s \leq n\}$ is a finite basic open cover of $f(B)$ becauce $\forall s \leq m \; \exists t \leq n \; (f(\overline{U}_{i_s}) \subset U_{k_s} \subset U_{j_t})$. Because $\forall t \leq n \; \exists s \leq m \; (U_{k_s} \subset U_{j_t})$ we have also that $\forall t \leq n \; (U_{j_t} \bigcap f(B) \neq \emptyset)$. This means that $\{U_{j_s} : s \leq n\}$ is a finite basic open cover of $f(B)$ which intersects $f(B)$.

On the other hand, suppose that $\{U_{j_t} : t \leq n\}$ is a finite basic open cover of $f(B)$ which intersects $f(B)$. For any $x \in A$, there is a $t \leq n$ such that $f(x) \in U_{j_t}$. By the continuity of f and the definition of δ_{XY}, there are $i, k \in \omega$ such that $x \in U_i \; \& \; f(U_i) \subset U_k \subset U_{j_t}$. This means that

$$\Gamma = \{U_i : i \in \omega \; \& \; \exists k \in \omega \; \exists t \leq n \; (\sharp i¢k\sharp \sqsubset p \; \& \; U_k \subset U_{j_t})\}$$

is a basic open cover of B. By the compactness of B, there is a finite subcover $\{U_{i_s}\}_{s \leq m} \subseteq \Gamma$ for some $m \in \omega$ which intersects B. That is, the string $j_0¢j_1¢ \cdots ¢j_n$ satisfies above conditions $1 - 4$ for some $i_0, \ldots i_m$ and $k_0, \ldots k_m$, hence will be enumerated finally by $M(p, q)$. That is, $M(p, q)$ enumerates all finite basic open cover of $f(B)$ which intersects $f(B)$.

Thus $\kappa_Y(f_M(p, q)) = f(B) = F(f, B)$, that is, the evaluation function F is $(\delta_{XY}, \kappa_X, \kappa_Y)$-computable. □

Lemma 6. *Let (X, d, A, α) be a computable metric space. Then the distance function $\tilde{d} : X \times K_X \to \mathbb{R}$ defined by $\tilde{d}(x, B) = \min\{d(x, y) : y \in B\}$ is $(\rho_X, \kappa_X, \rho_R)$-computable.*

Proof Suppose that $x = \rho_X(p_1)$ with $p_1 = \sharp i_0 \sharp i_1 \sharp i_2 \sharp \cdots$ and $B = \kappa_X(p_2)$. From p_2, we can effectively construct a sequence p_2' such that

$$p_2' = \sharp i_0^0 ¢ \cdots ¢ i_{m_0}^0 \sharp i_0^1 ¢ \cdots ¢ i_{m_1}^1 \sharp \cdots \ \&$$
$$\forall n (\sharp i_0^n ¢ \cdots ¢ i_{m_n}^n \sharp \sqsubset p_2 \ \& \ \forall j \leq m_n (\mathrm{rd}(i_j^n) < 2^{-(n+1)}).$$

Furthermore we can construct effectively two sequences $\{s_n^i\}_{n \in \omega}$ (for $i = 0, 1$) of natureal numbers which satisfy, for any n,

$$\nu_Q(s_n^1) < \nu_Q(s_{n+1}^1) < \min_{j \leq m_n} d(\alpha\pi_1(i_n), \alpha\pi_1(i_j^n)) - 2^{-n} \ \&$$
$$\min_{j \leq m_n} d(\alpha\pi_1(i_n), \alpha\pi_1(i_j^n)) + 2^{-n} < \nu_Q(s_{n+1}^2) < \nu_Q(s_n^2).$$

Let $r_n^i = \nu_Q(s_n^i)$ for all $n \in \omega$ and $i = 1, 2$. It is not difficult to see that

$$\forall n (r_n^1 < r_{n+1}^1 < \tilde{d}(x, B) < r_{n+1}^2 < r_n^2) \ \& \ \lim_{n \to \infty} (r_n^2 - r_n^1) = 0.$$

Thus we can construct a Turing machine M such that $f_M(p_1, p_2) = \sharp s_0^1 ¢ s_0^2 \sharp s_1^1 ¢ s_1^2 \sharp s_2^1 ¢ s_2^2 \sharp \cdots$. So $\tilde{d}(x, B) = \rho_R(f_M(p_1, p_2))$. Hence $\tilde{d}$ is a $(\rho_X, \kappa_X, \rho_R)$-computable function. □

We are now able to formulate our main theorem precisely and to prove it.

Theorem 7. *Let (Y, d_Y, A_Y, α_Y) be a computable metric space, (X, d_X, A_X, α_X) be an effectively locally compact computable metric space, let $C(X, Y)$ be the set of all continuous functions from X to Y. Then there is a $(\delta_{XY}, \rho_X, \rho_R, \rho_R)$-computable function $\hat{\delta} : C(X, Y) \times X \times \mathbb{R}^+ \to \mathbb{R}^+$ such that, for any $(f, x, \epsilon) \in C(X, Y) \times X \times \mathbb{R}^+$,*

$$\forall y \in X \ (d_X(x, y) < \hat{\delta}(f, x, \epsilon) \Longrightarrow d_Y(f(x), f(y)) < \epsilon). \tag{5}$$

That is, there is a $(\delta_{XY}, \rho_X, \rho_R, \rho_R)$-computable global modulus of continuity of $C(X, Y)$.

Proof Let γ_X be the (ρ_X, ρ_R)-computable function which witnesses the effectively local compactness of X. Define at first a function $\beta : X \times \mathbb{R} \to K_X$ by

$$\beta(x, \delta) = \begin{cases} \overline{B}(x, \gamma_X(x)) & \text{if } \delta > \gamma_X(x), \\ \overline{B}(x, \delta) & \text{otherwise.} \end{cases}$$

Then $\beta(x, \delta)$ is a compact subset of X for any $x \in X$ and $\delta \in \mathbb{R}$. Given any $x = \rho_X(p)$ and $\delta = \rho_R(q)$, the sequences $\{\alpha_X p(n)\}_{n \in \omega}$ and $\{\nu_Q q(n)\}_{n \in \omega}$ are two Cauchy sequences of the corresponding spaces with the limits x and δ, respectively. Because γ_X is (ρ_X, ρ_R)-computable, there is a Turing machine N such that $\rho_X(f_N(p)) = \gamma_X(x)$. Let $p_1 = f_N(p)$. Without lost of generality, we can assume that,

1. $d_X(x, \alpha_X(p(n)) < 2^{-(n+3)}$;
2. $|\delta - \nu_Q(q(n))| < 2^{-(n+3)}$; and
3. $|\gamma_X(x) - \nu_Q(p_1(n))| < 2^{-(n+3)}$.

Let $x_n = \alpha_X p(n)$ and $r_n = \min\{\nu_Q(q(n)), \nu_Q(p_1(n))\} + 2^{-(n+2)}$. Then $\beta(x, \delta) \subseteq B(x_n, r_n)$ for all $n \in \omega$.

Define $i_n = p(n)$ and $j_n = \mu j\ (r_n < \nu_Q(j) < r_n + 2^{-(n+3)})$. It is not difficult to see that $\beta(x, \delta) \subseteq U_{\langle i_n, j_n \rangle}$ and

$$\begin{aligned} d_H(U_{\langle i_n, j_n\rangle}, \beta(x,\delta)) &< d_X(x, \alpha_X(i_n)) + |r_n - \min\{\gamma_X(x), \delta\}| + |\nu_Q(j_n) - r_n| \\ &< 2^{-n}, \end{aligned}$$

for any $n \in \omega$. This means that $\{U_{\langle i_n, j_n\rangle}\}$ is a basic open cover of $\beta(x, \delta)$ satisfying that $d_H(U_{\langle i_n, j_n\rangle}, \beta(x, \delta)) < 2^{-n}$. Now we can construct a Turing machine M such that $f_M(p, q)$ enumerates the sequence $q_1 := \sharp\langle i_0, j_0\rangle \sharp \langle i_1, j_1\rangle \sharp \langle i_2, j_2\rangle \sharp \cdots$. Then we have $\kappa'_X(f_M(p,q)) = \beta(x, \delta)$ for any $(p, q) \in \mathrm{dom}(\rho_X) \times \mathrm{dom}(\rho_R)$. That is, β is $(\rho_X, \rho_R, \kappa'_X)$-computable. By Proposition 2, β is also $(\rho_X, \rho_R, \kappa_X)$-computable.

Define a function $h : C(X, Y) \times X \times \mathbb{R}^+ \to \mathbb{R}^+$ by

$$h(f, x, \delta) = \max\{d_Y(f(x), y) : y \in f\beta(x, \delta)\}.$$

By the $(\rho_X, \rho_R, \kappa_X)$-computability of β and Lemma 5, function F defined by $F(f, x, \delta) = f\beta(x, \delta)$ is $(\delta_{XY}, \rho_X, \rho_R, \kappa_Y)$-computable. Hence, by Lemma 6, h is a $(\delta_{XY}, \rho_X, \rho_R, \rho_R)$-computable function.

Note that $h(f, x, \delta)$ is also nondecreasing on δ. So $g(f, x, \delta) := h(f, x, \delta) + \delta$ is an on δ strictly increasing $(\delta_{XY}, \rho_X, \rho_R, \rho_R)$-computable function. Then we can define a $(\delta_{XY}, \rho_X, \rho_R, \rho_R)$-computable function $\hat{\delta}_1 : C(X, Y) \times X \times \mathbb{R}^+ \to \mathbb{R}^+$ by $\hat{\delta}_1(f, x, \epsilon) := \iota\delta(\epsilon = g(f, x, \delta))$, where "$\iota$" is the "minimal value" operator.

Finally, we define $\hat{\delta} : C(X, Y) \times X \times \mathbb{R}^+ \to \mathbb{R}^+$ by

$$\hat{\delta}(f, x, \epsilon) = \begin{cases} \hat{\delta}_1(f, x, \epsilon/2) & \text{if } \hat{\delta}_1(f, x, \epsilon/2) \le \gamma_X(x), \\ \gamma_X(x) & \text{otherwise.} \end{cases}$$

Similar to the classical proofs of the results on the computability of inverse function and patching function, (see e.g., [7]), it is easy to see that $\hat{\delta}$ is $(\delta_{XY}, \rho_X, \rho_R, \rho_R)$-computable. By the definitions of h and g, we have, for any $\delta \le \gamma_X(x)$, that

$$f(\overline{B}(x, \delta)) \subseteq \overline{B}(f(x), h(f, x, \delta)) \subseteq \overline{B}(f(x), g(f, x, \delta)).$$

Because $\hat{\delta}(f, x, \epsilon) \le \gamma_X(x)$ and $g(f, x, \hat{\delta}_1(f, x, \epsilon)) = \epsilon$ for any $\epsilon \ge 0$, it follows that

$$\forall \epsilon \ge 0 (f(\overline{B}(x, \hat{\delta}(f, x, \epsilon)) \subseteq f(\overline{B}(x, \hat{\delta}_1(f, x, \epsilon/2)) \subseteq \overline{B}(f(x), \epsilon/2)).$$

That is

$$\forall x' \in X (d_X(x, x') < \hat{\delta}(f, x, \epsilon) \Longrightarrow d_Y(f(x), f(x')) < \epsilon).$$

So the function $\hat{\delta}$ satisfies the theorem. $\square$

References

1. R. Engelking *General Topology*, Heldermann Verlag, Berlin, 1989.
2. Ker-I Ko *Complexity Theory of Real unctions*, Birkhäuser, Berlin, 1991.
3. Ch. Kreitz & K. Weihrauch Theory of representations, *Theoret. Comput. Sci.* 38(1985), 35–53.
4. Ch. Kreitz & K. Weihrauch Compactness in constructive analysis, *Annals of Pure and Applied Logic* 36(1987), 29-38.
5. B. A. Kushner *Lectures on Constructive Mathematical Analysis.* Translations of Mathematical Monographs, Vol. 60, Amer. Math. Soc. 1985.
6. E. Michael Continuous selections I, *Ann. of Math.* 63(1956), no. 2, 361–382.
7. M. Pour-El & J. Richards *Computability in Analysis and Physics.* Springer-Verlag, Berlin, Heidelberg, 1989.
8. D. Repovš & P. V. Semenov An application of the theory of selections in analysis. in *Proc. Int. Conf. Topol.* (Trieste, 1993), G. Gentili, Ed., Rent. Ist. Mat. Univ. Trieste 25(1993), 441-446. Abstract in *Abstr. Amer. Math. Soc.* 14(1993), 393, No. 93T-54-66.
9. H. Jr. Rogers *Theory of Recursive Functions and Effective Computability.* McGraw-Hill Book Company, 1967
10. W. Rudin *Real and Complex Analysis.* McGraw-Hill Book Company, 1974.
11. K. Weihrauch Type-2 recursion theory. *Theoret. Comput. Sci.* 38(1985), 17–33.
12. K. Weihrauch *Computability.* EATCS Monographs on Theoretical Computer Secience Vol. 9, Springer-Verlag, Berlin, Heidelberg, 1987.
13. K. Weihrauch Computability on computable metric spaces. *Theoret. Comput. Sci.* 113(1993), 191–210.
14. K. Weihrauch *Effektive Analysis.* Lecture Notes for Corresponding Course, Fern-Universität Hagen, 1994.
15. K. Weihrauch A foundation of computable analysis. *Computability and Complexity in Analysis*, pp25–40, *Informatik-Berichte* Nr. 190, FernUniversität Hagen, 1995
16. K. Weihrauch & Ch. Kreitz Representations of the real numbers and of the open subsets of the real numbers, *Annals of Pure and Applied Logic* 35(1987), 247-260.
17. K. Weihrauch & X. Zheng Computability on continuous, lower semi-continuous and upper semi-continuous real functions. *COCOON'97.* Shanghai, China, August 1997.

Proof Principles for Datatypes with Iterated Recursion*

ULRICH HENSEL
Inst. Theor. Inf., TU Dresden
D-01062 Dresden,
Germany.
hensel@tcs.inf.tu-dresden.de

BART JACOBS
Dep. Comp. Sci., Univ. Nijmegen,
P.O. Box 9010, 6500 GL Nijmegen,
The Netherlands.
bart@cs.kun.nl

Abstract. Data types like trees which are finitely branching and of (possibly) infinite depth are described by iterating initial algebras and terminal coalgebras. We study proof principles for such data types in the context of categorical logic, following and extending the approach of [14, 15]. The technical contribution of this paper involves a description of initial algebras and terminal coalgebras in total categories of fibrations for lifted "datafunctors". These lifted functors are used to formulate our proof principles. We test these principles by proving some elementary results for four kinds of trees (with finite or infinite breadth or depth) using the proof tool PVS.

1 Introduction

Algebras and coalgebras are of well-established importance in computer science, notably in the theory of datatypes, where especially *initial* algebras and *terminal* coalgebras play a distinguished rôle. Over the past decade there is more and more interest in the logic associated with initial algebras and terminal coalgebras. It is known for a long time [10] that induction is the appropriate proof method to reason about functions which are defined on initial algebras. More recently, bisimulation is recognised as the appropriate proof method to reason about (definable) functions taking values in terminal coalgebras [1, 29, 28]. In this context it became clear that there is an equivalent "binary" induction principle, stating that all congruences[2] on an initial algebra contain the equality relation (*i.e.* are reflexive). This induction principle for relations is completely dual to the bisimulation proof principle, which says that every bisimulation on a terminal coalgebra is contained in the equality relation. Unary predicates on coalgebras are also important, namely as "invariants" [17], or as "subcoalgebras" [30].

* This paper was written during a visit of Ulrich Hensel to the Computing Science Institute of the University of Nijmegen.

[2] We mean a relation which is closed under the constructors of the algebra, but which is not necessarily an equivalence relation. For example, a congruence $R \subseteq \mathbb{N} \times \mathbb{N}$ on the natural numbers satisfies $R(0,0)$ and $R(x,y) \Rightarrow R(S(x),S(y))$ for all $x, y \in \mathbb{N}$.

This (recent) interest in the logic of initial algebras and terminal coalgebras (see for example [6, 2, 22, 26, 7, 28, 9, 8, 12, 27]) is also motivated by the increasing use of powerful and expressive proof tools, like PVS, ISABELLE, HOL, LARCH, LEGO, COQ *etc.*, in which users may introduce their own inductively (and sometimes also coinductively) defined type. And, this is of most relevance, the tool[3] then automatically generates several standard definitions and results for such types, like "map" definitions (or functoriality) and (co)inductive proof principles. In practice the available (theoretical) possibilities are not yet fully exploited by such tools.

The concepts of initial algebra and terminal coalgebra form the basis for the experimental programming language CHARITY [5, 4]. It exploits initiality and terminality as the explicit definition schemas and demonstrates the remarkable power of these notions. CHARITY only involves a term language for writing programs, but no logic to reason about these programs.

It is the aim of this paper to contribute to the logical theory behind inductively and coinductively defined types, by developing proof principles for types which are obtained by iterated mixing of initial and terminal fixed points. Ideally, this should form a basis for a suitable blend of programming languages like CHARITY and proof tools like PVS, combining specification, programming and verification.

Interestingly, we expect that the theory that we are about to develop will (eventually) have practical applications. But the actual development of this theory takes place at a very abstract and remote level, using (technical) notions and techniques from categorical logic. Once we have formalised the relevant notions at a suitably abstract level, the proof principles that we seek take standard formulations (see Definition 6). In the present paper we only develop the technical details, and illustrate their usefulness in several typical examples. Space restrictions do not allow us to introduce all relevant notions from categorical logic, and we refer the interested reader to [3, II, 8.5] or [15, 20, 21] (or to the forthcoming book [18]) for background information.

This paper is organised as follows. To start, we identify the kind of "datafunctors" that we shall study, and present several examples of trees that will be used later. Then, in Section 3 we show how to obtain a fixed point for a functor on a total category of a fibration by using fixed points in the base category and in the fibre categories. This extends standard approaches to obtain such global structure [13]. Subsequently in Section 4 we formulate our proof principles as preservation properties, and establish their validity in the presence of comprehension and quotients (following [14, 15]). We pay special attention to handling mixed proof obligations in Section 5, and finally illustrate the use of these principles for four kinds of trees (with finite or infinite breadth or depth), each involving a composition operation. We prove in each case, using PVS, that composition with the empty tree yields the identity. To do so, we have to formalise our proof principles in (a suitable theory in) PVS.

[3] There is considerable difference between the various tools in the extend to which they do this.

Throughout, the data types and their proof principles involve various kinds of fixed points. In this (version of the) paper we are not concerned with the actual existence of these fixed points (which is a non-trivial topic in itself), but only with the derived formal properties.

2 Datafunctors

In this section we introduce the kind of functors that we will consider. These functors are built up from multiple arguments with constants, finite products and coproducts and, most importantly, with fixed points arising both from initial algebras and from terminal coalgebras. This allows us to iterate data type constructions, as in the examples below.

Definition 1. For a sufficiently complete and cocomplete category $\mathbb{C}$, the class of **datafunctors on** $\mathbb{C}$ is the least class of functors $\mathbb{C}^n \to \mathbb{C}$ (for $n \in \mathbb{N}$) containing:

1. the projection functors $\Pi_i: \mathbb{C}^n \to \mathbb{C}$ for $1 \leq i \leq n$, the constant functors $A: \mathbf{1} = \mathbb{C}^0 \to \mathbb{C}$ for $A \in \mathbb{C}$, and the (binary) product and coproduct functors $\times: \mathbb{C} \times \mathbb{C} \to \mathbb{C}$, $+: \mathbb{C} \times \mathbb{C} \to \mathbb{C}$;
2. the composite $S \circ (T_1, \ldots, T_m): \mathbb{C}^n \to \mathbb{C}$, for datafunctors $T_i: \mathbb{C}^n \to \mathbb{C}$ and $S: \mathbb{C}^m \to \mathbb{C}$;
3. the initial algebra carrier $\mu X. T(-, X): \mathbb{C}^n \to \mathbb{C}$, and also the terminal coalgebra carrier $\nu X. T(-, X) : \mathbb{C}^n \to \mathbb{C}$, for a datafunctor $T: \mathbb{C}^{n+1} \to \mathbb{C}$.

The completeness and cocompleteness which is assumed in this definition refers to the existence of the initial algebras and terminal coalgebras mentioned in point 3. The class of datafunctors we use is akin to the one in [19], but does not involve "shape". Also, at this stage, we do not consider the concept of "strength" (see [5, 23, 16], although it forms an essential part of the complete story of these functors.

Explicitly, the functor $\mu X. T(-, X): \mathbb{C}^n \to \mathbb{C}$ in the third clause maps a sequence of objects $\vec{Y} \in \mathbb{C}^n$ to the carrier $\mu X. T(\vec{Y}, X) \in \mathbb{C}$ of the initial algebra $\alpha_{\vec{Y}}: T(\vec{Y}, \mu X. T(\vec{Y}, X)) \xrightarrow{\cong} \mu X. T(\vec{Y}, X)$ of the functor $T(\vec{Y}, -): \mathbb{C} \to \mathbb{C}$. And it maps a sequence of morphisms $\vec{f}: \vec{Y} \to \vec{Z}$ in $\mathbb{C}^n$ to the unique morphism of algebras $\mu X. T(\vec{f}, X): \mu X. T(\vec{Y}, X) \to \mu X. T(\vec{Z}, X)$ in the following diagram.

$$\begin{array}{ccc}
T(\vec{Y}, \mu X. T(\vec{Y}, X)) & \xrightarrow{T(id, \mu X. T(\vec{f}, X))} & T(\vec{Y}, \mu X. T(\vec{Z}, X)) \\
 & & \downarrow T(\vec{f}, id) \\
\alpha_{\vec{Y}} \downarrow \cong & & T(\vec{Z}, \mu X. T(\vec{Z}, X)) \\
 & & \cong \downarrow \alpha_{\vec{Z}} \\
\mu X. T(\vec{Y}, X) & \xrightarrow[\mu X. T(\vec{f}, X)]{} & \mu X. T(\vec{Z}, X)
\end{array}$$

The terminal coalgebra functor $\nu X. T(-, X): \mathbb{C}^n \to \mathbb{C}$ is defined in a dual manner. Often we simply write μT and νT for $\mu X. T(-, X)$ and $\nu X. T(-, X)$. For an algebra $f: T(\vec{Y}, Z) \to Z$ we sometimes write $\mathsf{reduce}(f): \mu X. T(\vec{Y}, X) \to Z$ for the unique homomorphism of algebras resulting from initiality. Similarly, we write $\mathsf{coreduce}(g): Z \to \nu X. T(\vec{Y}, X)$ for the unique homomorphism of coalgebras induced by a coalgebra $g: Z \to T(\vec{Y}, Z)$. Hence the map $\mu X. T(\vec{f}, X)$ in the above diagram is $\mathsf{reduce}(\alpha_{\vec{Z}} \circ T(\vec{f}, id))$.

Example 1. (i) The polynomial functor

$$F_A : \mathbb{C} \times \mathbb{C} \longrightarrow \mathbb{C} \qquad \text{given by} \qquad (Y, X) \mapsto (A \times Y \times X) + 1$$

is a data functor built up with rules 1. and 2. in Definition 1. It will form the basis for the next four examples.

(ii) The functor $\mathsf{List}(A \times -) = \mu X. F_A(-, X): \mathbb{C} \to \mathbb{C}$ sends a (parameter) object Y to the carrier of the initial $F_A(Y, -)$-algebra $\mathsf{List}(A \times Y)$, with constructors $\mathsf{nil}: 1 \to \mathsf{List}(A \times Y)$ and $\mathsf{cons}: (A \times Y) \times \mathsf{List}(A \times Y) \to \mathsf{List}(A \times Y)$. The elements of $\mathsf{List}(A \times Y)$ are generated by these constructors and can thus be seen as finite sequences (of pairs).

(iii) The functor $\mathsf{Colist}(A \times -) = \nu X. F_A(-, X): \mathbb{C} \to \mathbb{C}$ sends a (parameter) object Y to the carrier of the terminal $F_A(Y, -)$-coalgebra $\mathsf{Colist}(A \times Y)$, with destructor $\mathsf{next}: \mathsf{Colist}(A \times Y) \to ((A \times Y) \times \mathsf{Colist}(A \times Y)) + 1$. The elements of $\mathsf{Colist}(A \times Y)$ are seen as finite or infinite sequences of elements of $A \times Y$.

(iv) A signature functor for finitely branching trees is

$$\mathsf{Tree}(A) : \mathbb{C} \longrightarrow \mathbb{C} \qquad \text{sending} \qquad X \mapsto \mathsf{List}(A \times X).$$

A $\mathsf{Tree}(A)$-algebra is a morphism $\mathsf{List}(A \times X) \to X$. It constructs a new node in X out of a list of subtrees in $\mathsf{List}(A \times X)$ which are all labeled with a value from A. The initial $\mathsf{Tree}(A)$-algebra contains all A-labeled finitely branching trees with finite depth. Note that the empty list of subtrees corresponds to a leaf. In the sequel we write the initial algebra as $\mathsf{br}: \mathsf{List}(A \times \mathsf{FTreeF}(A)) \xrightarrow{\cong} \mathsf{FTreeF}(A)$.

A $\mathsf{Tree}(A)$-coalgebra, is given by a morphism $X \to \mathsf{List}(A \times X)$ which decomposes a value in X into a list of successor states paired with a "visible" output in A. The carrier of the terminal $\mathsf{Tree}(A)$-coalgebra contains all A-labeled, finitely branching trees with possibly infinite depth. We use the same name br as before for the structure map of the terminal coalgebra $\mathsf{br}: \mathsf{FTreeI}(A) \xrightarrow{\cong} \mathsf{List}(A \times \mathsf{FTreeI}(A))$.

(v) The signature functor for possibly infinitely branching trees is

$$\mathsf{Cotree}(A) : \mathbb{C} \longrightarrow \mathbb{C} \qquad \text{given by} \qquad X \mapsto \mathsf{Colist}(A \times X).$$

The initial $\mathsf{Cotree}(A)$-algebra $\mathsf{ITreeF}(A)$ contains all possibly infinitely branching trees with finite depth. And the terminal $\mathsf{Cotree}(A)$-coalgebra $\mathsf{ITreeI}(A)$ additionally contains those with infinite depth. We use br to denote both the structure maps of these fixed points.

Notice that the examples (iv) and (v) describe the four combinatorial possibilities of finite versus possibly infinite in breadth or depth for trees, written as FTreeF, FTreeI, ITreeF and ITreeI. These four examples will reappear in Section 6.

3 Initial algebras and terminal coalgebras in total categories

From the perspective of categorical logic, the datafunctors from the previous section are functors $\mathbb{B}^n \to \mathbb{B}$ on a base category $\mathbb{B}$ of types and terms. In order to reason about such functors, one needs a fibration on top of this base category, where the objects in the fibres are seen as predicates on the underlying objects (see [18] for more information about this perspective). Appropriate reasoning principles are obtained by lifting the data functors on the base category to "logical" functors on the total category, by induction on the structure of the data functors. Essentially the same structure (of products and coproducts) is used in the total categories (derived from logical structure in fibre categories), see [14, 15] for more details. In this section we develop the theory which is required to extend this approach to data functors involving iterated fixed points. It will be the categorical contribution of this paper. The extension will involve an explanation of how to obtain fixed points for functors on total categories from fixed points in base categories, plus fixed points in the fibres. In later sections this theory will be applied to the particular situations that we are interested in.

Our starting point is the following situation. We have fibrations $\begin{smallmatrix}\mathbb{E}\\ \downarrow p\\ \mathbb{B}\end{smallmatrix}$ and $\begin{smallmatrix}\mathbb{D}\\ \downarrow q\\ \mathbb{A}\end{smallmatrix}$ together with a functor $T: \mathbb{A} \times \mathbb{B} \to \mathbb{B}$ between base categories, and over T a fibred functor $H: \mathbb{D} \times \mathbb{E} \to \mathbb{E}$ between total categories, in a (commuting) square:

$$\begin{array}{ccc} \mathbb{D} \times \mathbb{E} & \xrightarrow{H} & \mathbb{E} \\ {\scriptstyle q \times p}\downarrow & & \downarrow{\scriptstyle p} \\ \mathbb{A} \times \mathbb{B} & \xrightarrow[T]{} & \mathbb{B} \end{array} \qquad (1)$$

In this situation we think of p as the fibration which captures the logic that we are working in, and of q as a "parameter fibration"; often q will be p^n, for some $n \in \mathbb{N}$. The functor T between the base categories is a datafunctor, and H between total categories is a certain lifting of T. We shall assume that for each object $A \in \mathbb{A}$, the functor $T(A, -): \mathbb{B} \to \mathbb{B}$ has an initial algebra and a terminal coalgebra, written as

$$T(A, \mu T(A)) \xrightarrow[\cong]{\alpha_A} \mu T(A) \qquad \text{and} \qquad \nu T(A) \xrightarrow[\cong]{\beta_A} T(A, \nu T(A))$$

We thus get two functors $\mu T, \nu T: \mathbb{A} \to \mathbb{B}$, as described after Definition 1. Our aim in this section is to construct similar functors $\mu H, \nu H: \mathbb{D} \to \mathbb{E}$ over $\mu T, \nu T$

(respectively). They will be constructed via two auxiliary functors between fibres: for $X \in \mathbb{D}_A$, we define

$$\begin{array}{ccc} \mathbb{E}_{\mu T(A)} \xrightarrow{H_X} \mathbb{E}_{\mu T(A)} & & \mathbb{E}_{\nu T(A)} \xrightarrow{H^X} \mathbb{E}_{\nu T(A)} \\ Y \longmapsto (\alpha_A^{-1})^* (H(X,Y)) & & Y \longmapsto (\beta_A)^* (H(X,Y)) \end{array} \tag{2}$$

The following result from [15] will play an important rôle.

Lemma 2. *A natural transformation $\sigma: SU \Rightarrow UT$ in a situation*

$$\begin{array}{ccc} \mathbb{A} & \xrightarrow{S} & \mathbb{A} \\ U\uparrow & \overset{\sigma}{\Rightarrow} & \uparrow U \\ \mathbb{B} & \xrightarrow[T]{} & \mathbb{B} \end{array} \quad \textit{induces a functor} \quad \begin{array}{c} \mathrm{Alg}(S) \\ \uparrow \mathrm{Alg}(U) \\ \mathrm{Alg}(T) \end{array}$$

by $(TX \xrightarrow{f} X) \mapsto (SUX \xrightarrow{\sigma_X} UTX \xrightarrow{Uf} UX)$.

And if σ is an isomorphism, then a right adjoint G to U

$$\begin{array}{c} \mathbb{A} \\ U \left(\dashv \right) G \\ \mathbb{B} \end{array} \quad \textit{induces a right adjoint} \quad \begin{array}{c} \mathrm{Alg}(S) \\ \mathrm{Alg}(U) \left(\dashv \right) \mathrm{Alg}(G) \\ \mathrm{Alg}(T) \end{array}$$

where the functor Alg(G) *arises from $\tau: TG \Rightarrow GS$, the adjoint transpose of $UTG \cong SUG \overset{S\varepsilon}{\Longrightarrow} S$.*

Dually, a natural transformation $\tau: UT \Rightarrow SU$ gives a functor CoAlg(U): CoAlg$(T) \to$ CoAlg(S) *between categories of coalgebras. Furthermore, if this τ is an isomorphism, then, a left adjoint $F \dashv U$ induces a left adjoint* CoAlg$(F) \dashv$ CoAlg(U). □

We now turn to initial algebras and terminal coalgebras of functors $H(X, -)$ as above.

Theorem 3. *Consider the situation described in (1) above, with auxiliary functors H_X and H^X as in (2). Assuming that the initial algebras and terminal coalgebras in fibre categories in (i) and (ii) below exist, we can define for each $X \in \mathbb{D}$ over $A \in \mathbb{A}$ an initial algebra $\mu H(X)$ over $\mu T(A)$ and a terminal coalgebra $\nu H(X)$ over $\nu T(A)$ for the functor $H(X, -): \mathbb{E} \to \mathbb{E}$ as follows.*

(i) Let $\mu H(X)$ be the carrier of the initial algebra

$$\begin{array}{ccc} H_X(\mu H(X)) & \xrightarrow[\cong]{a_X} & \mu H(X) \\ \| & & \\ (\alpha_A^{-1})^* H(X, \mu H(X)) & & \end{array}$$

of the functor H_X *in the fibre category* $\mathbb{E}_{\mu T(A)}$. *It gives rise to an* $H(X,-)$-*algebra*

$$H(X,\mu H(X)) \dashrightarrow (\alpha_A^{-1})^* H(X,\mu H(X)) \xrightarrow{a_X} \mu H(X)$$

over the initial $T(A,-)$-*algebra* $\alpha_A\colon T(A,\mu T(A)) \xrightarrow{\cong} \mu T(A)$, *where the dashed arrow is the unique one over* α_A *for which the composite* $H(X,\mu H(X)) \to (\alpha_A^{-1})^* H(X,\mu H(X)) \to H(X,\mu H(X))$ *is the identity. This* $H(X,-)$-*algebra is the initial one.*

(ii) Assume now that p *is a bifibration,* i.e. *that substitution functors* u^* *have left adjoints* $\coprod_u$, *and let* $\nu H(X)$ *be the carrier of the terminal* H^X-*coalgebra in the fibre* $\mathbb{E}_{\nu T(A)}$:

$$\nu H(X) \xrightarrow[\cong]{b_X} H^X(\nu H(X)) = \beta_A^* H(X,\nu H(X))$$

It yields as terminal $H(X,-)$-*coalgebra the map:*

$$\nu H(X) \xrightarrow{b_X} \beta_A^* H(X,\nu H(X)) \longrightarrow H(X,\nu H(X))$$

over the terminal $T(A,-)$-*coalgebra* $\beta_A\colon \nu T(A) \xrightarrow{\cong} T(A,\nu T(A))$.

In case the fibration p in this result is a fibred preorder modelling some logic involving provability, rather than some type theory with proof-objects, the initial algebras and terminal coalgebras in the fibres are simply least and greatest fixed points of monotone functions. We shall see such examples later. But we present this result in its general—non-preordered—form.

Proof. (i) Assume an algebra $f\colon H(X,Y) \to Y$ in $\mathbb{E}$, say over $u\colon T(A,B) \to B$ in $\mathbb{B}$—with $f'\colon H(X,Y) \to u^*(Y)$ as the vertical part of f. Initiality of the algebra $\alpha_A\colon T(A,\mu T(A)) \xrightarrow{\cong} \mu T(A)$ gives a unique map of algebras $v = \mathsf{reduce}(u)\colon \mu T(A) \to B$ with $v \circ \alpha_A = u \circ T(id_A, v)$. Using that H is a fibred functor, we can construct an H_X-algebra $g\colon H_X(v^*(Y)) \to Y$ on $v^*(Y) \in \mathbb{E}_{\mu T(A)}$ as composite:

$$\begin{array}{ccc}
(\alpha_A^{-1})^* H(X, v^*(Y)) & \xrightarrow{\cong} & (\alpha_A^{-1})^* T(id_A, v)^* H(X,Y) \\
 & & \Big\downarrow (\alpha_A^{-1})^* T(id_A, v)^*(f') \\
 & & (\alpha_A^{-1})^* T(id_A, v)^* u^*(Y) \xrightarrow[\cong]{} v^*(Y)
\end{array}$$

By initiality of $\mu H(X)$ we get a unique (vertical) map $h\colon \mu H(X) \to v^*(Y)$ with $h \circ a_X = H_X(h) \circ g$. The required mediating $H(X,-)$-algebra map is then the composite $\mu H(X) \to v^*(Y) \to Y$ (over v) of h and the cartesian map $v^*(Y) \to Y$.

(ii) For an arbitrary coalgebra $f\colon Y \to H(X,Y)$ over $u\colon B \to T(A,B)$—with vertical part $f'\colon Y \to u^*H(X,Y)$—we get by terminality a unique map $v = \mathsf{coreduce}(u)\colon B \to \nu T(A)$ in $\mathbb{B}$ with $\beta_A \circ v = T(id_A, v) \circ u$. We can define a functor $K\colon \mathbb{E}_B \to \mathbb{E}_B$ by $Z \mapsto u^*H(X,Z)$. It yields a diagram commuting up-to-isomorphism:

$$\begin{array}{ccc} \mathbb{E}_B & \xrightarrow{\quad K \quad} & \mathbb{E}_B \\ {\scriptstyle v^*}\uparrow & \cong & \uparrow{\scriptstyle v^*} \\ \mathbb{E}_{\nu T(A)} & \xrightarrow[\quad H^X \quad]{} & \mathbb{E}_{\nu T(A)} \end{array}$$

since

$$\begin{aligned} Kv^*(Y) = u^*H(X, v^*(Y)) &\cong u^*T(id_A, v)^*H(X,Y) \\ &\cong v^*\beta_A^*H(X,Y) = v^*H^X(Y). \end{aligned}$$

The left adjoint $\coprod_v \dashv v^*$ now gives by Lemma 2 (ii) a left adjoint to the induced functor $\mathrm{CoAlg}(v^*)\colon \mathrm{CoAlg}(H^X) \to \mathrm{CoAlg}(K)$. As a consequence, this functor $\mathrm{CoAlg}(v^*)$ preserves terminal coalgebras. Hence the terminal K-algebra is the composite

$$v^*(\nu H(X)) \xrightarrow[\cong]{v^*(b_X)} v^*H^X(\nu H(X)) \xrightarrow{\cong} K(v^*(\nu H(X)))$$

The vertical part $f'\colon Y \to u^*H(X,Y) = K(Y)$ of our original map f is a K-coalgebra, and thus gives rise to a unique (vertical) map of K-coalgebras $g\colon Y \to v^*(\nu H(X))$. Composing this g with the cartesian map $v^*(\nu H(X)) \to \nu H(X)$ yields the required map of $H(X,-)$-coalgebras $Y \to \nu H(X)$ (over v). □

Proposition 4. *In the situation of the previous theorem, assume that the substitution functors u^* of the fibration p have both a left adjoint $\coprod_u$ and a right adjoint $\prod_u$. The resulting functors $\mu H, \nu H\colon \mathbb{D} \to \mathbb{E}$ are then fibred functors in diagrams:*

$$\begin{array}{ccc} \mathbb{D} & \xrightarrow{\quad \mu H \quad} & \mathbb{E} \\ {\scriptstyle q}\downarrow & & \downarrow{\scriptstyle p} \\ \mathbb{A} & \xrightarrow[\quad \mu T \quad]{} & \mathbb{B} \end{array} \qquad \begin{array}{ccc} \mathbb{D} & \xrightarrow{\quad \nu H \quad} & \mathbb{E} \\ {\scriptstyle q}\downarrow & & \downarrow{\scriptstyle p} \\ \mathbb{A} & \xrightarrow[\quad \nu T \quad]{} & \mathbb{B} \end{array}$$

Proof. Assume a morphism $w\colon B \to A$ in $\mathbb{A}$ and an object $X \in \mathbb{D}_A$. We get two morphisms $\mu T(w)\colon \mu T(B) \to \mu T(A)$ and $\nu T(w)\colon \nu T(B) \to \nu T(A)$ in $\mathbb{B}$, determined by the equations

$$\mu T(w) \circ \alpha_B = \alpha_A \circ T(w, \mu T(w)) \quad \text{and} \quad \beta_A \circ \nu T(w) = T(w, \nu T(w)) \circ \beta_B.$$

We also get two isomorphisms

$$\begin{array}{ccc} \mathbb{E}_{\mu T(B)} & \xrightarrow{H_{w^\bullet(X)}} & \mathbb{E}_{\mu T(B)} \\ {\scriptstyle \mu T(w)^*}\uparrow & \cong & \uparrow{\scriptstyle \mu T(w)^*} \\ \mathbb{E}_{\mu T(A)} & \xrightarrow[H_X]{} & \mathbb{E}_{\mu T(A)} \end{array} \qquad \begin{array}{ccc} \mathbb{E}_{\nu T(B)} & \xrightarrow{H^{w^*(X)}} & \mathbb{E}_{\nu T(B)} \\ {\scriptstyle \nu T(w)^*}\uparrow & \cong & \uparrow{\scriptstyle \nu T(w)^*} \\ \mathbb{E}_{\nu T(A)} & \xrightarrow[H^X]{} & \mathbb{E}_{\nu T(A)} \end{array}$$

Since

$$\begin{aligned} H_{w^\bullet(X)}(\mu T(w))^*(Y) &= (\alpha_B^{-1})^* H(w^*(X), (\mu T(w))^*(Y)) \\ &\cong (\alpha_B^{-1})^* T(w, \mu T(w))^*(H(X,Y) \\ &\cong (\mu T(w))^*(\alpha_A^{-1})^* H(X,Y) \\ &= (\mu T(w))^* H_X(Y). \\ H^{w^*(X)}(\nu T(w))^*(Y) &= \beta_B^* H(w^*(X), (\nu T(w))^*(Y)) \\ &\cong \beta_B^* T(w, \nu T(w))^* H(X,Y) \\ &\cong (\nu T(w))^* \beta_A^* H(X,Y) \\ &= (\nu T(w))^* H^X(Y). \end{aligned}$$

By Lemma 2, the two adjoints $(\mu T(w))^* \dashv \prod_{\mu T(w)}$ and $\coprod_{\nu T(w))} \dashv (\nu T(w))^*$ give rise to right and left adjoints:

$$\begin{array}{c} \mathrm{Alg}(H_{w^\bullet(X)}) \\ \mathrm{Alg}((\mu T(w))^*)\uparrow \dashv \downarrow \\ \mathrm{Alg}(H_X) \end{array} \qquad \begin{array}{c} \mathrm{CoAlg}(H^{w^*(X)}) \\ \downarrow \dashv \uparrow \mathrm{CoAlg}((\nu T(w))^*) \\ \mathrm{CoAlg}(H^X) \end{array}$$

Hence the functor $\mathrm{Alg}((\mu T(w))^*)\colon \mathrm{Alg}(H_X) \to \mathrm{Alg}(H_{w^\bullet(X)})$ preserves initial objects, and the functor $\mathrm{CoAlg}((\nu T(w))^*)\colon \mathrm{CoAlg}(H^X) \to \mathrm{CoAlg}(H^{w^*(X)})$ preserves terminal objects. In particular, this means that

$$(\mu T(w))^*(\mu H(X)) \cong \mu H(w^*(X)) \quad \text{and} \quad (\nu T(w))^*(\nu H(X)) \cong \nu H(w^*(X))$$

showing that μH and νH are fibred functors as required. □

Remark. As already mentioned, one is often interested in preorder fibrations describing some logic. Then the initial H_X-algebra and terminal H^X-coalgebra in Theorem 3 in the *preorder* fibre categories $\mathbb{E}_{\mu T(A)}$ and $\mathbb{E}_{\nu T(A)}$ (respectively) can be computed as least and greatest fixed points of the monotone functions H_X, H^X. The standard formulas are:

$$\mu H(X) = \bigwedge\{y \mid H_X(y) \leq y\} \qquad \text{and} \qquad \nu H(X) = \bigvee\{y \mid y \leq H^X(y)\}.$$

This leads to the familiar properties: $H_X(y) \leq y \Rightarrow \mu H(X) \leq y$ and $y \leq H^X(y) \Rightarrow y \leq \nu H(X)$. In this preorder setting, the latter implication can be strengthened to the so-called *strong coinduction* rule (see *e.g.* [27, 3.2]):

$$y \leq H^X(y) \vee \nu H(X) \qquad \text{implies} \qquad y \leq \nu H(X).$$

This rule is easily derived, since the assumption yields $y \vee \nu H(X) \leq H^X(y \vee \nu H(X))$.

4 Relational proof principles for datafunctors

In the previous section we have studied a fibred functor H over a functor T between base categories of fibrations. Here we shall instantiate this situation by taking T to be a datafunctor and H a "lifting" $\mathrm{Rel}(T)$ of T to a fibred category of relations $\mathrm{Rel}(\mathbb{E})$. Formally, this category $\mathrm{Rel}(\mathbb{E})$ is the total category of a fibration $\begin{smallmatrix}\mathrm{Rel}(\mathbb{E})\\ \downarrow\\ \mathbb{B}\end{smallmatrix}$ of relations, which is obtained from a fibration $\begin{smallmatrix}\mathbb{E}\\ \downarrow\\ \mathbb{B}\end{smallmatrix}$ by change-of-base (or pullback) along the functor $\mathbb{B} \to \mathbb{B}$ given by $A \mapsto A \times A$. Throughout this section we shall assume that the fibration $\begin{smallmatrix}\mathbb{E}\\ \downarrow\\ \mathbb{B}\end{smallmatrix}$ admits the structure required for the constructions that we perform. In particular we shall assume that there is an equality functor $\mathrm{Eq}: \mathbb{B} \to \mathrm{Rel}(\mathbb{E})$, were $\mathrm{Eq}(A) = \coprod_\delta(1A)$ with $1: \mathbb{B} \to \mathbb{E}$ a terminal object functor for the fibration.

We shall define for a datafunctor $T: \mathbb{B}^n \to \mathbb{B}$ a lifting $\mathrm{Rel}(T): \mathrm{Rel}(\mathbb{E})^n \to \mathrm{Rel}(\mathbb{E})$ by induction on the structure of T, adapting the approach of [14, 15]. That is, if T is a projection functor $\mathbb{B}^n \to \mathbb{B}$, then $\mathrm{Rel}(T): \mathrm{Rel}(\mathbb{E})^n \to \mathrm{Rel}(\mathbb{E})$ is the same projection; if T is a constant $A: \mathbf{1} \to \mathbb{B}$, then $\mathrm{Rel}(T)$ is the associated equality constant $\mathrm{Eq}(A) = \coprod_\delta(1A): \mathbf{1} \to \mathrm{Rel}(\mathbb{E})$; if T is $\times$ or $+$ from $\mathbb{B} \times \mathbb{B}$ to $\mathbb{B}$, then $\mathrm{Rel}(T)$ is $\times$ or $+$ from $\mathrm{Rel}(\mathbb{E}) \times \mathrm{Rel}(\mathbb{E})$ to $\mathrm{Rel}(\mathbb{E})$—given by canonical formulas $R \times S = (\pi \times \pi)^*(R) \wedge (\pi' \times \pi')^*(S)$, and $R + S = \coprod_{\kappa \times \kappa}(R) \vee \coprod_{\kappa' \times \kappa'}(S)$, using $\wedge$ and $\vee$ for products and coproducts in the fibres; if T is $S \circ (T_1, \ldots, T_m)$, then $\mathrm{Rel}(T)$ is $\mathrm{Rel}(S) \circ (\mathrm{Rel}(T_1), \ldots, \mathrm{Rel}(T_n))$; and finally, if T is μS or νS, then $\mathrm{Rel}(T)$ is $\mu\mathrm{Rel}(S)$ or $\nu\mathrm{Rel}(S)$, using the fixed points of fibred functors as described in the previous section. Under mild requirements, the lifted functors $\mathrm{Rel}(T)$ are fibred over T (using Proposition 4).

The following definition formalises appropriate interaction between T and $\mathrm{Rel}(T)$. The property that is defined can be proved for all non-recursive data functors, see [15], but in the presence of recursion it has to be required explicitly.

Definition 5. A datafunctor $T: \mathbb{B}^n \to \mathbb{B}$ is said to *respect equality* if there is a (canonical) natural isomorphism

$$\mathrm{Eq} \circ T \overset{\cong}{\Longrightarrow} \mathrm{Rel}(T) \circ \mathrm{Eq}^n.$$

(The canonicity of this isomorphism depends on the structure of T, but does not really play a rôle in what follows.)

A functor $T: \mathbb{B}^{n+1} \to \mathbb{B}$ respecting equality induces by lemma 2 functors between categories of algebras and of coalgebras, namely

$$\begin{array}{ccc} \mathrm{Alg}(T(\vec{X}, -)) & & \mathrm{CoAlg}(T(\vec{X}, -)) \\ \mathrm{Alg}(\mathrm{Eq}) \big\downarrow & \text{and} & \big\downarrow \mathrm{CoAlg}(\mathrm{Eq}) \\ \mathrm{Alg}(\mathrm{Rel}(T)(\mathrm{Eq}^n(\vec{X}), -)) & & \mathrm{CoAlg}(\mathrm{Rel}(T)(\mathrm{Eq}^n(\vec{X}), -)) \end{array}$$

for each tuple $\vec{X} \in \mathbb{B}^n$. These functors both use the isomorphism in the above definition, but in different directions. The first functor Alg(Eq) sends an algebra $\alpha: T(\vec{X}, Y) \to Y$ to the composite

$$\mathrm{Rel}(T)(\mathrm{Eq}^n(\vec{X}), \mathrm{Eq}(Y)) \xrightarrow{\cong} \mathrm{Eq}(T(\vec{X}, Y)) \xrightarrow{\mathrm{Eq}(\alpha)} \mathrm{Eq}(Y)$$

And the second functor CoAlg(Eq) sends a coalgebra $\beta: Y \to T(\vec{X}, Y)$ to

$$\mathrm{Eq}(Y) \xrightarrow{\mathrm{Eq}(\beta)} \mathrm{Eq}(T(\vec{X}, Y)) \xrightarrow{\cong} \mathrm{Rel}(T)(\mathrm{Eq}^n(\vec{X}), \mathrm{Eq}(Y))$$

In terms of these functors Alg(Eq) and CoAlg(Eq) we define induction and coinduction principles in a logic of a fibration (essentially as in [14, 15]).

Definition 6. Consider a fibration $\begin{smallmatrix}\mathbb{E}\\ \downarrow\\ \mathbb{B}\end{smallmatrix}$ with a datafunctor $T: \mathbb{B}^n \times \mathbb{B} \to \mathbb{B}$ respecting equality on its base category, together with the associated lifted functor $\mathrm{Rel}(T): \mathrm{Rel}(\mathbb{E})^n \times \mathrm{Rel}(\mathbb{E}) \to \mathrm{Rel}(\mathbb{E})$. The fibration satisfies

1. the *induction principle w.r.t.* T if for all objects $\vec{X} \in \mathbb{B}^n$ the functor

$$\mathrm{Alg}(T(\vec{X}, -)) \xrightarrow{\mathrm{Alg}(\mathrm{Eq})} \mathrm{Alg}(\mathrm{Rel}(T)(\mathrm{Eq}^n(\vec{X}), -))$$

preserves initial objects.

2. the *coinduction principle w.r.t.* T if, similarly, for all $\vec{X} \in \mathbb{B}^n$ the functor

$$\mathrm{CoAlg}(T(\vec{X}, -)) \xrightarrow{\mathrm{CoAlg}(\mathrm{Eq})} \mathrm{CoAlg}(\mathrm{Rel}(T)(\mathrm{Eq}^n(\vec{X}), -))$$

preserves terminal objects.

Lemma 7. *If* $\begin{smallmatrix}\mathbb{E}\\ \downarrow\\ \mathbb{B}\end{smallmatrix}$ *satisfies induction/coinduction* w.r.t. $T: \mathbb{B}^n \times \mathbb{B} \to \mathbb{B}$ *then both* $\mu T, \nu T : \mathbb{B}^n \to \mathbb{B}$ *respect equality.*

Proof. For each $\vec{X} \in \mathbb{B}$ the induction principle tells us that the carrier of the initial algebra $\mu\mathrm{Rel}(T)(\mathrm{Eq}^n(\vec{X}))$ of the functor $\mathrm{Rel}(T)(\mathrm{Eq}^n(\vec{X}), -)$ is $\mathrm{Eq}(\mu T(\vec{X}))$. As a result, there is a (canonical) isomorphism $\mu\mathrm{Rel}(T)(\mathrm{Eq}^n(\vec{X})) \cong \mathrm{Eq}(\mu T(\vec{X}))$ between two initial objects. The same holds in the coalgebraic case. □

The next result ensures the validity of induction and coinduction principles in the presence of comprehension and quotients, like in [15].

Theorem 8. *If our fibration admits both comprehension and quotients, then it satisfies the induction and coinduction principles w.r.t. all data functors.*

Proof. Recall that $\mathrm{Eq} = \coprod_\delta \circ 1$. The existence of comprehension as a right adjoint to 1 makes Eq a left adjoint. In the presence of quotients Eq becomes, by definition, a right adjoint. If a datafunctor T respects equality, then, by Lemma 2 the functor Alg(Eq) is a left and CoAlg(Eq) is a right adjoint preserving initial and terminal objects respectively. By induction on the structure of the datafunctors T, one establishes that T respects equality and therefore that induction and coinduction principles hold *w.r.t.* T. □

5 Handling mixed proof obligations

In this section we focus on the special case where we have an initial algebra α or a terminal coalgebra β of the form

$$\nu F(A) \xrightarrow[\cong]{\alpha} A \qquad \text{and} \qquad B \xrightarrow[\cong]{\beta} \mu F(B)$$

involving initial algebras and terminal coalgebras

$$\nu F(A) \xrightarrow{\cong} F(A, \nu F(A)) \qquad \text{and} \qquad F(B, \mu F(B)) \xrightarrow{\cong} \mu F(B)$$

where F is a datafunctor $\mathbb{C} \times \mathbb{C} \to \mathbb{C}$. For convenience, we assume that F has only two parameters instead of $n+1$, since there is no fundamental difference. The problem with the above α and β is that the fixed points $\nu F(A)$ and $\mu F(B)$ are "on the wrong side".

The same inconvenience occurs at a logical level. We shall explain this in a setting where we have a logical preorder structure $\vdash$ representing entailment, instead of a type theoretical (non-preorder) structure $\to$ describing explicit proof-terms. In this setting we assume that both the induction and coinduction principles hold *w.r.t* F.

We first consider the second situation with terminal coalgebra $\beta\colon B \xrightarrow{\cong} \mu F(B)$. Assume that we have a relation R on B, and that we wish to show by coinduction that R is contained in the equality relation $\mathrm{Eq}(B)$ on B (*i.e.* that $R \vdash \mathrm{Eq}(B)$). By Definition 6 we know that $\mathrm{Eq}(B)$ is the terminal coalgebra of the lifted functor $\mathrm{Rel}(\mu F)$, so it suffices to show that R carries a $\mathrm{Rel}(\mu F)$ coalgebra structure $R \to \mathrm{Rel}(\mu F)(R)$ over β: such a coalgebra results in a unique coalgebra homomorphism $R \to \mathrm{Eq}(B)$ over $\mathsf{coreduce}(\beta) = id$, see Theorem 3 (and its proof). It thus yields $R \vdash \mathrm{Eq}(B)$ as required (since we assumed the fibres to be preorders).

Hence in order to prove $R \vdash \mathrm{Eq}(B)$ we seek a coalgebra structure $R \to \mathrm{Rel}(\mu F)(R)$ over β. Equivalently, an entailment $R \vdash (\beta \times \beta)^*(\mathrm{Rel}(\mu F))(R)$ over B, or an entailment $\coprod_{\beta\times\beta}(R) \vdash \mathrm{Rel}(\mu F)(R)$ over $\mu F(B)$. Since $\mu F(B)$ is an initial algebra, we may try to prove the latter by induction. We can make this explicit by unraveling the left hand side $\coprod_{\beta\times\beta}(R)$. It can be described as

$$\begin{aligned}&\textstyle\coprod_{\beta\times\beta}(R)\\ &= \textstyle\coprod_{\pi'\times\pi'}\left(\pi^*(\beta\times id)^*\mathrm{Eq}(\mu F(B)) \wedge \pi'^*(\beta\times id)^*\mathrm{Eq}(\mu F(B)) \wedge (\pi\times\pi)^*(R)\right).\end{aligned}$$

Hence, what we have to prove is an entailment over $\mu F(B)$:

$$\begin{aligned}&\textstyle\coprod_{\pi'\times\pi'}(\pi^*(\beta\times id)^*\mathrm{Eq}(\mu F(B)) \wedge \pi'^*(\beta\times id)^*\mathrm{Eq}(\mu F(B)) \wedge (\pi\times\pi)^*(R))\\ &\quad \vdash \mathrm{Rel}(\mu F)(R).\end{aligned}$$

It is equivalent to an entailment over $B \times \mu F(B)$:

$$\begin{aligned}&\pi^*(\beta\times id)^*\mathrm{Eq}(\mu F(B)),\ \pi'^*(\beta\times id)^*\mathrm{Eq}(\mu F(B))\\ &\quad \vdash (\pi'\times\pi')^*\left((\pi\times\pi)^*(R) \Rightarrow \mathrm{Rel}(\mu F)(R)\right),\end{aligned}$$

where the comma on the left hand side of the turnstile describes the meet $\wedge$, and where $\Rightarrow$ on the right hand side is the associated implication (or exponent). Such an entailment can be proved by a "double" induction over the initial algebra $\mu F(B)$. Describing the precise details of this in categorical terms requires us to formally deal with strength at a logical level (see also [15]), leading us too far from the main points of the present paper. Also, we skip these details because this presents no problem in a proof assistant like PVS.

We turn to the initial algebra $\alpha: \nu F(A) \xrightarrow{\cong} A$, described in the beginning of this section. Suppose now that we wish to prove that a relation R on A is reflexive[4]. We can try to do so by induction, since A is an inductive type. This means that we wish to show $\mathrm{Eq}(A) \vdash R$ in the fibre over $A \times A$, using the binary induction principle. Since $\mathrm{Eq}(A)$ is by Definition 6 the initial algebra of $\mathrm{Rel}(\nu F)$, it suffices to show that R carries a $\mathrm{Rel}(\nu F)$-algebra $\mathrm{Rel}(\nu F)(R) \to R$ over α. Equivalently, this involves proving an entailment

$$\mathrm{Rel}(\nu F)(R) \vdash (\alpha \times \alpha)^*(R) \tag{3}$$

over $\nu F(A)$. In this case we do not know of a general way of massaging this goal into an equivalent obligation which we can prove by coinduction. But, in the special case where the right hand side is of suitable form, there is more we can do.

Let us assume that the right hand side $(\alpha \times \alpha)^*(R)$ of (3) is a conjunction of implications with substituted equations on terminal coalgebras as conclusions, *i.e.* that

$$(\alpha \times \alpha)^*(R) = \bigwedge_i S_i \Rightarrow (u_i \times v_i)^*(\mathrm{Eq}(Z_i))$$

where u_i, v_i are parallel maps $\nu F(A) \to Z_i$ to the carrier of some terminal coalgebra $\gamma_i: Z_i \xrightarrow{\cong} T_i(Z_i)$. The above proof obligation (3) is then equivalent to the collection of proof obligations

$$\mathrm{Rel}(\nu F)(R) \vdash S_i \Rightarrow (u_i \times v_i)^*(\mathrm{Eq}(Z_i)).$$

Or, equivalently, to

$$\coprod\nolimits_{u_i \times v_i}(\mathrm{Rel}(\nu F)(R) \wedge S_i) \vdash \mathrm{Eq}(Z_i).$$

The latter can be proved by coinduction (on the terminal coalgebras $\gamma_i: Z_i \xrightarrow{\cong} T_i(Z_i)$, namely, by proving that we have $\mathrm{Rel}(T_i)$-coalgebra structures on the left hand side $\coprod_{u_i \times v_i}(\mathrm{Rel}(\nu F)(R) \wedge S_i)$ given by entailments

$$\coprod\nolimits_{u_i \times v_i}(\mathrm{Rel}(\nu F)(R) \wedge S_i) \vdash (\gamma_i \times \gamma_i)^*(\mathrm{Rel}(T_i)(\coprod\nolimits_{u_i \times v_i}(\mathrm{Rel}(\nu F)(R) \wedge S_i))).$$

These two approaches for handling successive fixed points of different kinds are used in the next section.

---

[4] Often, the relation R will be of the form $\coprod_\delta(P)$ for a predicate P on A. Showing that R is reflexive is then the same as showing that P holds on A (since $\coprod_\delta(P)(x, y) \Leftrightarrow x = y \wedge P(x)$). In fact, in this preorder situation the case $R = \coprod_\delta(P)$ describes the general case: showing $\mathrm{Eq}(A) \vdash R$ is then equivalent to showing $\mathrm{Eq}(A) \vdash R \wedge \mathrm{Eq}(A)$. But by Frobenius $R \wedge \mathrm{Eq}(A) = R \wedge \coprod_\delta(1A) = \coprod_\delta(\delta^*(R) \wedge 1(A)) = \coprod_\delta(\delta^*(R))$, so that we may take as predicate $P = \delta^*(R)$.

6 Proof principles in use

Having described the general proof principles for iterated datafunctors in the previous sections we proceed with the presentation of some sample proofs. This should enable the reader to grasp the differences between pure induction and coinduction on the one hand and the iterated mixing of them on the other.

We define a composition operation $(x, y) \mapsto \mathsf{comp}(x, y)$ for each of the tree examples from Section 2, by appending a tree y at all leaves of the tree x. A leaf, hereby, is a tree which has no subtrees. This composition operation can be defined for all types of trees. We then prove, using appropriate proof principles for the different types of trees, that appending the empty tree τ yields the given tree itself, *i.e.* that $\mathsf{comp}(x, \tau) = x$. This is a non–trivial matter to prove formally.

These proofs are carried out using the specification and verification tool PVS [25, 24]. PVS is a tool which comprises a specification language based on higher order logic with predicate subtyping and dependent types, an interactive theorem prover and several tools for handling modular specifications, pretty printing, and generating readable LaTeX specification and proof files.

The formalisation of the employed proof principles involves the specification of coinductive sequences (colists) and the full range of our example trees. These specifications have some value in themself because they illustrate how PVS (or a similar tool) could (in future versions) incorporate iterated data structures and their proof principles.

We closely follow the abstract categorical development in this paper. It is an indispensable road map for our implementation in PVS. PVS itself supplies inductive (non iterated) abstract datatypes including the required lists and coproducts. Colists, however, have to be specified explicitly by providing the structure map next, describing the notion of homomorphism, and axiomatising terminality, as described in [11]. The (co)inductive trees are then formalised in a similar style in terms of the underlying (co)lists.

Relations (objects in the total category) are defined using predicates or, equivalently, characteristic functions. A fibre is a poset category with relations on a common type as objects and implication as (vertical) morphisms. The logic of PVS provides for the logical connectives yielding the appropriate fibred structure. An endofunctor on a fibre is then an implication preserving mapping from relations to relations (on the same type). We define general theories for fixed points of such endofunctors. Having introduced these basic definitions we formalise the specific endofunctors H_R and H^R from section 3 for lists and colists. Then we require the existence of fixed points in the fibres by suitable axioms and obtain the relational lifting.

The bisimulation and induction rules are introduced in terms of these fixed points (the polynomial liftings are "programmed" directly). PVS has a notion of comprehension via predicate subtypes and quotients can be implemented, so that the induction and coinduction principle can be proved within PVS. For all datatypes we provide strength, functorial properties, and the unrolling of the proof rules as used below.

The following subsections explain the sample proofs in more detail. We use a notion of a goal oriented proof tree starting with the goal sequence at the bottom and developing the proof in a bottom-up fashion. Moreover we simplify the injections for the coproduct in the functor $F(Y,X) = F_A(Y,X) = A \times Y \times X + 1$ from Example 1 by writing a triple $(a,y,x) \in A \times Y \times X$ for an element given by the left inclusion and $* \in 1$ for the right inclusion. The lifting $\mathrm{Rel}(F)$ of F to a (binary) functor on relations is then described as follows. For a relation R on Y and a relation S on X,

$$\begin{aligned}&\mathrm{Rel}(F)(R,S) \subseteq (A \times Y \times X + 1) \times (A \times Y \times X + 1)\\ &\quad = \{\langle (a,y,x),(a',y',x')\rangle \mid a = a' \text{ and } R(y,y') \text{ and } S(x,x')\} \cup \{\langle *,*\rangle\}.\end{aligned}$$

For this functor F we consider the initial algebra and terminal coalgebra

$$\mu F = \mu X.\, F(-,X) = \mathsf{List}(A \times -) \quad \text{and} \quad \nu F = \nu X.\, F(-,X) = \mathsf{Colist}(A \times -),$$

with structure maps

$$A \times Y \times \mathsf{List}(A \times Y) + 1 \xrightarrow[\cong]{[\mathsf{cons},\,\mathsf{nil}]} \mathsf{List}(A \times Y)$$

$$\mathsf{Colist}(A \times Y) \xrightarrow[\cong]{\mathsf{next}} A \times Y \times \mathsf{Colist}(A \times Y) + 1$$

The associated functors on the fibres—as in (2)—are as follows. For R a relation on Y, and S a relation on $\mathsf{List}(A \times Y)$,

$$\begin{aligned}&\mathrm{Rel}(F)_R(S) \subseteq \mathsf{List}(A \times Y) \times \mathsf{List}(A \times Y)\\ &\quad = \{\langle \mathsf{cons}(a,x,\ell),\mathsf{cons}(b,y,k)\rangle \mid a = b \text{ and } R(x,y) \text{ and } S(\ell,k)\} \cup \{\langle \mathsf{nil},\mathsf{nil}\rangle\}.\end{aligned}$$

And for S a relation on $\mathsf{Colist}(A \times Y)$,

$$\begin{aligned}&\mathrm{Rel}(F)^R(S) \subseteq \mathsf{Colist}(A \times Y) \times \mathsf{Colist}(A \times Y)\\ &\quad = \{\langle \ell,k\rangle \mid \mathsf{next}(\ell) = (a,x,\ell'), \mathsf{next}(k) = (b,y,k') \text{ with}\\ &\qquad\qquad a = b \text{ and } R(x,y) \text{ and } S(\ell',k')\} \cup \{\langle \ell,k\rangle \mid \mathsf{next}(\ell) = \mathsf{next}(k) = *\}.\end{aligned}$$

By definition, there are equalities

$$\begin{aligned}\mathrm{Rel}(\mu F)(R) &= (\mu\mathrm{Rel}(F))(R) = \mu(\mathrm{Rel}(F)_R)\\ \mathrm{Rel}(\nu F)(R) &= (\nu\mathrm{Rel}(F))(R) = \nu(\mathrm{Rel}(F)^R).\end{aligned}$$

Induction for $\mathsf{List}(A \times Y)$ takes the form of an equality (between initial algebras):

$$\mu(\mathrm{Rel}(F)_{\mathrm{Eq}(Y)}) = \mathrm{Eq}(\mathsf{List}(A \times Y)).$$

And coinduction for $\mathsf{Colist}(A \times Y)$ amounts to:

$$\nu(\mathrm{Rel}(F)^{\mathrm{Eq}(Y)}) = \mathrm{Eq}(\mathsf{Colist}(A \times Y)).$$

Below we will consider the following two initial algebras and two terminal coalgebras.

$$\mathsf{br}\colon \mathsf{List}(A \times \mathsf{FTreeF}(A)) \xrightarrow{\cong} \mathsf{FTreeF}(A)$$
$$\mathsf{br}\colon \mathsf{Colist}(A \times \mathsf{ITreeF}(A)) \xrightarrow{\cong} \mathsf{ITreeF}(A)$$
$$\mathsf{br}\colon \mathsf{FTreeI}(A) \xrightarrow{\cong} \mathsf{List}(A \times \mathsf{FTreeI}(A))$$
$$\mathsf{br}\colon \mathsf{ITreeI}(A) \xrightarrow{\cong} \mathsf{Colist}(A \times \mathsf{ITreeI}(A)$$

The associated induction and coinduction principles then take the following form.

Induction for $\mathsf{FTreeF}(A)$:	$\mu\mathrm{Rel}(\mu F) = \mathrm{Eq}(\mathsf{FTreeF})$
Induction for $\mathsf{ITreeF}(A)$:	$\mu\mathrm{Rel}(\nu F) = \mathrm{Eq}(\mathsf{ITreeF})$
Coinduction for $\mathsf{FTreeI}(A)$:	$\nu\mathrm{Rel}(\mu F) = \mathrm{Eq}(\mathsf{FTreeI})$
Coinduction for $\mathsf{ITreeI}(A)$:	$\nu\mathrm{Rel}(\nu F) = \mathrm{Eq}(\mathsf{ITreeI})$.

We formalise these principles in PVS (via axioms), and use them to obtain the results in the next four Subsections 6.1–6.4. This formalisation makes use of our earlier formalisation of colists (or sequences), see [11].

We emphasise that these principles can be generated automatically from the form of the functor (or signature) describing the relevant datatype. The formulation of our proof principles is entirely syntax-driven, and is thus suitable for incorporation in a proof-tool.

6.1 Finitely branching trees of finite depth

Finitely branching trees with finite depth are purely inductive. We briefly sketch the definition of composition and the required proof. In order not to complicate matters unnecessarily we avoid the usage of strength in inductive definitions and proofs. Therefore the composition for inductive trees (finite depth) here and in subsection 6.2 is unary by fixing the second argument.

Composition is defined inductively as $\mathsf{comp}_y = \mathsf{reduce}(\mathsf{comp\text{-}struct}_y)$ where

$$\mathsf{comp\text{-}struct}_y : \mathsf{List}(A \times \mathsf{FTreeF}(A)) \to \mathsf{FTreeF}(A);$$
$$\ell \mapsto \begin{cases} y & \text{if } \ell = \mathsf{nil} \\ \mathsf{br}(\ell) & \text{else} \end{cases}$$

The empty tree τ is constructed from the empty list as $\mathsf{br}(\mathsf{nil})$. In order to prove the equality $\mathsf{comp}_\tau(x) = x$ we show that $R = \{(x,x) \mid, \mathsf{comp}_\tau(x) = x\}$ contains the equality relation on trees. Using that $\mathrm{Rel}(\mu F)(R)$ is a least fixed point of $\mathrm{Rel}(F)_R$ we obtain

$$\cfrac{\cfrac{\mathrm{Rel}(F)_R((\mathsf{br} \times \mathsf{br})^*(R)) \vdash (\mathsf{br} \times \mathsf{br})^*(R)}{\mathrm{Rel}(\mu F)(R) \vdash (\mathsf{br} \times \mathsf{br})^*(R)}}{\mathrm{Eq} \vdash R}$$

The upper obligation amounts to two cases, namely $R(\mathsf{br}(\mathsf{nil}), \mathsf{br}(\mathsf{nil}))$ and

$$R(x,y) \wedge R(\mathsf{br}(\ell), \mathsf{br}(k)) \vdash R(\mathsf{br}(\mathsf{cons}((a,x),\ell)), \mathsf{br}(\mathsf{cons}((a,y),k))).$$

For our particular relation $R = \{(x,x) \mid, \mathsf{comp}_{\tau}(x) = x\}$ the first case yields

$$\mathsf{comp}_{\tau}(\mathsf{br}(\mathsf{nil})) = \mathsf{br}(\mathsf{nil})$$

and the second becomes

$$\begin{aligned} & x = y \wedge \mathsf{comp}_{\tau}(x) = x \wedge \mathsf{br}(\ell) = \mathsf{br}(k) \wedge \mathsf{comp}_{\tau}(\mathsf{br}(\ell)) = \mathsf{br}(\ell) \\ & \quad \vdash \mathsf{br}(\mathsf{cons}((a,x),\ell)) = \mathsf{br}(\mathsf{cons}((a,y),k)) \\ & \qquad \wedge \mathsf{comp}_{\tau}(\mathsf{br}(\mathsf{cons}((a,x),\ell))) = \mathsf{br}(\mathsf{cons}((a,x),\ell)). \end{aligned}$$

This can be easily discharged using above definitions and the fact that br is an isomorphism.

6.2 Possibly infinitely branching trees of finite depth

Here we expect a mixed argument for our proof. According to the structure of $\mathsf{ITreeF}(A)$ we start by induction, the inner proof is then performed by coinduction.

Composition is defined inductively as in section 6.1 but it involves colists instead of lists.

$$\begin{aligned} & \mathsf{comp\text{-}struct}_y : \mathsf{Colist}(A \times \mathsf{ITreeF}(A)) \to \mathsf{ITreeF}(A); \\ & \ell \mapsto \begin{cases} y & \text{if } \mathsf{next}(\ell) = * \\ \mathsf{br}(\ell) & \text{else} \end{cases} \end{aligned}$$

The empty tree is constructed from the empty colist. The relation to be proved reflexive is again $R = \{(x,x) \mid, \mathsf{comp}_{\tau}(x) = x\}$, a predicate transformed into a relation. $\mathsf{ITreeF}(A)$ yields the following obligation

$$\frac{\mathrm{Rel}(\nu F)(R) \vdash (\mathsf{br} \times \mathsf{br})^*(R)}{\mathrm{Eq} \vdash R}$$

Fortunately we can simplify $(\mathsf{br} \times \mathsf{br})^*(R)$ to a conjunction of substituted equalities over colists as explained in section 5.

$$\begin{aligned} (\mathsf{br} \times \mathsf{br})^*(R) &= \{(\ell,k) \mid, \mathsf{br}(\ell) = \mathsf{br}(k) \wedge \mathsf{comp}_{\tau}(\mathsf{br}(\ell)) = \mathsf{br}(\ell)\} \\ &= \{(\ell,k) \mid, \ell = k \wedge \mathsf{br}^{-1}(\mathsf{comp}_{\tau}(\mathsf{br}(\ell))) = k\} \\ &= \mathrm{Eq}(\mathsf{Colist}(A \times \mathsf{ITreeF}(A))) \\ &\qquad \wedge ((\mathsf{br}^{-1} \circ \mathsf{comp}_{\tau} \circ \mathsf{br}) \times id)^*(\mathrm{Eq}(\mathsf{Colist}(A \times \mathsf{ITreeF}(A)))) \end{aligned}$$

This yields the obligations over $\mathsf{Colist}(A \times \mathsf{ITreeF}(A))$:

$$\mathrm{Rel}(\nu F)(R) \vdash \mathrm{Eq} \quad \text{and} \quad \mathrm{Rel}(\nu F)(R) \vdash ((\mathsf{br}^{-1} \circ \mathsf{comp}_{\tau} \circ \mathsf{br}) \times id)^*(\mathrm{Eq})$$

which can be discharged by coinduction on colists, based on the fact that equality $\mathrm{Eq}(\mathsf{Colist}(A \times \mathsf{ITreeF}(A))$ on $A \times \mathsf{ITreeF}(A)$ is the terminal $\mathrm{Rel}(F)^{\mathrm{Eq}}$-coalgebra. The first obligation amounts to

$$\mathrm{Rel}(\nu F)(R) \vdash (\mathsf{next} \times \mathsf{next})^*(\mathrm{Rel}(F)(\mathrm{Eq}(A \times \mathsf{ITreeF}(A)), \mathrm{Rel}(\nu F)(R)))$$

which is trivially true as $R(x, y)$ implies $x = y$. The second one can be massaged as follows, for $f = \mathsf{br}^{-1} \circ \mathsf{comp}_\tau \circ \mathsf{br}$.

$$\cfrac{\cfrac{\cfrac{\mathrm{Rel}(\nu F)(R) \vdash (f \times id)^*(\mathrm{Rel}(F)^{\mathrm{Eq}}(\coprod_{f\times id}(\nu\mathrm{Rel}(F)^R)))}{\coprod_{f\times id}(\mathrm{Rel}(\nu F)(R)) \vdash \mathrm{Rel}(F)^{\mathrm{Eq}}(\coprod_{f\times id}(\mathrm{Rel}(\nu F)(R)))}}{\coprod_{f\times id}(\mathrm{Rel}(\nu F)(R)) \vdash \mathrm{Eq}}\ \begin{pmatrix}\text{coinduction}\\ \text{for Colist}\end{pmatrix}}{\mathrm{Rel}(\nu F)(R) \vdash (f \times id)^*(\mathrm{Eq})}$$

The upper sequence of this proof tree remains to be verified. Suppose $(\ell, k) \in \nu\mathrm{Rel}(F)^R = \mathrm{Rel}(\nu F)(R)$ and $\mathsf{next}(\ell) = \mathsf{next}(k) = *$ then

$$\mathsf{next}(\mathsf{br}^{-1}(\mathsf{comp}_\tau(\mathsf{br}(\ell)))) = \mathsf{next}(\mathsf{br}^{-1}(\tau)) = *$$

and therefore (ℓ, k) satisfies the right hand side of the sequence. Suppose now $\mathsf{next}(\ell) = (a, x, \ell')$ and $\mathsf{next}(k) = (b, y, k')$ then $f(\ell) = \mathsf{Colist}(id \times \mathsf{comp}_\tau)(\ell)$ and

$$\mathsf{next}(f(\ell)) = (a, \mathsf{comp}_\tau(x), \mathsf{Colist}(id \times \mathsf{comp}_\tau)(\ell'))$$

But by assumption we have $a = b$, $\mathsf{comp}_\tau(x) = x = y$, and

$$(\mathsf{Colist}(id \times \mathsf{comp}_\tau)(\ell'), k') \in \textstyle\coprod_{f\times id}(\nu\mathrm{Rel}(F)^R)).$$

ensuring that (ℓ, k) satisfies the right hand side of the sequence.

6.3 Finitely branching trees of possibly infinite depth

Composition for $\mathsf{FTreel}(A)$ needs to be defined coinductively, while the required structure map is defined inductively on lists.

$$\begin{array}{l}\mathsf{comp\text{-}struct} : \mathsf{FTreel}(A) \times \mathsf{FTreel}(A) \to \mathsf{List}(A \times (\mathsf{FTreel}(A) \times \mathsf{FTreel}(A)));\\ (x, y) \mapsto \begin{cases}\mathsf{merge\text{-}middle}(\mathsf{br}(y), x) & \text{if } \mathsf{br}(x) = \mathsf{nil}\\ \mathsf{merge\text{-}last}(\mathsf{br}(x), y) & \text{else}\end{cases}\end{array}$$

The functions $\mathsf{merge\text{-}middle} = \mathsf{List}(id \times \mathsf{c}) \circ \theta$ and $\mathsf{merge\text{-}last} = \theta$ make use of the *strength* $\theta : \mathsf{List}(A) \times B \to \mathsf{List}(A \times B)$[5], the commutativity of products $\mathsf{c} : A \times B \to B \times A$, and of the functor properties of the List–datatype. The empty tree τ is obtained by applying the inverse of the destructor br to the empty list nil.

The proof of our equation $\mathsf{comp}(x, \tau) = x$ requires a mixed argument. We start off by coinduction, then we need to switch to induction on lists. The coinduction principle for $\mathsf{FTreel}(A)$ states that the equation holds if there exists a relation R on trees such that $R(\mathsf{comp}(x, \tau), x)$ and

$$R \vdash (\mathsf{br} \times \mathsf{br})^*(\mathrm{Rel}(\mu F)(R)) \quad \text{or equivalently} \quad \textstyle\coprod_{\mathsf{br}\times\mathsf{br}}(R) \vdash \mathrm{Rel}(\mu F)(R).$$

[5] Defined be induction.

The latter obligation lies in the fibre over lists and reads

$$\forall \ell, k : \mathsf{List}(A \times \mathsf{FTreel}(A)), x, y : \mathsf{FTreel}(A).$$
$$R(x,y) \wedge \mathsf{br}(x) = \ell \wedge \mathsf{br}(y) = k \Rightarrow \mathrm{Rel}(\mu F)_R(\ell, k).$$

We take $R = \{(\mathsf{comp}(x,\tau), x) \mid x \in \mathsf{FTreel}(A)\}$ and proceed with induction on the lists ℓ and k. If both ℓ and k are nil then unfolding the fixed point $\mathrm{Rel}(\mu F)(R) = \mu\mathrm{Rel}(F)_R \cong \mathrm{Rel}(F)_R(\mu\mathrm{Rel}(F)_R)$ does the job. If either $\ell = \mathsf{nil}$ and $k = \mathsf{cons}(b, y', k')$ or vice versa then for all pairs (x, y) with $\mathsf{br}(x) = \ell$ $\mathsf{br}(y) = k$ the relation $R(x, y)$ does not hold because this would require $x = \mathsf{comp}(y, \tau)$. But we have $\mathsf{br}(x) = \mathsf{nil}$ and $\mathsf{br}(\mathsf{comp}(y,\tau)) \neq \mathsf{nil}$. For the case $\ell = \mathsf{cons}(a, x', \ell')$ and $k = \mathsf{cons}(b, y', k')$ the induction hypothesis holds for ℓ' and k' Assume $R(x, y)$ for all $\mathsf{br}(x) = \mathsf{cons}(a, x', \ell')$ and $\mathsf{br}(y) = \mathsf{cons}(b, y', k')$. It remains to show that $a = b$, $R(x', y')$, and $\mu\mathrm{Rel}(F)_R(\ell', k')$ but

$$\mathsf{br}(\mathsf{comp}(y,\tau)) = \mathsf{cons}((b, \mathsf{comp}(y',\tau)), \mathsf{List}(id \times \mathsf{comp})(\mathsf{merge\text{-}last}(k', \tau)))$$

which shows the first two cases. For $\mu\mathrm{Rel}(F)_R(\ell', k')$, we use the induction hypothesis and have to show, that there is a pair (x, y) in R such that $\mathsf{br}(x) = \ell'$ and $\mathsf{br}(y) = k'$. The inverse of br applied to ℓ' and k' does the job as

$$\ell' = \mathsf{tl}(\mathsf{br}(x)) = \mathsf{tl}(\mathsf{br}(\mathsf{comp}(y,\tau))) = \mathsf{List}(id \times \mathsf{comp})(\mathsf{merge\text{-}last}(k', \tau))$$

and therefore $\ell' = \mathsf{br}(\mathsf{comp}(\mathsf{br}^{-1}(k'), \tau))$.

6.4 Infinitely branching—infinite depth

The composition for the datatype of possibly infinite trees with possibly infinite branching is defined coinductively.

$$\mathsf{comp\text{-}struct}_t : \mathsf{ITreel}(A) \times \mathsf{ITreel}(A) \to \mathsf{Colist}(A \times (\mathsf{ITreel}(A) \times \mathsf{ITreel}(A)));$$
$$(x, y) \mapsto \begin{cases} \mathsf{merge\text{-}middle}(\mathsf{br}(y), x) & \text{if } \mathsf{next}(\mathsf{br}(x)) = * \\ \mathsf{merge\text{-}last}(\mathsf{br}(x), y) & \text{else} \end{cases}$$

The functions merge-middle and merge-last resemble those for the finitely branching case, but make use of the strength and functorial properties of Colist (which are of coinductive nature). The required composition is then defined as $\mathsf{comp} = \mathsf{coreduce}(\mathsf{comp\text{-}struct})$. The empty tree is defined via the inverse of br and the empty colist ϵ, namely $\tau = \mathsf{br}^{-1}(\epsilon)$. In order to prove $\mathsf{comp}(x, \tau) = x$ it suffices to show that $R = \{(\mathsf{comp}(x,\tau), x) \mid, x \in \mathsf{ITreel}(A)\} \subseteq \mathsf{ITreel}(A) \times \mathsf{ITreel}(A)$ is a bisimulation *w.r.t.* the lifted datatype functor, *i.e.*

$$\frac{R \vdash (\mathsf{br} \times \mathsf{br})^* \mathrm{Rel}(\nu F)(R)}{R \vdash \mathrm{Eq}}$$

The relational lifting of the colist functor $\mathrm{Rel}(\nu F)(R)$ is by definition the fixed point $\nu\mathrm{Rel}(F)^R$. Therefore

$$\frac{\dfrac{\dfrac{R \vdash (\mathsf{br}\times\mathsf{br})^*(\mathrm{Rel}(F)^R(\coprod_{\mathsf{br}\times\mathsf{br}}(R)))}{\coprod_{\mathsf{br}\times\mathsf{br}}(R) \vdash \mathrm{Rel}(F)^R(\coprod_{\mathsf{br}\times\mathsf{br}}(R))}}{\coprod_{\mathsf{br}\times\mathsf{br}}(R) \vdash \nu\mathrm{Rel}(F)^R}}{R \vdash (\mathsf{br}\times\mathsf{br})^*(\nu\mathrm{Rel}(F)^R)}$$

by coinduction. The upper line spells out as:

$$R \vdash \left\{ (x,y) \middle| \begin{array}{l} \mathsf{next}(\mathsf{br}(x)) = \mathsf{next}(\mathsf{br}(y)) = * \vee \\ \mathsf{next}(\mathsf{br}(x)) = (a,t,\ell) \wedge \mathsf{next}(\mathsf{br}(y)) = (b,s,k) \\ \quad \wedge\, a = b \wedge R(t,s) \wedge \coprod_{\mathsf{br}\times\mathsf{br}}(R)(\ell,k) \end{array} \right\}$$

Suppose now $R(x,y)$, *i.e.* $x = \mathsf{comp}(y,\tau)$ and, for the upper case, $\mathsf{next}(\mathsf{br}(y)) = *$ (y is the empty tree). Then

$$\begin{aligned} \mathsf{next}(\mathsf{br}(\mathsf{comp}(y.\tau))) &= \mathsf{next}(\mathsf{Colist}(id \times \mathsf{comp})(\mathsf{comp\text{-}struct}(y,\tau))) \\ &= \mathsf{next}(\mathsf{Colist}(id \times \mathsf{comp})(\epsilon)) \\ &= * \end{aligned}$$

For the lower case assume that $\mathsf{next}(\mathsf{br}(y)) = (b,s,k)$. Then, by applying the definitions,

$$\mathsf{next}(\mathsf{br}(\mathsf{comp}(y,\tau))) = (b, \mathsf{comp}(s,\tau), \mathsf{Colist}(id \times \mathsf{comp})(\mathsf{merge\text{-}last}(k,\tau)))$$

As, by assumption, $R(\mathsf{comp}(s,\tau), s)$ is valid, it remains to prove the last clause, namely the membership

$$\begin{aligned} &(\mathsf{Colist}(id \times \mathsf{comp})(\mathsf{merge\text{-}last}(k,\tau)), k) \\ &\quad \in \{(\ell,k) \mid, \exists(x,y).R(x,y) \wedge \mathsf{br}(x) = \ell \wedge \mathsf{br}(y) = k\}. \end{aligned}$$

So far we simply rewrote our definitions, the last obligation, however, requires a suitable instantiation of the existential quantifier. In our simple case there exists only the instantiation via the inverse of br, which yields the obligation

$$R(\mathsf{br}^{-1}(\mathsf{Colist}(id \times \mathsf{comp})(\mathsf{merge\text{-}last}(k,\tau))), \mathsf{br}^{-1}(k))$$

that is

$$\mathsf{Colist}(id \times \mathsf{comp})(\mathsf{merge\text{-}last}(k,\tau)) = \mathsf{br}(\mathsf{comp}(\mathsf{br}^{-1}(k),\tau))$$

which is shown by rewriting the right hand side.

7 Conclusion

We have shown how to obtain proof principles for datatypes with iterated recursion from a categorical analysis, and how to apply these principles using a modern proof tool.

References

1. P. Aczel. *Non-well-founded sets.* CSLI Lecture Notes 14, Stanford, 1988.
2. R.C. Backhouse, P. Chisholm, G. Malcolm, and E. Saaman. Do-it-yourself type theory. *Formal Aspects of Comp.*, 1:19–84, 1989.
3. F. Borceux. *Handbook of Categorical Algebra*, volume 50, 51 and 52 of *Encyclopedia of Mathematics.* Cambridge Univ. Press, 1994.
4. J.R.B. Cockett and T. Fukushima. About charity. Technical Report 92/480/18, Dep. Comp. Sci., Univ. Calgary, 1992.
5. J.R.B. Cockett and D. Spencer. Strong categorical datatypes II: A term logic for categorical programming. *Theor. Comp. Sci.*, 139:69–113, 1995.
6. Th. Coquand and Ch. Paulin. Inductively defined types. In P. Martin-Löf and G. Mints, editors, *COLOG 88 International conference on computer logic*, number 417 in Lect. Notes Comp. Sci., pages 50–66. Springer, Berlin, 1988.
7. P. Dybjer. Inductive families. *Formal Aspects of Comp.*, 6:440–465, 1994.
8. M.P. Fiore. A coinduction principle for recursive data types based on bisimulation. *Inf. & Comp.*, 127(2):186–198, 1996.
9. E. Giménez. Implementation of co-inductive types in Coq: an experiment with the Alternating Bit Protocol. In S. Berardi and M. Coppo, editors, *Types for Proofs and Programs*, number 1158 in Lect. Notes Comp. Sci., pages 135–152. Springer, Berlin, 1996.
10. J.A. Goguen, J. Thatcher, and E. Wagner. An initial algebra approach to the specification, correctness and implementation of abstract data types. In R. Yeh, editor, *Current Trends in Programming Methodology*, pages 80–149. Prentice Hall, 1978.
11. U. Hensel and B. Jacobs. Coalgebraic theories of sequences in PVS. Techn. Rep. CSI-R9708, Comput. Sci. Inst., Univ. of Nijmegen, 1997.
12. U. Hensel and D. Spooner. A view on implementing processes: Categories of circuits. In M. Haveraaen, O. Owe, and O.-J. Dahl, editors, *Recent Trends in Data Type Specification*, number 1130 in Lect. Notes Comp. Sci., pages 237–254. Springer, Berlin, 1996.
13. C. Hermida. Some properties of Fib as a fibred 2-category. *Journ. Pure & Appl. Algebra*, 1997, to appear.
14. C. Hermida and B. Jacobs. An algebraic view of structural induction. In L. Pacholski and J. Tiuryn, editors, *Computer Science Logic 1994*, number 933 in Lect. Notes Comp. Sci., pages 412–426. Springer, Berlin, 1995.
15. C. Hermida and B. Jacobs. Structural induction and coinduction in a fibrational setting. Full version of [14], 1996.
16. B. Jacobs. Parameters and parametrization in specification using distributive categories. *Fund. Informaticae*, 24(3):209–250, 1995.
17. B. Jacobs. Invariants, bisimulations and the correctness of coalgebraic refinements. Techn. Rep. CSI-R9704, Comput. Sci. Inst., Univ. of Nijmegen, 1997.
18. B. Jacobs. *Categorical Logic and Type Theory.* 1998, to appear.

19. B. Jay. Data categories. In M.E. Houle and P.Eades, editors, *Computing: The Australasian Theory Symposium Proceedings*, number 18 in Australian Comp. Sci. Comm., pages 21–28, 1996.
20. M. Makkai. The fibrational formulation of intuitionistic predicate logic I: completeness according to Gödel, Kripke, and Läuchli. Part 1. *Notre Dame Journ. Formal Log.*, 34(3):334–377, 1993.
21. M. Makkai. The fibrational formulation of intuitionistic predicate logic I: completeness according to Gödel, Kripke, and Läuchli. Part 2. *Notre Dame Journ. Formal Log.*, 34(4):471–499, 1993.
22. N.P. Mendler. Inductive types and type constraints in second-order lambda calculus. *Ann. Pure & Appl. Logic*, 51(1/2):159–172, 1991.
23. E. Moggi. Notions of computation and monads. *Inf. & Comp.*, 93(1):55–92, 1991.
24. S. Owre, S. Rajan, J.M. Rushby, N. Shankar, and M. Srivas. PVS: Combining specification, proof checking, and model checking. In R. Alur and T.A. Henzinger, editors, *Computer Aided Verification*, number 1102 in Lect. Notes Comp. Sci., pages 411–414. Springer, Berlin, 1996.
25. S. Owre, J.M. Rushby, N. Shankar, and F. von Henke. Formal verification for fault-tolerant architectures: Prolegomena to the design of PVS. *IEEE Trans. on Softw. Eng.*, 21(2):107–125, 1995.
26. Ch. Paulin-Mohring. Inductive definitions in the system Coq. Rules and properties. In M. Bezem and J.F. Groote, editors, *Typed Lambda Calculi and Applications*, number 664 in Lect. Notes Comp. Sci., pages 328–345. Springer, Berlin, 1993.
27. L.C. Paulson. Mechanizing coinduction and corecursion in higher-order logic. *Journ. of Logic and Computation*, 7:175–204, 1997.
28. A.M. Pitts. A co-induction principle for recursively defined domains. *Theor. Comp. Sci.*, 124(2):195–219, 1994.
29. J. Rutten and D. Turi. Initial algebra and final coalgebra semantics for concurrency. In J.W. de Bakker, W.P. de Roever, and G. Rozenberg, editors, *A Decade of Concurrency*, number 803 in Lect. Notes Comp. Sci., pages 530–582. Springer, Berlin, 1994.
30. J.J.M.M. Rutten. Universal coalgebra: a theory of systems. CWI Report CS-R9652, 1996.

When Do Datatypes Commute?

Paul Hoogendijk and Roland Backhouse

Department of Mathematics and Computing Science, Eindhoven University of Technology.
email: [paulh, rolandb] @win.tue.nl

Abstract. *Polytypic* programs are programs that are parameterised by type constructors (like List), unlike *polymorphic* programs which are parameterised by types (like Int). In this paper we formulate precisely the polytypic programming problem of "commuting" two datatypes. The precise formulation involves a notion of higher order naturality. We demonstrate via a number of examples the relevance and interest of the problem, and we show that all "regular datatypes" (tree-like datatypes that one can define in a functional programming language) do indeed commute according to our specification. The framework we use is the theory of allegories, a combination of category theory with the point-free relation calculus.

1 Polytypism

One of the most significant contributions to (re)usability of programs has been the notion of parametric polymorphism — first introduced by Strachey and later incorporated in the language ML by Milner [15]. In this paper we consider a problem that entails a higher level of parametricity than can normally be expressed by polymorphism. The problem is roughly stated in the title of the paper — "when do two datatypes commute?" — and an illustrative instance of two commuting dataypes is provided by the fact that a list of trees all of the same shape can always be transformed without loss of information to a tree of lists all of the same length.

The paper has three goals. First, we want to show that the problem is relevant and interesting. Second, we want to formulate the problem precisely and concisely. Third, we want to use this problem as a primer to a theory of higher order polymorphism that we are endeavouring to develop. It is not the purpose of this paper to provide a technical justification for all results claimed in the paper. A complete technical justification is given by Hoogendijk [7], refining earlier work of Backhouse, Doornbos and Hoogendijk [1].

Our commuting datatypes problem is an instance of what has recently been dubbed "polytypic programming" [9,13]. "Polytypic" programs distinguish themselves from polymorphic programs in that the parameter is a datatype like "list" or "tree" —a function from types to types— rather than a type like "integer", "list of integer" or "tree of string".

The emergence of polytypism as a viable research field has occurred gradually over a number of years. A landmark for us was the formulation by Malcolm [12] of a theorem expressing when two computations could be fused into one computation. Malcolm's fusion theorem was polytypic in that it was parameterised by a datatype and so could be instantiated in a variety of ways. Malcolm exploited the —polytypic— notion of a "catamorphism" and introduced the "banana bracket" notation which was popularised and extended to the —polytypic— notions of "anamorphisms" and "hylomorphisms" by Fokkinga, Meijer and Paterson [14]. (Malcolm referred to "promotion" rather than "fusion", that being the terminology used by Bird [2] at the time in his theory of lists.) Since then the theme of polytypism has been explored in a variety of ways. Several authors [3,9,13] have explored polytypic generalisations of existing programming problems, Doornbos [5] has developed a polytypic theory of program termination and the recently published book by Bird and De Moor [4] contains a wealth of material in which parameterisation by a datatype plays a central role.

Functional programmers have a well developed intuitive understanding of what it means for a function to be polymorphic. Being able to experiment with the notion by writing and executing polymorphic programs is clearly enormously beneficial to understanding. Nevertheless, an unequivocal formal semantics of "parametric polymorphism" is still an active area of research [11]. The situation with polytypism is much worse: the term is vague and probably understood in different ways by different authors. Moreover, experimental implementations of polytypism in functional programming languages [10] are only just beginning to get off the ground. The emphasis at this point in time is in showing the ubiquity of polytypism; a drawback is the ad hoc nature of some developments. To give one simple example: the "size" function for a datatype is often cited as a polytypic generalisation of the length of a list. But what is the appropriate notion of "size" for a tree — the number of nodes or, perhaps, the depth of the tree? Without a theoretical understanding of the notion of polytypism it is difficult to provide convincing arguments for one or the other choice.

This paper contributes to the theoretical foundations of polytypism, albeit tentatively. We draw inspiration from Reynolds' [18] and Plotkin's [17] seminal accounts of the semantics of parametric polymorphism. Roughly speaking, Reynolds and Plotkin showed that any parametrically polymorphic function satisfies a certain (di)naturality property that is derivable from the type of the function via so-called "logical relations". We turn this around and define the notion of commuting datatypes by requiring that a certain higher-order naturality property be satisfied. The framework we use for formalising such properties is the theory of allegories [6], a combination of category theory with the point-free relation calculus.

In the interests of greater understanding we approach the central topic of the paper slowly and deliberately. First we need to agree on what a datatype is. For this purpose we briefly summarise Hoogendijk and De Moor's [8] arguments. The next step is to present several illustrations of "commuting datatypes". One of these is a concrete example, concerning the transposition of matrices represented as lists of list, which we learnt from D.J. Lillie. A second is more abstract: we argue that Moggi's [16] notion of "strong" functor is an instance of the phenomenon "commuting datatypes". Armed with these examples we are able to proceed to a precise formalisation of the notion.

2 Allegories and Datatypes

A brief summary of this section is that our notion of a "datatype" is a "relator with membership" [8] and an appropriate framework for developing a theory of datatypes is the theory of allegories [6].

2.1 Parametric Polymorphism

To motivate these choices let us begin by giving a brief summary of Reynolds' [18] account of parametric polymorphism. (See also [11].) Suppose we have a polymorphic function f of type $T\alpha$ for all types α. That is, for each type A there is an instance f_A of type TA. Then parametricity of the polymorphism means that for any relation R of type $A \leftarrow B$ there is a relation TR of type $TA \leftarrow TB$ such that $(f_A, f_B) \in TR$.

In order to make the notion of parametricity completely precise, one has to be able to extend each type constructor T in the chosen programming language to a function $R \mapsto TR$ from relations to relations. The type requirement on this extension is that if $R : A \leftarrow B$ then $TR : TA \leftarrow TB$. This type requirement has of course exactly the same form as the type requirement on a functor and it has been known for a long time that datatypes are indeed functors. But just being a functor is probably much too weak a requirement to capture the notion of a datatype. Moreover, it seems to be difficult or clumsy to express non-deterministic properties in a strict categorical setting. An appropriate step to take, therefore, is to allegory theory [6] and the requirement that datatypes be "relators".

2.2 Allegories and Relators

Allegories An *allegory* is a category with additional structure, the additional structure capturing the most essential characteristics of relations. The additional axioms are as follows. First of all, arrows of the same type are ordered by the *partial order* $\subseteq$ and composition is monotonic with respect to this order. Secondly, for every pair of arrows $R, S : A \leftarrow B$, their *intersection* (*meet*) $R \cap S$ exists. Finally, for each arrow $R : A \leftarrow B$ its *converse* $R^{\cup} : B \leftarrow A$ exists.

The standard example of an allegory is Rel, the allegory with sets as objects and relations as arrows. With this allegory in mind, we refer henceforth to the arrows of an allegory as "relations".

Relators A *relator* is a monotonic functor that commutes with converse.

Two examples of relators have already been given. List is a unary relator, and product is a binary relator.

Division and Tabulation The allegory Rel has more structure than we have captured so far with our axioms. For instance, in Rel we can take arbitrary unions (joins) of relations. There are also two "division" operators, and Rel is "tabulated". In full, Rel is a unitary, tabulated, locally complete, division allegory. For full discussion of these concepts see [6] or [4]. Here we briefly summarise the relevant definitions.

We say that an allegory is *locally complete* if for each set $\mathcal{S}$ of relations of type $A \leftarrow B$, the union $\cup\mathcal{S} : A \leftarrow B$ exists and, furthermore, intersection and composition distribute over arbitrary unions. We use the notation $\perp\!\!\perp_{A,B}$ for the smallest relation of type $A \leftarrow B$ and $\top\!\!\top_{A,B}$ for the largest relation of the same type.

The existence of a largest relation for each pair of objects A and B is guaranteed by the existence of a "unit" object, denoted by 1. We say that object 1 is a *unit* if id_1 is the largest relation of its type and for every object A there exists a total relation $!_A : 1 \leftarrow A$. If an allegory has a unit then it is said to be *unitary*.

The most crucial consequence of the distributivity of composition over union is the existence of two so-called *division* operators "\" and "/". The interpretation of the factor $R\backslash T$ is

$$(b,c) \in R\backslash T \ \equiv\ \forall(a : (a,b) \in R : (a,c) \in T)\ .$$

The final characteristic of Rel is that it is "tabular". Formally, we say that an object C and a pair of functions $f : A \leftarrow C$ and $g : B \leftarrow C$ is a *tabulation* of relation $R : A \leftarrow B$ if

$$R = f \circ g^{\cup} \ \wedge\ f^{\cup} \circ f \cap g^{\cup} \circ g = id_C\ .$$

An allegory is said to be *tabular* if every relation has a tabulation.

If allegory $\mathcal{B}$ is tabular, a functor is monotonic iff it commutes with converse [4]. For this reason Bird and De Moor [4] define a relator to be a monotonic functor.

Domains In addition to the source and target of a relation it is useful to know their domain and range. The *domain* of a relation $R : A \leftarrow B$ is that subset $R>$ of id_B consisting of all y such that $(x,y) \in R$ for some x. The *range* of $R : A \leftarrow B$, which we denote by $R<$, is the domain of $R^{\cup}$. We use the names "domain" and "range" because we usually interpret relations as transforming "input" y on the right to "output" x on the left.

Pointwise Closed Classes of Relators We have already mentioned a few examples of relators. Of these, only product is primitive; the others are composite. In general, our concern is with establishing that certain *classes* of relators are commuting. That is, every pair of relators in the class commutes with each other. A requirement is that a class be sufficiently rich in the sense that it is closed under a number of composition operators. The composition operators that we consider indispensable are functional composition and tupling.

Little needs to be said about functional composition at this moment. It is easy to verify that the functional composition of two relators $F : \mathcal{A} \leftarrow \mathcal{B}$ and $G : \mathcal{B} \leftarrow \mathcal{C}$, which we denote by FG, is a relator. There is also an identity relator for each allegory $\mathcal{A}$, which we denote by Id leaving the specific allegory to be inferred from the context.

For our purposes, however, we need a variable-free mechanism for composing relators. This is achieved by making the arity of a relator explicit and introducing mechanisms for tupling and projection.

We consider a collection of allegories created by closing some base allegory $\mathcal{C}$ under the formation of finite cartesian products. An allegory in the collection is thus $\mathcal{C}^k$ where k, the *arity* of the allegory is either a natural number or $l*m$ where l is an arity and m is a number. Note that we identify $1*k$ and $k*1$ with k. The *arity* of a relator F is $k \leftarrow l$ if the target of F is $\mathcal{C}^k$ and its source is $\mathcal{C}^l$. We write $F : k \leftarrow l$ rather than the strictly correct $F : \mathcal{C}^k \leftarrow \mathcal{C}^l$. A relator with arity $1 \leftarrow 1$ is called an *endorelator* and a relator with arity $1 \leftarrow k$ for some k is called *single-valued.*

Given a number k and a number of relators F_i ($0 \leq i < k$) all of the same arity $l \leftarrow m$, the relators can be tupled in the obvious way to form a relator of arity $l*k \leftarrow m$. We denote the tupled relator by $\Delta(i{:}0 \leq i < k{:}\, F_i)$. Some variations on this notation are used. First, we often use F_k to abbreviate the mapping $(i{:}0 \leq i < k{:}\, F_i)$ in a tuple expression. That is, we abbreviate $\Delta(i{:}0 \leq i < k{:}\, F_i)$ to ΔF_k. Second, we sometimes use Δ as an infix operator —reduced slightly in size to avoid ambiguity—; thus, $F \vartriangle G$ is the relator that maps relation R to the pair of relations (FR, GR). Thirdly, when all the relators are equal to one and the same relator F we write simply ΔF ; this is the relator that given relation R makes k copies of FR to create a vector of length k. Finally, there are times when we need to make the implicit parameter k explicit. In such cases we add it as a subscript to Δ. In particular, we most often write $\Delta_k F$ in order to indicate clearly the amount of duplication of F.

Complementary to tupling is projection. For each number k and for each i, $0 \leq i < k$, we can define the relator $Proj_i$ that maps a k-tuple of relations R_0, $\ldots$, R_{k-1} to R_i . In the case that k is 2 we use the special notation Outl and Outr for the two projections. Note that the identity relator is a special case of a projection (the case $k = 1$).

Using tupling and projection we can define several other operations. The operation $\_^k$ can, of course, be extended to a functor. If F has arity $l \leftarrow m$ then $F^k \triangleq \Delta(i{:}0 \leq i < k{:}\, F Proj_i)$ has arity $l*k \leftarrow m*k$. Another relator transposes $l*k$ into $k*l$ (for integer arities l and k). We denote this relator by τ —irrespective of the dimensions l and k, relying on the context to determine what its dimensions are— . The definition of $\tau : k*l \leftarrow l*k$ is $\Delta_k \Delta_l(Proj_l Proj_k)$; it is the unique mapping such that for all matrices of single-valued relators $F_{i,j}$, where $0 \leq i < k$ and $0 \leq j < l$, one has $\tau(\Delta_k \Delta_l(F_{k,l})) = \Delta_l \Delta_k(F_{k,l})$. By composing $\_^k$ and τ we get a functor dual to $\_^k$; specifically, we define ${}^k F$ by ${}^k F = \tau F^k \tau$. Thus, for $F : l \leftarrow m$ we have $F^k : l*k \leftarrow m*k$ and ${}^k F : k*l \leftarrow k*m$.

Projection and tupling are connected by the law

$$H = \Delta F_k \equiv \forall(i{:}\, 0 \leq i < k{:}\, Proj_i H = F_i) \quad , \tag{1}$$

for all H and F.

Definition 1 (Pointwise Closed). A collection of relators is said to be *pointwise closed* with *base allegory* $\mathcal{C}$ if each relator in the collection has type $\mathcal{C}^k \leftarrow \mathcal{C}^l$ for some arities k and l, and the collection includes all projections and is closed under functional composition and tupling. □

Regular Relators The "regular relators" are those relators constructed from three primitive (classes of) relators by pointwise closure and induction.
For each object A in an allegory there is a relator K_A defined by $K_A R = id_A$. Such relators are called *constant* relators.
A *coproduct* of two objects consists of an object and two injection relations. The object is denoted by A+B and the two relations by $inl_{A,B} : A+B \leftarrow A$ and $inr_{A,B} : A+B \leftarrow B$. Having the functions inl and inr, we can define the *junc* operator: for all $R : C \leftarrow A$ and $S : C \leftarrow B$,

$$R \triangledown S \triangleq R \circ inl^{\cup}_{A,B} \cup S \circ inr^{\cup}_{A,B} \quad , \tag{2}$$

and the *coproduct relator*: for all $R : C \leftarrow A$ and $S : D \leftarrow B$

$$R+S \triangleq (inl_{C,D} \circ R) \triangledown (inr_{C,D} \circ S) \ .$$

A *product* of two objects consists of an object and two projection arrows. The object is denoted by A×B and the two arrows by $outl_{A,B} : A \leftarrow A \times B$ and $outr_{A,B} : B \leftarrow A \times B$. Having the projection functions outl and outr, we can define the *split* operator on relations: for all $R : A \leftarrow C$ and $S : B \leftarrow C$

$$R \vartriangle S \triangleq outl^{\cup}_{A,B} \circ R \cap outr^{\cup}_{A,B} \circ S \quad , \tag{3}$$

and the *product relator*: for all for $R : C \leftarrow A$ and $S : D \leftarrow B$,

$$R \times S \triangleq (R \circ outl_{A,B}) \vartriangle (S \circ outr_{A,B}) \ .$$

Tree relators are defined as follows. Suppose that relation $in : A \leftarrow FA$ is an initial F-algebra. That is to say, suppose that for each relation $R : B \leftarrow FB$ there exists a unique F–homomorphism to R from in. We denote this unique homomorphism by $(\![F;R]\!)$. Formally, $(\![F;R]\!)$ and in are characterized by the universal property that, for each relation $X : B \leftarrow A$ and each relation $R : B \leftarrow FB$,

$$X = (\![F;R]\!) \ \equiv \ X \circ in = R \circ FX \ . \tag{4}$$

Now, let ⊗ be a binary relator and assume that, for each A, $in_A : TA \leftarrow A \otimes TA$ is an initial algebra of $(A\otimes)$[1]. Then the mapping T defined by, for all $R : A \leftarrow B$,

$$TR = (\![A\otimes ; in_B \circ R \otimes id_{TB}]\!)$$

is a relator, *the tree relator induced by* ⊗.
Note that the function space constructor is not included in the class of regular relators.

[1] Here and elsewhere we use the section notation (A⊗) for the relator $\otimes(K_A \vartriangle Id)$.

2.3 Natural Transformations

Reynolds' characterisation of parametric polymorphism predicts that certain polymorphic functions are natural transformations. To see this it helps to re-express Reynolds' pointwise definition of the $\leftarrow$ operator in the following point-free form:

$$(f,g) \in R \leftarrow S \equiv f \circ S \subseteq R \circ g \quad .$$

Now consider, for example, the reverse function on lists, denoted here by rev. This has polymorphic type $\mathsf{List}A \leftarrow \mathsf{List}A$ for all A and so $(\mathsf{rev},\mathsf{rev}) \in \mathsf{List}R \leftarrow \mathsf{List}R$ for all relations R. That is, $\mathsf{rev} \circ \mathsf{List}R \subseteq \mathsf{List}R \circ \mathsf{rev}$ for all relations R. Similarly the function that makes a pair out of a single value, here denoted by fork, has type $A \times A \leftarrow A$ for all A, and so is predicted to satisfy the property $\mathsf{fork} \circ R \subseteq R \times R \circ \mathsf{fork}$ for all relations R.

The above properties of rev and fork are not natural transformation properties because they assert an inclusion and not an equality; they are sometimes called "lax" natural transformation properties. It so happens that the inclusion in the case of rev can be strengthened to an equality but this is certainly not the case for fork. This leads to three distinct notions of natural transformation. A collection of relations α indexed by objects (equivalently, a mapping α of objects to relations) is said to be *a natural transformation of type* $F \leftarrow G$, for relators F and G, iff $FR \circ \alpha_B = \alpha_A \circ GR$ for each $R : A \leftarrow B$. A collection of relations α is a natural transformation of type $F \hookleftarrow G$ iff $FR \circ \alpha_B \supseteq \alpha_A \circ GR$ for each $R : A \leftarrow B$; also, α is a natural transformations of type $F \hookrightarrow G$ iff $\alpha^{\cup} : G \hookleftarrow F$.

Since natural transformations of type $F \hookleftarrow G$ are the more common ones and, in fact agree with the categorical notion of natural transformation in the case that they are functions, we say that α is a *natural transformation* if $\alpha : F \hookleftarrow G$ and we say that α is a *proper* natural transformation if $\alpha : F \leftarrow G$.

The natural transformations studied in the computing science literature are predominantly (collections of) functions. In contrast, the natural transformations discussed in this paper are almost all non-functional either because they are partial or because they are non-deterministic (or both).

2.4 Membership and Fans

Since our goal is to use naturality properties to specify relations it is useful to be able to interpret what it means to be "natural". All interpretations of naturality that we know of assume either implicitly or explicitly that a datatype is a way of structuring information and, thus, that one can always talk about the information stored in an instance of the datatype. A natural transformation is then interpreted as a transformation of one type of structure to another type of structure that rearranges the stored information in some way but does no actual computations on the stored information. Doing no computations on the stored information guarantees that the transformation is independent of the stored information and thus also of the representation used when storing the information. Hoogendijk and De Moor have made this precise [8]. Their argument, briefly summarised here, is based on the thesis that a datatype is a relator with a membership relation.

Given a relator F, Hoogendijk and De Moor give an abstract definition of a *membership relation* mem for F. The interpretation of mem_A is the relation holding between a value and an F-structure whenever the value is stored in the F-structure.

The membership, mem, of relator $F : k \leftarrow l$ is a natural transformation. In the case of an endorelator, $\mathsf{mem} : \mathsf{Id} \hookleftarrow F$; in the general case of a relator of arity $k \leftarrow l$, the membership relation is a collection of arrows of arity $k * l \leftarrow l$ and type $(\Delta_k)^l \hookleftarrow \Delta_l F$. Hoogendijk and De Moor prove that mem and $\cap(\mathsf{mem} \setminus ((\Delta_k)^l)\mathsf{id})\Delta_l$ are the *largest* natural transformations of their types[2]. Moreover, supposing F and G are relators with memberships mem.F and mem.G respectively, they show that the largest natural transformation of type $F \hookleftarrow G$ is $\cap(\mathsf{mem.F} \setminus \mathsf{mem.G})$. (We refer the reader to [8] for proofs of all these properties in the case of endorelators, and to [7] in the general case.)

The insight that these properties give is that natural transformations between datatypes can only rearrange values; computation on the stored values or invention of new values is prohibited. A proper natural transformation $\alpha : F \leftarrow G$ has types $F \hookleftarrow G$ and $F \hookrightarrow G$. Consequently, a proper natural transformation copies values without loss or duplication.

The natural transformation $\cap(\mathsf{mem} \setminus ((\Delta_k)^l)\mathsf{id})\Delta_l$, which is the largest natural transformation of type $F\Delta_l \hookleftarrow \Delta_k$, is called the *canonical fan* of F. It transforms an arbitrary value into an F-structure by non-deterministically creating an F-structure and then copying the given value at all places in the structure. It plays a crucial role in the sequel. Rules for computing the canonical fan for all regular relators are as follows. (These are used later in the construction of "zips".)

$$\begin{aligned} \mathsf{fan.Proj} &= \mathsf{id} & \mathsf{fan.}\Delta F_k &= \Delta(\mathsf{fan.}F_k) \\ \mathsf{fan.}K_A &= \top\!\!\top_{A,-} & \mathsf{fan.FG} &= F(\mathsf{fan.G}) \circ \mathsf{fan.F} \\ \mathsf{fan.+} &= (\mathsf{id} \triangledown \mathsf{id})^{\cup} & \mathsf{fan.}\times &= \mathsf{id} \vartriangle \mathsf{id} \\ \mathsf{fan.T} &= (\![\mathsf{id}\otimes\, ; (\mathsf{fan.}\otimes)^{\cup}]\!)^{\cup} & & \text{(where T is the tree relator induced by } \otimes\text{).} \end{aligned}$$

3 Commuting Datatypes: Examples

In this section we want to argue that the notion that two datatypes "commute" is a common occurrence.

The best known example of a commutativity property is the fact that two lists of the same length can be mapped into single list of pairs. The function that performs this operation is known as the "zip" function to functional programmers. Zip commutes a pair of lists into a list of pairs.

---

[2] $\cap S$ denotes the intersection of the l elements of the vector of relations S. Division in a product allegory is componentwise division in the base allegory.

Another example is the function that commutes m lists each of length n into n lists each of length m. This function is also well known under the name *matrix transposition*. There is also a function that "broadcasts" a value to all elements of a list —thus

$$(a, [b_1, b_2, \ldots]) \mapsto [(a, b_1), (a, b_2), \ldots]$$

— . That is, the datatype an element of type A paired with (a list of elements of type B) is "commuted" to a list of (element of type A paired with an element of type B). More precisely, for each A, the family of broadcasts indexed by B is a natural transformation of type $\mathsf{List}(A\times) \hookleftarrow (A\times)\mathsf{List}$; the two datatypes being "commuted" are thus $(A\times)$ and List. This list broadcast is itself an instance of a subfamily of the operations that we discuss later. In general, a *broadcast* operation copies a given value to all locations in a given data structure.

A final example of a generalised zip would be the (polymorphic) operation that maps values of type $(A+B)\times(C+D)$ to values of type $(A\times C)+(B\times D)$, i.e. commutes a product of disjoint sums to a disjoint sum of products. A necessary restriction is that the elements of the input pair of values have the same "shape", i.e. both be in the left component of the disjoint sum or both be in the right component.

In general then, a *zip* operation transforms F-structures of G-structures to G-structures of F-structures. Typically, "zips" are partial since they are only well-defined on structures of the same shape. As we shall see, they may also be non-deterministic; that is, a "zip" is a relation that need not be a function. Finally, the arity of the two datatypes, F and G, need not be the same; for example, the classical zip function maps pairs of lists to lists of pairs, and pairing has arity $1 \leftarrow 2$ whereas list formation has arity $1 \leftarrow 1$.

3.1 Structure Multiplication

A good example of the beauty of the "zip" generalisation is afforded by what we shall call "structure multiplication". (This example we owe to D.J. Lillie [private communication, December 1994].) A simple, concrete example of structure multiplication is the following. Given two lists $[a_1, a_2, \ldots]$ and $[b_1, b_2, \ldots]$ form a matrix in which the (i,j)th element is the pair (a_i, b_j). We call this "structure multiplication" because the input type is the product $\mathsf{List}A \times \mathsf{List}B$ for some types A and B. It is also called cross-product.

The point we want to make is that there is an obvious generalisation of this procedure: replace $\mathsf{List}A$ by FA and $\mathsf{List}B$ by GB for some arbitrary relators F and G. Doing so leads to the realisation that every step involves a "zip" operation (i.e. commuting the order of a pair of datatypes). This is made explicit in the diagram below.

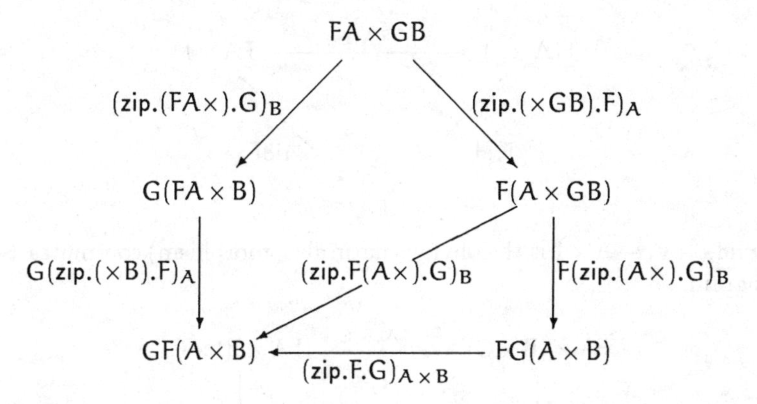

In order to make evident which datatypes are being "commuted" at each step, each arrow has been labelled by an expression involving a "zip" term. A "zip" takes the form zip.F.G for some datatypes F and G. In the absence of a formal specification (to be given later) one should interpret zip.F.G as a family of relations indexed by types such that $(\mathsf{zip.F.G})_A : \mathsf{GFA} \leftarrow \mathsf{FGA}$.

An additional edge has been added to the diagram to show the usefulness of generalising the notion of commutativity beyond just broadcasting; this additional inner edge shows how the commutativity of the diagram can be decomposed into smaller parts. Specifically, in order to show that the whole diagram commutes (in the standard categorical sense of commuting diagram) it suffices to show that two smaller diagrams commute. Specifically, the following two equalities must be established:

$$(\mathsf{zip.F(A\times).G})_B \;=\; (\mathsf{zip.F.G})_{A\times B} \circ \mathsf{F}(\mathsf{zip.(A\times).G})_B \tag{5}$$

$$(\mathsf{zip.F(A\times).G})_B \circ (\mathsf{zip.(\times GB).F})_A \;=\; \mathsf{G}(\mathsf{zip.(\times B).F})_A \circ (\mathsf{zip.(FA\times).G})_B \tag{6}$$

We shall in fact design our definition of "commuting datatypes" in such a way that these two equations are satisfied (almost) by definition. In other words, our notion of "commuting datatypes" is such that the commutativity of the above diagram is automatically guaranteed.

3.2 Strength

Several scientists have argued that the notion of functor is too general to capture the notion of a datatype as understood by programmers. Moggi [16] argues that the notion of "strength" is fundamental to computation, "strength" being defined as follows.

Definition 2 (Strength). A collection of arrows str is said to be a *strength* of relator F iff $\mathsf{str}_{A,B} : \mathsf{F}(A \times B) \hookleftarrow \mathsf{FA} \times B$ and $\mathsf{str}_{A,B}$ is a function that behaves coherently with respect to product in the following sense. First, the diagram

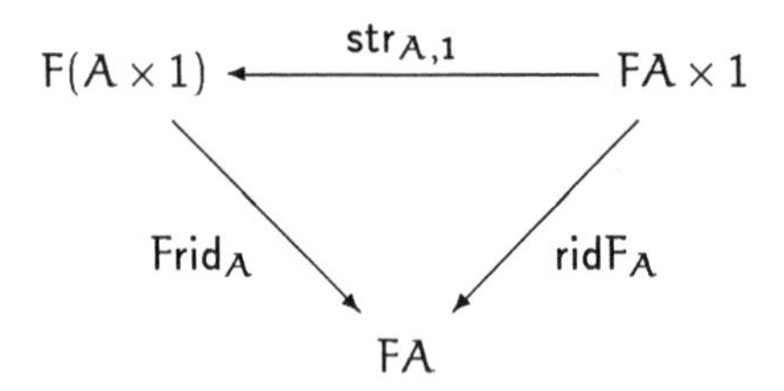

(where $\mathsf{rid}_A : A \leftarrow A \times 1$ is the obvious natural isomorphism) commutes. Second, the diagram

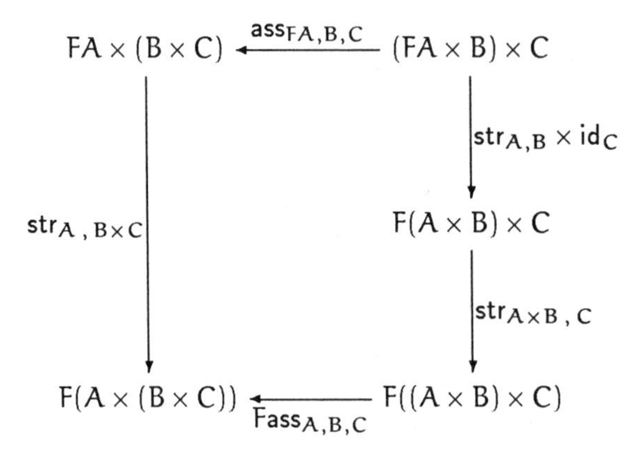

(where $\mathsf{ass}_{A,B,C} : A \times (B \times C) \leftarrow (A \times B) \times C$ is the obvious natural isomorphism) commutes as well. A relator that has at least one strength is said to be *strong*. □

Note that "strength" is what we have called a broadcast operation.

The type of the "strength" $\mathsf{str}_{A,B}$ of relator F is the same as that of $(\mathsf{zip}.(\times B).F)_A$, namely $F(A \times B) \leftarrow FA \times B$. We shall argue that, if F and the family of relators $(\times B)$ are included in a class of commuting relators, then any relation satisfying the requirements of $(\mathsf{zip}.(\times B).F)_A$ also satisfies the definition of $\mathsf{str}_{A,B}$.

In the first diagram there are two occurrences of the canonical isomorphism rid. In general, we recognise a projection of type $A \leftarrow A \times B$ as a broadcast where the parameter F is instantiated to K_A, the relator that is constantly A when applied to objects and is the identity on A when applied to arrows. Thus rid_A is $(\mathsf{zip}.(\times 1).K_A)_B$ for some arbitrary B. In words, rid_A commutes the relators $(\times 1)$ and K_A. Redrawing the first diagram above, using that all the arrows are broadcasts and thus zips, we get the following diagram.

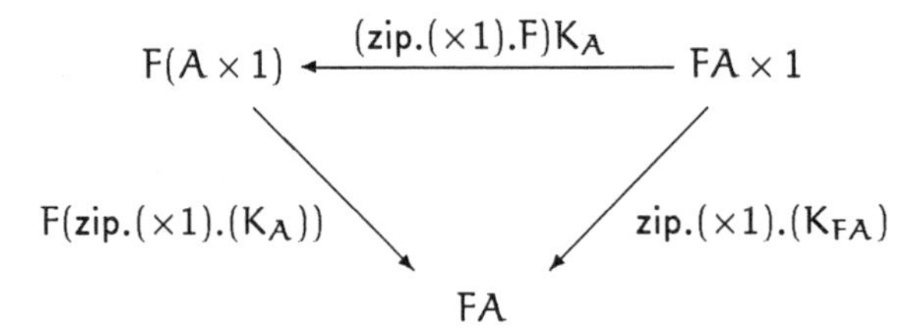

Expressed as an equation, this is the requirement that

$$\mathsf{zip}.(\times 1).(\mathsf{K}_{\mathsf{FA}}) \;=\; \mathsf{F}(\mathsf{zip}.(\times 1).(\mathsf{K}_{\mathsf{A}})) \circ (\mathsf{zip}.(\times 1).\mathsf{F})\mathsf{K}_{\mathsf{A}} \tag{7}$$

Now we turn to the second diagram in the definition of strength. Just as we observed that rid is an instance of a broadcast and thus a zip, we also observe that ass is a broadcast and thus a zip. Specifically, $\mathsf{ass}_{\mathsf{A,B,C}}$ is $(\mathsf{zip}.(\times\mathsf{C}).(\mathsf{A}\times))_{\mathsf{B}}$. Once again, every edge in the diagram involves a zip operation! That is not all. Yet more zips can be added to the diagram. For our purposes it is crucial to observe that the bottom left and middle right nodes —the nodes labelled $\mathsf{F}(\mathsf{A}\times(\mathsf{B}\times\mathsf{C}))$ and $\mathsf{F}(\mathsf{A}\times\mathsf{B})\times\mathsf{C}$— are connected by the edge $(\mathsf{zip}.(\times\mathsf{C}).\mathsf{F}(\mathsf{A}\times))_{\mathsf{B}}$.

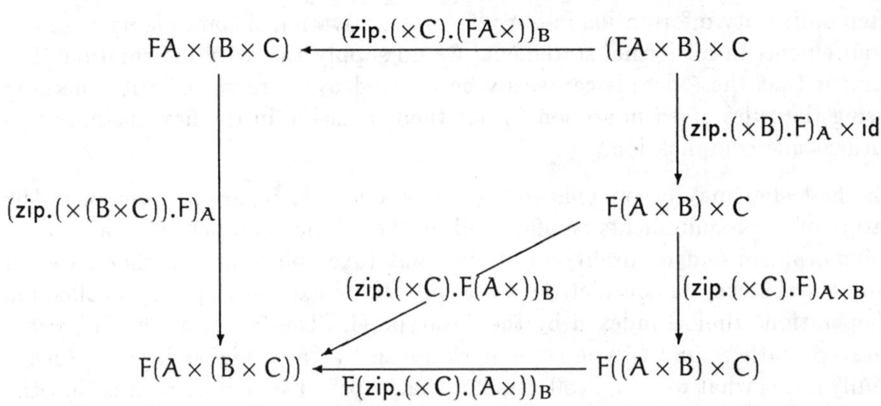

$$\begin{array}{ccc}
\mathsf{FA}\times(\mathsf{B}\times\mathsf{C}) & \xleftarrow{(\mathsf{zip}.(\times\mathsf{C}).(\mathsf{FA}\times))_{\mathsf{B}}} & (\mathsf{FA}\times\mathsf{B})\times\mathsf{C} \\
\Big\downarrow{\scriptstyle (\mathsf{zip}.(\times(\mathsf{B}\times\mathsf{C})).\mathsf{F})_{\mathsf{A}}} & & \Big\downarrow{\scriptstyle (\mathsf{zip}.(\times\mathsf{B}).\mathsf{F})_{\mathsf{A}}\times\mathsf{id}_{\mathsf{C}}} \\
 & & \mathsf{F}(\mathsf{A}\times\mathsf{B})\times\mathsf{C} \\
 & {\scriptstyle (\mathsf{zip}.(\times\mathsf{C}).\mathsf{F}(\mathsf{A}\times))_{\mathsf{B}}} & \Big\downarrow{\scriptstyle (\mathsf{zip}.(\times\mathsf{C}).\mathsf{F})_{\mathsf{A}\times\mathsf{B}}} \\
\mathsf{F}(\mathsf{A}\times(\mathsf{B}\times\mathsf{C})) & \xleftarrow[\mathsf{F}(\mathsf{zip}.(\times\mathsf{C}).(\mathsf{A}\times))_{\mathsf{B}}]{} & \mathsf{F}((\mathsf{A}\times\mathsf{B})\times\mathsf{C})
\end{array}$$

This means that we can decompose the original coherence property into a combination of two properties of zips. These are as follows. First, the lower triangle:

$$(\mathsf{zip}.(\times\mathsf{C}).\mathsf{F}(\mathsf{A}\times))_{\mathsf{B}} \;=\; \mathsf{F}(\mathsf{zip}.(\times\mathsf{C}).(\mathsf{A}\times))_{\mathsf{B}} \circ (\mathsf{zip}.(\times\mathsf{C}).\mathsf{F})_{\mathsf{A}\times\mathsf{B}} \tag{8}$$

Second, the upper rectangle:

$$(\mathsf{zip}.(\times(\mathsf{B}\times\mathsf{C})).\mathsf{F})_{\mathsf{A}} \circ (\mathsf{zip}.(\times\mathsf{C}).(\mathsf{FA}\times))_{\mathsf{B}} = (\mathsf{zip}.(\times\mathsf{C}).\mathsf{F}(\mathsf{A}\times))_{\mathsf{B}} \circ (\mathsf{zip}.(\times\mathsf{B}).\mathsf{F})_{\mathsf{A}}\times\mathsf{id}_{\mathsf{C}} \tag{9}$$

Note the strong similarity between (7) and (8). They are both instances of one equation parameterised by three different datatypes. There is also a similarity between these two equations and (5); the latter is an instance of the same parameterised equation after taking the converse of both sides and assuming that $\mathsf{zip.F.G} = (\mathsf{zip.G.F})^{\cup}$. Less easy to spot is the similarity between (6) and (9). As we shall see, however, both are instances of one equation parameterised again by three different datatypes except that (9) is obtained by applying the converse operator to both sides of the equation and again assuming that $\mathsf{zip.F.G} = (\mathsf{zip.G.F})^{\cup}$.

4 The Requirement

In this section we formulate precisely what we mean by two datatypes commuting.

Looking again at the examples above, the first step towards an abstract problem specification is clear enough. Replacing "list", "tree" etc. by "datatype F" the problem is to specify an operation zip.F.G for given datatypes F and G that maps FG-structures into GF-structures.

Note that the informal language we use here seems to imply that we consider only endo relators (relators of arity $1 \leftarrow 1$). If $F : m \leftarrow k$ and $G : n \leftarrow l$ then, to be perfectly precise, we should talk about mapping $({}^{n}F)(G^{k})$-structures to $(G^{m})({}^{l}F)$-structures.

Being able to handle relators of arbitrary arity and not restricting ourselves to endorelators is an important element of our development but nevertheless we often omit arity information in our informal motivation of some elements of our requirement. In all formal statements we do supply the arity information. The point is that these details can easily be inferred by a process of arity checking (using the rules given in section 2) but their inclusion in the first instance is a burdensome complication.

The first step may be obvious enough, subsequent steps are less obvious. The nature of our requirements is influenced by the relationship between parametric polymorphism and naturality properties but takes place at a higher level. In subsequent steps, we consider the datatype F to be fixed and specify a collection of operations zip.F.G indexed by the datatype G. (The fact that the index is a datatype rather than a type is what we mean by "at a higher level".) Such a family forms what we call a collection of "half-zips". The requirement is that the collection be "parametric" in G. That is, the elements of the family zip.F should be "logically related" to each other. The precise formulation of this idea leads us to three requirements on "half-zips". The symmetry between F and G, lost in the process of fixing F and varying G, is then restored by the simple requirement that a zip is both a half-zip and the converse of a half-zip.

4.1 Naturality Requirements

Our first requirement is that zip.F.G be natural. That is to say, its application to an FG-structure should not in any way depend on the values in that structure. Suppose that $F : m \leftarrow k$ and $G : n \leftarrow l$. Then we demand that

$$\mathsf{zip.F.G} : (G^{m})({}^{l}F) \leftarrow ({}^{n}F)(G^{k}) \ . \qquad (10)$$

Note that we require zip.F.G to be a *proper* natural transformation since for a zip operation on a structure no loss or duplication of values should occur.

Demanding naturality is not enough. The zips zip.F.G and zip.F.H should be "coherent" with all transformations $\alpha : G \leftarrow H$. That is to say, having both zips and α, there are two ways of transforming FH-structures into GF-structures as given by the following diagram.

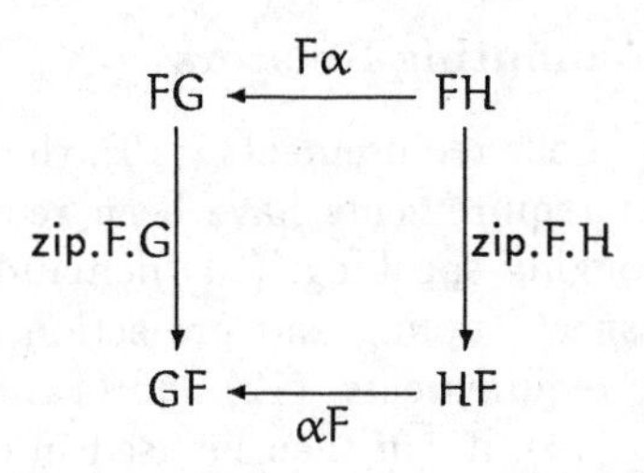

One might suppose that an equality is required, i.e. $\alpha F \circ \mathsf{zip.F.H} = \mathsf{zip.F.G} \circ F\alpha$ for all natural transformations $\alpha : G \leftarrow H$. But this requirement is too severe. Instead we impose the requirement:

$$\alpha F \circ \mathsf{zip.F.H} \subseteq \mathsf{zip.F.G} \circ F\alpha \quad \text{for all } \alpha : G \hookleftarrow H \ .$$

Including arity information, the formal statement of the requirement is that for all relators $F : m \leftarrow k$ and $G, H : n \leftarrow l$, and all $\alpha : G \hookleftarrow H$,

$$(\alpha^m)({}^l F) \circ \mathsf{zip.F.H} \subseteq \mathsf{zip.F.G} \circ ({}^n F)(\alpha^k) \ . \tag{11}$$

4.2 Pointwise Integrity

The variable-free mechanism we have introduced for "pointwise closing" a class of relators allows some freedom in the manner in which relators are composed. Our second set of requirements guarantee that this algebraic structure is respected by the mapping zip.F. We begin with tupling and projection. In view of arity considerations the obvious requirements are:

$$\mathsf{zip.F.G} = \tau\Delta_n(\mathsf{zip.F.Proj}_n G) \tag{12}$$

where n is the arity of the target of G —the zip of a tuple is the tuple of the zips— and

$$\mathsf{zip.F.Proj} = (\mathsf{id}^m)F(\mathsf{Proj}^k) \tag{13}$$

for each projection relator $\mathsf{Proj} : 1 \leftarrow l$, assuming $F : m \leftarrow k$.
In fact, (12) becomes redundant when we introduce requirement (14) on the composition of relators.
For our final requirement we consider the monoid structure of functors under composition. Fix functor F and consider the collection of zips, zip.F.G, indexed by (endo)functor G. Since the (endo)functors form a monoid it is required that the mapping zip.F is a monoid homomorphism. The demand is that, for all $F : m \leftarrow k$, $G : n \leftarrow l$ and $H : l \leftarrow o$,

$$\mathsf{zip.F.GH} = (G^m)(\mathsf{zip.F.H}) \circ (\mathsf{zip.F.G})(H^k) \ , \tag{14}$$

and, for $F : m \leftarrow k$ and the identity relator $\mathrm{Id} : l \leftarrow l$,

$$\mathsf{zip.F.Id} = (\mathsf{id}^{l*m})({}^l F) \ . \tag{15}$$

4.3 Half Zips and Commuting Relators

Apart from the very first of our requirements ((10), the requirement that zip.F.G be natural), all the other requirements have been requirements on the nature of the mapping zip.F. Roughly speaking, (11) demands that it be parametric, (12) and (13) that it respect tupling and projection, and (14) and (15) that it be functorial. Of these requirements, (12) and (15) are redundant. ((12) can be derived from (13) and (14); it can then be used in combination with (13) to derive (15).) We find it useful to bundle the (non-redundant set of) requirements together into the definition of something that we call a "half zip".

Definition 3 (Half Zip). Consider a fixed relator $F : m \leftarrow k$ and a pointwise closed class of relators $\mathcal{G}$. Then the members of the collection zip.F.G, where G ranges over $\mathcal{G}$, are called *half-zips* iff
(a) $zip.F.G : (G^m)(^lF) \leftarrow (^nF)(G^k)$, for each $G : n \leftarrow l$
(b) $zip.F.Proj = (id^m)F(Proj^k)$ for all $Proj : 1 \leftarrow l$,
(c) $zip.F.GH = (G^m)(zip.F.H) \circ (zip.F.G)(H^k)$ for all $G : n \leftarrow l$ and $H : l \leftarrow o$,
(d) $(\alpha^m)(^lF) \circ zip.F.H \subseteq zip.F.G \circ (^nF)(\alpha^k)$ for all $G, H : n \leftarrow l$ and $\alpha : G \hookleftarrow H$.
□

Note that for $F : m \leftarrow k$ and $G : n \leftarrow l$, zip.F.G and $(zip.G.F)^\cup$ do not have the same arity. The source and target arities are clearly related by matrix transposition, i.e. the relator τ. So, the general definition becomes:

Definition 4 (Commuting Relators). The half-zip zip.F.G is said to be a *zip* of (F, G) if there exists a half-zip zip.G.F such that

$$zip.F.G = \tau(zip.G.F)^\cup\tau$$

We say that datatypes F and G *commute* if there exists a zip for (F, G). □

5 Consequences

In this section we address two concerns. First, it may be the case that our requirement is so weak that it has many trivial solutions. We show that, on the contrary, the requirement has a number of consequences that guarantee that there are no trivial solutions. On the other hand, it could be that our requirement for datatypes to commute is so strong that it is rarely satisfied. Here we show that the requirement can be met for all regular datatypes. (Recall that the "regular" datatypes are the sort of datatypes that one can define in a conventional functional programming language.) Moreover, we can even prove the remarkable result that for the regular relators our requirement has a *unique* solution.

5.1 Shape Preservation

Zips are partial operations: zip.F.G should map F–structures of (G–structures of the same shape) into G–structures of (F–structures of the same shape). This requirement is, however, not explicitly stated in our formalisation of being a zip. It is nevertheless a consequence of that formal requirement. In particular a half zip always constructs G–structures of (F–structures of the same shape).

Suppose $F : k \leftarrow l$ and $G : m \leftarrow n$ are datatypes. Then, if fan.G is the canonical fan of G,

$$((\mathsf{fan}.G)^k)F = (\mathsf{zip}.F.G)((\Delta_n)^l) \circ ({}^m F)((\mathsf{fan}.G)^l) \ . \tag{16}$$

From equation (16) it also follows that the range of $(\mathsf{zip}.F.G)_1$ is the range of $(\mathsf{fan}.G)_{F1}$, i.e. arbitrary G-structures of which all elements are the same, but arbitrary, F-shape. It is (16) that often uniquely characterises zip.F.G.

5.2 Commuting relators

One reason why our requirements might have trivial solutions is that they are expressed in terms of lax natural transformations. Requiring properness of a natural transformation is stronger. The next lemma establishes a properness result for zips on commuting datatypes; it proves to be the key in showing that certain zips are unique.

Let ξ denote a class of commuting datatypes. Then for all relators $F : k \leftarrow l$, and $G, H : m \leftarrow n$ in ξ and all families of *functions* α such that $\alpha : G \leftarrow H$,

$$({}^m F)(\alpha^l) \circ \mathsf{zip}.H.F = \mathsf{zip}.G.F \circ (\alpha^k)({}^n F) \ . \tag{17}$$

Note that the lemma does not imply that the zips are themselves functional. On the face of it, the property stated in the lemma is quite weak.

5.3 All regular datatypes commute

We now come to the main result of this paper, namely, that all regular relators commute. Morever, for each pair of regular relators F and G there is a *unique* natural transformation zip.F.G satisfying our requirements.

The regular relators are constructed from the constant relators, product and coproduct by pointwise extension and/or the construction of tree relators. The requirement that zip.F.G and zip.G.F be each other's converse (modulo transposition) demands the following definitions:

$$\mathsf{zip}.\mathsf{Id}.G = \mathsf{id}G \tag{18}$$

$$\mathsf{zip}.\mathsf{Proj}.G = \mathsf{id}G(\mathsf{Proj}^k) \quad \text{for all } G : 1 \leftarrow k \text{ and all } \mathsf{Proj} : 1 \leftarrow l \tag{19}$$

$$\mathsf{zip}.\Delta F_k.G = \tau\Delta(\mathsf{zip}.F_k.G) \tag{20}$$

$$\mathsf{zip}.FG.H = (\mathsf{zip}.F.H)({}^k G) \circ F(\mathsf{zip}.G.H) \quad \text{for all } H : 1 \leftarrow k \tag{21}$$

The restriction to single-valued relators in these equations is made possible by the rule for $\mathsf{zip}.G.\Delta F_l$.

For the constant relators and product and coproduct, the zip function is uniquely characterised by (16). One obtains the following definitions, for all $\mathsf{G} : 1 \leftarrow k$:

$$\mathsf{zip.K_A.G} = (\mathsf{fan.G})(\mathsf{K_A}) \tag{22}$$
$$\mathsf{zip.+.G} = \mathsf{Ginl} \triangledown \mathsf{Ginr} \tag{23}$$
$$\mathsf{zip.\times.G} = (\mathsf{Goutl} \vartriangle \mathsf{Goutr})^{\cup} \tag{24}$$

Note that, in general, $\mathsf{zip.K_A.G}$ and $\mathsf{zip.\times.G}$ are not functions; moreover, the latter is typically partial. That is the right domain of $(\mathsf{zip.\times.G})_{(\mathsf{A,B})}$ is typically a proper subset of $\mathsf{GA} \times \mathsf{GB}$. Zips of datatypes defined in terms of these datatypes will thus also be non-functional and/or partial. Nevertheless, broadcast operations ("strengths") are always functional.

Tree relators are the last sort of relators in the class of regular relators. Let T be the tree relator induced by $\otimes$ as defined in section 2.2. Here the uniqueness of $\mathsf{zip.T.G}$ for all G is assured by (17) with α instantiated to in. One obtains:

$$\mathsf{zip.T.G} = (\!|\mathsf{id_G}\otimes ;\ \mathsf{G}(^k\mathsf{in}) \circ (\mathsf{zip.\otimes.G})(^k(\mathsf{Id}\vartriangle\mathsf{T}))|\!) \quad \text{for all } \mathsf{G} : 1 \leftarrow k. \tag{25}$$

5.4 Broadcast and Structure Multiplication, Again

In our motivation of commuting datatypes, we said that the requirements for structure multiplication and "strength" would be met "almost by definition". In this section we observe in what sense that is indeed the case.

The requirements for structure multiplication are given by equations (5) and (6); those for broadcasts by (7), (8) and (9).

We begin with (5), (7) and (8). Note that all of these correspond to triangular diagrams. All are instances or simple consequences of the compositionality requirement of zips, 3(c). This is easiest to see in the case of (8) since it suffices to make the substitutions $\mathsf{F,G,H} := (\times\mathsf{C}), \mathsf{F}, (\mathsf{A}\times)$. Next easiest to see is (7). Here the observation has to be made that $\mathsf{K_{FA}} = \mathsf{FK_A}$. Then make the substitutions $\mathsf{F,G,H} := (\times 1), \mathsf{F}, \mathsf{K_A}$. Finally, (5) is a combination of 3(c) and (4) with the substitutions $\mathsf{F,G,H} := \mathsf{G}, \mathsf{F}, (\mathsf{A}\times)$. Thus all three requirements are satisfied, by definition, if it can be shown that all the relators involved belong to a class of commuting relators. In particular, since the sections $(\times\mathsf{C})$ and $(\mathsf{A}\times)$ are regular relators, all the requirements are met if in each case F is a regular relator.

The remaining two requirements, (6) and (9), are instances of (17) and 3(d), respectively. This is less easy to see. The key is to observe that the broadcast α where $\alpha_\mathsf{B} = (\mathsf{zip.}(\times\mathsf{B}).\mathsf{F})_\mathsf{A}$ is a proper, functional natural transformation of type $\mathsf{F}(\mathsf{A}\times) \leftarrow (\mathsf{FA})\times$ for each regular relator F and each A. (Note that the functionality is a special property of broadcasts. As mentioned before, zips are typically partial and nondeterministic. Hoogendijk [7] proves that $(\mathsf{zip.}(\times\mathsf{B}).\mathsf{F})_\mathsf{A}$ is functional for all regular relators F.) Property (6) is then an instance of (17) after making the substitutions $\mathsf{F,G,H} := \mathsf{G}, \mathsf{F}(\mathsf{A}\times), (\mathsf{FA})\times$ and defining α as above. Property (9) is obtained from 3(d) using the substitutions $\mathsf{F,G,H},\alpha_\mathsf{B} := (\times\mathsf{C}), \mathsf{F}(\mathsf{A}\times), (\mathsf{FA})\times, (\mathsf{zip.}(\times\mathsf{B}).\mathsf{F})_\mathsf{A}$. This results in an inclusion

—not an equality— but every term is a broadcast, and thus a function, and inclusion of functions is equivalent to their equality. We conclude that (6) and (9) are also met provided that F and all sections of the form $(\times C)$ and $(A\times)$ are members of a class of commuting relators, and in particular if F is a regular relator.

6 Conclusion

Polytypism is a new concept in the repertoire of generic programming. In this paper we have made contributions to the theoretical and practical development of polytypism. First, we have provided evidence for the necessity of developing a theory of polytypism in a relational rather than a functional framework. Membership and fans can only be discussed at a metalevel in a functional framework and the fact that all regular relators commute is just not true in a functional framework since some of the transformations are necessarily nondeterministic. Second, we have demonstrated how to cope with non-endo relators thus overcoming a limitation of all other work in this field published to date that we know of (including our own). Third, we have illustrated a general approach to the specification of polytypic programs. Roughly summarised, the approach is to require that the class of programs is compositional with respect to the pointwise definition of datatypes, and that the class is "higher order natural" in the sense that it maps related datatypes to related datatypes (just as polymorphic functions map related objects to related objects). This is a major advance on our earlier work [1] in which the commuting requirement was substantially more operational in flavour and hence *ad hoc.*

Several challenges remain. Possibly the most important is that we have been unable to establish a general unicity property of the "zip" operators even though in every individual case that we have studied we can prove unicity. This suggests that our requirements can be made stronger and, in the process, yet simpler and more elegant.

Acknowledgement The diagrams were drawn with the aid of Paul Taylor's commutative diagrams package.

References

1. R. Backhouse, H. Doornbos, and P. Hoogendijk. Commuting relators. Available via World-Wide Web at `http://www.win.tue.nl/cs/wp/papers`, September 1992.
2. R. Bird. Lectures on constructive functional programming. In M. Broy, editor, *Constructive Methods in Computing Science*, pages 151–216. Springer-Verlag, 1989. NATO ASI Series, vol. F55.
3. R. Bird, O. de Moor, and P. Hoogendijk. Generic functional programming with types and relations. *J. of Functional Programming*, 6(1):1–28, January 1996.
4. R. S. Bird and O. de Moor. *Algebra of Programming.* Prentice-Hall International, 1996.

5. H. Doornbos and R. Backhouse. Reductivity. *Science of Computer Programming*, 26(1–3):217–236, 1996.
6. P. Freyd and A. Scedrov. *Categories, Allegories*. North-Holland, 1990.
7. P. Hoogendijk. *A Generic Theory of Datatypes*. PhD thesis, Department of Mathematics and Computing Science, Eindhoven University of Technology, 1997.
8. P. Hoogendijk and O. de Moor. What is a datatype? Technical Report 96/16, Department of Mathematics and Computing Science, Eindhoven University of Technology, 1996. Submitted to Science of Computer Programming. Available via World-Wide Web at `http://www.win.tue.nl/cs/wp/papers`.
9. J. Jeuring. Polytypic pattern matching. In *Conference Record of FPCA '95, SIGPLAN-SIGARCH-WG2.8 Conference on Functional Programming Languages and Computer Architecture*, pages 238–248, 1995.
10. J. Jeuring and P. Jansson. Polytypic programming. In J. Launchbury, E. Meijer, and T. Sheard, editors, *Proceedings of the Second International Summer School on Advanced Functional Programming Techniques*, pages 68–114. Springer-Verlag, 1996. LNCS 1129.
11. A. Jung (Editor). Domains and denotational semantics: History, accomplishments and open problems. *Bulletin of the European Association for Computer Science*, 59:227–256, June 1996.
12. G. Malcolm. Homomorphisms and promotability. In J. van de Snepscheut, editor, *Conference on the Mathematics of Program Construction*, pages 335–347. Springer-Verlag LNCS 375, 1989.
13. L. Meertens. Calculate polytypically! In H. Kuchen and S. D. Swierstra, editors, *Proceedings of the Eighth International Symposium PLILP '96 Programming Languages: Implementations, Logics and Programs*, volume 1140 of *Lecture Notes in Computer Science*, pages 1–16. Springer Verlag, 1996.
14. E. Meijer, M. Fokkinga, and R. Paterson. Functional programming with bananas, lenses, envelopes and barbed wire. In *FPCA91: Functional Programming Languages and Computer Architecture*, volume 523 of *LNCS*, pages 124–144. Springer-Verlag, 1991.
15. R. Milner. A theory of type polymorphism in programming. *J. Comp. Syst. Scs.*, 17:348–375, 1977.
16. E. Moggi. Notions of computation and monads. *Information and Computation*, 93(1):55–92, 1991.
17. G. D. Plotkin. Lambda-definability in the full type hierarchy. In J. Seldin and J. Hindley, editors, *To H.B. Curry: Essays on Combinatory Logic, Lambda Calculus and Formalism*. Academic Press, London, 1980.
18. J. Reynolds. Types, abstraction and parametric polymorphism. In R. Mason, editor, *IFIP '83*, pages 513–523. Elsevier Science Publishers, 1983.

A Calculus for Collections and Aggregates

Kazem Lellahi[1] and Val Tannen[2]

[1] LIPN, URA 1507 du CNRS,
Université de Paris 13, Institut Galilée,
93430 Villetaneuse, France
kl@ura1507.univ-paris13.fr

[2] CIS Department, University of Pennsylvania,
200 S. 33rd St, Philadelphia, PA 19104, USA
val@cis.upenn.edu

Abstract. We present a calculus that should play for database query languages the same role that the lambda calculus plays for functional programming. For the semantic foundations of the calculus we introduce a new concept: monads *enriched* with algebraic structure. We model collection types through enriched monads and aggregate operations through enriched monad algebras. The calculus derives program equivalences that underlie a good number of the optimizations used in query languages.

1 Motivation

This paper proposes a calculus for programming with collection data types and aggregate operations on such collections. From a philosophical perspective, such a calculus should play for database query languages the role that the lambda calculus plays for functional programming languages. From a practical perspective, such a calculus can form the foundation of an *intermediate language*: the surface syntax of queries gets translated into terms of the calculus and the equational theory of the calculus derives program equivalences used in optimization.

Collections are data types such as lists, sets, bags (multisets), trees. Aggregates are functions such as that which adds all the elements of a bag of integers or returns the largest number in a set. Collections and aggregates are essential components of the basic database paradigms: relational, object-oriented, and (most recently) object-relational. In recent years, the design of database query languages has started to finally benefit from advances in general programming language design. The industry standard, SQL, has evolved in the direction of more referential transparency and compositionality, especially in the versions used for object-relational databases A rich and challenging standard has been proposed for object-oriented databases: OQL [Cat96].

The programming language designer must be aware, however, of the particularities of query languages. The size of the data they are routinely manipulating is such that most programs (called queries in the trade) are simply not feasible unless subjected to sophisticated optimizations. Most of these will not be found among those routinely performed by compilers for general-purpose programming languages. There is a good reason for this: these optimizations are not sound if programs are allowed to diverge or have side-effects. It follows that query languages will have limited computing power and thus allow certain specialized optimizations. It also follows that the choice of query language constructs and the understanding of the identities that these constructs satisfy are essential. With the calculus presented in this paper we are trying to capture a good deal of these aspects, encompassing certainly the relational data model, the nested relational paradigm, and many features of the object-relational and object-oriented data models.

We propose two main contributions. First, we capture the principles of optimized programming with collections and their aggregates in an economical formalism that is suitable for study using the tools of the programming semantics community. Second, and a precondition for the first contribution, we develop a theory of *enriched* monads that provides the semantic foundation for the calculus. Along the way we emphasize a new role for (Eilenberg-Moore) monad algebras and monad morphisms in programming semantics, as well as an interesting connection with interchange laws.

2 Overview

In section 3 we give examples of the data types and operations we wish to model. We also introduce algebras for an endofunctor which will play a crucial role in the subsequent development. In section 4 we discuss, in categorical terms, the semantics of collections and aggregates, using monads and monad algebras. In section 5 we discuss, again semantically, three different ways of enriching the basic monad setting and show, surprisingly, that they are equivalent. The equivalence is exploited thoroughly in subsequent sections.

In section 6 we put it all together and extract a small set of primitives and axioms for the calculus that corresponds to one collection type. We give examples of optimizations. In section 7 we put together the calculi that correspond to collections that are distinct but related by conversions (such as duplicate removal from bags to sets).

3 Collections and Aggregates

Collections, in this treatment, are data types such as homogenous, finite sets, bags, lists, and trees with data on the leaves[3]. We always have collections of one element—*singletons*. We denote by $\mathsf{Coll}\,X$ the set of finite collections of a certain kind made out of elements in X. Examples: $\mathsf{Set}\ X$, $\mathsf{Bag}\ X$ and $\mathsf{List}\ X$.

Aggregates are operations of the form $\mathsf{agg} : \mathsf{Coll}\ A \longrightarrow A$ which compute an element of A out of each collection of elements of A. The aggregate of a singleton must of course equal its only element. Examples of aggregates that fit in our treatment are $\mathsf{agg}_{\mathsf{or}}, \mathsf{agg}_{\mathsf{and}} : \mathsf{Set\ bool} \longrightarrow \mathsf{bool}$ and $\mathsf{agg}_{\mathsf{max}}, \mathsf{agg}_{\mathsf{min}} : \mathsf{Set\ num} \longrightarrow \mathsf{num}$ for the Set data type, $\mathsf{agg}_{\mathsf{sum}} : \mathsf{Bag\ num} \longrightarrow \mathsf{num}$ for Bag ,

[3] A treatment of trees with data on the nodes would be useful too.

and $\mathsf{agg}_{\mathsf{concat}}$: List string $\rightarrow$ string for List . (For $\mathsf{agg}_{\mathsf{max}}$ and $\mathsf{agg}_{\mathsf{min}}$ to be defined on the empty set, we assume that num has a least element $-\infty$, and a greatest element $+\infty$.)

On the other hand, not all functions Coll $A \rightarrow A$ are aggregates. We explain in section 4 why the obvious function addup : Set num $\rightarrow$ num does not fit in our treatment. Query language designers however, have long remarked that this function is less than useful. For instance, $\mathsf{agg}_{\mathsf{sum}} \circ \mathsf{map}\ (\lambda x\,.\,1)$ computes the cardinality of bags while $\mathsf{addup} \circ \mathsf{map}\ (\lambda x\,.\,1)$ is always 1. That's why in practical query languages such as SQL and OQL the default semantics is bags, while sets can be requested explicitly using "select unique".

A bit more formally, an *algebraic*[4] *collection type* is a parameterized type that admits an inductive definition of the following form (in ML-like syntax):

```
datatype 'a Coll = sng of 'a | opcoll of 'a Coll * ...* 'a Coll | ...
```

with one or more finitary algebraic operations of various arities (including nullary) on collections as constructors (in addition to singleton). Moreover, the meaning of such types can be further constrained by equations between the algebraic constructors (otherwise we would never get sets or bags).

An interesting example is the *Boom hierarchy* of types (see for instance [Hoo94]). These types have two algebraic constructors, one nullary, one binary, in addition to the singleton:

```
datatype 'a Coll = sng of 'a | empty | comb of 'a Coll * 'a Coll
```

The hierarchy is obtained by imposing progressively richer sets of equational constraints. With no constraints we get binary trees. When empty is a unit for $\mathsf{comb}(\cdot,\cdot)$ we get binary trees with no empty subtree, call them *compact* binary trees. If we add to this the associativity of comb we get lists (built with append, not with cons). Further adding the commutativity of comb gives bags and we get sets by also adding the idempotence of comb. Notice that lists have a monoid structure, for bags the monoid is commutative and for sets it is moreover idempotent.

The semantics of algebraic collection types can be formalized with the help of *algebras for an endofunctor*. Fix a category, Base . In all our examples, Base is the category of sets and functions, but there seem to be no drawbacks and there may be advantages to seeking more generality. Let E be an endofunctor E : Base $\rightarrow$ Base . As usual, we denote the action of E on an arrow f by E f.

Definition 1. An E-*algebra* is simply a pair $(A,\ \mathsf{op} : \mathsf{E}\,A \rightarrow A)$. An E-algebra *homomorphism* between $(A,\ \mathsf{op})$ and (B, op') is an arrow $h : A \rightarrow B$ such that $h \circ \mathsf{op} = \mathsf{op}' \circ (\mathsf{E}\,h)$. We denote by E-Alg the category of E-algebras and their homomorphisms.

---

[4] This is to distinguish from other types that people often call "collection" or "bulk" types, eg., relations, finite mappings, queues, or arrays.

For the semantics of algebraic collection types we require that **Base** have finite (including nullary) products and we take $\mathsf{E}X = X \times \cdots \times X$. Then $\mathsf{op} : A \times \cdots \times A \to A$ is just a finitary algebraic operation on A. Several operations can be postulated concomitantly when **Base** has sums (coproducts). For instance, the structures in the Boom hierarchy are B-algebras where $\mathsf{B}X = \mathsf{unit} + X \times X$ (where unit is the nullary product—a *terminator*). Monoids and similar structures are B-algebras.

Let V_C be the class (variety) of E-algebras satisfying a given set Eq of equational constraints. For each X, the semantics of the algebraic collection type $\mathsf{Coll}\, X$ determined by V_C is the algebra $(\mathsf{Coll}\, X, \mathsf{opcoll})$ satisfying C *freely generated* by X via singletons. That is, for any other algebra A in V_C, and for any arrow $f : X \to A$ there is a unique homomorphism $h : \mathsf{Coll}\, X \to A$ such that $h \circ \mathsf{sng}_X = f$ (the semantics is a left *adjoint* to the forgetful functor $\mathsf{V}_C \to \mathsf{Base}$). We say that h is defined by *structural recursion*. For instance, it is easy to see that $\mathsf{Set}\, X$ is the commutative idempotent monoid freely generated by X via singletons.

This adjunction semantics suggests an immediate approach to designing query languages for algebraic collection types: give syntax for *structural recursion*. This works fine in the absence of equational constraints and it would amount to special cases of the recursion on inductive datatypes found in languages like ML. But it is cumbersome when we have equational constraints, because they need to be verified for the structural recursion construct to have meaning [BTS91]. Equations like associativity and even idempotence are powerful enough to encode undecidable, even non-r.e. properties [BTS91, SBT94]. A more subtle objection is that the adjunction semantics in itself does not suggest directly a purely equational theory, rather it offers an induction principle. This principle likely proves all the desired program equivalences, but this leaves the discovery of these equivalences to an ad-hoc process.

Instead, what we can do is extract out of the adjunction semantics constructs that are always defined, and characterize their properties through equations, which leads easily and directly to optimizations. It was already shown in [BBW92, BNTW95] that such a programme leads us naturally to consider the monads associated with the adjunctions. A calculus was already proposed in these papers, called the Nested Relational Calculus (NRC). Algebraic constructor operators such as empty set or binary set union were considered in NRC but their equational axiomatization was ad-hoc. Important expressiveness results concerning the sommation aggregate were obtained in [LW94b, LW94a]. Equational rewriting is used for some of these results, but the axiomatization is less than systematic. The properties of conversions between collections have apparently not been considered.

In this paper we model, as before, collections through monads. In addition we model aggregates through *monad algebras* and algebraic constructor operations *and* conversions between collections through *enrichments*.

4 Collections as Monads, Aggregates as Monad Algebras

4.1 Monads

We begin with the most commonly used definition of monads (often called *triples* [BW85, LS86]), using however a notation inspired by our domain of application: collections.

Definition 2 [Mac71]. A *monad* on Base is given by a functor Base $\rightarrow$ Base, whose action on objects we denote by $X \mapsto \mathsf{Coll}\, X$ and whose action on morphisms we denote, instead of the usual $\mathsf{Coll}\, f$, by $f : X \rightarrow Y \quad \mapsto \quad \mathsf{map}\, f : \mathsf{Coll}\, X \rightarrow \mathsf{Coll}\, Y$ and two natural transformations $\mathsf{sng}_X : X \rightarrow \mathsf{Coll}\, X$ and $\mathsf{flatten}_X : \mathsf{Coll}\,\mathsf{Coll}\, X \rightarrow \mathsf{Coll}\, X$ such that:

$$\mathsf{flatten}_X \circ (\mathsf{map}\ \mathsf{sng}_X) = \mathsf{id}_{\mathsf{Coll}\, X} \tag{1}$$

$$\mathsf{flatten}_X \circ \mathsf{sng}_{\mathsf{Coll}\, X} = \mathsf{id}_{\mathsf{Coll}\, X} \tag{2}$$

$$\mathsf{flatten}_X \circ (\mathsf{map}\ \mathsf{flatten}_X) = \mathsf{flatten}_X \circ \mathsf{flatten}_{\mathsf{Coll}\, X} \tag{3}$$

Examples. $\mathsf{Coll}\, X$ is the set of finite sets (bags, lists, compact binary trees, binary trees, ternary trees, "2-3" trees, etc.) with elements from X. sng builds a singleton set (bag, etc.), map f applies f to each element of a set, and flatten is the union of a set of sets. We denote the first three of these monads by Set, Bag, and List.

For applications to the theory of programming languages, Moggi [Mog91] has found more suitable a succinct definition of monads formalized by Manes [Man76] and inspired by Kleisli's work [Kle65]:

Definition 3 [Man76]. A *(Kleisli) monad* on Base is given by a function mapping any object X of Base to an object $\mathsf{Coll}X$ also of Base, a family of morphism $\mathsf{sng}_X : X \rightarrow \mathsf{Coll}X$, and the "extension" operation on arrows $p : X \rightarrow \mathsf{Coll}\, Y \mapsto \mathsf{ext}\ p : \mathsf{Coll}\, X \rightarrow \mathsf{Coll}\, Y$ such that ($p : X \rightarrow \mathsf{Coll}\, Y$, $q : Y \rightarrow \mathsf{Coll}\, Z$): :

$$\mathsf{ext}\ \mathsf{sng}_X = \mathsf{id}_{\mathsf{Coll}\, X} \tag{4}$$

$$(\mathsf{ext}\ p) \circ \mathsf{sng}_X = p \tag{5}$$

$$(\mathsf{ext}\ q) \circ (\mathsf{ext}\ p) = \mathsf{ext}\ ((\mathsf{ext}\ q) \circ p) \tag{6}$$

The two definitions are equivalent via the transformations:

$$\mathsf{map}\ f \mapsto \mathsf{ext}\ (\mathsf{sng}_Y \circ f) \tag{7}$$

$$\mathsf{flatten}_X \mapsto \mathsf{ext}\ \mathsf{id}_{\mathsf{Coll}\, X} \tag{8}$$

$$\mathsf{ext}\ p \mapsto \mathsf{flatten}_Y \circ (\mathsf{map}\ p) \tag{9}$$

In particular, the functoriality of map and the naturality of sng and flatten are all derivable from the three axioms of the Kleisli-Manes succinct presentation.

It is precisely the extension operation ext that arises naturally in the work on query languages mentioned above [BBW92, BNTW95]. It is shown there that together with sng, empty set, set union, pairing, projections, conditional and equality, the ext operation can express many useful database operations, in particular the relational and the nested relational algebra.[5].

4.2 Monad algebras

Let (Coll , map , sng, flatten) be a monad. The following is due to Eilenberg and Moore:

Definition4 [Mac71]. A *monad algebra* for the monad Coll is a Coll -algebra (definition 1) $(A,\ \mathsf{agg}_A : \mathsf{Coll}\ A \rightarrow A)$ such that

$$\mathsf{agg}_A \circ \mathsf{sng}_A = \mathsf{id}_A \tag{10}$$

$$\mathsf{agg}_A \circ (\mathsf{map}\ \mathsf{agg}_A) = \mathsf{agg}_A \circ \mathsf{flatten}_A \tag{11}$$

A *homomorphism* of monad algebras is just a homomorphism of Coll -algebras. We denote by MonAlg the category of monad algebras and their homomorphisms.

Note. Tagging agg with A only is an abuse of notation since one may have two different monad algebra structures on the same object A of Base . This will be corrected in the calculus (section 6) while here we try to keep a simple and intuitive notation. In the same spirit we will sometimes refer here to the entire monad algebra as A.

An immediate remark is that from (2) and (3) it follows that flatten is a particular case of agg, that is, $(\mathsf{Coll}\ X, \mathsf{flatten}_X)$ is a monad algebra . In fact, it turns out that the forgetful functor from MonAlg to Base has a left adjoint, with $(\mathsf{Coll}\,X, \mathsf{flatten}_X)$ being the free monad algebra generated by X (via $\mathsf{sng}_X : X \rightarrow \mathsf{Coll}\,X$). The monad canonically associated to this adjunction is Coll again. This property is precisely why monad algebras were introduced, by Eilenberg and Moore [EM65] who managed thus to show that any monad comes out of an adjunction.

Another useful remark is that (11) states that $\mathsf{agg}_A : \mathsf{Coll}\ A \rightarrow A$ itself is a homomorphism.

Examples. For the monad Set , the aggregate operations $\mathsf{agg}_{\mathsf{or}}, \mathsf{agg}_{\mathsf{and}} : \mathsf{Set\ bool} \rightarrow \mathsf{bool}$ and $\mathsf{agg}_{\mathsf{max}}, \mathsf{agg}_{\mathsf{min}} : \mathsf{Set\ num} \rightarrow \mathsf{num}$ form monad algebras [6]. A monad algebra for Bag is given by $\mathsf{agg}_{\mathsf{sum}} : \mathsf{Bag\ num} \rightarrow \mathsf{num}$ and a monad algebra for List is given by $\mathsf{agg}_{\mathsf{concat}} : \mathsf{List\ string} \rightarrow \mathsf{string}$.

Note that disjunction, conjunction, max, and min aggregation are also monad algebras for Bag and List and sum aggregation is also a monad algebra for List (we explore inter-collection relationships in section 7) but sum aggregation is *not* a monad algebra for Set . Indeed, (11) fails: if we apply the left hand side to $\{\{1,2\},\{1,3\}\}$ we get 7 while the right hand side produces 6. We see that the

[5] These formalisms, as well as the calculus developed in section 6 correspond in fact to *strong* monads [Mog91]

[6] Yes, lattices come to mind and this is not accidental—given the correspondence between commutative idempotent monoids and semilattices.

problem comes from the fact that addition is not an idempotent operation while union is.

A Kleisli-style definition for monad algebras We have already remarked that $\mathsf{flatten}$ is an instance of agg. We generalize the extension operation ext that produces collections, hence elements of *free* monad algebras, to an "aggregate extension" operation that produces elements of an arbitrary monad algebra:

Definition 5. A *(Kleisli-style)monad algebra* for the monad Coll consists of an object A of Base and an operation on arrows $r : X \to A \;\mapsto\; \mathsf{agext}\ r : \mathsf{Coll} X \to A$ such that $(p : W \to \mathsf{Coll}\ X)$:

$$(\mathsf{agext}\ r) \circ \mathsf{sng}_X = r \tag{12}$$

$$(\mathsf{agext}\ r) \circ (\mathsf{ext}\ p) = \mathsf{agext}\ ((\mathsf{agext}\ r) \circ p) \tag{13}$$

Similar to an earlier remark, it follows from (5,6) that $(\mathsf{Coll} Y, \mathsf{ext}\ )$ is a monad algebra, that is, ext is a particular case of agext.

The two definitions of monad algebras are equivalent via transformations that generalize (8) and (9):

$$\mathsf{agg}_A \mapsto \mathsf{agext}\ \mathsf{id}_A \tag{14}$$

$$\mathsf{agext}\ r \mapsto \mathsf{agg}_A \circ (\mathsf{map}\ r) \tag{15}$$

Homomorphisms of monad algebras can also be characterized in terms of the Kleisli-style definition:

Proposition 6. *$h : A \to B$ is a homomorphism of monad algebras if and only if:*

$$h \circ (\mathsf{agext}\ r) = \mathsf{agext}\ (h \circ r) \qquad \forall r : X \to A \tag{16}$$

It follows that for any $r : X \to A$, $\mathsf{agext}\ r : \mathsf{Coll}\ X \to A$ is a homomorphism of monad algebras, by (13). In particular, $\mathsf{ext}\ p$, $\mathsf{map}\ f$, agg and $\mathsf{flatten}$ are also homomorphisms. Moreover, the agext construct is intimately connected to the Eilenberg-Moore result. Indeed, let $r : X \to A$, and let $h : \mathsf{Coll} X \to A$ be a homomorphism such that $h \circ \mathsf{sng}_X = r$. By (16), $h \circ (\mathsf{ext}\ \mathsf{sng}_X) = \mathsf{agext}\ (h \circ \mathsf{sng}_X)$ hence $h = \mathsf{agext}\ r$. Therefore:

Theorem 7 The Eilenberg-Moore Adjunction. *For any monad algebra, for any $r : X \to A$, $\mathsf{agext}\ r : \mathsf{Coll}\ X \to A$ is the only homomorphism of monad algebras such that $(\mathsf{agext}\ r) \circ \mathsf{sng}_X = r$.*

4.3 Adding products

The typed calculus that we introduce in section 6 has terms that can have several free variables. Relating such a calculus to its categorical semantics usually requires that Base have finite products, an assumption that we now make. For simplicity, we only introduce notation for binary products, denoted $X \times Y$. Denote by $\pi_1 : X \times Y \to X, \pi_2 : X \times Y \to Y$ the two corresponding projections, and by $\langle f, g \rangle : Z \to X \times Y$ the pairing of arrows $f : Z \to X, g : Z \to Y$. It is straightforward to extend everything (including the calculus) assuming also a nullary products, i.e., a terminator.

Now this extension provides us with an interesting opportunity for modeling "horizontal fusion" of aggregates. Indeed, it is known that the category MonAlg is "as complete as" Base , more precisely, the forgetful functor from MonAlg to Base creates limits [BW85] p117. Therefore, MonAlg has finite products too, and the product of (A, agext_A) and (B, agext_B) is $(A \times B, \mathsf{agext}_{A\times B})$ where $(s : X \to A \times B)$:

$$\mathsf{agext}_{A\times B}\ s = \langle \mathsf{agext}_A\ (\pi_1 \circ s), \mathsf{agext}_B\ (\pi_2 \circ s)\rangle \tag{17}$$

Therefore, to product-aggregate a collection of pairs, project on each component and aggregate separately the resulting collections. By the way, there may be other interesting aggregates (monad algebra structures) that one can define on $A \times B$, but they are not the categorical product. We are interested in fact in using this the other way: fusing different agext's over the same collection into one product agext .

Note. (17) is equivalent, in terms of the original definition of monad algebras, to

$$\mathsf{agg}_{A\times B} = \langle \mathsf{agg}_A \circ (\mathsf{map}\ \pi_1), \mathsf{agg}_B \circ (\mathsf{map}\ \pi_2)\rangle \tag{18}$$

It will be useful to observe that this equation is in turn equivalent to the conjunction of

$$\pi_1 \circ \mathsf{agg}_{A\times B} = \mathsf{agg}_A \circ (\mathsf{map}\ \pi_1) \tag{19}$$

$$\pi_2 \circ \mathsf{agg}_{A\times B} = \mathsf{agg}_B \circ (\mathsf{map}\ \pi_2) \tag{20}$$

which state that the two projections are homomorphisms of monad algebras.

Conclusion for this section: Remembering that ext is a particular case of agext we conclude that *monads and their algebras can be axiomatized in terms of* sng *and* agext *using only (4), (12) and (13). For products add (17).* These are the first ingredients for the calculus introduced in section 6.

5 Enriched Monads and Monad Algebras

The general theory of monads accounts explicitely only for the singleton collection constructor. We wish now to formalize the presence of constructor operations such as binary union of sets or nullary empty set as well as corresponding finitary algebraic operations on monad algebras, such as binary disjunction and nullary false on bool or binary addition and nullary 0 on num . Surprisingly, we have found that with a little extra generality we can treat finitary algebraic operations and *conversions* between collection types (see section 7) in the same semantic framework [7].

Assume an endofunctor $\mathsf{E} : \mathsf{Base} \to \mathsf{Base}$ and denote by E-Alg the category of E-algebras and their homomorphisms. Our first example is when Base has finite (possibly nullary) products and $\mathsf{E}X = X \times \cdots \times X$. Then, E-algebras $\mathsf{E}A \to A$ are just finitary algebraic operations on A. Our second example assumes that E is another monad and uses its monad algebras (which are, in particular, E-algebras).

We present *three different ways* to *enrich* a monad Coll and its monad algebras with structure related to the endofunctor E. These three definitions turn out to be equivalent (theorem 11) which suggests that they capture a rather robust concept.

In the **first** kind of enrichment, we postulate an additional E-algebra structure on each monad algebra for the monad Coll. There are several possibilities for the accompanying *coherence* conditions but they can also be shown to be equivalent:

Definition 8. An E-*algebra enrichment* of the monad Coll (and its monad algebras) is an E-algebra structure $(A, \mathsf{op}_A : \mathsf{E}A \to A)$ on each monad algebra (A, agg_A) such that any one of the following equivalent (Proposition!) coherence conditions hold:

(i) Any homomorphism of monad algebras is also a homomorphism for the corresponding E-algebra structures. In other words, the correspondence $\mathsf{agg}_A \mapsto \mathsf{op}_A$ is a functor $\mathsf{MonAlg} \to \mathsf{E}$-$\mathsf{Alg}$ that commutes with the underlying forgetful functors $\mathsf{MonAlg} \to \mathsf{Base}$ and E-$\mathsf{Alg} \to \mathsf{Base}$.

(ii) $\mathsf{map}\ f$ and agg_A are homomorphisms for the corresponding E-algebra structures.

(iii) For each $r : X \to A$, $\mathsf{agext}_A\ r$ is a homomorphism for the corresponding E-algebra structures.

Notes. Condition (i) is equivalent to asking that op be a natural transformation from $\mathsf{E} \circ \mathsf{U}$ to U where U is the forgetful functor from monad algebras to Base. In (ii), asking that $\mathsf{map}\ f$ be a homomorphism is the same as asking that $\mathsf{op}_{\mathsf{Coll}\,X}$ be a natural transformation from $\mathsf{E} \circ \mathsf{Coll}$ to Coll.

The third coherence condition is the one we will use for our calculus. Let us give a special name to the E-algebra structure on the free monad algebras: $\mathsf{opcoll}_X \stackrel{\text{def}}{=} \mathsf{op}_{\mathsf{Coll}\,X}$ (we shall reuse this name below in a different but equivalent definition of enrichment). With this, condition (iii) becomes:

$$(\mathsf{agext}\ r) \circ \mathsf{opcoll}_X = \mathsf{op}_A \circ (\mathsf{E}\,(\mathsf{agext}\ r)) \qquad \text{where} \qquad r : X \to A \quad (21)$$

Products. Like MonAlg, and with the same construction (minus checking the monad algebra laws), E-Alg has finite products when Base has them.

[7] Although, as we explain later, it is impractical to try to achieve the same generality in the calculus.

Interestingly, it *already* follows from the its definition that the E-algebra enrichment functor MonAlg $\rightarrow$ E-Alg preserves products on the nose. Indeed, we have remarked that the projections are homomorphisms for the monad algebra structures, hence also for the corresponding E-algebra structure, which entails

$$\mathsf{op}_{A\times B} = \langle \mathsf{op}_A \circ (\mathsf{E}\,\pi_1), \mathsf{op}_B \circ (\mathsf{E}\,\pi_2)\rangle \qquad (22)$$

The **second** kind of enrichment we consider is enriching as above but only the free monad algebras (for us—the collection structures), that is, postulating only opcoll instead of op. This generalizes Wadler's notion of *ringad* [WT91].

The **third** kind of enrichment was considered in [Lel94a, Lel94b] and it consists simply of a natural transformation $\mathsf{tconv}_X : \mathsf{E}X \rightarrow \mathsf{Coll}\,X$. The intuition comes from our primary example: $\mathsf{E}X = X \times \cdots \times X$. In this case $\mathsf{tconv}(x_1, \ldots, x_n)$ is the collection built out of the elements of the tuple $(x_1, \ldots, x_n)$ (eg. if the monad is Set the ordering and the duplicates are lost).

We present the equivalence between the three kinds of enrichments in a slightly more general context, in which we postulate not only algebraic operations but also possible laws that such operations may obey (for instance, we may enrich Set monad algebras with commutative idempotent monoid structures). The intention again is to convince that the notions of enrichment we consider are quite robust. Such laws could be added to the equational theory of the calculus in section 6, in conjunction with specifying particular collection types, such as Set .

Definition 9. An *enriching variety* (for lack of a better name!) is a class $\mathcal{EV}$ of E-algebras that is closed under the following property: for any homomorphism of E-algebras $h : A \rightarrow B$ that has a right inverse in Base , $h \circ g = \mathsf{id}_B$[8], if A is in $\mathcal{EV}$ then B is also in $\mathcal{EV}$.

This is not an unfamiliar condition. Algebraic collection types (section 3) arise in equationally defined varieties of algebraic structures and all such varieties have this property [9]. In addition, it is straightforward to verify the following (another robustness check, which will be useful in section 7):

Proposition 10. *If* E *is a monad, then the class of* E*-algebras which are monad algebras is an enriching variety.*

Our main result:

Theorem 11. *Fix a monad and its monad algebras (notation as above). Let $\mathcal{EV}$ be an enriching variety (definition 9). The following three notions of $\mathcal{EV}$-*enrichment *are equivalent:*

(i) *An* E*-algebra enrichment (see Definition 8) such that each* E*-algebra (A, op_A) belongs to $\mathcal{EV}$.*

[8] Such an h is often called a *retract* or a *split epi.*

[9] In fact, this allows us to avoid the explicit use of coequalizers, which otherwise play a central role for monads, as Beck's theorem shows [Mac71, BW85].

(ii) *An* E*-algebra structure* $(\mathsf{Coll}\,X,\ \mathsf{opcoll}_X : \mathsf{E}\,\mathsf{Coll}\,X \rightarrow \mathsf{Coll}\,X)$ *on each free monad algebra, belonging to* $\mathcal{EV}$ *and such that* $\mathsf{map}\ f$ *and* $\mathsf{flatten}$ *are homomorphisms of* E*-algebras (*$\mathsf{map}\ f$ *is a homomorphism also means that* opcoll_X *is a natural transformation).*

(iii) *A family of arrows* $\mathsf{tconv}_X : \mathsf{E}\,X \rightarrow \mathsf{Coll}X$ *which is a natural transformation, i.e.,*

$$(\mathsf{map}\ f) \circ \mathsf{tconv}_X = \mathsf{tconv}_Y \circ (\mathsf{E}\,f) \qquad \textit{where} \qquad f : X \rightarrow Y$$

and such that all E*-algebras of the form* $(\mathsf{Coll}X, \mathsf{flatten}_X \circ \mathsf{tconv}_{\mathsf{Coll}\,X})$ *belong to* $\mathcal{EV}$*.*

Proof Sketch. We will at least give the passages between the three notions and explain how the closure property of the enriching variety is used. (ii) is really a fragment of (i): $\mathsf{opcoll}_X \stackrel{\mathrm{def}}{=} \mathsf{op}_{\mathsf{Coll}\,X}$, as above. The passage from (ii) to (iii) is given by $\mathsf{tconv}_X \stackrel{\mathrm{def}}{=} \mathsf{opcoll}_X \circ (\mathsf{E}\,\mathsf{sng}_X)$. Naturality is immediate, moreover

$$\mathsf{flatten}_X \circ \mathsf{tconv}_{\mathsf{Coll}\,X} = \mathsf{flatten}_X \circ \mathsf{opcoll}_{\mathsf{Coll}\,X} \circ (\mathsf{E}\,\mathsf{sng}_{\mathsf{Coll}\,X}) =$$

$$\mathsf{opcoll}_X \circ (\mathsf{E}\,\mathsf{flatten}_X) \circ (\mathsf{E}\,\mathsf{sng}_{\mathsf{Coll}\,X}) = \mathsf{opcoll}_X \circ (\mathsf{E}\,\mathsf{id}_{\mathsf{Coll}\,X} = \mathsf{opcoll}_X$$

Finally, the passage from (iii) to (i) is given by $\mathsf{op}_A \stackrel{\mathrm{def}}{=} \mathsf{agg}_A \circ \mathsf{tconv}_A$. One can check that any homomorphism of monad algebras is also a homomorphism for the corresponding E-algebra structures. It remains to show that (A, op_A) belongs to $\mathcal{EV}$. This follows from the fact that $\mathsf{agg}_A : \mathsf{Coll}\,A \rightarrow A$ is homomorphism of monad algebras which has a right inverse, sng_A, because the E-algebra structure on $\mathsf{Coll}\,A$, $\mathsf{op}_{\mathsf{Coll}\,A} = \mathsf{agg}_{\mathsf{Coll}\,A} \circ \mathsf{tconv}_{\mathsf{Coll}\,A} = \mathsf{flatten}_A \circ \mathsf{tconv}_{\mathsf{Coll}\,A}$ belongs to $\mathcal{EV}$. □

Anticipating the treatment in section 7, we point out here that in the case when E is a monad itself and $\mathcal{EV}$ is its class of monad algebras (cf. proposition 10) the conditions on tconv in theorem 11 (iii) turn out to be equivalent to tconv being a monad morphism (proposition 13).

Conclusion for this section: Now consider an algebraic collection type and the monad associated with its adjunction semantics. It is easy to see that this monad can be enriched in the sense discussed above with the algebraic operations corresponding to the non-singleton constructors of the algebraic collection type. This is our main example. Remembering that opcoll is particular case of op, we conclude that *enrichments to monads and their algebras can be axiomatized in terms of* op *using only (21). For products, add (17). No need to add (22), it follow from the rest.* We now have all the ingredients for the basic version of our calculus (one collection type). Another example of enrichment is given by the conversions of section 7.

6 The Calculus

Using the concept of enriched monad and its monad algebras as semantic foundation, we develop in this section a calculus of operations for *one* collection type and its aggregates. Extensions to multiple collections and product aggregates are developed in sections 7 and 4.3.

In the calculus we wish to avoid explicit higher-order functions, so all the expressions will denote "elements" of the objects of Base . This is in the spirit of [Mog91, BNTW95] and also of practical query languages. Expressions will have free variables however, and we can bind variables in the principal construct (that corresponds to aggregate extension), so at least the intensional effect of functions as arguments is achieved.

This is a typed calculus and the types are of two kinds Base and MonAlg . We have *type constants* for each kind, such as $\mathsf{or}, \mathsf{and}, \mathsf{max} : \mathsf{MonAlg}$ and $\mathsf{bool}, \mathsf{num} : \mathsf{Base}$. We have the following constructions on types:

$$\frac{\sigma : \mathsf{Base}}{\mathsf{Coll}\,\sigma : \mathsf{Base}} \qquad \frac{\sigma : \mathsf{Base}}{\mathsf{W}\,\sigma : \mathsf{MonAlg}}$$

$$\frac{\sigma_1 : \mathsf{Base} \quad \sigma_2 : \mathsf{Base}}{\sigma_1 \times \sigma_2 : \mathsf{Base}} \qquad \frac{\alpha_1 : \mathsf{MonAlg} \quad \alpha_2 : \mathsf{MonAlg}}{\alpha_1 \times \alpha_2 : \mathsf{MonAlg}}$$

In order to state the typing rules, we *define* an auxiliary meta-function U from MonAlg to Base : $\mathsf{U}(\mathsf{W}\,\sigma) \stackrel{\text{def}}{=} \mathsf{Coll}\,\sigma$, $\quad \mathsf{U}(\alpha_1 \times \alpha_2) = \mathsf{U}\,\alpha_1 \times \mathsf{U}\,\alpha_2$ and, depending on the semantics, links between the MonAlg type constants and the Base type constants, for instance $\mathsf{U}\,\mathsf{or} = \mathsf{bool}$, $\mathsf{U}\,\mathsf{and} = \mathsf{bool}$ or $\mathsf{U}\,\mathsf{max} = \mathsf{num}$.

The terms of the calculus are introduced together with their typing rules in figure 1. We omit the explanation of the (standard) syntax for typing rules. By convention, $\sigma : \mathsf{Base}$ and $\alpha : \mathsf{MonAlg}$. Note that the variables only range over Base types. Note also that in $\Theta_\alpha x \in e_1 \,.\, e_2$, x is a bound variable whose scope is e_2 (but *not* e_1). The usual conventions about bound variables apply.

$$\frac{}{\Gamma, x : \sigma \vdash x : \sigma} \qquad \frac{\Gamma \vdash e : \sigma}{\Gamma \vdash \mathsf{sng}(e) : \mathsf{Coll}\,\sigma}$$

$$\frac{\Gamma, x : \sigma \vdash e : \mathsf{U}\,\alpha \quad \Gamma \vdash S : \mathsf{Coll}\,\sigma}{\Gamma \vdash \Theta_\alpha x \in S \,.\, e : \mathsf{U}\,\alpha} \qquad \frac{\Gamma \vdash e_1, \ldots, e_m : \mathsf{U}\,\alpha}{\Gamma \vdash \mathsf{op}_\alpha(e_1, \ldots, e_m) : \mathsf{U}\,\alpha}$$

$$\frac{\Gamma \vdash e_1 : \sigma_1 \quad \Gamma \vdash e_2 : \sigma_2}{\Gamma \vdash (e_1, e_2) : \sigma_1 \times \sigma_2} \qquad \frac{\Gamma \vdash e : \sigma_1 \times \sigma_2}{\Gamma \vdash \pi_1 e : \sigma_1 \quad \Gamma \vdash \pi_2 e : \sigma_2}$$

Fig. 1. Typing rules

The equational axioms are in figure 2. The following abbreviations are used in the axioms:

$$\Theta x \in S \,.\, R \stackrel{\text{def}}{=} \Theta_{\mathsf{W}\tau} x \in S \,.\, R \qquad \mathsf{opcoll}(S_1, \ldots, S_m) \stackrel{\text{def}}{=} \mathsf{op}_{\mathsf{W}\sigma}(S_1, \ldots, S_m)$$

(proj$_i$)	$\pi_i(e_1, e_2) = e_i$
(surjprod)	$(\pi_1 e, \pi_2 e) = e$
(sng)	$\Theta\, x \in S \,.\, \mathsf{sng}(x) = S$
(β)	$\Theta_\alpha x \in \mathsf{sng}(e) \,.\, e' = e'[e/x]$
(assoc)	$\Theta_\alpha x \in (\Theta\, y \in R \,.\, S) \,.\, e = \Theta_\alpha y \in R \,.\, (\Theta_\alpha x \in S \,.\, e)$
(hom)	$\Theta_\alpha x \in \mathsf{opcoll}(S_1, \ldots, S_m) \,.\, e = \mathsf{op}_\alpha(\Theta_\alpha x \in S_1 \,.\, e, \ldots, \Theta_\alpha x \in S_m \,.\, e)$
(aggprod)	$\Theta_{\alpha_1 \times \alpha_2} x \in S \,.\, e = (\Theta_{\alpha_1} x \in S \,.\, \pi_1 e, \Theta_{\alpha_2} x \in S \,.\, \pi_2 e)$

Fig. 2. Axioms

It is worth writing down the calculus version of (22):

$$\mathsf{op}_{\alpha_1 \times \alpha_2}((e_{11}, e_{21}), \ldots, (e_{1m}, e_{2m})) = (\mathsf{op}_{\alpha_1}(e_{11}, \ldots, e_{1m}), \mathsf{op}_{\alpha_2}(e_{21}, \ldots, e_{2m}))$$

and noting that, just as (22) holds in the enriched monad formalization, this equation follows from the axioms in figure 2.

One can give a formal semantics for this calculus in terms of the categorical constructs presented in sections 4 and 5, along the lines of [Mog91] or of [BNTW95], provided that Coll is a *strong* monad. This is well understood so we omit the development. Not to loose sight of the intuition, we note that if **Base** is sets and functions then (confusing syntax and semantics for a moment):

$$\Theta_\alpha x \in S \,.\, e = (\mathsf{agext}\ (\lambda x.e))S$$

where agext corresponds to the monad algebra denoted by α.

We also explain briefly the choice of syntax for representing enrichments (section 5) in the calculus. The problem is that it is not clear how to express "elements" of type $\mathsf{E}X$ when E is an arbitrary endofunctor. So for the one-monad calculus presented in this section we have decided to consider only the particular case when the enrichments are finitary algebraic operations. We also consider without loss of generality only one algebraic operation, since for several operations the enrichments functor can be simply defined as a sum and the various laws can be just added up. In section 7 we consider multiple monads and conversion enrichments, for whose calculus we provide a different kind of syntax. Finally, we have seen in theorem 11 that the enrichments of the free algebras can actually define the enrichments on any monad algebras. This seems to encumbers the notation however, so we found it more useful for understanding optimizations to postulate constructs and axioms for enrichments of arbitrary monad algebras.

We should also explain tagging the terms with types. Polymorphism and type inference for such calculi is an interesting topic but it is not addressed in this paper. Instead, our goal is to provide just enough type annotation so that typable terms have unique types (and unique type derivations), modulo the simple equational constraints that link type constants. Hence the type tag in $\Theta_\alpha x \in S \,.\, e$. Notice however that we didn't tag the construct with σ, because it can be derived from the type of S. Similarly, when $\alpha = \mathsf{W}\,\tau$, we do not need to tag the construct at all because τ can be derived from the type of e. Similarly also for opcoll (except when $m = 0$!).

The reader must have noticed that this is not a presentation of *all* the monad algebras for Coll , just of the free ones and as many additional ones as we postulate through type constants. Indeed, including a mechanism for "user-defined" or "programmed" aggregates runs into the same problems as user-defined structural recursion on algebraic collection types with equational constraints [BTS91, SBT94]: checking the axioms is undecidable, not even r.e.[10]

Examples. Consider the bag monad. To make the notation more suggestive, write $\biguplus x \in S \,.\, R$ instead of $\Theta\, x \in S \,.\, R$. Consider the type constants num : Base , $\mathsf{sum} : \mathsf{MonAlg}$ with $\mathsf{U}\,\mathsf{sum} = \mathsf{num}$. Write $\sum x \in S \,.\, e$ instead of $\Theta_{\mathsf{sum}} x \in S \,.\, e$. As is the case for all the collection types in the Boom hierarchy, the bag monad and its algebras can be enriched with two algebraic operations, one binary (ignored in these examples), and one nullary. Write empty for the nullary opcoll abbreviation and 0 for the nullary $\mathsf{op}_{\mathsf{sum}}$[11]

First we use a somewhat concrete query to show that (assoc) generalizes the idea of *vertical loop fusion* [Won94]. We shall also need a conditional, but adding it to the calculus is straightforward. Let B be a bag of pairs of numbers and suppose that we want the sum of all the first components of the pairs in the bag whose second components satisfy a given predicate $P(\cdot)$. An efficient answer to this problem is of course the expression $\sum z \in B \,.\, \mathsf{if}\ P(\pi_2 z)\ \mathsf{then}\ \pi_1 z\ \mathsf{else}\ 0$. But a more "modular" mind may well think as follows: first select the pairs whose second component satisfies the predicate, producing a bag B' and then add up the first components of the pairs in B':

$$\sum x \in (\biguplus y \in B \,.\, \mathsf{if}\ P(\pi_2 y)\ \mathsf{then}\ \mathsf{sng}(y)\ \mathsf{else}\ \mathsf{empty}) \,.\, \pi_1 x$$

By applying the (assoc) axiom, followed by a standard conditional transformation, and then by the (hom) and (β) axioms, we actually obtain the efficient answer given above.

---

[10] This kind of extensibility may well be desirable, and could be present in practical languages (in fact, why not user-defined collection types?). But programmers must use it at their own risks: if the structure they programmed is not an aggregate or not a collection type then the optimizer may make mathematically invalid transformations.

[11] We should have tagged empty with a type, but for these examples, the types is easily inferred from the mathematical context.

Our second example shows that (aggprod) realizes a form of *horizontal loop fusion*, quite a bit more general than that considered in [Won94]. Suppose we wish to compute the average of a bag of numbers. The immediate answer is

$$\mathsf{div}\,(\textstyle\sum x \in B\,.\,x, \sum x \in B\,.\,1)$$

A naive implementation would traverse B twice. If access to B is expensive, we are better off applying (aggprod) (right to left!) which produces

$$\mathsf{div}\; \Theta_{\mathsf{sum}\times\mathsf{sum}} x \in B\,.\,(x, 1)$$

To make this truly useful in practice, one may have to detect, using a common subexpression algorithm, loops that iterate over the same collection in different positions in an expression and replace them with projections from the result of a product aggregate. While we can certainly do horizontal fusion without justifying it with products of monad algebras (!), the value of this formalization lies in being able to mix this optimization together with others, all based on rewritings modulo a uniform equational theory.

7 Multiple Collections and Conversion Enrichments

Query languages are seldom based on only one collection type. SQL works with both bags and sets, while object-oriented query languages can in addition manipulate lists. Moreover, these languages use explicit or implicit conversions from lists to bags (ordering removal) and from bags to sets (duplicate removal)[12].

We shall see that such conversions also behave like enrichments, although in this case we enrich a monad (and its algebras), with monad algebra structures corresponding to another monad! When we put the two monads together, the axioms satisfied by the conversions provide nice compatibility conditions between the corresponding constructs of the monads.

Moreover, each monad (sets, bags, etc.) is separately enriched by finitary algebraic operations. We will look at the situation when these enrichments are compatible with each other modulo the conversions.

Definition 12 [BW85]. Let Coll^1 and Coll^2 be two monads on the same category Base (rest of the notation as in section 4, with corresponding superscript 1 or 2). A *morphism of monads* is a family of arrows $\mathsf{conv}_X : \mathsf{Coll}^1 X \to \mathsf{Coll}^2 X$ such that $(f : X \to Y)$:

$$(\mathsf{map}^2\, f) \circ \mathsf{conv}_X = \mathsf{conv}_Y \circ (\mathsf{map}^1\, f) \tag{23}$$
$$\mathsf{conv}_X \circ \mathsf{sng}^1_X = \mathsf{sng}^2_X \tag{24}$$
$$\mathsf{conv}_X \circ \mathsf{flatten}^1_X = \mathsf{flatten}^2_X \circ (\mathsf{map}^2\, \mathsf{conv}_X) \circ \mathsf{conv}_{\mathsf{Coll}^1 X} \tag{25}$$

[12] See the SQL and OQL "select unique" constructs.

The first condition says that conv is a natural transformation and the third condition is clear when we notice that $(\mathsf{map}^2\ \mathsf{conv}_X) \circ \mathsf{conv}_{\mathsf{Coll}^1 X}$ (by naturality, the same as $\mathsf{conv}_{\mathsf{Coll}^2 X} \circ (\mathsf{map}^1\ \mathsf{conv}_X)$) is the horizontal composition of conv with itself [Mac71].

In the spirit of the succinct presentations of section 4 and keeping in mind the intended formalization for the calculus, we remark that (23) and (25) can be replaced by a single axiom ($p : X \to \mathsf{Coll}^1\ Y$):

$$(\mathsf{ext}^2\ (\mathsf{conv}_Y \circ p)) \circ \mathsf{conv}_X = \mathsf{conv}_Y \circ (\mathsf{ext}^1\ p) \tag{26}$$

Proposition 13. *A natural transformation* $\mathsf{Coll}^1 \to \mathsf{Coll}^2$ *gives an* MonAlg^1 *-enrichment (as defined in theorem 11) of* Coll^2 *if and only if it is a morphism of monads.*

Proof sketch. We have already shown (proposition 10) that the class MonAlg^1 of monad algebras for Coll^1 is an enriching variety. Now we can put to use the third characterization of enrichments ((iii) in theorem 11). Using the properties of morphisms of monads, one can check that the Coll^1-algebra ($\mathsf{Coll}^2\ X$, $\mathsf{flatten}^2_X \circ \mathsf{conv}_{\mathsf{Coll}^2 X}$) is an algebra for the monad Coll^1. The converse is straightforward. □

Now all the development in the proof of theorem 11 applies here. Given a monad algebra for Coll^2, $\mathsf{agg}^2_A : \mathsf{Coll}^2\ A \to A$, we obtain a monad algebra for Coll^1 on the same Base object A: $\mathsf{agg}^1_A \stackrel{\text{def}}{=} \mathsf{agg}^2_A \circ \mathsf{conv}_A : \mathsf{Coll}^1\ A \to A$, whose corresponding aggregate extension operation can be defined as $\mathsf{agext}^1\ r \stackrel{\text{def}}{=} \mathsf{agext}^2 \circ \mathsf{conv}_X$.

Suppose now that we have an enriching variety $\mathcal{EV}$, an $\mathcal{EV}$-enrichment of Coll^1, $\mathsf{tconv}^1_X : \mathsf{E}\,X \to \mathsf{Coll}^1\ X$, and a morphism of monads from Coll^1 to Coll^2, $\mathsf{conv}_X : \mathsf{Coll}^1 X \to \mathsf{Coll}^2 X$. It follows from theorem 11 that these two enrichments *compose*, yielding an $\mathcal{EV}$-enrichment of Coll^2, $\mathsf{tconv}^2_X \stackrel{\text{def}}{=} \mathsf{conv}_X \circ \mathsf{tconv}^1_X : \mathsf{E}\,X \to \mathsf{Coll}^2\ X$. We shall assume that this is indeed the $\mathcal{EV}$-enrichment that we desire, as is the case, for instance, with the conversions from lists to bags, or from bags to sets.

If (A, agg^2_A) is monad algebra for Coll^2 we can derive two E-algebra structures on the object A of Base : $\mathsf{op}^2_A \stackrel{\text{def}}{=} \mathsf{agg}^2_A \circ \mathsf{tconv}^2_A$ corresponding to the Coll^2 monad algebra structure and $\mathsf{op}^1_A \stackrel{\text{def}}{=} \mathsf{agg}^1_A \circ \mathsf{tconv}^1_A$ corresponding to the Coll^1 monad algebra structure $\mathsf{agg}^1_A \stackrel{\text{def}}{=} \mathsf{agg}^2_A \circ \mathsf{conv}_A$. But $\mathsf{tconv}^2_X = \mathsf{conv}_X \circ \mathsf{tconv}^1_X$ insures that these are the same.

Looking to extend our calculus, we will have to axiomatize this situation in terms of $\mathsf{opcoll}^1_X : \mathsf{E}\,\mathsf{Coll}^1\ X \to \mathsf{Coll}^1\ X$ rather than in terms of tconv^1. Of course, opcoll^2 is expressible in terms of opcoll^1, but at the price of some lack of elegance:

$$\mathsf{opcoll}^2_X \stackrel{\text{def}}{=} \mathsf{flatten}^2_X \circ \mathsf{conv}_{\mathsf{Coll}^2 X} \circ \mathsf{opcoll}^1_{\mathsf{Coll}^2 X} \circ (\mathsf{E}\,\mathsf{sng}^1_{\mathsf{Coll}^2 X})$$

So we prefer to postulate both opcoll^2 and opcoll^1, and link them by observing that

$$\mathsf{conv}_X \circ \mathsf{tconv}^1_X = \mathsf{tconv}^2_X \quad \longleftrightarrow \quad \mathsf{conv}_X \circ \mathsf{opcoll}^1_X = \mathsf{opcoll}^2_X \circ (\mathsf{E}\,\mathsf{conv}_X)$$

To build a calculus for multiple collections, we put together the calculi for Coll^1 and Coll^2 as defined in section 6 (taking care to keep the same Base) and we add one term construct:

$$\frac{\sigma : \mathsf{Base} \qquad S : \mathsf{Coll}^1\,\sigma}{\mathsf{conv}(S) : \mathsf{Coll}^2\,\sigma}$$

as well as three new axioms, see figure 3.

$$\mathsf{conv}(\mathsf{sng}^1(x)) = \mathsf{sng}^2(x)$$

$$\mathsf{conv}(\Theta^1\, x \in S\,.\,R) = \Theta^2\, x \in \mathsf{conv}(S)\,.\,\mathsf{conv}(R)$$

$$\mathsf{conv}(\mathsf{opcoll}^1(S_1,\ldots,S_m)) = \mathsf{opcoll}^2(\mathsf{conv}(S_1),\ldots,\mathsf{conv}(S_m))$$

Fig. 3. Conversion axioms

According to the preceding discussion, the calculus behaves *as if* we had a new type construct that associates to each $\alpha : \mathsf{MonAlg}^2$ an $\mathsf{conv}(\alpha) : \mathsf{MonAlg}^1$ such that $\mathsf{U}^1\mathsf{conv}(\alpha) = \mathsf{U}^2\alpha$. But we do not need to add such a construct because the corresponding aggregate extension and algebraic enrichment operations are already definable, as we saw: $\Theta^1_{\mathsf{conv}(\alpha)} x \in S\,.\,e \stackrel{\text{def}}{=} \Theta^2_\alpha x \in \mathsf{conv}(S)\,.\,e$ and $\mathsf{op}^1_{\mathsf{conv}(\alpha)}(e_1,\ldots,e_m) \stackrel{\text{def}}{=} \mathsf{op}^2_\alpha(e_1,\ldots,e_m)$

However, during calculations it will be useful to keep these abbreviations in mind, knowing that the fact that they satisfy the axioms of the calculus as if they were among its primitives is provable from the axioms already given.

8 Related and Further Work, Conclusions

The use of monads for collection types is inspired by the work of Moggi, Wadler, Trinder and Watt [Mog91, Wad92, Tri91, WT91]. The calculi considered here and in [BNTW95] are similar to those considered by Moggi for general monads [Mog91]. Mulry [Mul92, Mul92] has used monad algebras in his treatment of semantics for partial datatypes and fixpoints. His recent results on liftings appear to be related to our notion of monad enrichment [Mul97], a connection that should be investigated. Recent work by Filinski on the semantics of effects and continuations using monads [Fil96] shows intriguing similarities with our use of monad algebras, a connection that should also be investigated. Another interesting question is whether the correspondence between finitary algebraic theories and finitary (finite rank) monads (see for example [Bor94]) together with the monad semantics for collections can be used for a general understanding of algebraic collection types (section 3). There are other interesting developments that concern primarily category theorists and that we will present elsewhere, such as connections with tripleability, free monads, and sequi-enriched Kleisli categories [Mac71, BW85]. Because of the space limitations, we must leave for another paper some results about certain equations that hold only for certain collections types. There are also interesting connections with the Bird-Meertens formalism [Bac88, MFP91] where we borrowed the notion of Boom hierarchy. In particular, the recent work on *deforestation*, the elimination of intermediate data structures [GLP93, LS95] is closely related to the loop fusion optimizations we have discussed. These connections should be the subject of further work. Fegaras and Maier [FM95] consider monoid types which are similar to our monad algebra types but quite a bit more particular. Their monoid comprehension calculus can be translated in our calculus, and it captures some of the equivalences derived by our axiomatization, but apparently not all of them.

For the record, we regard this calculus as a basis for validating some of the program equivalences used in query optimizations. This does not yet constitute a query optimizer. Indeed, for some of these axioms, it is not clear at all which orientation, left to right or right to left constitutes an optimization. For others, the orientation is obvious most of the time, because they eliminate intermediate data structures but there are surprising exceptions. To settle these problems, a *cost model* that takes into account the size and the retrieval latency of the data is indispensable. It should also be said that there are query optimizations about which the theory of monads has apparently nothing to say, for instance optimizations that depend on constraints on the data such as the existence of inverse relationships. This paper does not attempt to address such optimizations or cost models.

The main thrust now is to study the calculus we have proposed as an equationally axiomatized (collection) type theory. This is where the methods developed for programming languages in general will prove valuable. A number of relevant results, especially strong normalization properties, have been already obtained by Limsoon Wong [Won93, Won94] (see also [LW94b, LW94a, Di 96]).

In conclusion, we tried to provide a succinct but expressive calculus that can serve as an intermediate language for the optimization of database query languages. Most of the work is of foundational nature—choosing between possible axiomatizations, seeking economy of formalism. The benefit we expect is that we will understand, and thus apply, optimizations in their full generality.

Acknowledgements. We are heavily indebted to Peter Buneman and Limsoon Wong for sharing numerous ideas. Thanks to Scott Harker for a perceptive remark about converting sets to bags. Kazem Lellahi is grateful to University of Pennsylvania's CIS Department and Institute for Research in Cognitive Science while Val Tannen is similarly grateful to LRI, Université de Paris-Sud: the hospitality of both institutions has made this collaboration possible. Val Tannen was partially supported by grants NSF CCR-90-57570 and CCR-92-16122, ARO DAAH04-93-G0129, and DARPA NOOO14-94-1-1086.

Apology. The authors are painfully aware of having written this entire paper without drawing a single commutative diagram. (Lots of them in our working notes though!) This is only due to the time and space constraints of conference submission and we intend to remedy it in an expanded version.

References

[Bac88] R. C. Backhouse. An exploration of the Bird-Meertens formalism. Technical Report CS 8810, Groningen University, The Netherlands, 1988.

[BBW92] Val Breazu-Tannen, Peter Buneman, and Limsoon Wong. Naturally embedded query languages. In J. Biskup and R. Hull, editors, *LNCS 646: Proceedings of 4th International Conference on Database Theory, Berlin, Germany, October, 1992*, pages 140–154. Springer-Verlag, October 1992.

[BNTW95] Peter Buneman, Shamim Naqvi, Val Tannen, and Limsoon Wong. Principles of programming with complex objects and collection types. *Theoretical Computer Science*, 149(1):3–48, September 1995.

[Bor94] F. Borceux. *Handbook of categorical algebra, Vol.2.* Cambridge University Press, 1994.

[BTS91] V. Breazu-Tannen and R. Subrahmanyam. Logical and computational aspects of programming with Sets/Bags/Lists. In *LNCS 510: Proceedings of 18th International Colloquium on Automata, Languages, and Programming, Madrid, Spain, July 1991*, pages 60–75. Springer Verlag, 1991.

[BW85] Michael Barr and Charles Wells. *Toposes, Triples, and Theories.* Springer-Verlag, New York, 1985.

[Cat96] R. G. G. Cattell, editor. *The Object Database Standard: ODMG-93.* Morgan Kaufmann, San Mateo, California, 1996.

[Di 96] R. Di Cosmo. On the power of simple diagrams. In *LNCS 1103: Rewriting Techniques and Applications, New Brunswick, NJ, July 1996*, Berlin, July 1996. Springer-Verlag.

[EM65] S. Eilenberg and J. C. Moore. Adjoint functors and triples. *Illinois Journal of Mathematics*, 9:381–398, 1965.

[Fil96] A. Filinski. *Controlling effects.* PhD thesis, School of Computer Science, Carnegie Mellon University, Pittsburgh, PA 15213, May 1996. Available as CMU Technical Report CMU-CS-96-119.

[FM95] Leonidas Fegaras and David Maier. Towards an effective calculus for object query languages. In *Proceedings of ACM SIGMOD International Conference on Management of Data*, pages 47–58, San Jose, California, May 1995.

[GLP93] A. Gill, J. Launchbury, and S. Peyton Jones. A short cut to deforestation. In *Proceedings of Conference on Functional Programming Languages and Computer Architecture*, pages 223–232, 1993.

[Hoo94] P. Hoogendijk. Relational programming laws in the Boom hierarchy of types. Technical report, Eindhoven University of Technology, The Netherlands, 1994.

[Kle65] H. Kleisli. Every standard construction is induced by a pair of adjoint functors. *Proceedings of the American Mathematical Society*, 16:544–546, 1965.

[Lel94a] K. Lellahi. Towards a characterization of bulk types. Technical Report 94-01, Université Paris 13, LIPN, 1994.

[Lel94b] K. Lellahi. Type de collection et monades. In *Actes des Journées Cateégories, Algèbres, Esquisses et neo-esquisses*, Caen, 1994.

[LS86] J. Lambek and P. J. Scott. *Introduction to Higher Order Categorical Logic*, volume 7 of *Cambridge Studies in Advanced Mathematics*. Cambridge University Press, London, 1986.

[LS95] J. Launchbury and T. Sheard. Warm fusion: deriving build-catas from recursive definitions. In *Proceedings of Conference on Functional Programming Languages and Computer Architecture*, pages 314–323, 1995.

[LW94a] Leonid Libkin and Limsoon Wong. Aggregate functions, conservative extension, and linear orders. In Catriel Beeri, Atsushi Ohori, and Dennis E. Shasha, editors, *Proceedings of DBPL-4, New York, August 1993*, pages 282–294. Springer-Verlag, January 1994.

[LW94b] Leonid Libkin and Limsoon Wong. Some properties of query languages for bags. In Catriel Beeri, Atsushi Ohori, and Dennis E. Shasha, editors, *Proceedings of DBPL-4, New York, August 1993*, pages 97–114. Springer-Verlag, January 1994.

[Mac71] S. MacLane. *Categories for the Working Mathematician*. Springer-Verlag, Berlin, 1971.

[Man76] Ernest G. Manes. *Algebraic Theories*, volume 26 of *Graduate Texts in Mathematics*. Springer-Verlag, Berlin, 1976.

[MFP91] Erik Meijer, Maartens Fokkinga, and Ross Paterson. Functional programming with bananas, lenses, envelopes, and barbed wires. In J. Hughes, editor, *LNCS 523: 5th ACM Conference on Functional Languages and Computer Architecture*, pages 124–144. Springer-Verlag, August 1991.

[Mog91] Eugenio Moggi. Notions of computation and monads. *Information and Computation*, 93:55–92, 1991.

[Mul92] P. Mulry. Monads and algebras in the semantics of partial data types. *Theoretical Computer Science*, 99:141–155, 1992.

[Mul97] P. Mulry, May 1997. Private communication.

[SBT94] Dan Suciu and Val Breazu-Tannen. A query language for NC. In *Proceedings of 13th ACM Symposium on Principles of Database Systems*, pages 167–178, Minneapolis, Minnesota, May 1994. See also UPenn Technical Report MS-CIS-94-05.

[Tri91] P. W. Trinder. Comprehensions, a query notation for DBPLs. In *Proceedings of 3rd International Workshop on Database Programming Languages, Nahplion, Greece*, pages 49–62. Morgan Kaufmann, August 1991.

[Wad92] Philip Wadler. Comprehending monads. *Mathematical Structures in Computer Science*, 2:461–493, 1992.

[Won93] Limsoon Wong. Normal forms and conservative properties for query languages over collection types. In *Proceedings of 12th ACM Symposium on Principles of Database Systems*, pages 26–36, Washington, D. C., May 1993. See also UPenn Technical Report MS-CIS-92-59.

[Won94] Limsoon Wong. *Querying Nested Collections*. PhD thesis, Department of Computer and Information Science, University of Pennsylvania, Philadelphia, PA 19104, August 1994. Available as University of Pennsylvania IRCS Report 94-09.

[WT91] David A. Watt and Phil Trinder. Towards a theory of bulk types. Fide Technical Report 91/26, Glasgow University, Glasgow G12 8QQ, Scotland, July 1991.

Lifting

Anna Bucalo and Giuseppe Rosolini*

Dipartimento di Matematica, via Dodecaneso 35, 16146 Genova, Italy

Abstract. The problem we consider is that of recognizing a lifting functor in a category, possibly lacking part of the usual structure required to define the partial maps. We solve it in the general case of a category with a terminal object; this indicates that a lifting functor does not depend on having pullbacks in the category. An interesting characterization of lifting is obtained when we specialize our result to categories with finite products.
This approach to partiality seems new in the literature on categories of partial maps, although it is certainly implicit in the computational point of view, and it relates directly to the question of modularity.

1 Introduction

The categorical aspects of partiality have been studied in a variety of fashions, see *e.g.* [1–3, 6–9, 12, 15, 16]. In all cases but [6], the structure of a category of partial maps is taken as primitive, and a characterization is provided in terms of a suitably defined category of "total" maps. Moreover, additional structure on the category is assumed, usually in the form of monoidal structure with special properties—some reasons for this are explained in [16]. The characterization allows the possibility of extending the category of total maps to a topos, and there one obtains representability of partial maps, see [13] which reports on the quoted result from the less easily available [17].

The approach of [6] considers different data for a notion of partiality: instead of taking the category of partial maps as primitive, it considers a category of total maps **C** together with a notion of "undefinedness" as expressed by an endofunctor $L: \mathbf{C} \longrightarrow \mathbf{C}$ which one should think of as *adding* an element to an object to denote divergence. A simple example is $\mathbf{C} = \mathbf{Pos}$, the category of posets and monotone functions, and $L = (-)_\perp$, the functor which adds a new least element to a poset.

The endofunctor L is constructed in [6] by having monomorphisms in the locally small category **C** representing the subobjects which are domains of definition for the category of partial maps to be introduced. These form a collection $\mathcal{M}$ of monos of **C**, closed under inverse images (in the sense that a pullback of a mono in $\mathcal{M}$ along any map always exists, and is in $\mathcal{M}$). In the example on **Pos**, the collection $\mathcal{M}$ consists of those order-reflecting monomorphisms whose image is upward closed. Then the functor L is completely determined as LC represents

* Research supported by MURST 40% and by HCM project 'Typed lamba-calculus'

partial maps into C defined on monos in $\mathcal{M}$, *i.e.* spans $X \xleftarrow{m} A \xrightarrow{g} C$ such that the left leg m is in $\mathcal{M}$, indicated by a hooked arrow. Closure under inverse images is exactly what is required to obtain a functor $\mathrm{Ptl}_{\mathcal{M}}(-, C): \mathbf{C}^{\mathrm{op}} \longrightarrow \mathbf{Set}$ of partial maps into a given codomain C, and LC is a representing object for this functor.

To see this in more detail, let $LC \xleftarrow{\eta_C} E_C \xrightarrow{\varpi_C} C$ be the universal element for the representation. The following properties are well-known:

1. $\eta_C: E_C \longrightarrow LC$ is in $\mathcal{M}$, hence has all inverse images.
2. L extends immediately to an endofunctor on $\mathbf{C}$. It preserves all existing connected limits, as follows directly from the representation.
3. $\varpi_C: E_C \longrightarrow C$ is iso if and only if $\mathrm{id}: C \longrightarrow C$ is in $\mathcal{M}$. (Assuming id_C is in $\mathcal{M}$, one sees that ϖ_C is a retraction by considering the span $C \xleftarrow{\mathrm{id}} C \xrightarrow{\mathrm{id}} C$. Then it is easy to see that ϖ_C is monic. The converse follows from 1.)
4. In case all identities are in $\mathcal{M}$, one sees $\eta: \mathrm{Id}_{\mathbf{C}} \longrightarrow L$ is a cartesian natural transformation—a natural transformation is *cartesian* when all naturality squares are pullbacks.
5. The following are equivalent:
   (i) The class $\mathcal{M}$ is closed under composition,
   (ii) there is a (necessarily unique) map $m: L^2 1 \longrightarrow L1$ such that the diagram

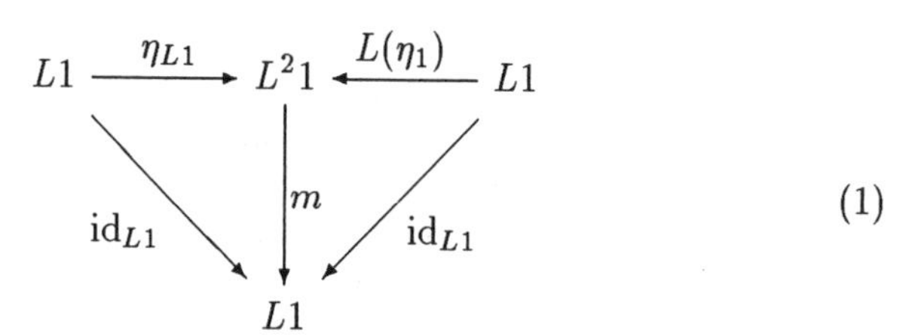

(1)

   commutes,
   (iii) there is a natural transformation $\mu: L^2 \longrightarrow L$ such (L, η, μ) is a monad.
   Under such circumstances, μ (as well as m) is uniquely determined, and it is cartesian.
6. It follows from 5 that the span $L1 \xleftarrow{L!} L^2 C \xrightarrow{\mu_C} LC$ is jointly monic.

The representation takes the pleasing form that the fill-in, in the diagram below, is in natural bijection with the left-upper pair of maps (up to a commuting iso)

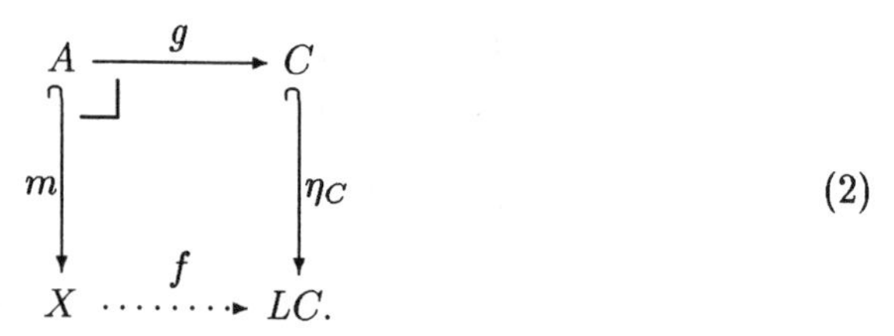

(2)

When $\mathcal{M}$ satisfies (any of) the equivalent conditions in 3 and in 5, one says that $\mathcal{M}$ is a class of *admissible* monomorphisms. A pointed endofunctor (L, η), which is determined by the representation of partial maps on monos in $\mathcal{M}$ as in (2), is usually called a *lifting* (for $\mathcal{M}$-partiality).

The problem we want to consider is that of recognizing a lifting functor on a category even if the category lacks part of the structure required to define the representation: no class of admissible monos is given in advance, nor is the category required to have pullbacks. We shall solve it in the general case of a category with a terminal object; this indicates that a lifting functor does not depend on the category having enough pullbacks. The significance of pullbacks is that, from the pointed endofunctor, one obtains the collection of admissible monomorphisms, but this is assured only under existence of pullbacks of the maps of the form η_C. Our main result is that every category with a lifting can be fully embedded into one with an admissible class $\mathcal{M}$ of monos, such that the maps $\eta_C: C \longrightarrow LC$ are universal with respect to $\mathcal{M}$-partial maps. In particular, we obtain an interesting characterization of lifting when we specialize our result to categories with finite products.

This approach to partiality does not appear in any of the literature on categories of partial maps since the central rôle was always played by the *Kleisli category* of (the monad on) L: the category of total maps was only instrumental, and was definable from that of partial maps (when this came with sufficient structure). On the other hand, it is certainly implicit in the point of view proposed by Moggi in [13], and it relates directly to the question of modularity in that context, see [14]. Once the structure of a lifting is known, one knows how to mix that with other computational structures.

For an example to illustrate why one might be interested in a lifting functor without an admissible class of monomorphisms, consider the category **1CPO** of directed complete posets with least element and Scott-continuous maps. Adding a new least element is (probably the most commonly studied example of) a lifting as one sees immediately when **1CPO** is embedded in the supercategory **CPO** of cpo's (not necessarily with a least element) and Scott-continuous maps.

Another example is the category of enumerated sets, [5, 4]: an *enumerated set* is a surjection $e: N \twoheadrightarrow A$ from the set of natural numbers. A map $f: e \longrightarrow e'$ from the enumerated set $e: N \twoheadrightarrow A$ to $e': N \twoheadrightarrow A'$ is a function $f: A \longrightarrow A'$ such that there is a recursive function g such that

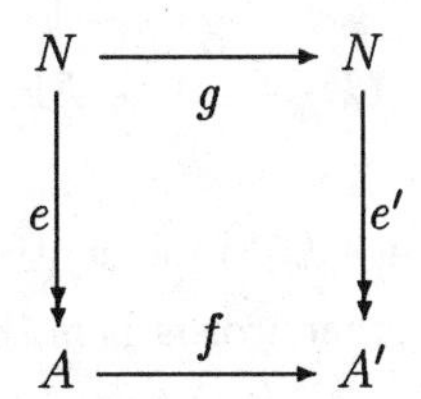

commutes. The category **EN** is obtained using standard set-theoretic composition.

Eršov noted that the operation of adding an *undefined* element $N \cup \{\uparrow\}$ to denote divergence for partial recursive functions extends to enumerated sets: $Le: N \twoheadrightarrow A \cup \{\uparrow\}$ first computes the nth partial recursive function on input 0, then uses the output m to list the element $e(m) \in A$.

In Sect. 2 we review some known properties of the monad obtained from the representation of partial maps on admissible monomorphisms. In Sect. 3 we define a new notion of lifting on a category **C**, assuming **C** has a terminal object, and we prove a representation theorem introducing a certain category $\mathbf{C_t}$ satisfying a universal property with respect to categories with a lifting representing partial maps. In Sect. 4 we show that, when a category has finite products, a lifting can be recognized by algebraic conditions on the pointed endofunctor. We draw some comparisons with other work. In Sect. 5 we sketch some applications to free constructions and to the case of toposes.

2 Properties on Lifting as a Representation

In this section, we review some properties of the monad obtained from the representation of partial maps on admissible monomorphisms. For the purpose of discussion, we suppose that **C** is a category with a terminal object 1 and $L: \mathbf{C} \longrightarrow \mathbf{C}$ is a functor on **C**, and that a collection $\mathcal{M}$ of admissible monos is given on **C** such that LC represents partial maps into C defined on admissible monos.

Note that L determines the admissible monomorphisms as the pullbacks of $\eta_1: 1 \longrightarrow L1$ along maps $C \longrightarrow L1$.

There are two equivalent ways to define the category of partial maps. They are both standard constructions: one is the Kleisli category $\mathbf{C}_L$ on the monad L, the other is determined as the quotient category $\mathrm{Ptl}_{\mathcal{M}}$—by taking the poset reflection of each hom—of the bicategory of those spans $C \xleftarrow{m} A \xrightarrow{f} B$ such that the left leg m is an admissible mono, see *e.g.* [13, 16] for explicit descriptions.

By the universal property of η_C, an admissible mono $m: A \hookrightarrow C$ determines two maps from C: one is the classifying map of m, the unique map filling in

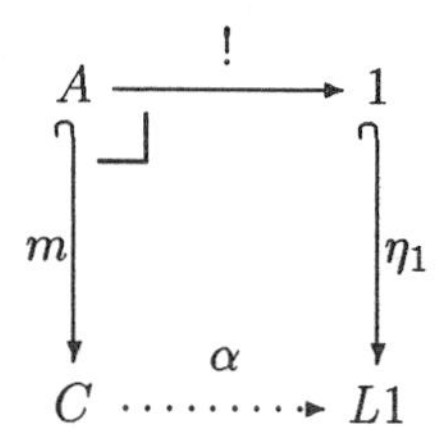

so as to make it a pullback.

Another is the map $a: C \longrightarrow LC$ representing the partial map given by m itself as $C \xleftarrow{m} A \xrightarrow{m} C$, in other words it is the unique map filling in

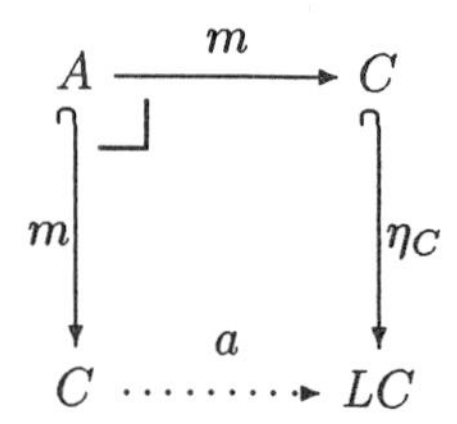

so as to make it a pullback.

It is useful to understand the behaviour of these maps in the Kleisli category $\mathbf{C}_L$. As we look at more than one category structure on the arrows in $\mathbf{C}$, we shall use an index $\longrightarrow_L$ and $\circ_L$ to indicate an arrow and composition in the Kleisli category.

The map $\alpha: C \longrightarrow_L 1$ is an object of the comma category $\mathbf{C}_L/1$—recall that 1 is a terminal object in $\mathbf{C}_L$ exactly when L is the identity.

The map $a: C \longrightarrow_L C$ is an idempotent on the object C in $\mathbf{C}_L$. Note that $\alpha = L! \circ a$. Moreover, supposing $b: C \longrightarrow_L C$ is the map obtained as representation of another partial map of the form $C \xleftarrow{n} B \xrightarrow{n} C$, one has that, taking their composites as endomaps on C in $\mathbf{C}_L$

$$b \circ_L a = a \circ_L b. \tag{3}$$

The explicit condition in $\mathbf{C}$ becomes, of course, $\mu_C \circ Lb \circ a = \mu_C \circ La \circ b$. Moreover the composite in (3) is the idempotent obtained from the admissible mono $m \cap n$, the intersection of m and n: this is defined as the diagonal from P to C in the pullback

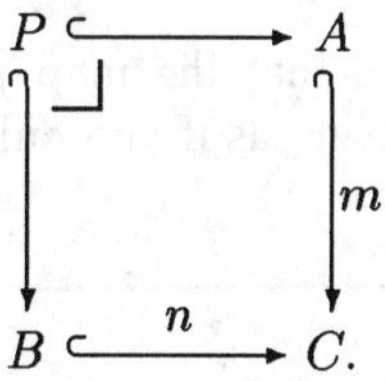

$$\begin{array}{ccc} P & \hookrightarrow & A \\ \downarrow & \lrcorner & \downarrow{\scriptstyle m} \\ B & \xhookrightarrow{n} & C. \end{array}$$

Finally, if $g: A \longrightarrow A'$ is a map in $\mathbf{C}$ between the domains of the admissible monos $m: A \hookrightarrow C$ and $m': A' \hookrightarrow C'$, then consider the fill-in

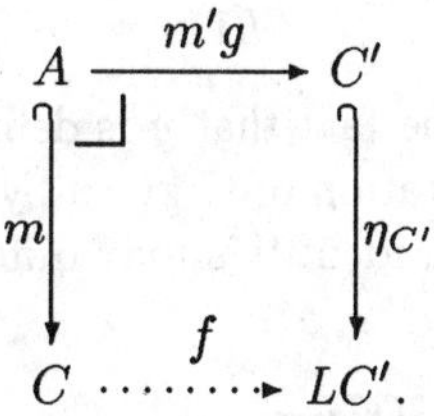

$$\begin{array}{ccc} A & \xrightarrow{m'g} & C' \\ {\scriptstyle m}\downarrow & \lrcorner & \downarrow{\scriptstyle \eta_{C'}} \\ C & \overset{f}{\cdots\cdots\rightarrow} & LC'. \end{array}$$

The condition that f is obtained from a map actually between A and A' is the commutativity of

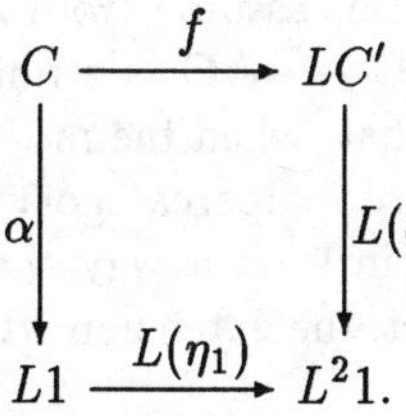

$$\begin{array}{ccc} C & \xrightarrow{f} & LC' \\ {\scriptstyle \alpha}\downarrow & & \downarrow{\scriptstyle L(\alpha')} \\ L1 & \xrightarrow{L(\eta_1)} & L^2 1. \end{array}$$

Since the pair $L1 \xleftarrow{L!} L^2 1 \xrightarrow{\mu_1} L1$ is jointly monic (see 6 of section 1), this can be restated equivalently as two commuting diagrams. These are usefully read in $\mathbf{C}_L$ as

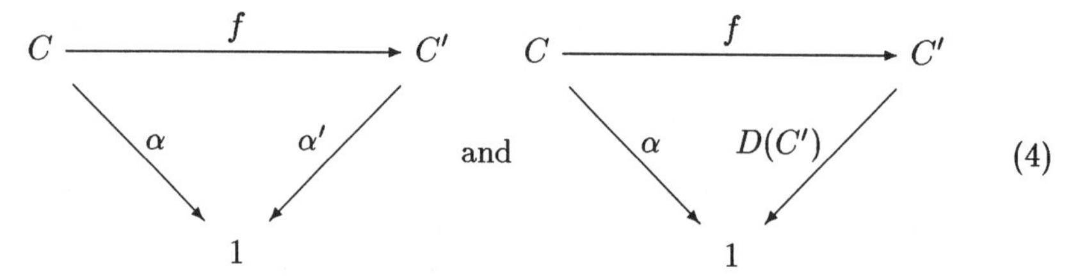

where $D(C') = \eta_1 \circ ! = L! \circ \eta_{C'}$ is the map classifying the maximal admissible subobject of C'.

Summing up, each admissible mono $m: A \hookrightarrow C$ in $\mathbf{C}$ determines an object $\alpha: C \longrightarrow_L 1$ of $\mathbf{C}_L/1$: in fact,

Proposition 1. *There is a bijection between isomorphism classes of admissible monos and objects of* $\mathbf{C}_L/1$. *Every map* $g: A \longrightarrow A'$ *in* $\mathbf{C}$ *determines a map* $f: \alpha \longrightarrow_L \alpha'$ *in* $\mathbf{C}_L/1$ *such that*

$$\alpha = D(C') \circ_L f. \tag{5}$$

Note that equation (5) states that the map f of $\mathbf{C}$ gives rise to a map, in $\mathbf{C}_L/1$, from α into $D(C')$, and it holds if and only if the diagram

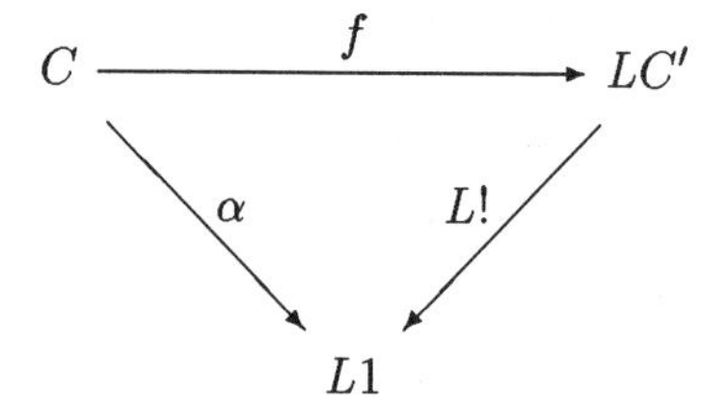

commutes in $\mathbf{C}$: it expresses the fact that g is defined *totally* on A.

Also, in general, the presentation of $\mathbf{C}$ given by Proposition 1 does not result in a subcategory of $\mathbf{C}_L/1$ since identities may differ.

3 A New Notion of Lifting

In this section we introduce a more general notion of lifting than that based on admissible monos. To this purpose, assume from now on that $\mathbf{C}$ is a category with a terminal object 1 and $L: \mathbf{C} \longrightarrow \mathbf{C}$ is a functor on $\mathbf{C}$ which underlies a monad (L, η, μ). We shall prove that, when the monad satisfies suitable algebraic conditions, $\mathbf{C}$ can be extended to a supercategory which comes equipped with a class $\mathcal{M}$ of admissible monos in such a way that the maps $\eta_C: C \longrightarrow LC$ represent $\mathcal{M}$-partial maps. In fact, the extension will be the "best" such, the free solution to the problem.

Let $\mathbf{C}_t$ consist of the same objects as $\mathbf{C}_L/1$, and of those arrows $f: \alpha \longrightarrow \alpha'$ in $\mathbf{C}_L/1$ such that f also defines a map from α to $D(C')$ as in (4).

The standard notation, which we employ, is slightly misleading: a map in a comma category is actually a triple (f, α, α') generating a certain commutative triangle. So the condition selecting the appropriate arrows in $\mathbf{C}_t$ is not categorical. And we have carefully chosen the words introducing objects and arrows of the graph $\mathbf{C}_t$, because it may not define a *category.* Composition is inherited from $\mathbf{C}_L/1$, but identities need not exist. For instance, the identity $\eta_C: C \longrightarrow_L C$ in $\mathbf{C}_L$ determines an identity in $\mathbf{C}_t$ only on the object $D(C)$.

Definition. Let (L, η, μ) be a cartesian monad on $\mathbf{C}$, in the sense that both natural transformations η and μ are cartesian. Let $\mathbf{C}_t$ be the graph defined as above. The monad (L, η, μ) is a *lifting* if

- L preserves existing pullbacks,
- the pair $L1 \xleftarrow{L!} L^2 1 \xrightarrow{\mu_1} L1$ is jointly monic,
- there are maps in $\mathbf{C}$ which define identity maps in $\mathbf{C}_t$ (*i.e.* $\mathbf{C}_t$ is a category),
- any two of these with the same source in $\mathbf{C}$ (hence, the same target) commute in $\mathbf{C}_L$, *i.e.* satisfy (3), and their composition is an identity.

For an object $\alpha: C \longrightarrow L1$ of $\mathbf{C}_t$ (a map in $\mathbf{C}$), we write $a: C \longrightarrow LC$ for the map which defines the identity on α in $\mathbf{C}_t$. Arrows and composition in $\mathbf{C}_t$ will be denoted as $\longrightarrow_t$ and $\circ_t$.

The assignment $C \longmapsto D(C) = \eta_1 \circ !$ (like in Sect. 2) extends to a functor $D: \mathbf{C} \longrightarrow \mathbf{C}_t$ using the left adjoint into the Kleisli category: one defines the action on maps as

$$(f: C \longrightarrow C') \longmapsto \eta_{C'} f = L(f) \eta_C .$$

Theorem 2. *The functor $D: \mathbf{C} \longrightarrow \mathbf{C}_t$ is full and faithful, and preserves the terminal object. Moreover, D preserves all limits which exist in $\mathbf{C}$ and are preserved by $L: \mathbf{C} \longrightarrow \mathbf{C}$. In particular, D preserves pullbacks; hence all existing finite products.*

Proof. Full faithfulness follows from cartesianness of η. As to preservation of the terminal object, note the unique map from an object $\alpha: C \longrightarrow_L 1$ into $D(1) = (\mathrm{id}_{L1}: L1 \longrightarrow_L 1)$ must be α itself. Also the rest is simple diagram chasing. □

Theorem 3. *The category $\mathbf{C}_t$ has a class of admissible monomorphisms for which the maps $\eta_C: C \longrightarrow LC$ represent partial maps. The full embedding $D: \mathbf{C} \longrightarrow \mathbf{C}_t$ is universal with that property.*

Proof. This is straightforward. The identity $a: \alpha \longrightarrow_t \alpha$, for $\alpha: C \longrightarrow_L 1$, determines a mono $a: \alpha \longrightarrow_t D(C)$. Moreover, it is immediate to see that, for b the identity on $\beta: C \longrightarrow_L 1$,

$$a \leq b \text{ if and only if } a = b \circ_L a = a \circ_L b$$

since, if $f: \alpha \longrightarrow_t \beta$ is such that $b \circ_L f = a$, then $f = a$ because b is the identity on β. Define the admissible monos as those maps of the form $a: \alpha \hookrightarrow_t \beta$ where also $\beta: C \longrightarrow_L 1$.

Representation is given by the subobject $D(\eta_C): D(C) \longrightarrow_t D(LC)$. Note that it is isomorphic to $L\eta_C: L(!_C) \longrightarrow_t D(LC)$.

Universality of D means that, given any category **A** with a class of admissible monos and a functor $F: \mathbf{C} \longrightarrow \mathbf{A}$ such that $F(\eta_C: C \longrightarrow LC)$ is the representation of partial maps into $F(C)$ for every object C in **C**, there is an extension $G: \mathbf{C}_t \longrightarrow \mathbf{A}$ which preserves the representing pullbacks, and the extension is unique (up to a unique natural isomorphism) with this property. The proof is easy. □

In the case of the category **lCPO**, the extension $\mathbf{lCPO}_t$ is **CPO**, the category of directed complete posets. In the case of the category **EN** of enumerated sets, the extension $\mathbf{EN}_t$ has just one initial object added.

Since $D: \mathbf{C} \longrightarrow \mathbf{C}_t$ is a universal full embedding with respect to the notion of lifting, it is possible to deduce diagrammatic properties for the monad components in **C**. For instance, the diagram

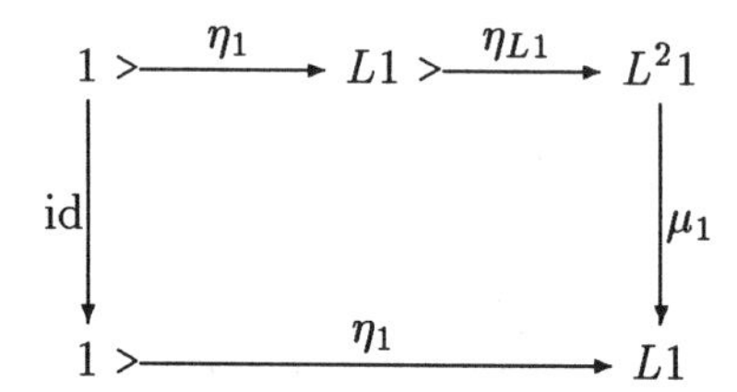

is a pullback in **C**, as follows from 5 of Sect. 1, since D reflects pullbacks by Theorem 2.

4 An Application to Categories with Finite Products

We reconsider the conditions on the pointed endofunctor $L: \mathbf{C} \longrightarrow \mathbf{C}$ under the further assumption that **C** has binary products. The increased definability allows us to give a more compact presentation for a lifting as defined in Sect. 3. In particular, the monad inherits a strength: a natural transformation

$$s_{C,C'}: C \times L(C') \longrightarrow L(C \times C')$$

which internalizes the action of the functor L as well as that of η and μ, see [11]. For instance, the two triangles which appear in the statemente of Prop. 4 internalize the facts that L preserves identities and that η is a strong natural transformation, see [10] for a quick recap on strong monads and related notions.

An application of the full embedding of Theorem 3 gives the following result, which is standard for a functor which represents partial maps.

Proposition 4. *Suppose that* **C** *is a category with finite products and* (L, η, μ) *is a lifting on* **C**. *Then there is a unique natural transformation*

$$s_{C,C'}: C \times L(C') \longrightarrow L(C \times C')$$

making the following diagrams commute, q denoting second projections,

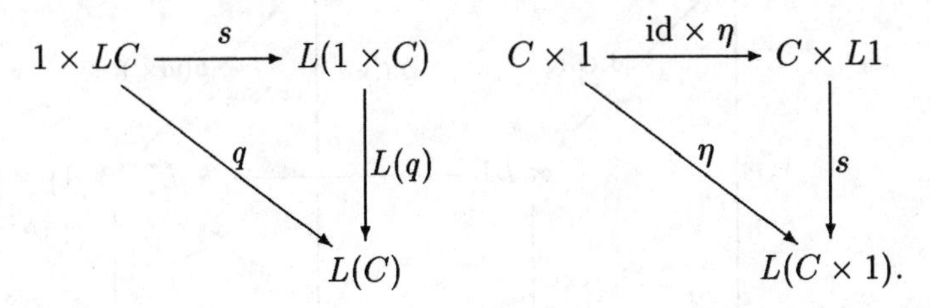

Proof. We employ the full embedding $D: \mathbf{C} \longrightarrow \mathbf{C}_t$: for simplicity, we identify $\mathbf{C}$ with its image in D. Define $s_{C,C'}: C \times L(C') \longrightarrow L(C \times C')$ as the unique map filling in the diagram

$$\begin{array}{ccc} C \times C' & \xrightarrow{\mathrm{id}_{C \times C'}} & C \times C' \\ {\scriptstyle \mathrm{id}_C \times \eta_{C'}} \downarrow & & \downarrow {\scriptstyle \eta_{C \times C'}} \\ C \times L(C') & \xrightarrow{s_{C,C'}} & L(C \times C') \end{array} \tag{6}$$

It is straightforward to check that s is a natural transformation and the triangles in the statement commute.

As for uniqueness, note first that pullback pasting gives the first commutative triangle in the statement for a general pair C, C'

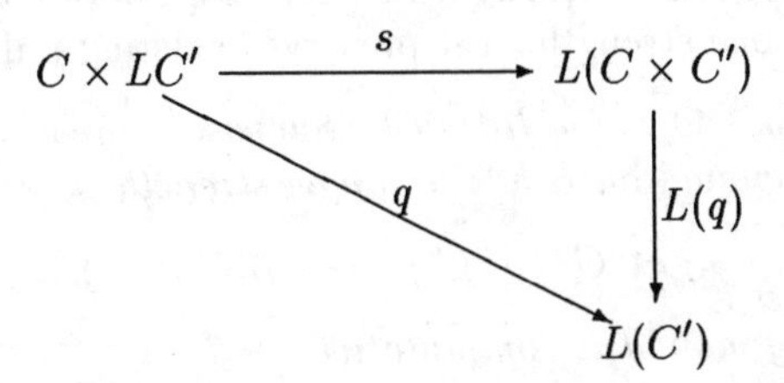

because s is natural and $q = q \circ (! \times \mathrm{id})$. It follows that s is cartesian in the second component. Thus the bottom face in the cube drawn in Fig. 1 is a pullback. So the component of the transformation at a pair C, C' is completely determined by those at the pairs $1, C'$ and $C, 1$ since, in the cube drawn in Fig. 1, also the two side faces are pullbacks, hence so is the top by pasting. Since the left diagram in the statement defines s on the pairs $1, C$, it remains to show that the value at a pair $C, 1$ is uniquely determined. This is so because the commutative diagram

$$\begin{array}{ccc} C \times 1 & \xrightarrow{\mathrm{id}} & C \times 1 \\ {\scriptstyle \mathrm{id} \times \eta} \downarrow & & \downarrow {\scriptstyle \eta} \\ C \times L(1) & \xrightarrow{s} & L(C \times 1) \end{array}$$

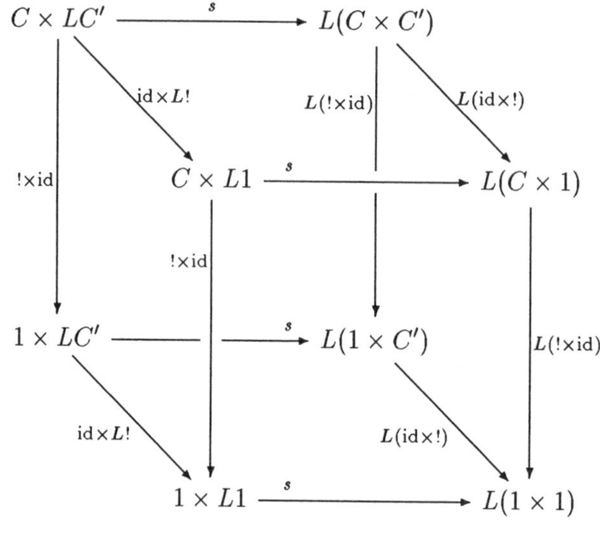

Figure 1

is a pullback. Indeed, given any span $C \times L(1) \xleftarrow{\langle f,x \rangle} X \xrightarrow{\langle g,! \rangle} C \times 1$ which makes the square with s and η commute, this must have $x = \eta!$ by prolonging the square with $L(q)$. Hence $f = g$ because η is monic. □

Note that the proof of uniqueness in Prop. 4 shows that any pointed endofunctor has at most one strength, if it preserves existing pullbacks.

Corollary 5. *Suppose* **C** *has a lifting* L. *Suppose moreover that* **C** *has binary products. Then the monad on* L *has a unique strength*

$$s_{C,C'}: C \times L(C') \longrightarrow L(C \times C').$$

Moreover, the strong monad is commutative.

Proof. It is straightforward to check commutativity of the various diagrams which s, η and μ must satisfy by chasing them in $\mathbf{C}_t$. For instance, to see that the monad is commutative one must check that the two possible ways to connect $LC \times LC'$ to $L(C \times C')$ do not differ. More precisely, one must check that the diagram commutes. □

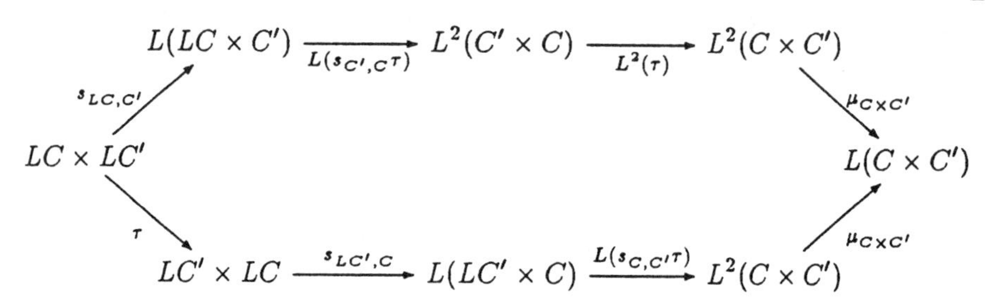

Theorem 6. *Let* $\mathbf{C}$ *be a category with finite products. Suppose* $L: \mathbf{C} \longrightarrow \mathbf{C}$ *is an endofunctor, and* $\eta: \mathrm{Id} \longrightarrow L$, $\mu: L^2 \longrightarrow L$, *and* s *are natural transformations. Then the following are equivalent*

(i) (L, η, μ) *is a lifting, and* s *is its strength,*

(ii) L *preserves existing pullbacks,* (L, η, μ) *is a cartesian (strong) commutative monad, and the following diagram commutes*

$$\begin{array}{ccc} LX & \xrightarrow{L\eta} & L^2X \\ {\scriptstyle \langle \mathrm{id}, L! \rangle}\downarrow & & \uparrow{\scriptstyle Lp} \\ LX \times L1 & \xrightarrow{s} & L(LX \times 1). \end{array} \tag{7}$$

Proof. (i) $\Rightarrow$ (ii) requires us to check commutativity of (7) which is trivial in $\mathbf{C}_t$. By Theorem 2 the diagram commutes also in $\mathbf{C}$.

(ii) $\Rightarrow$ (i) is obtained by mimicking the construction of the "domains" in [3, 16] to produce the identities in $\mathbf{C}_t$. Let $\alpha: C \longrightarrow L1$ be an object in $\mathbf{C}_t$; define a as the composite

$$C \xrightarrow{\langle \mathrm{id}, \alpha \rangle} C \times L1 \xrightarrow{s} L(C \times 1) \xrightarrow{L(p)} LC$$

where p denotes first projections. It is easy to check that maps of this form commute with each other. □

5 Conclusions

The characterization obtained by the representation theorem allows us to see that the category $\mathbf{\Delta}$ of finite non-empty ordinals is the initial category with a *non-trivial* lifting, where non-triviality is expressed in a positive form asserting that "there are undefined values": there is a cartesian natural transformation $\flat: 1 \longrightarrow L$ from the constant functor evaluated at the terminal object into L.

Also, the characterization of a lifting on a (small) category $\mathbf{C}$ with finite products given in Theorem 6(ii) explains why the extension to a category of sheaves on $\mathbf{C}$ of the lifting functor is possible.

On a last note, we would like to point out that it seems possible to obtain the free completion with lifting and products of a category, by first adding lifting, then products, and finally picking a ***certain*** full subcategory of the idempotent splitting. There is the problem of characterizing these idempotents.

Acknowledgements

Marcelo Fiore and John Power provided very useful comments on a preliminary version.

References

1. A. Carboni. Categorie di mappe parziali. *Riv. Mat. Univ. Parma*, 9:373–377, 1984.
2. P.L. Curien and A. Obtułowicz. Partiality and cartesian categories. *Inform. and Comput.*, 80:50–95, 1989.
3. R. Di Paola and A. Heller. Dominical categories: recursion theory without elements. *Journ. Symb. Logic*, 57:594–635, 1986.
4. Ju.L. Eršov. Theorie der Numerierungen. *Zeitschrift fur Math. Log.*, 19(4):289–388, 1973.
5. Ju.L. Eršov. Model C of partial continuous functionals. In R.O. Gandy and J.M.E. Hyland, editors, *Logic Colloquium '77*, pages 455–467. North Holland Publishing Company, 1977.
6. M.P. Fiore. *Axiomatic Domain Theory in Categories of Partial Maps.* Cambridge University Press Distinguished Dissertations in Computer Science, 1996.
7. M. Grandis. Cohesive categories and manifolds. *Ann. Mat. Pura Appl. (4)*, 157:199–244, 1990.
8. M. Grandis. Cohesive categories and measurable operators. *Rend. Accad. Naz. Sci. XL Mem. Mat. (5)*, 14(1):195–234, 1990.
9. H.J. Hoenke. On partial algebras. *Col. Math. Soc. J. Bolyai*, 29:373–412, 1977.
10. Bart Jacobs. Semantics of weakening and contraction. *Ann. Pure Appl. Logic*, 69(1):73–106, 1994.
11. A. Kock. Strong functors and monoidal monads. *Arch. Math.*, 23:113–120, 1972.
12. E. Moggi. *The partial-lambda calculus.* Ph.D. Thesis, University of Edinburgh, 1988.
13. E. Moggi. Notions of computations and monads. *Inform. and Comput.*, 79:95–130, 1992.
14. A.J. Power. Modularity in denotational semantics. to appear in *E.N.T.C.S.* **6**, 1997.
15. M. Proietti. Connections between partial maps categories and tripos theory. In *Category theory and computer science (Edinburgh, 1987)*, volume 283 of *Lecture Notes in Comput. Sci.*, pages 254–269. Springer, Berlin, 1987.
16. E.P. Robinson and G. Rosolini. Categories of partial maps. *Inform. and Comput.*, 79:95–130, 1988.
17. G. Rosolini. *Continuity and effectiveness in topoi.* D.Phil. Thesis, University of Oxford, 1986.

General Synthetic Domain Theory – A Logical Approach (Extended abstract)

B. REUS
Institut für Informatik
Ludwig-Maximilians-Universität
Oettingenstr. 67
D-80538 München
reus@informatik.uni-muenchen.de

Th. STREICHER
Fachbereich 4 Mathematik
TH Darmstadt
Schlossgartenstr. 7
D-64289 Darmstadt
streicher@mathematik.th-darmstadt.de

Abstract. Synthetic Domain Theory (SDT) is a version of Domain Theory where "all functions are continuous". In [14, 12] there has been developed a logical and axiomatic version of SDT which is special in the sense that it captures the essence of Domain Theory *à la Scott* but rules out other important notions of domain.

In this article we will give a *logical* and *axiomatic* account of *General Synthetic Domain Theory* (GSDT) aiming to grasp the structure common to all notions of domain as advocated by various authors. As in [14, 12] the underlying logic is a sufficiently expressive version of *constructive type theory*. We start with a few basic axioms giving rise to a core theory on top of which we study various notions of predomains as well-complete and replete S-spaces [9], define the appropriate notion of domain and verify the usual *induction principles*.

1 Introduction

In various lectures starting from end of the 70's Dana Scott has promoted the idea of using intuitionistic higher order logic or set theory as an adequate logical framework for axiomatising Domain Theory as an *Extensional Theory of Computation* where *all functions between domains are continuous* and even *computable*. Moreover, there are *intended models* for this situation, namely the various *realisability models*.

In [15] the basic notion of Synthetic Domain Theory was introduced, based on previous work of Scott and Mulry: the "r.e. subobject classifier" Σ providing the collection of propositions equivalent to some Σ^0_1-sentence. Using Σ one may *define* for every intuitionistic set a relation of "observational inequality" corresponding to the "information ordering" of Classical Domain Theory (CDT). This line of research was taken up again by W. Phoa in his Ph.D. Thesis [11] where he gave a detailed account of Domain Theory in the Effective Topos. Whereas these approaches were model-based, in 1989 J.M.E. Hyland [6] and P. Taylor [21] independently introduced two (essentially) equivalent formulations of a synthetic notion of domain, namely the so-called *Σ-replete objects*. This property of Σ-repleteness can be shown to be equivalent to closure under "all generalised limit processes", cf. [20]. In "traditional" accounts of SDT it often is not clear whether definitions and theorems have to be understood internally or externally[1].

In contrast the starting point of our *Logical Approach to SDT* is that *an axiomatic theory of domains has to be developed in a purely logical way* if one wants it to provide a *logical framework for the verification of functional programs.* Of course, this does not at all mean to avoid categorical notions and diagrams but, rather, to understand them as *formulated in the internal language.* For this purpose one needs a *full internal subcategory* Set of the ambient category which is a model for the type theory in use. Such a full internal subcategory provides an appropriate *type-theoretic universe* Set. Although we will often use diagrammatic language we emphasize that *all diagrams have to be understood internally as living in the universe Set.*

Using such a universe Set in this paper we develop elementary domain theory in an appropriate type theory concentrating on the verification of the usual induction principles for canonical fixpoints, for which purpose we introduce a new notion of admissible predicate. We also show how to obtain canonical solutions of domain equations if the universe Set is impredicative thereby following a suggestion of A. Simpson. Most proofs are omitted due to space restrictions, a full version, however, is in preparation.

2 General Synthetic Domain Theory

The basic idea of synthetic domain theory à la Scott is that the set Σ^X of Σ-predicates on a set X provides sort of a topology on X. But already the requirement that Σ-predicates are closed under finite disjunction gives rise to the function $\lambda u, v{:}\Sigma.\, u \vee v : \Sigma \times \Sigma \longrightarrow \Sigma$ which *cannot be computed sequentially* (as it is not even stable). Thus, there is need for a *more general theory* which does not rule out interesting models but, nevertheless, allows one to develop basic domain theory as needed for the verification of functional programs.

A fairly general categorical axiomatisation has been developed by G. Rosolini in [16]. His presentation, however, is not free from external arguments and his axioms do not seem to be strong enough for deriving the usual induction principles for canonical fixpoints.

A detailed account of domain theory in *general realisability models* has been given by John Longley and Alex Simpson in [9]. Their exposition is *purely model based* and refrains from developing their theory axiomatically. In [17] Simpson gives a logical treatment of well-complete sets based on intuitionistic set theory instead of type theory. (This may be more in accordance with mainstream mathematics but makes rather strong ontological assumptions.)

Our logical approach to *General Synthetic Domain Theory* (GSDT) is, of course, based on and inspired by the above mentioned work. The main purpose of our logical approach is to demonstrate that the *consequent use of type-theoretic language* facilitates a *simpler* presentation avoiding complicated external category-theoretic arguments as in [16] as well as complicated explicit

[1] see also the remark after Definition 4.2

constructions in realisability models as in [9]. This is also the case for [17] but we think that our type theoretic formulation is even simpler and more suitable for formalisation, see [12].

3 The Logic

The logic underlying our formulation of SDT is the Extended Calculus of Constructions (ECC) as described in [10] and implemented within the LEGO system. The Extended Calculus of Constructions is a constructive type theory with an *impredicative* universe Prop of propositions together with an infinite hierarchy of *predicative* universes $\mathsf{Type}_0 \in \mathsf{Type}_1 \in \ldots \in \mathsf{Type}_n \in \ldots$ where $\mathsf{Prop} \in \mathsf{Type} := \mathsf{Type}_0$. This hierarchy of universes is *cumulative* in the sense that we also have $\mathsf{Prop} \subseteq \mathsf{Type}_0 \subseteq \mathsf{Type}_1 \subseteq \ldots \subseteq \mathsf{Type}_n \subseteq \ldots$.

For our purposes we need the following extension of ECC.

Axiom 1 *The universe Prop of propositions is "proof-irrelevant", i.e.*

$$\forall P{:}\mathit{Prop}.\, \forall p, q{:}P.\; p = q$$

Axiom 2 *There is a universe Set $\in$ Type with Prop $\subseteq$ Set but* **not** *Prop $\in$ Set. Furthermore, Set contains the inductive type $\mathbb{N}$ of natural numbers.*

In the following the word "set" refers to objects of the universe Set.

Axiom 3 *Functions between sets are extensional, i.e.*

$$\forall A{:}\mathit{Set}.\, \forall B{:}A \to \mathit{Set}.\, \forall f, g : \Pi x{:}A.\, B(x).\; (\forall a{:}A.\, f\, a = g\, a) \Rightarrow f = g$$

Axiom 4 *Functional relations are required to be traced by functions, i.e. we require the Axiom of Unique Choice (AUC)*

$$\begin{aligned} &\Pi A : \mathit{Set}.\, \Pi B : A \to \mathit{Set}.\, \Pi P : \Pi x{:}A.\, B(x) \to \mathit{Prop}. \\ &\quad (\forall x{:}A.\, \exists!\, y{:}B(x).\, P\, x\, y) \to \Sigma f : \Pi x{:}A.\, B(x).\, \forall a{:}A.\, P\, a\, (f\, a) \end{aligned}$$

The universe Set may be impredicative but need not.

It has been shown in [19, 10] that ECC can be interpreted in the category of realisability sets over an arbitrary partial combinatory algebra A. For interpreting the extension above in arbitrary realisability models one has to interpret the universe Set as the collection of partial equivalence relations on A and Prop as the sub-collection of those partial equivalence relations R with $A_{/R}$ containing at most one element. See [12] for detailed description of these models and the verifications of Axioms 3 and 4.

Notice that our extension of type theory is closely related to the internal language of toposes, see e.g. [11]. But it is different from topos logic in the following two aspects. On the one hand in a topos there need not exist a universe Set on the other hand our type theory does not assume that equivalent propositions are always equal. It turns out that subsequently this principle will never be needed.

This is good so since for the case of impredicative universe Set it is unknown whether it is consistent with our other assumptions.

Notice that due to the assumption that the universe Prop is proof-irrelevant we have subset types available as in topos logic. If A is a set and $P : A \to \mathsf{Prop}$ is a predicate then the corresponding subset is given by

$$\{ x \in A \mid P(x) \} \equiv \Sigma x{:}A.\, P(x)$$

whose inclusion into A is given by projection on the first component, i.e. the term $\lambda z : \{ x \in A \mid P(x) \}.\, \pi_0(z)$, which is one-to-one as for $z, z' \in \{ x \in A \mid P(x) \}$ we have $\pi_1(z) = \pi_1(z')$ provided $\pi_0(z) = \pi_0(z')$ as $P(\pi_0(z)) = P(\pi_0(z'))$ contains at most one element due to Axiom 1.

Following the practice of topos theory we will use informal set-theoretic notation instead of the formal type theoretic language which makes sense as we have subset types and set comprehension available. Thus by *abus de langage* we (almost always) omit the inclusion maps $\pi_0 : \{ x \in A \mid P(x) \} \longrightarrow A$ for subset types. The only restriction compared to the internal language of a topos is that one must not consider equivalent predicates as equal.

Furthermore, all diagrams below have to be understood as living in some of the universes Set, $\mathsf{Type}_0, \ldots, \mathsf{Type}_n, \ldots$ considered as *full internal subcategories* of the ambient category establishing a model of the type theory in use.

4 $\mathcal{L}$-completeness and Separatedness

4.1 $\mathcal{L}$-completeness

Many notions of completeness can be formulated quite elegantly (cf. Section 6) via the following notion of *orthogonality*, originally introduced in [4].

Definition 4.1 *A map $f : X \longrightarrow Y$ is called* orthogonal *to a map $g : Z \longrightarrow U$, abbreviated as $f \perp g$, iff for all $h : X \longrightarrow Z$ and $k : Y \longrightarrow U$ with $k \circ f = g \circ h$ there is a unique map $\alpha : Y \longrightarrow Z$ (called "fill-in") with $\alpha \circ f = h$ and $g \circ \alpha = k$, i.e. the diagram*

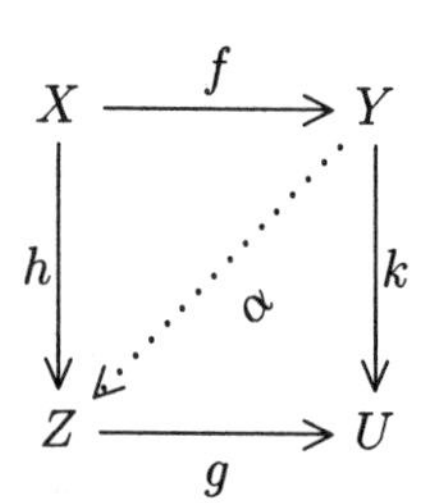

commutes.
If U is terminal then we write $f \perp Z$ for $f \perp !_Z$. That means that any $h : X \longrightarrow Z$ uniquely extends along f to a map $\overline{h} : Y \longrightarrow Z$, i.e. $h = \overline{h} \circ f$.

Using the notion of orthogonality we define an abstract notion of completeness relative to a class $\mathcal{L}$ of maps in some universe.

Definition 4.2 *Let $\mathcal{L}$ be a class of maps, i.e. a predicate on*

$$\Sigma X, Y : U.\, Y^X$$

where U is one of the universes $\mathsf{Set} \subseteq \mathsf{Type}_0 \subseteq \mathsf{Type}_1 \ldots$. A type A is called $\mathcal{L}$-complete iff $e \bot A$ for all $e \in \mathcal{L}$ (i.e. $(X, Y, e) \in \mathcal{L}$ where $e : X \to Y$).
A map m is called $\mathcal{L}$-closed *iff $e \perp m$ for all $e \in \mathcal{L}$.*

This general pattern will later be instantiated by specific choices of $\mathcal{L}$ in order to obtain particular notions of predomain as e.g. complete or replete sets.

Remark: Note that in the rest of this paper we will mostly consider the case $U = \mathsf{Set}$ and thus speak about $\mathcal{L}$-complete sets. In principle, however, we could also use any other of the universes Type_n instead of Set.

In any case, the important point is that *our type theory allows us to quantify over $\mathcal{L}$* as $\mathcal{L}$ is contained in some universe[2].

Theorem 4.1 *$\mathcal{L}$-complete sets enjoy the following closure properties:*

(i) Dependent products preserve $\mathcal{L}$-completeness.

(ii) Let A be $\mathcal{L}$-complete and $m : P \rightarrowtail A$ be a subobject of A. Then P is $\mathcal{L}$-complete iff m is $\mathcal{L}$-closed.

(iii) Let $f_1, f_2 : A \longrightarrow B$ be maps between $\mathcal{L}$-complete sets. If $m : E \rightarrowtail A$ is an equaliser of f_1 and f_2 then E is also $\mathcal{L}$-complete.

4.2 S-spaces

Now we define the notion of *S-space* as generalisation of T_0-spaces where points are determined uniquely by its open neighbourhoods represented by continuous maps to *Sierpinski space* S. This notion has been considered by Rosolini and later by W. Phoa [11] in the particular context of the effective topos choosing for S the "r.e.-subobject-classifier" Σ.

In this subsection we just assume S to be an arbitrary set, i.e. $S \in \mathsf{Set}$.

Notation. We write $S(\_)$ for S^{-} and $S^n(\_)$ for the n-times application of the functor $S(\_)$.

Definition 4.3 *For any type X we define $\varepsilon_X : X \longrightarrow S^2(X)$ as $\varepsilon_X(x) = \lambda p{:}S(X).\, p(x)$.*

---

[2] Notice that the collection of *all* S-isos does not form an internal collection unless one restricts to S-isos in some universe U. Typically, one takes Set for U, *c.f.* Section 6.3. However, if the ambient category is a topos there need not exist such a universe and, therefore, repleteness cannot be expressed by a predicate expressible in the internal language.

Remark Notice that ε is the unit of the so-called "continuation monad" $S^2(\_)$ and therefore a natural transformation from the identity functor to $S^2(\_)$ which restricts to any type universe Type_i or Set.

Definition 4.4 *A set $X \in$ Set is called an S-*space *or S-*separable *iff ε_X is monic.*

The following representation theorem is useful in many places, e.g. for showing closure properties of S-spaces.

Theorem 4.2 *A set X is an S-space iff it is a subobject of $S(I)$ for some $I \in$ Set.*

Proof. "$\Rightarrow$": If $\varepsilon_X : X \longrightarrow S^2(X)$ is monic then X is a subobject of $S^2(X)$.
"$\Leftarrow$": Suppose $m : X \rightarrowtail S(I)$ is monic. By naturality of ε we get that

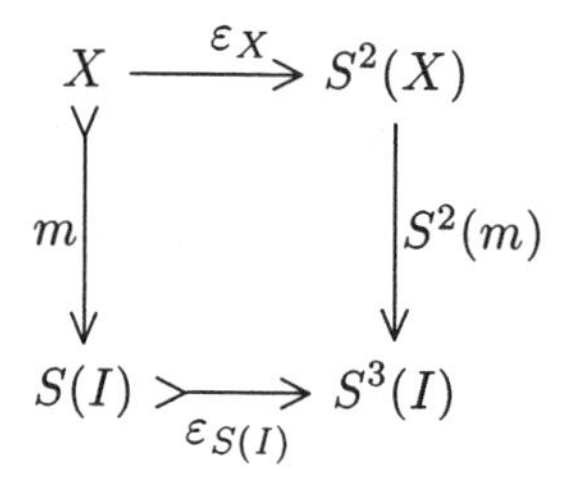

where $\varepsilon_{S(I)}$ is split monic since $S(\varepsilon_X) \circ \varepsilon_{S(X)} = \mathsf{id}_{S(X)}$ (as can be verified by straightforward computation). Thus, ε_X is monic and X is is an S-space.

A representation theorem for $\mathcal{L}$-complete S-spaces follows from the representation theorem for S-spaces.

Corollary 4.3 *A set is an $\mathcal{L}$-complete S-space iff it is an $\mathcal{L}$-closed subobject of some power of S by a set.*

S-spaces have the following closure properties.

Theorem 4.4 *If $A \in$ Set and $B : A \longrightarrow$ Set with $B(x)$ an S-space for all $x \in A$ then $\Pi x{:}A.\, B(x) \in$ Set is an S-space, i.e. S-spaces are closed under products of families indexed over a set.*

Moreover, if Set *is an impredicative universe then S-spaces are closed under products of* arbitrary *families.*

Furthermore, S-spaces are closed under equalisers.

5 Initial and Terminal Lift-algebra

5.1 The Dominance S

Up to now we have made no assumptions about S. Now we will assume that S is a set of propositions providing a notion of "well-behaved" subobject as introduced originally by G. Rosolini in [15] for the case of toposes under the name *dominance.*

As opposed to toposes in our case equivalent propositions need not be equal. Accordingly, we cannot consider S simply as a subset of Prop. Instead, we postulate an element $\top \in S$ giving rise to a map $\mathsf{def} := \lambda u{:}S.\,(u = \top) : S \to \mathsf{Prop}$ required to reflect equivalence to equality, i.e. that $u = v \in S$ iff the propositions $u = \top$ and $v = \top$ are equivalent. Thus, via def the set S may be considered "as a subobject of Prop *modulo equivalence*".

Axiom 5 *There is a distinguished $S \in \mathsf{Set}$ together with distinguished elements $\bot, \top \in S$ such that*

$$\neg(\bot = \top)$$

and

$$\forall u, v{:}S.\, [u = \top \Leftrightarrow v = \top] \Rightarrow u = v$$

i.e. the map $\mathsf{def} := \lambda u{:}S.\,(u = \top) : S \to \mathsf{Prop}$ reflects logical equivalence in Prop to equality in S.

Moreover, S is required to be a dominance, *i.e. for any $u \in S$ and $v : \mathsf{def}(u) \to S$ there exists a (necessarily unique) $u \angle v \in S$ with*

$$\mathsf{def}(u \angle v) \Leftrightarrow \exists p{:}\mathsf{def}(u).\, \mathsf{def}(v(p))$$

providing a "dependent conjunction" on S.

Remark: Notice that $\angle$ provides us with an operation $\wedge_S : S \times S \longrightarrow S$ given by $u \wedge_S v := u \angle\ \lambda p{:}\mathsf{def}(u).\ v$ satisfying $\mathsf{def}(u \wedge_S v) \Leftrightarrow \mathsf{def}(u) \wedge \mathsf{def}(v)$. Therefore we write $u \wedge v$ for $u \wedge_S v$.

Some useful notions are gathered in the following definition.

Definition 5.1 *A map $e : X \longrightarrow Y$ in Set is called S-*iso *iff $e \perp S$. Accordingly, a map e in Set is called S-*epi *iff $S(e)$ is monic, i.e.*

$$p \circ e = q \circ e \Rightarrow p = q$$

for all $p, q \in Y \longrightarrow S$.

*A mono $m : X \rightarrowtail Y$ is called S-*mono *or S-*subobject of Y *iff there exists a (necessarily unique) map $\chi : Y \longrightarrow S$ with $\mathsf{def}(\chi(y)) \Leftrightarrow \exists x{:}X.\, y = m(x)$ for all $y \in Y$ called* classifying map for m.

Remark: Notice that from the assumption that $\top \in S$ is a dominance it follows that S-monos are closed under composition and pullbacks along arbitrary maps in Set.

5.2 Lifting

Now we will construct a *lifting operation* from the dominance allowing one to classify partial maps with domain of definition given by an S-valued predicate. Note that – just as for the notion of $\mathcal{L}$-completeness – we could consider lifting on different universes, but for the purposes of this paper it suffices to have a lifting operation on the universe Set.

Definition 5.2 *The lifting functor* $L : \mathsf{Set} \to \mathsf{Set}$ *is defined as follows. For any* $A \in \mathsf{Set}$

$$LA \triangleq \Sigma u{:}S.\,(\mathit{def}(u) \longrightarrow A)$$

and for any $f : A \longrightarrow B$ *in* Set

$$L(f) \triangleq \lambda z{:}LA.\,\langle \pi_0(z), \lambda p{:}\mathit{def}(\pi_0(z)).\, f(\pi_1(z)(u))\rangle$$

Moreover, let η *be the natural transformation from the identity functor on* Set *to* L *given by*

$$\eta_A \triangleq \lambda x{:}A.\,\langle \top, \lambda p{:}\mathit{def}(\top).\, x\rangle$$

for all $A \in \mathsf{Set}$.

Notice that $L0 \cong 1$ and $L1 \cong S$ when 0 is an empty set and 1 is a "terminal" set containing precisely one element, say $*$.

Convention : In the sequel we adopt the convention of not distinguishing between $a \in A \in \mathsf{Set}$ and the function $1 \longrightarrow A$ sending $*$ to a.

Definition 5.3 *For any set* A *let* $\bot_A \in LA$ *be defined as*

$$\bot_A \triangleq \langle \bot, \lambda p{:}\mathit{def}(\bot).\, ?_A(p)\rangle$$

where $?_A$ *is the unique map* $\mathit{def}(\bot) \cong 0 \longrightarrow A$ *and let*

$$c_A \triangleq [\bot_A, \eta_A] : 1 + A \longrightarrow LA$$

called covering *map of* LA.

Remark: Notice that for any set A the map $\eta_A : A \longrightarrow LA$ *classifies* partial maps into A whose domain of definition is given by an S-subobject. More precisely, for any $B \in \mathsf{Set}$, $p : B \longrightarrow S$, $B' \triangleq \{\, x \in B \mid \mathsf{def}(p(x)) \,\}$ and $f : B' \longrightarrow A$ there exists a *unique classifying* map $\chi_f : B \longrightarrow LA$ such that the following diagram is a pullback

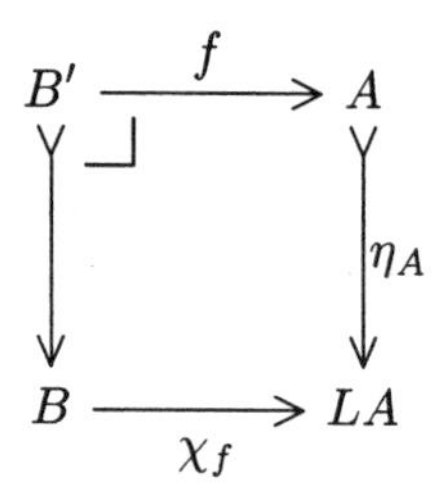

where χ_f is given explicitly as $\chi_f(b) = \langle\, p(b), \lambda u{:}\mathsf{def}(p(b)).\, f(\langle b, u\rangle)\,\rangle$.

5.3 Initial L-Algebra and Terminal L-Coalgebra

Following [21, 7] one may show the existence of initial and terminal L-algebras in Set.

Theorem 5.1 *In Set there exists an initial L-algebra $\phi : L\omega \longrightarrow \omega$ and a terminal L-coalgebra $v : \overline{\omega} \longrightarrow L\overline{\omega}$ where $\overline{\omega}$ is a subobject of $S(\mathbb{N})$ and ω is the least sub-L-algebra of v^{-1}.*

As ϕ and v are known to be isomorphisms, c.f. [3], the following definitions make sense.

Definition 5.4 *The maps*

$$\sigma \triangleq \phi \circ \eta_\omega : \omega \longrightarrow \omega$$

$$\overline{\sigma} \triangleq v^{-1} \circ \eta_{\overline{\omega}} : \overline{\omega} \longrightarrow \overline{\omega}$$

are called successor maps *on ω and $\overline{\omega}$, respectively.*

The unique L-algebra morphism from ϕ to v^{-1} is called $\iota : \omega \longrightarrow \overline{\omega}$ satisfying $v \circ \iota \circ \phi = L\iota$.

From these definitions and naturality of η it is obvious that $\overline{\sigma} \circ \iota = \iota \circ \sigma$.

Definition 5.5 *Whenever $\alpha : LA \longrightarrow A$ is an L-algebra let $\mathsf{step}_\alpha : \mathbb{N} \longrightarrow A$ be the unique map satisfying*

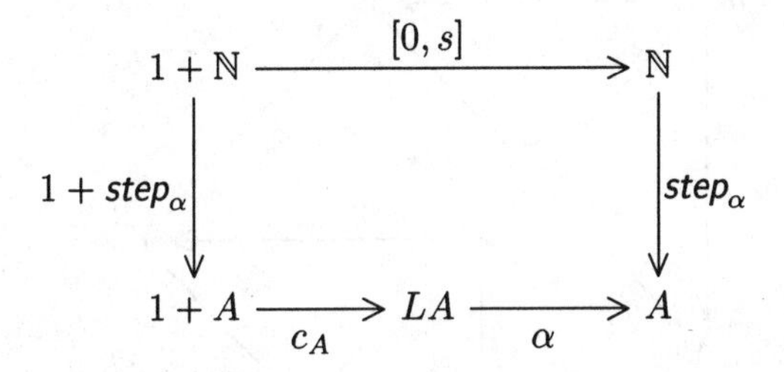

where $s(n) = n + 1$. The uniqueness of step_α follows from the fact that $[0, s]$ by definition is the initial algebra for the functor $1 + \_$.

We write step for $\mathsf{step}_\phi : \mathbb{N} \longrightarrow \omega$ and $\overline{\mathsf{step}}$ for $\mathsf{step}_{v^{-1}} : \mathbb{N} \longrightarrow \overline{\omega}$ and, therefore, $\overline{\mathsf{step}} = \iota \circ \mathsf{step}$.

Next we define an object $\infty \in \overline{\omega}$ corresponding intuitively to the "limit" of the sequence $(\mathsf{step}(n))_{n \in \mathbb{N}}$ in $\overline{\omega}$. It will play a key role in the construction of canonical fixpoints for endofunctions on domains (cf. [16]).

Definition 5.6 *Let $\infty : 1 \longrightarrow \overline{\omega}$ be the unique L-coalgebra morphism from η_1 to v.*

Lemma 5.2 *The object ∞ is a fixpoint of $\overline{\sigma}$, i.e. $\overline{\sigma} \circ \infty = \infty$, and, moreover, ∞ is an equaliser of $\mathsf{id}_{\overline{\omega}}$ and $\overline{\sigma}$.*

5.4 Closure under Lifting

We first give a necessary condition for a class $\mathcal{L}$ of maps in Set such that $\mathcal{L}$-complete sets are closed under lifting.

Lemma 5.3 *If $\mathcal{L}$ is a class of maps in* **Set** *such that $\mathcal{L}$-complete sets are closed under lifting then $e \perp S$ for all e in $\mathcal{L}$.*

Proof. As 1 is $\mathcal{L}$-complete and $\mathcal{L}$-complete sets are closed under lifting it follows that $S \cong L1$ is $\mathcal{L}$-complete i.e. $e \perp S$ for all e in $\mathcal{L}$.

Maps e with $e \perp S$ were originally introduced by Martin Hyland in [6] and called S-isos. As closure under lifting is sort of a minimal requirement for a class of predomains given as $\mathcal{L}$-complete sets for some $\mathcal{L}$ the previous lemma tells us that one may restrict attention to classes $\mathcal{L}$ of S-isos. A detailed explanation of S-isos as "generalised limit process" has been given in [20]. Thus one may consider a class $\mathcal{L}$ of S-isos as a *class $\mathcal{L}$ of generalised limit processes.*

The following lemma will be crucial for giving a necessary and sufficient characterisation of those $\mathcal{L}$ for which $\mathcal{L}$-complete sets are closed under lifting.

Lemma 5.4 *If $\mathcal{L}$ is a class of S-isos and $A \in$* **Set** *then $L\,A$ is $\mathcal{L}$-complete iff $m^*e \perp A$ for all $e \in \mathcal{L}$ and S-monos m.*

Proof. The proof idea is illustrated by the following diagram

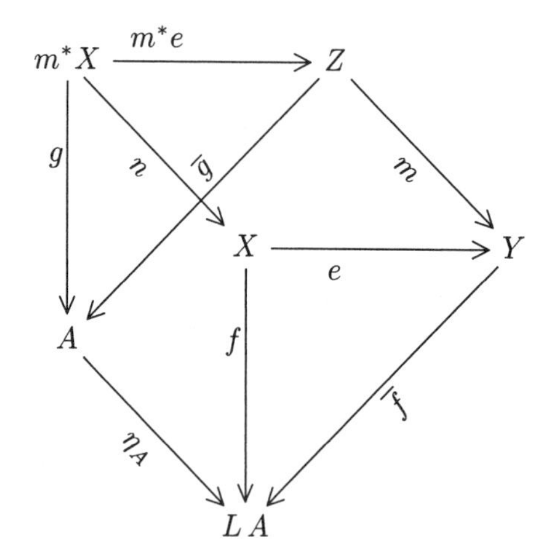

where all squares are pullback squares.

Corollary 5.5 *If $\mathcal{L}$ contains only S-isos then A is $\mathcal{L}$-complete if $L\,A$ is $\mathcal{L}$-complete.*

Proof. If $L\,A$ is $\mathcal{L}$-complete then by Lemma 5.4 it follows that $m^*e \perp A$ for all $e : X \longrightarrow Y$ in $\mathcal{L}$ and S-subobjects m of Y. As all identity maps are S-monos we get that $e \perp A$ for all e in $\mathcal{L}$, i.e. that A is $\mathcal{L}$-complete.

Furthermore, we have the following lemma.

Lemma 5.6 *S-isos are stable under pullbacks along S-monos.*

From this we get a sufficient condition for $\mathcal{L}$-complete sets being closed under lifting.

Theorem 5.7 *If $\mathcal{L}$ is a class of S-isos stable under pullbacks along S-monos then the $\mathcal{L}$-complete sets are closed under lifting.*

Finally, we observe that S-spaces are closed under lifting.

Lemma 5.8 *The S-spaces are closed under lifting.*

6 Predomains

In this section we study several notions of predomain arising from different choices of the set $\mathcal{L}$ of generalised limit processes.

6.1 Complete Sets

As known from classical domain theory it is not sufficient to consider partial orders for interpreting recursive definitions of objects. Instead, one has to require that any ω-chain has a supremum, i.e. a *limit*.

In the vein of being "order-free" we have replaced partially ordered sets by S-spaces. Next we will give an "order-free" analogue of the requirement that ascending chains have limits.

Instead of $\mathbb{N}$-indexed ascending chains in an S-space A we consider ω-chains in A, i.e. maps $a : \omega \longrightarrow A$ where ω is the initial L-algebra. Although for any such $a : \omega \longrightarrow A$ one may obtain an $\mathbb{N}$-indexed ascending chain $a \circ \mathsf{step} : \mathbb{N} \longrightarrow A$ there exist even realisability models where some ascending chains $c : \mathbb{N} \longrightarrow A$ cannot be obtained as $a \circ \mathsf{step}$ from some $a : \omega \longrightarrow A$ (cf. [9]).

Definition 6.1 *A set A is called* complete *iff $\iota \perp A$, i.e. for any $a : \omega \longrightarrow A$ there is a unique $\overline{f} : \overline{\omega} \longrightarrow A$ with $\overline{f} \circ \iota = f$.*

Thus, the complete sets are the $\mathcal{L}$-complete sets for $\mathcal{L}$ containing precisely the map $\iota : \omega \longrightarrow \overline{\omega}$ and therefore satisfy the following closure properties.

Theorem 6.1 *If A is a set and $B : A \longrightarrow$ **Set** is a family of complete sets (S-spaces) then $\Pi x{:}A.\, B(x)$ is a complete set (S-space), too.*

Moreover, if **Set** *is an impredicative universe then complete sets (S-spaces) are closed under arbitrary dependent products.*

Complete sets (S-spaces) are closed under equalisers.

In order to ensure that S itself is complete, however, we have to postulate

Axiom 6 $\iota \perp S$.

Remark: Our definition of "completeness" is an internal version of the external notion of completeness introduced by Longley and Simpson [9] for realisability models. Notice also that the above definition of "completeness" is nothing but a weakening of Hyland and Taylor's notion of "replete object" [6, 21] (cf. also Section 6.3). Thus completeness in the above sense is a minimal closure property whereas repleteness is a maximal closure property.

6.2 Well-complete Sets

The main problem with complete S-spaces is that they are not (known to be) closed under lifting. Therefore one has to look for a more restrictive notion which nevertheless satisfies all the closure properties of complete sets and S-spaces.

A minimal restriction in order to obtain closure under lifting is the notion of *well-completeness* as suggested by [8, 9].

Definition 6.2 *A set A is called* well-complete *iff its lifting $L\,A$ is orthogonal to ι, i.e. iff $L\,A$ is complete.*

From Corollary 5.5 we immediately get that

Lemma 6.2 *Any well-complete set A is complete.*

The following Lemma identifies well-complete as an instance of the notion of $\mathcal{L}$-completeness.

Lemma 6.3 *A set A is well-complete iff $m^*\iota \perp A$ for all S-subobjects m of $\overline{\omega}$, i.e. iff A is $\mathcal{L}$-complete for $\mathcal{L} \triangleq \{ m^*\iota \mid m \ S\text{-subobject of } \overline{\omega} \}$.*

Proof. Immediate by Lemma 5.4.

Therefore, we get the usual desired closure properties.

Theorem 6.4 *Well-complete sets (S-spaces) are closed under equalisers, lifting, and dependent products of families indexed over a set. Moreover, if* Set *is impredicative then well-complete sets (S-spaces) are closed under arbitrary dependent products.*

Lemma 6.5 *S is well-complete, i.e. $\iota \perp L\,S$.*

Proof. $L\,S$ is a retract of $S \times S$ which can be seen by establishing the retraction maps $e : L\,S \longrightarrow S \times S$ such that $e : \langle u, f \rangle \mapsto \langle u, u \angle f \rangle$ and $p : S \times S \longrightarrow L\,S$ such that $p : \langle u, v \rangle \mapsto \langle u, \lambda p{:}\mathsf{def}(u).\, v \rangle$.
By Axiom 6 and the closure properties of complete sets it follows that $L\,S$ must be complete too.

Note that by Lemma 6.2 the above Lemma is even equivalent to Axiom 6. Accordingly, we get the following representation theorem for well-complete S-spaces.

Theorem 6.6 *A set is a well-complete S-space iff it is a well-complete subobject of some power of S.*

6.3 S-replete Sets

The notion of S-replete set has been introduced originally in [6] and [21] in the context of a more restrictive axiomatic setting tailored towards a domain theory *á la Scott.* Nevertheless, their definition makes sense also in our more general framework.

Definition 6.3 *A set A is called (S-)*replete *iff $e \perp A$ for all S-isos e.*

Obviously, this is again an instance of $\mathcal{L}$-completeness with $\mathcal{L}$ being the collection of *all* S-isos. This observation is crucial for obtaining the subsequent closure properties which follow from Theorem 4.1, Lemma 5.6, and Theorem 5.7.

Theorem 6.7 *The S-replete sets are closed under lifting, dependent products of families indexed over a set, and equalisers. If* **Set** *is impredicative then S-replete sets are closed under arbitrary dependent products.*

Finally, we observe that S-repleteness is the most restrictive notion of predomain containing S.

7 Domains

The well-complete and S-replete *S-spaces* studied so far are "synthetic" analogues of "classical" *predomains*, i.e. partial orders closed under limits of ω-chains. As in classical domain theory for construction of canonical fixpoints one has to restrict attention to *domains*, i.e. predomains having a least element. Next we give an "order-free" reformulation of the notion of domain.

Definition 7.1 *A* focal L-algebra *is a map $\alpha : LA \longrightarrow A$ satisfying $\alpha \circ \eta_A = id_A$. The element $\alpha(\perp_A)$ will be referred to as $\perp_\alpha$.*

If $\alpha : LA \longrightarrow A$ and $\beta : LB \longrightarrow B$ are focal L-algebras then a homomorphism from α to β *is a map $h : A \longrightarrow B$ with $h \circ \alpha = \beta \circ Lh$.*

Notice that *a priori* focal L-algebra structure on a set need not be unique. For S-spaces, however, uniqueness of focal L-algebra structure is guaranteed by the following two axioms.

Axiom 7 *For any S-space A the map $c_A = [\perp_A, \eta_A] : 1 + A \longrightarrow LA$ is an S-epi.*

Intuitively, that means that for any S-space A maps from LA to S are uniquely determined by their behaviour on the image of η_A and the element $\perp_A$.

Axiom 8 *If $f(\perp) = \top$ for some $f : S \longrightarrow S$ then $f(u) = \top$ for all $u \in S$.*

So $\perp$ is the least element of S w.r.t. "information ordering" (*c.f.* Def. 9.2).

We next show that a focal L-algebra structure on an S-space A gives rise to a least element of A w.r.t. "information ordering".

Lemma 7.1 *Let $\alpha : L\,A \longrightarrow A$ be a focal L-algebra structure on an S-space A. If $p \in S(A)$ with $p(\bot_\alpha) = \top$ then $p(a) = \top$ for all $a \in A$.*

Proof. Let $p \in S(A)$ with $p(\bot_\alpha) = \top$. Let $a \in A$. Define $h_a : S \longrightarrow LA$ as $h_a(u) \triangleq \langle u, \lambda p : \mathsf{def}(u).a\rangle$ for which we have $h_a(\bot) = \bot_A$ and $h_a(\top) = \eta_A(a)$. Thus, we have

$$(p \circ \alpha \circ h_a)(\bot) = p(\alpha(\bot_A)) = p(\bot_\alpha) = \top$$

$$(p \circ \alpha \circ h_a)(\top) = p(\alpha(\eta_A(a))) = p(a)$$

which by Axiom 8 implies that $p(a) = \top$.

Now we get uniqueness of focal L-algebra structure for S-spaces.

Theorem 7.2 *There is at most one focal L-algebra structure on an S-space A.*

Proof. Suppose α_1 and α_2 are focal L-algebra structures on an S-space A. As A is an S-space and c_A is S-epic by Axiom 7 it suffices to show that $\alpha_1 \circ c_A = \alpha_2 \circ c_A$. As $\alpha_1 \circ \eta_A = \mathsf{id}_A = \alpha_2 \circ \eta_A$ it even suffices to show that $\alpha_1(\bot_A) = \alpha_2(\bot_A)$. But by the previous Lemma 7.1 we have for all $p \in S(A)$ that $p(\alpha_1(\bot_A)) = \top$ iff $p(\alpha_2(\bot_A)) = \top$. As A is an S-space this implies $\alpha_1(\bot_A) = \alpha_2(\bot_A)$.

Theorem 7.3 *Focal S-spaces are closed under dependent products, equalisers of L-algebra morphisms and under lifting.*

Proof. By straightforward construction of the required structure maps and direct verification. The focal structure on $L\,A$ is given by $\mu_A : L\,L\,A \longrightarrow L\,A$ classifying the partial map $(\eta_{LA} \circ \eta_A, \mathsf{id}_A)$.

Remark. The map μ_A can be defined for arbitrary sets A and together with η gives rise to the so-called *lifting monad* (L, η, μ) on Set.

An appropriate notion of *domain* is now given by *well-complete focal S-spaces* or the full subcategory of *replete focal S-spaces*.

8 Fixpoints

For any focal L-algebra $\alpha : L\,A \longrightarrow A$ over a complete A any map $f : A \longrightarrow A$ admits a *canonical fixpoint* which can be constructed *à la Kleene* analogous to classical domain theory *c.f.* [1, 16]. Since completeness encompasses well-completeness and S-repleteness this can be applied to all notions of domains considered in this paper.

Lemma 8.1 *Let $\alpha : LA \longrightarrow A$ be a focal L-algebra. For any $f : A \longrightarrow A$ there is a canonical map $\mathsf{kl}_{\alpha,f} : \omega \longrightarrow A$ called* Kleene chain of f (w.r.t. α), *such that*

$$\mathsf{kl}_{\alpha,f}(\mathit{step}(n)) = f^n(\bot_\alpha)$$

for all $n \in \mathbb{N}$.
Accordingly, if A is complete for the unique map $\overline{\mathsf{kl}_{\alpha,f}} : \overline{\omega} \longrightarrow A$ with $\mathsf{kl}_{\alpha,f} = \overline{\mathsf{kl}_{\alpha,f}} \circ \iota$ it holds that $\overline{\mathsf{kl}_{\alpha,f}}(\overline{\mathit{step}}(n)) = f^n(\bot_\alpha)$ for all $n \in \mathbb{N}$.

Theorem 8.2 *([16, 1])*
Let A be a complete set and $\alpha : LA \longrightarrow A$ be a focal L-algebra structure on A. Then for any endomap $f : A \longrightarrow A$ the element

$$\mathit{fix}_\alpha(f) \triangleq \overline{\mathsf{kl}_{\alpha,f}}(\infty) \in A$$

is a fixpoint of f, *i.e.* $f(\mathit{fix}_\alpha(f)) = \mathit{fix}_\alpha(f)$.

9 Domain-Theoretic Induction Principles

For the purposes of *program verification* one needs proof principles allowing one to prove properties of canonical fixpoint such as *Fixpoint Induction*, *Computational Induction* and *Park Induction* in Classical Domain Theory.

We now will derive "synthetic" analogues of these domain-theoretic induction principles which is a somewhat delicate task in GSDT as one has to avoid reference to order-theoretic notions.

As in classical domain theory induction principles are valid only for a certain class of so-called *admissible* predicates. Classically, a predicate P on a cpo A is called *admissible* iff P is closed under suprema of ascending chains. The straightforward "synthetic" analogue of "classical" admissibility would be to require that P is complete, i.e. orthogonal to ι (because the very idea of the synthetic approach is to replace ascending $\mathbb{N}$-indexed chains by ω-chains).

Alas, this straightforward choice would enforce a *reformulation* already of fixpoint induction as illustrated by the following theorem.

Theorem 9.1 *("Synthetic" Fixpoint Induction)*
Let A be complete, $f : A \longrightarrow A$ and $\alpha : LA \longrightarrow A$ be a focal L-algebra. If $m : P \rightarrowtail A$ is complete and also a subalgebra of f_α, i.e. $f_\alpha \circ L(m)$ factors through m, then $\overline{\mathsf{kl}_{\alpha,f}}$ factors through m and therefore $\mathit{fix}_\alpha(f) \in P$.

That P is a sub-L-algebra of f_α explicitly reads as follows

$$\forall u{:}S.\forall a : \mathsf{def}(u) \longrightarrow A.\, (\forall p : \mathsf{def}(u).P(a(p))) \Rightarrow P(\alpha(\langle u, f \circ a \rangle))$$

which would be equivalent to the premiss of classical fixpoint induction if it were sufficient to consider only the cases $u = \top$ and $u = \bot$. But such case analysis is not available as the proposition $\forall u{:}S.\, u = \top \vee u = \bot$ is not valid in the Kleene realisability model which is sort of a standard model guiding our intuitions [3].

We will next give an improved definition of "admissible predicate" for which fixpoint induction and the other induction principles will be valid in their classical form.

[3] However, the weaker statement $\forall u{:}S.\, \neg\neg(u = \top \vee u = \bot)$ is valid in the Kleene realisability model and this guarantees the validity of $P(\bot) \wedge P(\top) \Rightarrow \forall u{:}S.\, P(u)$ provided the predicate P is $\neg\neg$*-closed, i.e.* $\forall u{:}S.\, \neg\neg P(u) \Rightarrow P(u)$. Of course, this is only useful if sufficiently many predicates P on S are $\neg\neg$-closed which is the case in the Kleene realisability model and has been exploited axiomatically in the first author's Thesis [12, 13] by postulating *Markov's Principle.*

9.1 Admissible Predicates

Classically, an admissible predicate on predomain A is a subset $P \subseteq A$ such that for any ω-chain $a : \omega \longrightarrow A$ its image is contained in P iff the image of $a \circ \mathsf{step} = \overline{a} \circ \overline{\mathsf{step}}$ is contained in P, *i.e.* iff P contains all "finite approximations" $a(\mathsf{step}(n))$. This can be reformulated as an orthogonality condition as follows.

Definition 9.1 *A subobject $m_P : P \rightarrowtail A$ of a completes et A is called* admissible *iff $\overline{\mathsf{step}} \perp m_P$.*

The relationship between completeness and admissibility is given by the following lemma whose proof is straightforward.

Lemma 9.2 *A subobject $m_P : P \rightarrowtail A$ of a complete set A is admissible iff P is complete and $\mathsf{step} \perp m_P$.*

For equality on complete S-spaces being admissible we need the following theorem whose proof is essentially due to Alex Simpson, *c.f.* [17], where, however, it has been used for quite a different purpose.

Theorem 9.3 *The maps $\mathsf{step} : \mathbb{N} \longrightarrow \omega$ and $\overline{\mathsf{step}} : \mathbb{N} \longrightarrow \overline{\omega}$ are S-epic.*

Proof. As $S(\iota)$ is an isomorphism and $\overline{\mathsf{step}} = \iota \circ \mathsf{step}$ it is sufficient to show that step is an S-epi.

For this purpose let $m : P \rightarrowtail \omega$ be the greatest subobject of ω through which step factors by an S-epi. Let $e : \mathbb{N} \longrightarrow P$ be the unique map with $\mathsf{step} = m \circ e$. Of course, e is an S-epi. Notice that

$$\phi \circ L(\mathsf{step}) \circ c_{\mathbb{N}} = \phi \circ c_\omega \circ (1 + \mathsf{step}) = \mathsf{step} \circ [0, s]$$

as $L(\mathsf{step}) \circ c_{\mathbb{N}} = c_\omega \circ (1+\mathsf{step})$ by naturality of c and $\phi \circ c_\omega \circ (1+\mathsf{step}) = \mathsf{step} \circ [0, s]$ by definition of step.

Now consider the following diagram

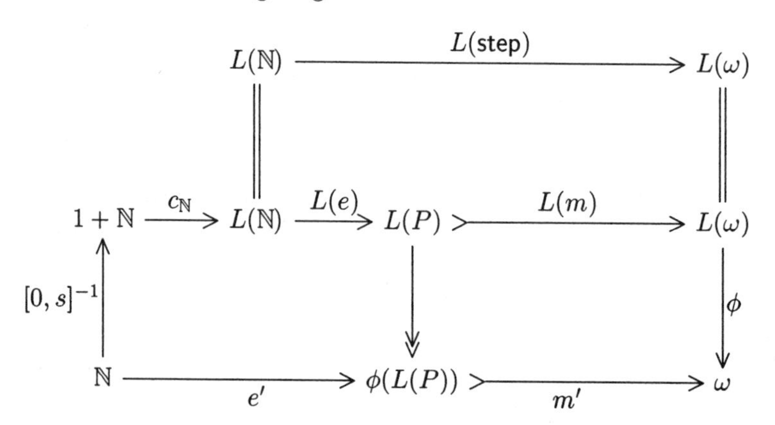

where m' is the image of $\phi \circ L(m)$ and e' is the unique map with $\mathsf{step} = m' \circ e'$.

By Axiom 7 it follows that $c_{\mathbb{N}}$ and $L(e)$ are S-epic (the latter as $L(e) \circ c_{\mathbb{N}} = c_P \circ e$ and both $c_{\mathbb{N}}$ and $c_P \circ e$ are S-epic). Thus, the map e' is an S-epi as it arises as a composition of S-epis. Therefore, m' is contained in m from which it follows that P is a sub-L-algebra of ϕ via m. As ϕ is an initial L-algebra the L-algebra morphism m is an isomorphism and therefore step is an S-epi.

Now we can establish the usual closure properties for admissible predicates.

Theorem 9.4 *Equality on complete S-spaces is admissible.*

Proof. Let A be a complete S-space. We have to show that the equality on A as given by the subobject $\delta_A \triangleq \langle \mathsf{id}_A, \mathsf{id}_A \rangle$ is orthogonal to $\overline{\mathsf{step}}$.
Suppose $f = \langle f_1, f_2 \rangle : \overline{\omega} \longrightarrow A \times A$ and $g : \mathbb{N} \longrightarrow A$ with $f \circ \overline{\mathsf{step}} = \delta_A \circ g$. But then $f_1 \circ \overline{\mathsf{step}} = g = f_2 \circ \overline{\mathsf{step}}$ entailing $f_1 = f_2$ as A is an S-space and $\overline{\mathsf{step}}$ is an S-epi by Theorem 9.3.

Well known "abstract nonsense" about orthogonality classes gives rise to the following closure properties of admissible predicates.

Theorem 9.5 *Admissible predicates are closed under arbitrary intersections and, therefore, in particular under conjunction and universal quantification. They are also stable under substitution.*

9.2 Computational Induction and Fixpoint Induction

Now one can prove validity of computational induction for admissible predicates.

Theorem 9.6 *(Computational Induction)*
Let A be a complete set and $\alpha : LA \longrightarrow A$ a focal L-algebra. Then for any map $f : A \longrightarrow A$ and admissible predicate P on A it holds that $P(\mathsf{fix}_\alpha(f))$ provided $\forall n{:}\mathbb{N}.\, P(f^n(\bot_\alpha))$.

Proof. By Lemma 8.1 we know that $\overline{\mathsf{kl}_{\alpha,f}}(\overline{\mathsf{step}}(n)) = f^n(\bot_\alpha)$. Thus, by assumption $\forall n{:}\mathbb{N}.\, P(f^n(\bot_\alpha))$ the map $\overline{\mathsf{kl}_{\alpha,f}} \circ \overline{\mathsf{step}}$ factors through $m_P : P \rightarrowtail A$ and, therefore, by admissibility of P the map $\overline{\mathsf{kl}_{\alpha,f}}$ factors through m_P. Thus, in particular $\mathsf{fix}_\alpha(f) = \overline{\mathsf{kl}_{\alpha,f}}(\infty) \in P$.

Fixpoint Induction follows from Computational Induction by the usual argument.

Theorem 9.7 (Fixpoint Induction for Admissible Predicates)
Let A be a complete set and $\alpha : LA \longrightarrow A$ be a focal L-algebra. Then for any map $f : A \longrightarrow A$ and admissible predicate P on A it holds that $P(\mathsf{fix}_\alpha(f))$ provided $P(\bot_\alpha)$ and $\forall x : A.\, P(x) \Rightarrow P(f(x))$.

9.3 Park Induction

In Classical Domain Theory the principle of "Park induction"[4] says that the canonical fixpoint of a (continuous) endofunction $f : A \longrightarrow A$ on a domain A is the least pre-fixpoint of f. Of course, the notions "least" and "pre-fixpoint" refer to the *"information ordering"* $\sqsubseteq_A$ *on* A to be defined next.

Definition 9.2 *For any set* A *the binary relation*

$$x \sqsubseteq_A y \;:\Leftrightarrow \forall p : S(A).\ p(x) = \top \Rightarrow p(y) = \top$$

is called information preorder *on* A.

Obviously, $\sqsubseteq_A$ is reflexive and transitive. Moreover, $\sqsubseteq_A$ is antisymmetric, i.e. a partial order, iff A is an S-space.

Of course, every function between sets is monotonic in the sense that it preserves the information preorder. Furthermore, we have the following two easy lemmas where the second follows immediately from Theorem 9.5 by the logical form of its definition.

Lemma 9.8 *Let* A *be a complete set and* $a \in A$ *then the predicate* $\sqsubseteq_A$ *is an admissible binary predicate on* A.

Now validity of Park induction for complete S-spaces can be proved as usual.

Theorem 9.9 *(Park Induction)*
Let A *be a complete focal* S*-space. If* $f : A \longrightarrow A$ *and* $a \in A$ *with* $f(a) \sqsubseteq_A a$ *then* $\mathsf{fix}(f) \sqsubseteq_A a$ *where* $\mathsf{fix}(f)$ *is* $\mathsf{fix}_\alpha(f)$ *for the unique focal* L*-algebra structure* α *on* A.

Proof. Let $a \in A$ with $f(a) \sqsubseteq a$. By Theorem 9.5 $P(x) \equiv x \sqsubseteq_A a$ is an admissible predicate on A. By Lemma 7.1 the domain A has a least element $\bot := \alpha(\bot_A)$, i.e. $P(\bot)$. As f is monotonic we get that $P(x)$ implies $P(f(x))$ for all $x \in A$. Now, by fixpoint induction it follows that $P(\mathsf{fix}_\alpha)$. Thus, we have $\mathsf{fix}(f) \sqsubseteq_A a$.

10 Domain Equations

Recursive types are given by domain equations $D \cong F(D, D)$ where F is a (necessarily internal) functor $F : \mathcal{C}^{op} \times \mathcal{C} \longrightarrow \mathcal{C}$ and $\mathcal{C}$ is the (necessarily internal) category of domains and L-algebra morphisms, i.e. *strict* maps[5].

---

[4] Park induction is useful for proving that a recursively defined function $f \triangleq \mathsf{fix}(\Phi) : A \longrightarrow B$ diverges for some argument $a \in A$, i.e. that $f(a) = \bot_B$. Usually, this is achieved by exhibiting a function $g : A \longrightarrow B$ with $\Phi(g) \sqsubseteq_{B^A} g$ and $g(a) = \bot_B$.

[5] If A and B are S-spaces with focal algebra structure α and β, respectively, then $f : A \longrightarrow B$ is an algebra morphism, i.e. $f \circ \alpha = \beta \circ L\,f$, iff $f \circ \alpha \circ c_A = \beta \circ L\,f \circ c_A$ which in turn is equivalent to $f(\bot_\alpha) = \bot_\beta$.

Following a suggestion of A. Simpson (personal communication, March 1997) one may establish the existence of solutions of domain equations provided the *universe* **Set** *is impredicative.*

First one shows that $\mathcal{C}$ is *algebraically compact*, i.e. for any functor $F : \mathcal{C} \longrightarrow \mathcal{C}$ there exists an an initial F-algebra α whose inverse is a terminal F-coalgebra. By Freyd's [3] the category $\mathcal{C}^{op} \times \mathcal{C}$ is algebraically compact, too. Thus, any internal mixed-variant functor $F : \mathcal{C}^{op} \times \mathcal{C} \longrightarrow \mathcal{C}$ admits a canonical solution $D \cong F(D, D)$ by taking the initial/terminal algebra for the functor $F^{\S} : \mathcal{C}^{op} \times \mathcal{C} \longrightarrow \mathcal{C}^{op} \times \mathcal{C}$ where $F^{\S}(-_1, -_2) = \langle F(-_2, -_1), F(-_1, -_2)\rangle$.

Algebraic compactness of $\mathcal{C}$ can be shown as follows. Let $F : \mathcal{C} \longrightarrow \mathcal{C}$ be a functor. As $\mathcal{C}$ is small complete the category of F-algebras is small complete, too. The initial F-algebra is constructed as the least sub-algebra of the weakly initial F-algebra $\phi : FP \longrightarrow P$ where $P := \Pi_{\alpha:FA \longrightarrow A} A$ and for any F-algebra $\alpha : FA \longrightarrow A$ the α-st component of ϕ is given by $\pi_\alpha \circ \phi = \alpha \circ F(\pi_\alpha)$. Thus, we have $\pi_\alpha : \phi \longrightarrow \alpha$. This construction makes *essential use of impredicativity of* **Set** *as ϕ is defined by quantification over* all *F-algebras including ϕ itself.* Now let $\alpha : FA \longrightarrow A$ be an initial F-algebra. For showing that α^{-1} is a terminal F-coalgebra (*c.f.* [3]) let $\beta : B \longrightarrow FB$ be some F-coalgebra. An F-coalgebra morphism $h : B \longrightarrow A$ from α^{-1} to β is obtained as the canonical solution of the fixpoint equation $h = \alpha \circ Fh \circ \beta$ (which exists as the set $[B \circ\!\!\longrightarrow A]$ of strict maps from B to A is a domain). It remains to show that this $h : \beta \longrightarrow \alpha^{-1}$ is unique as an F-coalgebra morphism. Let e be the canonical solution of $e = \alpha \circ Fe \circ \alpha^{-1}$. By initiality of α any of its endomorphisms is equal to id_A and, therefore, $e = \mathsf{id}_A$. Now if $k : \beta \longrightarrow \alpha^{-1}$ is an F-coalgebra morphism then we get $h = e \circ k$ by fixpoint induction. Thus, we get $h = k$.

Such "cheating" is not possible in classical domain theory due to the absence of a non-trivial impredicative universe **Set**. Instead, canonical solutions of domain equations are constructed explicitly as inverse limits, *c.f.* [18]. This classical construction can be performed also in a special version of SDT suitable for Domain Theory á la Scott, *c.f.* [12] for a proof checked mechanically within the LEGO system. In general SDT, however, this is not possible anymore as not every ascending $\mathbb{N}$-indexed chain appears as restriction of an ω-chain. However, in [17] A. Simpson has suggested to avoid this problem by considering *inverse limits of ω-diagrams* instead of $\mathbb{N}$-diagrams. It remains to be checked whether his method will work in our type-theoretic framework as well.

11 Markov's Principle and its Consequences

In the intended models of our theory, namely the *realisability models*, for all types X equality is $\neg\neg$-closed, i.e. $\forall x, y{:}X. \neg\neg(x = y) \Rightarrow x = y$. For the purposes of GSDT it suffices to assume that equality on S is $\neg\neg$-closed which is equivalent to def being $\neg\neg$-closed. For the case when S is the collection of all Σ^0_1-propositions this axiom is known as *Markov's Principle*, a terminology we adopt also for our more general situation.

In presence of Markov's Principle one may easily verify the following powerful proof principles used intrinsically in [12, 13].

Lemma 11.1 *In presence of Markov's Principle for any type X and $\neg\neg$-closed predicate P on $L\,X$ we have that $\forall z : LX.\,P(z)$ iff $P(\bot_X)$ and $\forall x : X.P(\eta_X(x))$.*

The following Theorem shows that Axiom 7 can be derived from Markov's Principle.

Theorem 11.2 *Under Markov's Principle $p = q$ is equivalent to $p \circ \eta_X = q \circ \eta_X$ and $p(\bot_X) = q(\bot_X)$ for all types X and maps $p, q : LX \longrightarrow S$.*

Proof. Follows from Lemma 11.1 by taking the predicate $\lambda z{:}L\,X.\,p(z) = q(z)$ which is $\neg\neg$-closed by Markov's Principle.

12 Conclusion and Future Work

We have presented a logical approach to general SDT avoiding any reference to external notions. Any realisability structure as considered in [9] provides a model of the logic and axioms employed in this paper. However, it is not clear how the sheaf models of Fiore and Rosolini [2] can be considered as models of our theory as it is unknown whether they admit an appropriate universe Set (though there cannot exist an *impredicative* universe Set !).

In [12] extensional S-spaces have been studied as an axiomatic analogue of the *extensional pers* of Freyd et. al. [5], however, under the stronger assumptions of [6, 21] and Markov's Principle. It seems to be worthwhile investigating to which extent the theory of extensional S-spaces can be developed under the weaker assumptions of GSDT.

Furthermore, there is need for a more detailed account of domain equations when Set is not impredicative.

Acknowledgments.
We wish to thank Alex Simpson, Pino Rosolini, Eugenio Moggi, Marcelo Fiore, and everybody else on the SDT-mailing-list for fruitful discussions. Especially we want to thank A. Jung for pointing out some unnecessary detours in an earlier draft of this paper. Two anonymous referees gave valuable comments. Paul Taylor's diagram macros are acknowledged.

References

1. R.L. Crole and A.M. Pitts. New foundations for fixpoint computations: Fix-hyperdoctrines and the fix-logic. *Information and Computation*, 98:171–210, 1992.
2. M. Fiore and G. Rosolini. Two models of synthetic domain theory. 1996. To appear in the Freyd Festschrift.
3. P. Freyd. Remarks on algebraically compact categories. In *Applications of Categories in Computer Science*, volume 177 in Notes of the London Mathematical Society, 1992.

4. P. Freyd and G. Kelly. Categories of continuous functors. *Journal of Pure and Applied Algebra*, 2:169–191, 1972.
5. P. Freyd, P. Mulry, G. Rosolini, and D. Scott. Extensional PERs. *Information and Computation*, 98:211–227, 1992.
6. J.M.E. Hyland. First steps in synthetic domain theory. In A. Carboni, M.C. Pedicchio, and G. Rosolini, eds., *Proc. of the 1990 Como Category Theory Conference*, volume 1488 of *Lecture Notes in Mathematics*, pages 131–156, Berlin, 1991. Springer.
7. M. Jibladze. A presentation of the initial lift-algebra. *JPAA*, 116(2–3):199–220, 1997.
8. J.R. Longley. *Realizability Toposes and Language Semantics.* PhD thesis, University of Edinburgh, 1994.
9. J.R. Longley and A.K. Simpson. A uniform account of domain theory in realizability models. To appear in the special edition of *Math.Str.Comp.Sc.* for *LDPL* 1995.
10. Z. Luo. *An Extended Calculus of Constructions.* PhD thesis, University of Edinburgh, 1990. Available as report ECS-LFCS-90-118.
11. W.K. Phoa. *Domain Theory in Realizability Toposes.* PhD thesis, University of Cambridge, 1990. Also available as report ECS-LFCS-91-171, University of Edinburgh.
12. B. Reus. *Program Verification in Synthetic Domain Theory.* PhD thesis, Ludwig-Maximilians-Universität München, 1995. Shaker Verlag, Aachen, 1996.
13. B. Reus. Synthetic domain theory in type theory: Another logic of computable functions. In J. Grundy, J. von Wright and J. Harrison, eds., *TPHOLs'96*, volume 1125 of *LNCS*, pages 363–381. Springer, 1996.
14. B. Reus and T. Streicher. Naive Synthetic Domain Theory – a logical approach. Draft, September 1993.
15. G. Rosolini. *Continuity and effectiveness in topoi.* PhD thesis, University of Oxford, 1986.
16. G. Rosolini. Notes on synthetic domain theory. Draft, August 1995.
17. A.K. Simpson. Domain theory in intuitionistic set theory. Draft, 1996.
18. M.B. Smyth and G.D. Plotkin. The category-theoretic solution to recursive domain equations. *SIAM Journal of Computing*, 11:761–783, 1982.
19. T. Streicher. *Semantics of Type Theory, Correctness, Completeness and Independence Results.* Birkhäuser, 1991.
20. T. Streicher. Inductive Construction of Repletion. To appear in *Applied Categorical Structures*, 1997.
21. P. Taylor. The fixed point property in synthetic domain theory. In *6th Symp. on Logic in Computer Science*, pages 152–160, Washington, 1991. IEEE Computer Soc. Press.

Lecture Notes in Computer Science

For information about Vols. 1–1229

please contact your bookseller or Springer-Verlag

Vol. 1230: J. Duncan, G. Gindi (Eds.), Information Processing in Medical Imaging. Proceedings, 1997. XVI, 557 pages. 1997.

Vol. 1231: M. Bertran, T. Rus (Eds.), Transformation-Based Reactive Systems Development. Proceedings, 1997. XI, 431 pages. 1997.

Vol. 1232: H. Comon (Ed.), Rewriting Techniques and Applications. Proceedings, 1997. XI, 339 pages. 1997.

Vol. 1233: W. Fumy (Ed.), Advances in Cryptology — EUROCRYPT '97. Proceedings, 1997. XI, 509 pages. 1997.

Vol 1234: S. Adian, A. Nerode (Eds.), Logical Foundations of Computer Science. Proceedings, 1997. IX, 431 pages. 1997.

Vol. 1235: R. Conradi (Ed.), Software Configuration Management. Proceedings, 1997. VIII, 234 pages. 1997.

Vol. 1236: E. Maier, M. Mast, S. LuperFoy (Eds.), Dialogue Processing in Spoken Language Systems. Proceedings, 1996. VIII, 220 pages. 1997. (Subseries LNAI).

Vol. 1238: A. Mullery, M. Besson, M. Campolargo, R. Gobbi, R. Reed (Eds.), Intelligence in Services and Networks: Technology for Cooperative Competition. Proceedings, 1997. XII, 480 pages. 1997.

Vol. 1239: D. Sehr, U. Banerjee, D. Gelernter, A. Nicolau, D. Padua (Eds.), Languages and Compilers for Parallel Computing. Proceedings, 1996. XIII, 612 pages. 1997.

Vol. 1240: J. Mira, R. Moreno-Díaz, J. Cabestany (Eds.), Biological and Artificial Computation: From Neuroscience to Technology. Proceedings, 1997. XXI, 1401 pages. 1997.

Vol. 1241: M. Akşit, S. Matsuoka (Eds.), ECOOP'97 – Object-Oriented Programming. Proceedings, 1997. XI, 531 pages. 1997.

Vol. 1242: S. Fdida, M. Morganti (Eds.), Multimedia Applications, Services and Techniques – ECMAST '97. Proceedings, 1997. XIV, 772 pages. 1997.

Vol. 1243: A. Mazurkiewicz, J. Winkowski (Eds.), CONCUR'97: Concurrency Theory. Proceedings, 1997. VIII, 421 pages. 1997.

Vol. 1244: D. M. Gabbay, R. Kruse, A. Nonnengart, H.J. Ohlbach (Eds.), Qualitative and Quantitative Practical Reasoning. Proceedings, 1997. X, 621 pages. 1997. (Subseries LNAI).

Vol. 1245: M. Calzarossa, R. Marie, B. Plateau, G. Rubino (Eds.), Computer Performance Evaluation. Proceedings, 1997. VIII, 231 pages. 1997.

Vol. 1246: S. Tucker Taft, R. A. Duff (Eds.), Ada 95 Reference Manual. XXII, 526 pages. 1997.

Vol. 1247: J. Barnes (Ed.), Ada 95 Rationale. XVI, 458 pages. 1997.

Vol. 1248: P. Azéma, G. Balbo (Eds.), Application and Theory of Petri Nets 1997. Proceedings, 1997. VIII, 467 pages. 1997.

Vol. 1249: W. McCune (Ed.), Automated Deduction – CADE-14. Proceedings, 1997. XIV, 462 pages. 1997. (Subseries LNAI).

Vol. 1250: A. Olivé, J.A. Pastor (Eds.), Advanced Information Systems Engineering. Proceedings, 1997. XI, 451 pages. 1997.

Vol. 1251: K. Hardy, J. Briggs (Eds.), Reliable Software Technologies – Ada-Europe '97. Proceedings, 1997. VIII, 293 pages. 1997.

Vol. 1252: B. ter Haar Romeny, L. Florack, J. Koenderink, M. Viergever (Eds.), Scale-Space Theory in Computer Vision. Proceedings, 1997. IX, 365 pages. 1997.

Vol. 1253: G. Bilardi, A. Ferreira, R. Lüling, J. Rolim (Eds.), Solving Irregularly Structured Problems in Parallel. Proceedings, 1997. X, 287 pages. 1997.

Vol. 1254: O. Grumberg (Ed.), Computer Aided Verification. Proceedings, 1997. XI, 486 pages. 1997.

Vol. 1255: T. Mora, H. Mattson (Eds.), Applied Algebra, Algebraic Algorithms and Error-Correcting Codes. Proceedings, 1997. X, 353 pages. 1997.

Vol. 1256: P. Degano, R. Gorrieri, A. Marchetti-Spaccamela (Eds.), Automata, Languages and Programming. Proceedings, 1997. XVI, 862 pages. 1997.

Vol. 1258: D. van Dalen, M. Bezem (Eds.), Computer Science Logic. Proceedings, 1996. VIII, 473 pages. 1997.

Vol. 1259: T. Higuchi, M. Iwata, W. Liu (Eds.), Evolvable Systems: From Biology to Hardware. Proceedings, 1996. XI, 484 pages. 1997.

Vol. 1260: D. Raymond, D. Wood, S. Yu (Eds.), Automata Implementation. Proceedings, 1996. VIII, 189 pages. 1997.

Vol. 1261: J. Mycielski, G. Rozenberg, A. Salomaa (Eds.), Structures in Logic and Computer Science. X, 371 pages. 1997.

Vol. 1262: M. Scholl, A. Voisard (Eds.), Advances in Spatial Databases. Proceedings, 1997. XI, 379 pages. 1997.

Vol. 1263: J. Komorowski, J. Zytkow (Eds.), Principles of Data Mining and Knowledge Discovery. Proceedings, 1997. IX, 397 pages. 1997. (Subseries LNAI).

Vol. 1264: A. Apostolico, J. Hein (Eds.), Combinatorial Pattern Matching. Proceedings, 1997. VIII, 277 pages. 1997.

Vol. 1265: J. Dix, U. Furbach, A. Nerode (Eds.), Logic Programming and Nonmonotonic Reasoning. Proceedings, 1997. X, 453 pages. 1997. (Subseries LNAI).

Vol. 1266: D.B. Leake, E. Plaza (Eds.), Case-Based Reasoning Research and Development. Proceedings, 1997. XIII, 648 pages. 1997 (Subseries LNAI).

Vol. 1267: E. Biham (Ed.), Fast Software Encryption. Proceedings, 1997. VIII, 289 pages. 1997.

Vol. 1268: W. Kluge (Ed.), Implementation of Functional Languages. Proceedings, 1996. XI, 284 pages. 1997.

Vol. 1269: J. Rolim (Ed.), Randomization and Approximation Techniques in Computer Science. Proceedings, 1997. VIII, 227 pages. 1997.

Vol. 1270: V. Varadharajan, J. Pieprzyk, Y. Mu (Eds.), Information Security and Privacy. Proceedings, 1997. XI, 337 pages. 1997.

Vol. 1271: C. Small, P. Douglas, R. Johnson, P. King, N. Martin (Eds.), Advances in Databases. Proceedings, 1997. XI, 233 pages. 1997.

Vol. 1272: F. Dehne, A. Rau-Chaplin, J.-R. Sack, R. Tamassia (Eds.), Algorithms and Data Structures. Proceedings, 1997. X, 476 pages. 1997.

Vol. 1273: P. Antsaklis, W. Kohn, A. Nerode, S. Sastry (Eds.), Hybrid Systems IV. X, 405 pages. 1997.

Vol. 1274: T. Masuda, Y. Masunaga, M. Tsukamoto (Eds.), Worldwide Computing and Its Applications. Proceedings, 1997. XVI, 443 pages. 1997.

Vol. 1275: E.L. Gunter, A. Felty (Eds.), Theorem Proving in Higher Order Logics. Proceedings, 1997. VIII, 339 pages. 1997.

Vol. 1276: T. Jiang, D.T. Lee (Eds.), Computing and Combinatorics. Proceedings, 1997. XI, 522 pages. 1997.

Vol. 1277: V. Malyshkin (Ed.), Parallel Computing Technologies. Proceedings, 1997. XII, 455 pages. 1997.

Vol. 1278: R. Hofestädt, T. Lengauer, M. Löffler, D. Schomburg (Eds.), Bioinformatics. Proceedings, 1996. XI, 222 pages. 1997.

Vol. 1279: B. S. Chlebus, L. Czaja (Eds.), Fundamentals of Computation Theory. Proceedings, 1997. XI, 475 pages. 1997.

Vol. 1280: X. Liu, P. Cohen, M. Berthold (Eds.), Advances in Intelligent Data Analysis. Proceedings, 1997. XII, 621 pages. 1997.

Vol. 1281: M. Abadi, T. Ito (Eds.), Theoretical Aspects of Computer Software. Proceedings, 1997. XI, 639 pages. 1997.

Vol. 1282: D. Garlan, D. Le Métayer (Eds.), Coordination Languages and Models. Proceedings, 1997. X, 435 pages. 1997.

Vol. 1283: M. Müller-Olm, Modular Compiler Verification. XV, 250 pages. 1997.

Vol. 1284: R. Burkard, G. Woeginger (Eds.), Algorithms — ESA '97. Proceedings, 1997. XI, 515 pages. 1997.

Vol. 1285: X. Jao, J.-H. Kim, T. Furuhashi (Eds.), Simulated Evolution and Learning. Proceedings, 1996. VIII, 231 pages. 1997. (Subseries LNAI).

Vol. 1286: C. Zhang, D. Lukose (Eds.), Multi-Agent Systems. Proceedings, 1996. VII, 195 pages. 1997. (Subseries LNAI).

Vol. 1287: T. Kropf (Ed.), Formal Hardware Verification. XII, 367 pages. 1997.

Vol. 1288: M. Schneider, Spatial Data Types for Database Systems. XIII, 275 pages. 1997.

Vol. 1289: G. Gottlob, A. Leitsch, D. Mundici (Eds.), Computational Logic and Proof Theory. Proceedings, 1997. VIII, 348 pages. 1997.

Vol. 1290: E. Moggi, G. Rosolini (Eds.), Category Theory and Computer Science. Proceedings, 1997. VII, 313 pages. 1997.

Vol. 1292: H. Glaser, P. Hartel, H. Kuchen (Eds.), Programming Languages: Implementations, Logigs, and Programs. Proceedings, 1997. XI, 425 pages. 1997.

Vol. 1294: B.S. Kaliski Jr. (Ed.), Advances in Cryptology — CRYPTO '97. Proceedings, 1997. XII, 539 pages. 1997.

Vol. 1295: I. Prívara, P. Ružička (Eds.), Mathematical Foundations of Computer Science 1997. Proceedings, 1997. X, 519 pages. 1997.

Vol. 1296: G. Sommer, K. Daniilidis, J. Pauli (Eds.), Computer Analysis of Images and Patterns. Proceedings, 1997. XIII, 737 pages. 1997.

Vol. 1297: N. Lavrač, S. Džeroski (Eds.), Inductive Logic Programming. Proceedings, 1997. VIII, 309 pages. 1997. (Subseries LNAI).

Vol. 1298: M. Hanus, J. Heering, K. Meinke (Eds.), Algebraic and Logic Programming. Proceedings, 1997. X, 286 pages. 1997.

Vol. 1299: M.T. Pazienza (Ed.), Information Extraction. Proceedings, 1997. IX, 213 pages. 1997. (Subseries LNAI).

Vol. 1300: C. Lengauer, M. Griebl, S. Gorlatch (Eds.), Euro-Par'97 Parallel Processing. Proceedings, 1997. XXX, 1379 pages. 1997.

Vol. 1302: P. Van Hentenryck (Ed.), Static Analysis. Proceedings, 1997. X, 413 pages. 1997.

Vol. 1303: G. Brewka, C. Habel, B. Nebel (Eds.), KI-97: Advances in Artificial Intelligence. Proceedings, 1997. XI, 413 pages. 1997. (Subseries LNAI).

Vol. 1304: W. Luk, P.Y.K. Cheung, M. Glesner (Eds.), Field-Programmable Logic and Applications. Proceedings, 1997. XI, 503 pages. 1997.

Vol. 1305: D. Corne, J.L. Shapiro (Eds.), Evolutionary Computing. Proceedings, 1997. X, 313 pages. 1997.

Vol. 1308: A. Hameurlain, A M. Tjoa (Eds.), Database and Expert Systems Applications. Proceedings, 1997. XVII, 688 pages. 1997.

Vol. 1310: A. Del Bimbo (Ed.), Image Analysis and Processing. Proceedings, 1997. Volume I. XXI, 722 pages. 1997.

Vol. 1311: A. Del Bimbo (Ed.), Image Analysis and Processing. Proceedings, 1997. Volume II. XXII, 794 pages. 1997.

Vol. 1312: A. Geppert, M. Berndtsson (Eds.), Rules in Database Systems. Proceedings, 1997. VII, 213 pages. 1997.

Vol. 1314: S. Muggleton (Ed.), Inductive Logic Programming. Proceedings, 1996. VIII, 397 pages. 1997. (Subseries LNAI).

Vol. 1315: G. Sommer, J.J. Koenderink (Eds.), Algebraic Frames for the Perception-Action Cycle. Proceedings, 1997. VIII, 395 pages. 1997.